how
to
know
the

beetles

The **Pictured Key Nature Series** has been published since 1944 by the Wm. C. Brown Company. The series was initiated in 1937 by the late Dr. H. E. Jaques, Professor Emeritus of Biology at Iowa Wesleyan University. Dr. Jaques' dedication to the interest of nature lovers in every walk of life has resulted in the prominent place this series fills for all who wonder **"How to Know."**

John F. Bamrick and Edward T. Cawley
Consulting Editors

The Pictured Key Nature Series

How to Know the

- **AQUATIC INSECTS,** Lehmkuhl
- **AQUATIC PLANTS,** Prescott
- **BEETLES,** Arnett-Downie-Jaques, Second Edition
- **BUTTERFLIES,** Ehrlich
- **FALL FLOWERS,** Cuthbert
- **FERNS AND FERN ALLIES,** Mickel
- **FRESHWATER ALGAE,** Prescott, Third Edition
- **FRESHWATER FISHES,** Eddy-Underhill, Third Edition
- **GILLED MUSHROOMS,** Smith-Smith-Weber
- **GRASSES,** Pohl, Third Edition
- **IMMATURE INSECTS,** Chu
- **INSECTS,** Bland-Jaques, Third Edition
- **LICHENS,** Hale, Second Edition
- **LIVING THINGS,** Jaques, Second Edition
- **MAMMALS,** Booth, Third Edition
- **MITES AND TICKS,** McDaniel
- **MOSSES AND LIVERWORTS,** Conard-Redfearn, Third Edition
- **NON-GILLED FLESHY FUNGI,** Smith-Smith
- **PLANT FAMILIES,** Jaques
- **POLLEN AND SPORES,** Kapp
- **PROTOZOA,** Jahn, Bovee, Jahn, Third Edition
- **SEAWEEDS,** Abbott-Dawson, Second Edition
- **SEED PLANTS,** Cronquist
- **SPIDERS,** Kaston, Third Edition
- **SPRING FLOWERS,** Cuthbert, Second Edition
- **TREES,** Miller-Jaques, Third Edition
- **TRUE BUGS,** Slater-Baranowski
- **WEEDS,** Wilkinson-Jaques, Third Edition
- **WESTERN TREES,** Baerg, Second Edition

how to know the beetles

Second Edition

Ross H. Arnett, Jr.
Siena College

N.M Downie, Professor Emeritus
Purdue University

and the late H.E. Jaques

Boston, Massachusetts Burr Ridge, Illinois Dubuque, Iowa
Madison, Wisconsin New York, New York San Francisco, California St. Louis, Missouri

WCB/McGraw-Hill

A Division of The McGraw-Hill Companies

Library of Congress Catalog Card Number: 79-55045

ISBN 0-697-04776-8

Printed in the United States of America

Contents

Contents

Preface

This new edition, although intended for the beginner or amateur, has been made useful to those many entomologists, not specialists in the Coleoptera, by the addition of all of the families of beetles found in the United States and Canada. Representative species of each have been added. Some changes have been made in the species of other families as well, so that the book is useful to those who have occasion to make identifications of beetles, either to determine whether or not the beetle is a pest, or perhaps for an ecological survey. Most of the pest species are now included.

It has always been our belief that a key should be easy to use. The picture-keys make no attempt to account for every exception in each couplet. However, to make it accurate for the species in this book, it has been necessary to have some family names come out in more than one place. These are cross-referenced in the index. The keys then do not reflect the classification of the order. However, a separate list of families and their common names is a new feature found at the end of the book.

This book describes several hundred of the species of beetles most likely to be seen by the general student of insect life. They are pictured, keyed, and the recognition features given for each. Obviously, this is a small part of the approximately 29,000 species of beetles living in Canada and the United States. To include all of these species in one book is impossible. Therefore, we have indicated the approximate (or actual) number of species in each family, and with this is usually a brief note about the species not included. Therefore, if the beetle that you have does not seem to fit well into the picture key and is not described in the text, set it aside until you can get help from a specialist, or until you have collected and identified the approximately 1500 species (900 of which are illustrated) in this book. Then you will be ready for some of the specialized literature available. A brief list of such publications appears at the end of the next section.

Many individuals specialize in the collection of beetles. In 1947 one of us (Arnett) founded a separate journal, *The Coleopterists Bulletin*, to provide a means of communication among this growing body of amateurs and professionals. It is now the official organ of *The Coleopterists Society;* membership is open to anyone interested in beetles. All major libraries subscribe to this journal and the current address of the secretary may be obtained from any recent copy.

The illustrations used in this second edition are mostly those used by Jaques in the first edition. Some new ones have been added and acknowledged in their proper place. The user will find a slight inconsistency in the numbering of the figures. In a few cases it has been necessary to use letters instead of numbers (see figure 177) and in another case (Fig. 187) it was necessary to number the parts of the figure instead of using letters.

The system of classification used in this book is that adopted by the Board of the North American Beetle Fauna Project of which one of the authors (Arnett) is the director. The family numbers correspond to those used in the *Checklist of the Beetle of Canada, United States, Mexico, Central America, and the West Indies*, a series published by the North American Beetle Fauna Project.

We are grateful to several of our colleagues for reading portions of the manuscript, particularly Dr. Richard L. Jacques, Jr., of Fairleigh Dickinson University, Rutherford, New Jersey. Most of the hard work was done years ago by Professor H. E. Jaques. He in turn received considerable help from the Coleopterists of the United States Department of Agriculture and the United States National Museum of Natural History. At that time the entire list of names used in this book was circulated among those specialists. Therefore there is no doubt in our mind that the list actually represents the common species. Our major task has been to update the names and to add to the text. Some parts of the book have been further reviewed by specialists. Their help is acknowledged where appropriate. For all of this help, past and present, we are very grateful.

About Beetles

Beetles, like all insects, are covered with a coat of leathery or horny material called the cuticle. This material, which is very complex chemically, forms the exoskeleton. In addition, most beetles have their front wings modified into wing covers composed of the same material. These wing covers protect the hind wings which are folded beneath them when not in use. It is the wings covers, or elytra, that give beetles the Greek Latinized name, Coleoptera. This feature along with their complete metamorphosis (i.e., egg, larva, pupa, and adult stages of their life cycle), chewing mouthparts (at least in almost all species), and their usually compact bodies, is characteristic of this, the largest order (in number of species) of all the insect orders, and for that matter, of any other group of animals of plants.

Figure 1 Every fifth living thing is a beetle.

The most recent estimates of the number of insect species is about one million, of which beetles comprise nearly a third, or over 300,000 species. If compared with the plant kingdom, there are about the same number of species. This means that beetles represent a fifth (Fig. 1) of all living organisms. Many more species are described each year.

Specimens of beetles are easily caught, using a variety of methods, some of which are described later in this section. When properly mounted, they make an interesting display and offer endless opportunities for study. Many are brightly colored, all have interesting shapes, and best of all, the great diversity of habit and habitats of the many species gives an endless challenge to learn more about their lives and their contributions to the many ecosystems in which they live.

Most beetles are of no economic importance as insect pests and none is dangerous to handle alive unless you deliberately place a finger in its powerful mandibles or squeeze it (the blister beetles and some others exude a fluid that irritates human skin, and may cause blisters). However, a few species are of great importance as pests of forests, grains, vegetables, flowers, and stored products. Although insect collections are not generally considered "stored products," some beetles do not appreciate these scientific "graveyards" and they (certain species of Dermestidae, or skin beetles) attack and severely damage collections unless precautions are taken.

The order Coleoptera is only one of 29 orders comprising the class Insecta, phylum Arthropoda. Classification details of the phylum and class may be found in various encyclopedias, and in other books in this series. For our purposes here, it is important that you be able to recognize a beetle, because unless you have a beetle in front of you, the keys and illustrations in this book will be useless.

Most, but not all beetles have elytra (Fig. 2) which cover the entire abdomen. Some beetles with short elytra may be confused with earwigs, order Dermaptera. But earwigs almost always have fairly long, rather stout pinchers at the *end of the abdomen.* No beetle has this feature. A beginner might confuse some of the true bugs, order Hemipera, with beetles. The true bugs have very fine, long, piercing mouthparts used for sucking plant juices or blood. Those mouthparts project backward and would never be confused with the chewing mouthparts of beetles. (A few very rare, and very small beetles have sucking mouthparts; none of these are known to occur in the United States. Some species of blister beetles, Meloidae, have a sucking tube for nectar *in addition to* their chewing mouthparts.)

All beetles, and all plants and animals, have been given a scientific name (if the species has been collected and studied). The name is always composed of two Latin words, the

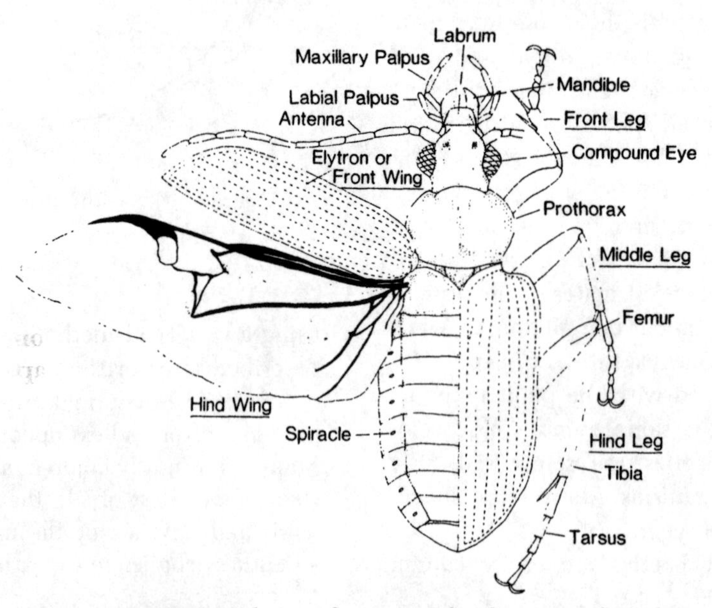

Figure 2 **Dorsal view of a common beetle naming the parts used in the identification keys.**

generic name, and the specific name. Together these names form the name of the species. The two words (or the initial of the genus) must always be used together when referring to a species. Some species also have common names in English. Those names are used in this book in addition to the scientific name. In some cases, a genus may be divided into two or more subgenera. If it is necessary to use these names, the subgeneric name is placed between the generic and the specific name set off by parentheses. Finally, in a few groups, subspecific names may be used. Those names follow the specific name. Finally, it is customary to cite the original describer's name after the specific or subspecific name. If there has been a change of names due to the discovery of additional information about the species so that it is necessary to assign the species to another genus, then the original author of the species has his name placed in parenthesis. An example of all of this is *Cicindela* (*Plectographa*) *suturalis hebraea* (Klug). If you do not understand the meaning of all of these parts of the name of this tiger beetle, reread the paragraph. Further details of nomenclature may be found in introductory textbooks on biology. Note, however, in most cases in this book and elsewhere, only the generic, specific, and authors name are used. Subgenera and subspecies are of concern primarily to the specialists.

BEETLE STRUCTURE

The size of the species of beetles vary from a minute species, less than 0.5 mm, that lives in the spore tubes of certain fungi, to the giants of tropical regions over 15 cm in length. It is estimated that it would take about 40 million of the smallest beetles to equal in weigh one of the largest! Such a great range in size of the species reflects only one aspect of the great species diversity of this order. There may be a considerable size range even within the species. Some adult beetles, for reasons unknown, perhaps nutritional, perhaps genetic, may be less than half the size of the largest specimens of the same species. This is not generally the case, however, and often two very similar species may be separated on the basis of their size range difference alone.

The exoskeleton (described in the introduction) not only serves as an anchor for muscles, but also gives each species its distinctive shape and contains the pigments that are responsible for the beetle's color. Often this skeleton is so hard that the beetle is protected from most predators. In fact, some beetles from the Old World deserts are so hard that it is necessary to use a sharp steel drill to bore a hole into them so that they may be mounted on an insect pin. This also means that a collection of these insects, once dried and protected by a fumigant, will stay in nearly lifelike condition indefinitely.

As for percentage of the total number of species of beetles, the majority would fall in the dark, drab color range. This is not true for some groups. Most Chrysomelidae (Leaf beetles), Cerambycidae (Longhorned beetles), Buprestidae (Metallic Woodboring beetles). Cicindelidae (Tiger beetles), and certain species of many other families are brightly colored and rival the butterflies in color patterns and variety of colors.

Modern classification of the beetles is based on many features including details of the external anatomy of the thorax, abdomen, including the organs of copulation, wing veins and folding patterns, and the shape of the legs and the number of tarsal segments on each. Much attention has been given to structures found in the larvae and this has been correlated with adult structures. The total analysis has resulted in a fairly stable family classification. However, this is much different than the features used in keys. Anatomical features in keys are only those that are easiest to see and at the same time enable the separation of families, genera and species. An accurate knowledge of the external anatomy of beetles aids keying for identification. Beginners will find the simple diagramatic sketches in the key more helpful if each part is compared with the beetle specimen before keying is attempted (See Figs. 2, 3, and 4). Many parts are explained and pictured in the glossary and often within the keys at the place that the term is used.

When viewed from above, beetles show three usually definitely separated sections: head, prothorax, and elytra. In addition some species have the tip of the abdomen exposed and others (e.g. Staphylinidae) show a considerable portion of the abdomen. All insects possess three main body regions: head, thorax, and abdomen. In beetles these regions are best seen when viewed ventrally. Dorsally, the thorax that is seen is only the *pronotum*, or dorsal surface of the first thoracic segment. The elytra, as previously explained are modified front wings, and these are the appendages of the mesothorax, or second thoracic segment. The small triangular piece that is usually exposed in the center between the prothorax and the elytra is the *scutellum*, a part of the mesothorax. When the beetle is flying, the elytra are usually held open so that the flying wings, which are appendages of the metathorax, or third thoracic segment, can function. Some members of the family Buprestidae do not open the elytra. Instead the flying wings are unfolded and protruded from under the elytra for flight. The elytra provide excellent protection of the wings and abdomen of the beetle but they are definitely in the way in flight. Therefore most beetles are rather slow, clumsy

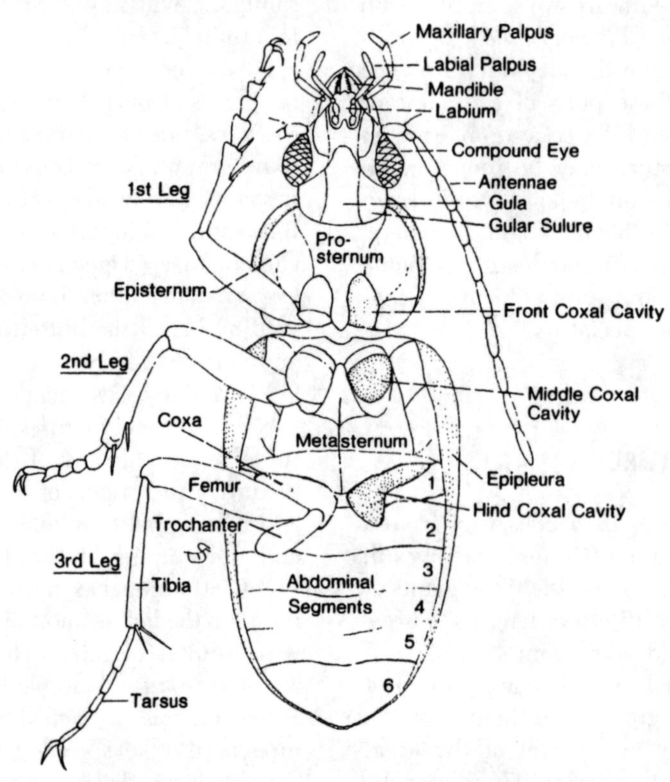

Figure 3 Ventral view of a common beetle (Carabidae) with the more important parts named.

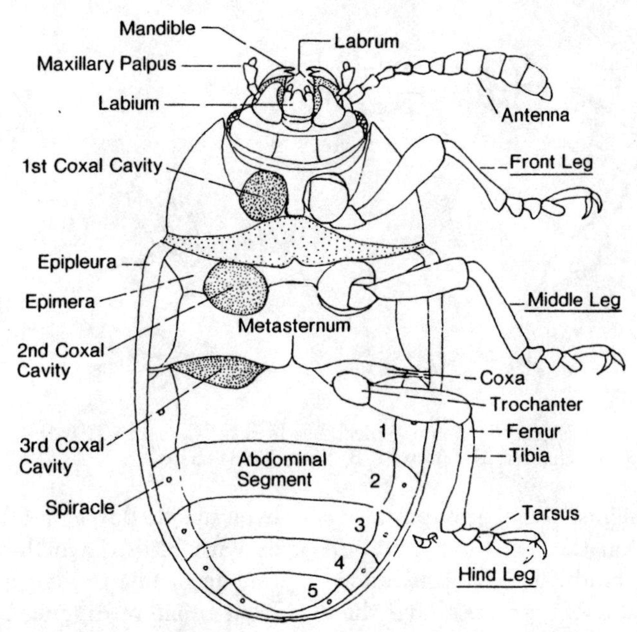

Figure 4 **Ventral view of a common beetle (Chrysomelidae) showing the more important parts.**

fliers. It was once thought that only the second pair of wings contributed to flying, but it is now known that in some species, at least, the elytra take some part in the flight. It is also known that beetles with the elytra removed cannot fly. A considerable number of beetle species lack flying wings. These flightless beetles usually have the elytra fused and hence are never opened (See some Tenebrionidae). Other beetles (see especially Buprestidae) have hard elytra that are locked in place by an intricate notched device. It is impossible to force these elytra apart to see the wings beneath unless they are first raised and then separated. The function of this system has never been determined experimentally. It may offer some additional protection to these wood-boring species.

In addition to the wings, the thorax has as appendages three pairs of legs, one pair per segment (Fig. 5). The legs are named according to the thoracic segment to which they are attached, i.e., forelegs, middlelegs, and hindlegs. The legs are attached to the thorax by a ball and socket joint, the basal (closest to body) segment of the leg forming the "ball". This is termed the *coxa*. The coxa rotates within the socket or *coxal cavity*. As with almost all insects, the coxa has attached to it the *trochanter*, usually a small triangular segment that is sometimes off center and appears to be at the side of the next segment, the *femur*. This is the part of the leg containing the muscles that operate the following segments. In some species the hind femur is enlarged to accommodate muscles necessary for jumping. The *tibia*, which may be the longest of the leg segments, follows the femur. The apical end usually has two or more heavy spines or spurs. The *tarsus*, or foot portion of the leg, is the most variable. The common number of segments is five, but some groups of beetles have less (never more) than this number. The most frequent modification is that the next to the last segment is greatly reduced so that it is very difficult to see. Therefore, there appear to be

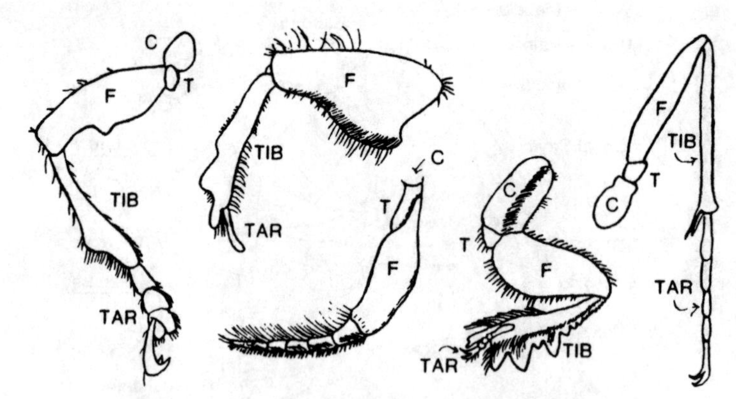

Figure 5 Some typical beetle legs. C, coxa; T, trochanter; F, femur; TIB, tibia; TAR, tarsus.

only four tarsal segments. Other groups have five tarsal segments on the front and middle pairs of legs, but the hind legs have this number reduced to four. Other groups have the segments further reduced in number. This feature is used extensively in identification keys. Attached to the end of the last tarsal segment are two claws, sometimes one. These too vary greatly in shape from species to species. Beneath one (usually the next to last) or more of the tarsal segments in many species is a pad of closely packed very fine and short setae. These pads combined with the claws enable the beetle to walk on very steep surfaces or even upside down on a branch or leaf. Beetles, as with birds, have their legs and feet modified according to their mode of life. The heavy front tibiae of digging beetles are armed with heavy spines to aid in this process; the tarsi of these insects are greatly reduced or even absent. Swimming modifications are apparent in the several water beetle families. Other modifications of the legs are discussed under the respective families.

The heavily armored head (Fig. 6) varies greatly in size, shape, and position. It is the head, of course, that bears the eyes, mouthparts, and the antennae. The shape and posi-

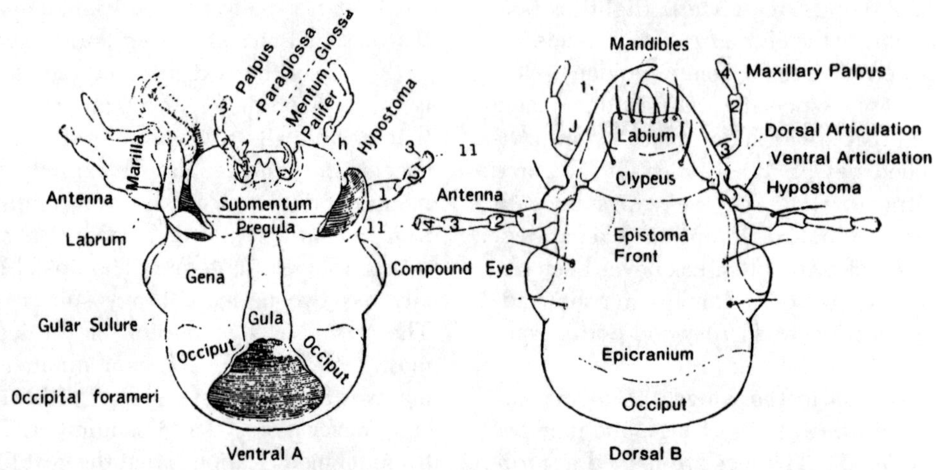

Figure 6 Head of a ground beetle, *Pterostichus*. A, ventral view; B, dorsal view. (After Hopkins, U.S. D.A.)

tion of the head as it is attached to the thorax is modified according to the type of food consumed and the insect's feeding habits. Therefore the mouthparts may be directed forward, downward, or extended outward at the end of a long beak (Fig. 7).

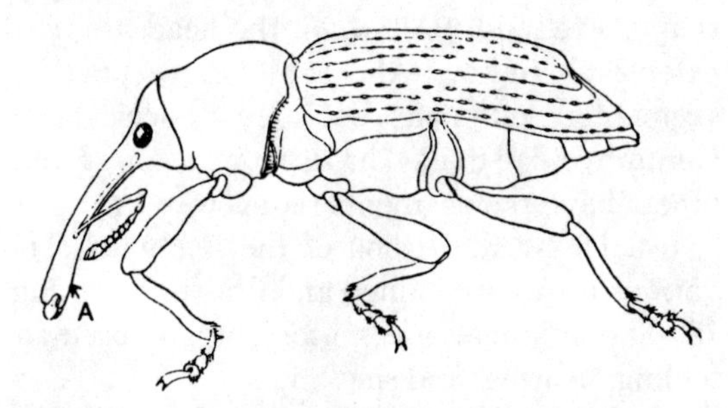

Figure 7 A common curculionid, *Pissodes fraseri*. A, mouth parts. (After Hopkins, U.S.D.A.)

Beetles chew their food by means of a pair of usually heavy jaws, the *mandibles,* which are aided by some accessory parts. The mandibles are modified in many ways according to the type of food eaten. Some are heavy for chewing leaves, others are long and thin for piercing soft-bodied prey, and still others are used to bore into wood, fruit, or grain. One large family, the Curculionidae, or weevils, has its mouthparts at the end of a long snout or beak (Fig. 7). Adults of this group of over 50,000 species pierce fruits, nuts, or seeds, eating out a deep hole into which is deposited one or more eggs. The surface of the hole grows over, sealing the egg inside. Soon the larva hatches and feeds within on the fruit, seed, or nut.

The mouthparts of beetles meet in a vertical plane instead of a horizontal plane as do ours. That changes the whole process of managing food. As already indicated, the mandibles do the heavy work. They are supplemented by a pair of *maxillae* operating just behind (or below) the mandibles (Fig. 8). Each member of the pair bears a toothed *lacinia* and a fingerlike *galea.* This is the structure that is elongated in certain species of Meloidae into a long sucking tube. On the outside of these parts

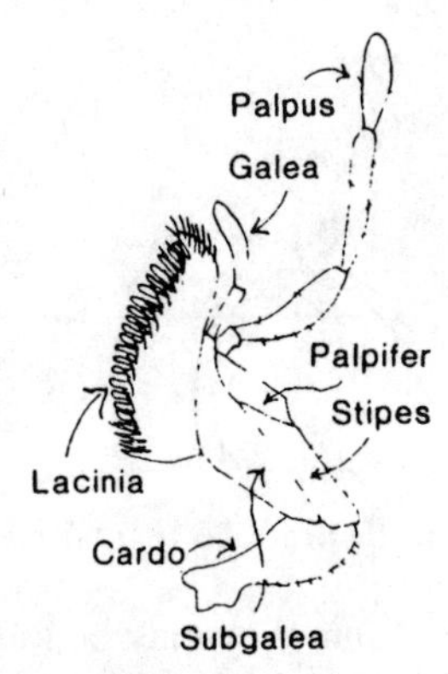

Figure 8 The maxilla with parts named.

is a *maxillary palpus* consisting of a few segments. This helps to support bits of food and probably functions mostly as a sensory structure for taste and smell. Still lower is a single *labium* or lower lip (Fig. 9) with two united *mentums* and on either side a *labial palpus* with structure and function much the same as the maxillary palpi. These palpi are often spoken of as palps.

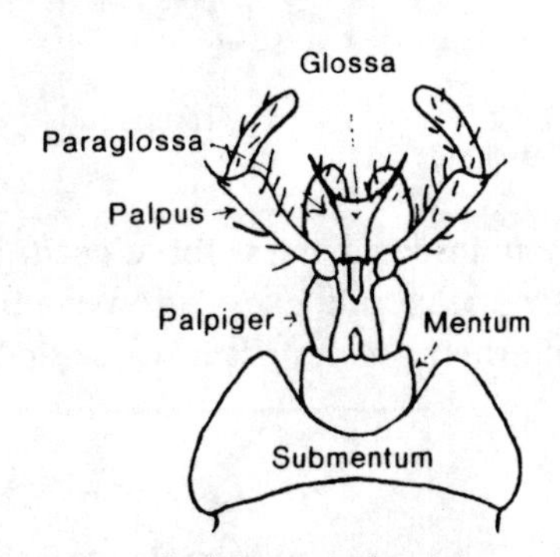

Figure 9 The labium with parts named.

A broad platelike structure is spread above all these mouth parts and functions with the labium to keep food in line with the mouth. This is called the upper lip or *labrum* (Fig. 10). At its base is a narrow supporting strip, the *clypeus.*

The compound eyes are located in various positions on the head. They are usually large and oval, but in some groups they are

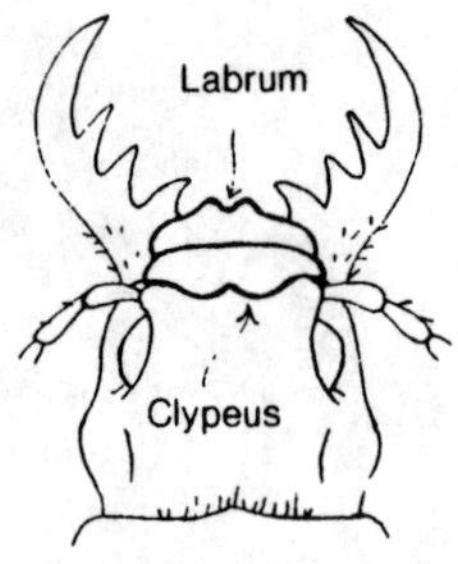

Figure 10 A beetle head showing labrum and clypeus.

small, and in others they may be kidney-shaped with the antennae arising from the inside curve of the eye. Sometimes the eyes are entirely divided (See Gyrinidae and some Cerambycidae) to form four eyes. A compound eye is composed of many eye units, each possessing a more or less six-sided lens (Fig. 11).

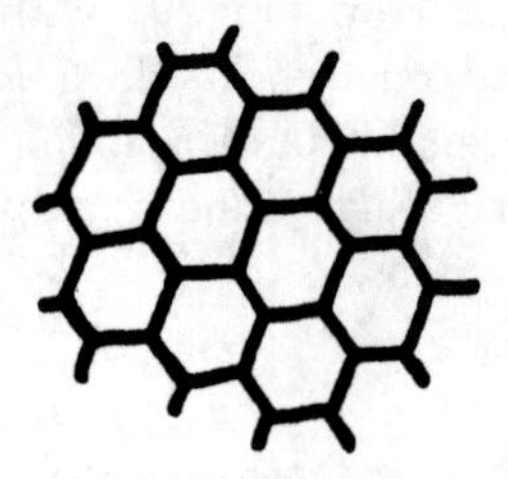

Figure 11 The lenses in the compound eyes of insects are six-sided.

Many insects possess three *ocelli* in addition to the compound eyes. A few beetles have ocelli, but their presence is rare (some Staphylinidae and Dermestidae, etc.). A few beetles that spend their entire lives in caves or in sand are wholly without eyes.

All insects always have only one pair of antennae. Beetles have the most diversified antennae of the class Insecta. Consequently their shape, size, and position on the head are used extensively in keys. Almost all beetles have 11-segmented antennae, but a few species (see Cerambycidae) may have more than 11 and others have the segments reduced to as few as 2, usually by the fusion of the segments. The various forms are named and illustrated in Fig. 12. The antennae are sensory organs used for feeling, hearing, and smelling.

The abdomen is composed of 10 segments; but usually only 5, or at most 8, segments are visible from below. The remaining segments are telescoped within the body and these parts are modified as egg-laying organs, *ovipositors* or male copulatory organs, *genitalia*. *Spiracles*, fairly large breathing pores, are arranged along the sides of the abdomen. These spiracles regulate the size of the opening of the *tracheae*, the internal breathing tubes. These openings are visible in some species, but most often they are hidden by the closed elytra. The upper part of each abdominal segment, the *tergum*, is usually membranous. The lower part, the *sternum*, is visible from beneath. In some species, the last abdominal tergum, the

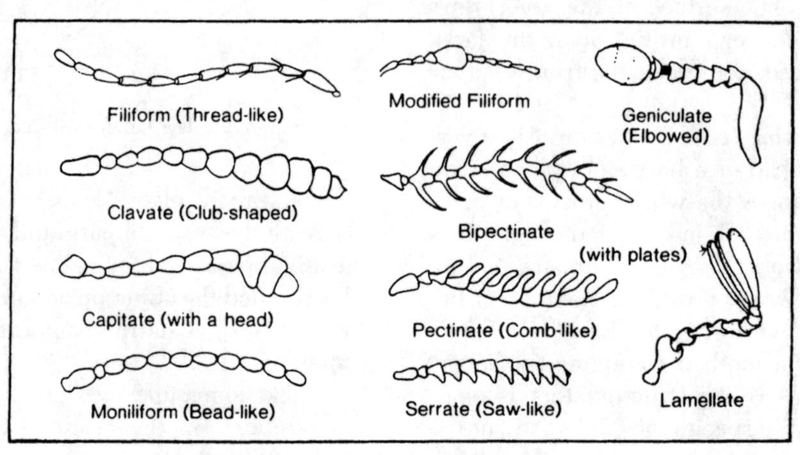

Figure 12 Some forms of beetle antennae.

pygidium is exposed. This usually occurs when the elytra are square at the ends, called *truncate,* appearing to be cut straight across. If a second abdominal segment is exposed as well, that segment is called the *propygidium.* The abdominal sterna (pl. of sternum) may have two or more segments fused and immovable. Sometimes the segments are distinct, and, in other cases, there is no evidence of the fusion.

Abdominal appendages are found in only one group, the males of the genus *Anaspis.* Even in this group the structures are small and apparently functionless; they are probably the vestiges of abdominal legs. The copulatory organs (genitalia), hidden within the abdomen except when in use, are variously modified. For many families these structures alone are used for the identification of species. This requires dissection and special preparation for small beetles. With larger beetles, the genitalia may be squeezed out and left to dry attached to the specimen.

USE OF KEYS

The preceding, short introduction to the anatomy of beetles points out many of the features that are useful in the identification of specimens. The keys that follow are for the most part self-explanatory. Every attempt has been made to make them useful. As pointed out previously, only about one in every twenty species of the beetles of the United States and Canada has been included. Therefore it is always possible that the specimen at hand is not included in this book. Again it is emphasized that the keys and descriptions should be carefully read and compared with the specimen. A misidentified specimen is useless and worse than no identification at all because it is misleading and may result in the spread of misinformation. Beginners should make tentative identifications and whenever possible have their specimens checked by specialists by contacting them through the Coleopterists Society or the North American Beetle Fauna Project. Once an accurate reference collection is assembled, it will be a lot easier to identify new material as it is collected.

Note that the keys have been constructed to eliminate the largest groups first. Be sure that you are using the correct part of the book. If the pictures look strange, go back and start over. Soon you will recognize the larger and major families and will be able to go directly to that part of the book to make identifications. To help you backtrack, the number in parenthesis following the couplet is the number that directed you to the couplet. Turn back and check for the correct alternative.

WHERE TO FIND BEETLES

The identification of beetles is both the key to the literature, i.e., gives a name under which is filed information about the beetles, and a means of recording new information. Therefore, when collecting beetles, it is important to keep a record of where they were caught.

Beetles occur almost everywhere. The brief list that follows is a handy guide to use when in the field so that "no stone is left unturned, and no nook and cranny is unexplored."

1. Crawling on ground
2. Beneath stones, wood, and leaves
3. In loose stones and gravel
4. In the soil in humus
5. Burrowing into the soil beneath humus
6. Along shore of bodies of water
7. In sea wrack
8. Swimming on water surface
9. Diving beneath surface
10. In or under dung
11. In or under carrion
12. In termite nests
13. In ant nests
14. In bee nests

15. In wasps nests
16. In bird nests
17. In mammal nests
18. In human habitations
19. Ectoparasites
20. In fungus
21. Feeding on leaves
22. Mining leaves
23. Boring in stems
24. Boring in roots
25. In soil feeding on roots
26. Feeding on flowers
27. Feeding on fruit
28. Feeding in seeds
29. Boring into wood
30. Under loose bark and in rotten wood
31. Feeding on sap oozing from plants
32. In or under decaying plant material
33. In stored plant products
34. In stored animal products
35. Flying around lights at night

Many collectors set up light traps (Fig. 14) using ultraviolet light. These may be obtained from many biological supply companies. Although the use of these traps, or the use of the lights and a white sheet to attract beetles (see next section) results in hundreds of specimens, little is learned about the habits and habitats of beetles. It is an excellent way to start and build up a reference collection.

HOW TO COLLECT, MOUNT, AND STORE BEETLES FOR STUDY

These brief pointers for starting a beetle collection are only for the beginner. Whole books have been written on this subject. As the collection grows, more equipment may be needed and more refined techniques may be used than those described here. The most important thing to remember about any insect collection is that it must have accurate data associated with it; it must be properly labeled. Secondly, it must be properly mounted if it is to have lasting values and allow others to study the specimens. Standard mounting methods must be used if you want the help of specialists. Finally, proper storage will preserve the specimens indefinitely so that specimens representing new information will always be available for study. At the very least, a new collection will provide new data on the distribution of the species, and it may record new biological information or even previously unknown species. So with the pleasures of collecting and forming a collection comes the responsibility of preserving the data it represents.

The instruments and materials needed to start a collection need not be elaborate. Some of the items may be constructed or obtained locally. Others may be purchased from one of the several biological supply houses. The following list of items is a starter. Their use is discussed in the following pages.

1. Several killing jars of a convenient size, e.g., tall olive jars with screw caps, not snap on or clampedged caps and large cork-stoppered test tubes. Sometimes a larger jar is useful for the very large beetles that may be attracted to lights at night.
2. Small vials with neoprene stoppers, corks, or screw caps (tightly sealed), with 70% rubbing alcohol (isopropyl alcohol).
3. Two nets (at least); one with light netting used to capture beetles flying; the other of heavy material to sweep across vegetation. An aquatic net is also desirable.
4. A white umbrella or similar device which is useful for catching beetles that drop from branches shaken or struck by a stick.
5. A pair of fine-pointed forceps to remove small beetles from decaying wood or from under bark; a heavier pair, knife, or screwdriver to help break up soil, wood, bark, decaying vegetation, etc. One may usually pick up large beetles by the fingers with less damage to the beetles than large forceps will cause (and rarely have

the fingers been damaged!). Forceps do aid in the capture of small beetles or beetles in places inaccessible to fingers.

6. One or more small camel's hair (or similar material) brushes for picking up extremely small beetles. This is done by moistening the tip of the brush in alcohol, touching it on the beetle, and then dipping the brush back into the alcohol in the catching jar.
7. An aspirator or suction bottle for removing small beetles from a collecting sheet, bark, etc. when otherwise impossible to capture.
8. A small notebook to record field observations.
9. Insect pins for mounting. These are of a standard length and several sizes are available. These must be purchased from a biological supply house because in this country, unlike Japan, insect collecting equipment is not sold in department stores.
10. A step block to regulate the height of the specimens and labels.
11. Good grade (rag) of heavy paper for the labels and mounting points.
12. Standard size insect boxes or drawers for storing specimens, or wooden cigar boxes as a start.
13. Knapsack to hold field gear.

Killing Jars

The best killing jars are those that use potassium cyanide as the killing agent. However, unless you are associated with a college or university it is difficult to obtain this material. Ethyl acetate, a liquid, is second best, followed by chloroform, or ether. The use of cyanide requires a jar constructed somewhat differently from those using liquids. Remember that all of these materials are extremely poisonous and must be handled with extreme care. They should never be used indoors and they must always be kept out of the reach of children. Mark each jar with POISON labels. Keep a first aid book handy in case of accidents and the phone number of the poison center of the area.

The cyanide jar (see Fig. 13 for type of jar) must have a tightly fitting screw cap. Jars with straight sides are best. Working outdoors on newspapers, place about 12 mm of cyanide crystals in the bottom of the jar. Tap this down with a stick with a flat end. Place about 7 cm

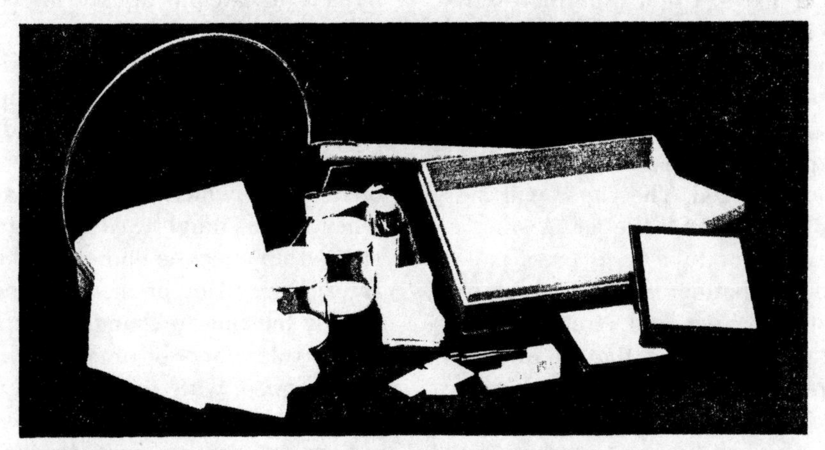

Figure 13 Basic collecting kit for beetle collectors. The insect net is unassembled to show its construction. Its net is folded and lying across the net ring. Killing jars and killing fluid is in the center of the photo. On the right are two types of mounting boxes; the larger has a solid cover; the two smaller ones, glass covers. In the center foreground are glassine envelopes with labels, packages of insect pins, forceps, and a step block.

of DRY sawdust over the cyanide. Press this down tightly (to about 4 to 5 cm). Then pour in about 12 mm of Plaster of Paris mixed with water so that the solution will just barely run. Tap the jar slightly to make the plaster of Paris settle smoothly across the top of the sawdust. Make sure that the sawdust is entirely covered by the plaster so that no bubble holes open up and give access to the sawdust. After the jars are complete, gather up the newspaper and any spilled cyanide and place these where no one will contact the material. Do not bring this into a building. Let the jars stand in the shade with the covers off for a few hours until the plaster is firmly set and the excess moisture has evaporated from the jar. As soon as the cyanide is moistened by the water from the plaster, the cyanide gas will be generated. They are extremely dangerous. To help prevent accidental breakage, each jar should be wrapped with adhesive tape around the base and near the top. Do not cover the entire jar as you will need to examine the contents to make sure that everything is dead.

A killing jar (or a large test tube with a cork stopper) using one of the liquid agents listed above (Fig. 13) may be made in the fashion described above. Use a little more sawdust in place of the cyanide. Some collectors like to place a cork upright on the sawdust and pour the plaster of Paris around the cork. After the plaster has set, the cork may be removed for easy charging of the jar with the killing agent and then replaced. The vapor will pass through the plaster and fill the jar. A simpler way to make a jar or tube is to press paper toweling into the bottom and then cut out pieces of blotting paper into circles slightly larger than the diameter of the jar or tube. These also are pressed into place to hold the paper. Jars constructed in this manner may be cleaned out periodically and reused. Once cyanide jars are "spent" they are useless and must be discarded into trash containers that will not be opened by someone else.

Most beetles may be collected directly into alcohol without damage. Alcohol preserved specimens have several advantages. They may be stored indefinitely in the alcohol (make sure to change the original alcohol for storage for it is diluted by the water in the specimens). They may be removed for mounting at any convenient time. Also they may be dissected at anytime without any further preparation. Dry killed specimens must be relaxed by placing them in a humid container (or in hot water) before they can be dissected without breakage.

The disadvantage of alcohol stored specimens is that they tend to become coated with a thin layer of fat that darkens their colors. Storage vials may also leak around the tops so that frequent checks must be made to make certain that the vials do not dry out. Some collectors put a small amount of glycerin in the storage alcohol (1-2%) which will keep the specimens moist if the jar does dry out.

Nets

Nets are of three types: aerial nets for catching insects in flight; sweeping or beating nets for taking insects resting on vegetation; and water nets for catching aquatic insects. These may be home built or purchased.

Aerial nets (Fig. 13) consist of a cloth bag, a metal ring 30-35 cm in diameter to hold the mouth of the bag open, and a handle to which the ring is attached. The ring is made of steel wire which will spring back into shape when used roughly. The metal ferrule is slipped up over the end of the handle to hold the ring in place or may be pushed back to release the ring to change bags. If the proper size ferrule cannot be obtained, the handle may be wrapped with wire. Usually tape is too weak to hold the net in place very long. Hose clamps are another possible alternative. Other types of construction plans are available or the handy person may easily devise his own plan of construction.

The net bag for an aerial net consists of

two parts: the light weight netting material (knitted in such a way that strands of fabric do not slip and allow a lot of little holes to open through which small beetles may escape) and a heavier muslin around the top to enclose the ring. The top part is double thick and cut so that it makes a band at least 5 cm around the ring. The end toward the handle is left open so that the ring may be inserted. The net bag should be at least twice as long as the diameter of the ring. This is so that the insects captured may be forced to the bottom and the ring of the net turned to close the opening and to hold them in place until they are removed and placed into the killing jar. The netting is, of course, sewn along the side and bottom. Then the net is turned inside out so that the edges of the seams are inside and away from twigs, branches, etc. upon which they might snag.

It takes practice to use skillfully an aerial net, capture a flying insect, and keep it in the net until it is removed for killing. Fortunately most beetles, once in the net, will close their elytra and walk on the netting from which they may be easily picked off.

The sweep net is constructed in the same manner except that it is made entirely of muslin. This will withstand the rough use of sweeping over vegetation consisting of all sorts of rough plants. After a few sweeps, the contents may be dumped into a small cage or similar container for sorting and preserving.

A water net may be made of cloth similar to the sweep net. However, the net frame should be shaped into a semicircle and be about two-thirds to one-half the diameter of the sweep net. The bag should be about the same length as the width of the net. However, a net welded together in the shape of a spoon with the frame covered with hardware cloth is much better. For many water beetles an ordinary kitchen strainer with a long handle attached is very useful.

The beating umbrella is used to catch insects that drop from vegetation when it is tapped or shaken. If a white umbrella is unobtainable, a net may be made by stretching a square of bleached muslin or canvas about 2/3rds of a meter square onto two sticks or small aluminum pipes. Slots for the sticks are sewn into each corner of the hemmed cloth. A third stick or pipe may be used as the beater. When not in use the apparatus may be disassembled for easy carrying and storage. For beating shrubs and tree limbs this sort of device is much better than the umbrella.

Traps

Four types of traps are used to catch beetles; pitfall, yellow pan, light, and Malaise. A pitfall trap is constructed from two tin cans, one fitting into the other. The larger is used as the bait container and the smaller one as the trap. The trap can has both the top and the bottom removed. Across the bottom is soldered some ordinarily fly screen. This prevents the beetles that are attracted to the bait from reaching the bait and becoming soiled or drowned. Both can be buried in the soil so that the top of the cans are even with the soil surface. A rain cover in the form of an aluminum tent open at each end is placed over the trap. Baits of various sorts may be used. Dead fish or frogs will attract carrion beetles; rotting fruit other beetles; dung, etc., still others. Traps such as these are subject to damage or destruction by foraging racoons and other small mammals.

Yellow pan traps are shallow enamel kitchen pans painted yellow. Such pans are buried in the soil at ground level. The pan is partially filled with colorless anti-freeze. Beetles and other insects walking across the field or forest floor will enter the trap and be killed and preserved in the fluid. Because the fluid does not evaporate, this type of trap may be left in place for long periods of time. However, it must be protected from the rain by an aluminum tent.

A Malaise trap is a barrier made of mosquito type netting, a green cloth used in tropical mosquito bars. The trap is erected so that it is in flyways in a forest, along a creek, path, or natural openings in the forest, or it may be used in open fields. The ordinary fly patterns of insects leads many bees, bugs, moths, butterflies, and beetles into the trap. The arrangement of the cloth is such that it leads the insect up to the peak of the trap into the collecting jar, which may contain alcohol or be a dry cyanide killing jar. These traps may be constructed, but they are fairly complicated and difficult to sew. Several designs are manufactured commercially but run to over several hundred dollars in cost.

The light trap (Fig. 14) requires electric power, either battery or house current. It is constructed of aluminum, bolted or soldered, or it may be plastic. The light is usually "blacklight" or low range ultraviolet. Insects, including beetles, have a different visible spectrum than human beings. The orange to red end of the spectrum is invisible, but the blue range of the spectrum attracts the insects. At the bottom of the trap is a container with a killing agent, usually alcohol. Beetles and many other insects are attracted to the light, flying around it, and soon drop down and are killed in the collecting jar. Many collectors place a baffle of coarse hardware cloth in the trap above the killing jar to screen out large moths. A blacklight may also be mounted in an ordinary fluorescent bulb holder and hung near a vertically suspended white sheet or the side of a white painted building. Then the insects are hand collected as they land on the sheet or wall. It should be noted that light traps work best on warm, humid, dark nights. They work best on summer nights in the South. One of the writers used them once in Northern Idaho with results that averaged about three insects per night. A word of caution: when using this system of collecting, protect your ears from moths by inserting a loose cotton plug in the ears or using an ear cover. Small insects often crawl into the ear canal, resulting in very painful experiences. If such happens, pour in some light table or baby oil that will immobilize the insect and allow for its removal with a pair of forceps.

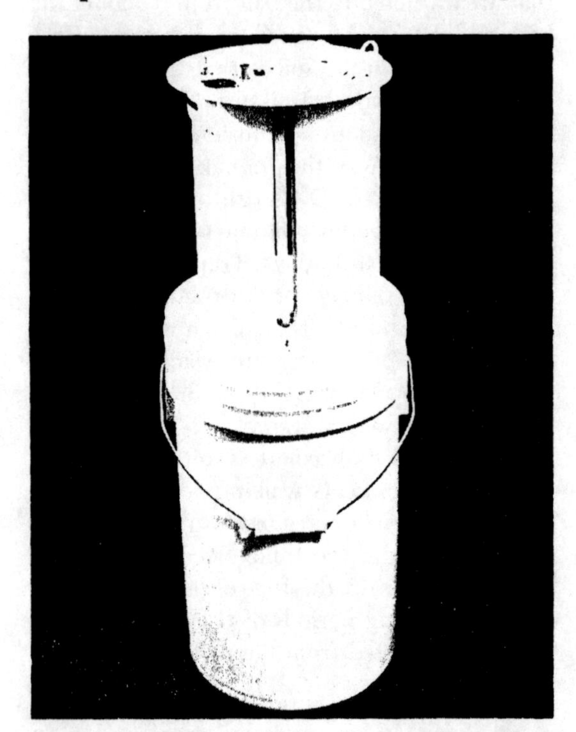

Figure 14 A portable light trap; may be used on house current or batteries. (Courtesy of Ward's Natural Science Establishment, Inc.)

Berlese Funnel

Several forms of the "Berlese" funnel trap are used to collect small beetles from soil litter. Such collecting is productive, even in mid-winter. Samples of leaf litter and similar materials are gathered into paper bags (labelled for source, etc.). The contents are placed into the funnel part of the separator. Heat from the light bulb suspended above the funnel drives the insects from the litter into the alcohol jar beneath the funnel. The hardware cloth inside the funnel keeps the litter itself from dropping into the jar. Some of the finer particles may drop through so the sample should be poured into a shallow dish (such as a Petri dish or Syracuse watch glass) and sorted with the aid of a low power dissecting microscope. Many of the small species of beetles are collected in this manner. Representative species of many of the so-called "rare" beetle families may turn up in these vials.

Mounting

Adult beetles are mounted on standard size (38 mm) insect pins. Although very fine pins are available, a Number 3 pin is suitable for all of the larger beetles (10 mm or over). Smaller, broad beetles may be pinned, but very small or narrow beetles should be glued to points. These are cut or punched out of heavy paper such as a good grade of drawing paper or heavy ledger paper. The beetles are attached to the point at the side of the thorax between the middle and hind legs. Care must be taken not to obscure the base of the legs because very important key characteristics are located there. For the larger beetles, the insect pin is carefully pushed through the body at the center of the right elytron. Make certain that the pin is placed so that the beetle is at right angles to the pin front and back as well as side to side (Fig. 15).

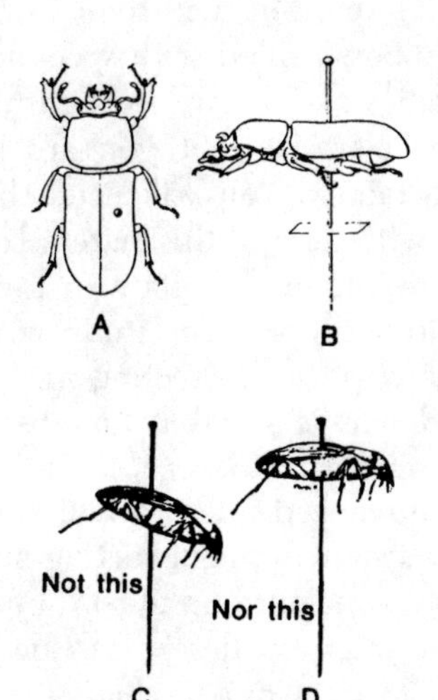

Figure 15 Pin beetles through right elytron as shown in *A*; regulate the height on the pin as shown in *B* to give room to grasp the pin with thumb and forefinger, unlike *C*. Specimens should not tip in front to back or side to side as in *D*.

Height on the pin is regulated by a pinning or step block (Fig. 13). This may be made from strips of fine grained wood, e.g., oak or hickory, 8 mm thick and 25 mm wide. These pieces are nailed together with brads and a very small vertical hole drilled in the center of each step. If hard wood is not used, thin pieces of aluminum may be nailed on the surface of each of the three steps. This will prevent "chewing" of the surface by the pins as they are thrust into the holes of the block.

The lower step of the block is used to regulate the height of the insect on the pin by inserting the head of the pin into the hole and adjusting the position of the beetle on the pin. This will insure that enough of the pin protrudes so that it may be grasped either by the fingers or with a pair of pinning forceps (especially made forceps obtainable from a biological supply company, Fig. 13). The top hole is used to regulate the height of points on the pin, and the middle step is used for locality labels. If more than one label is used, either as a set of locality and data labels, or data label and identification label, the lower step is again used to regulate the height of the lower label.

Paper points are used for mounting beetles too small to pin (Fig. 16). Usually it is necessary to bend the end of the point to fit the contour of the side of the beetle. The water soluble, milk base glues work very well. Be careful not to use too much.

As soon as the specimens are mounted, they should be labeled, taking care not to confuse data. Labels may be printed by hand using a fine point crow quill dip pen on heavy rag content paper. Labels should be as small as possible. Obviously, if there are a lot of specimens from the same locality, hand printed labels are impractical. Printed labels may be obtained by carefully typing the data on a typewriter with clean keys or with a carbon ribbon. Many labels are typed on a uniform sheet of typing paper and this sheet is taken to a printer. A printing plate is made (usually for an offset press) by reducing the page to about

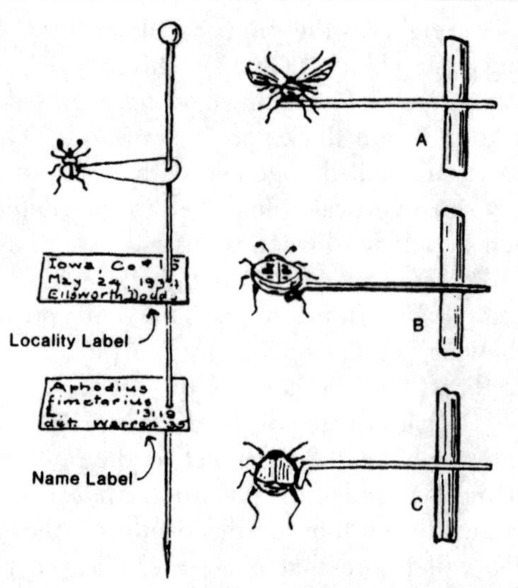

Figure 16 Small beetles may be glued to points cut from heavy paper; the position on the pin is shown in *A*; the method of gluing to point is shown in *B* and *C*.

1/3 the original, or to approximately 4 point type. Some biological supply companies offer this service and will have the labels set into type as well.

The data to be included on the labels are of two types: geographical and ecological. Often a collector will use two labels, one as a geographical label and the other for ecological or biological data. The locality or geographical label should have the name of the country, state or province, nearby town, and/or name of park, river, lake, etc. with greatly abbreviated directions. Note that the name of a city or town is not very exact. It implies that the collection was made within the city. This is seldom the case, so additional information is usually necessary. If the geographical label is to be used as a separate label, then this may be used by many persons who happen to collect at this site. If only one label is used, then either the date the collection was made is added, or space is left to print in the date with ink. Usually the collector's name is the bottom line of this label. The two label system is recommended because more precise locality data may be included. The second label gives the data and the collector's name. Additional details of how the specimen was collected, e.g., at ultraviolet light, and something about the collecting site are added. For example, the dominant plant cover, the type of habitat, etc. Some collectors also number their collections (lot numbers) to agree with numbered notes in their field notebook. This is a good practice, but never use only a lot number on a label. If for some reason the notes are lost or separated from the collection, the specimens are worthless as scientific objects. Specimens without data labels should be discarded. The old saying in museums is: "When in doubt, throw it out." Once the specimens are labeled, they are ready for sorting to species and arranging in the collection.

The purpose of the remainder of this book is to help you make identifications of your specimens to family, genus, and species. You will soon be able to sort specimens to family with relative ease. The key is arranged so that the most easily recognized families are keyed out first. The more obscure families are further on in the key. You will soon find that you have many boxes filled with specimens of the larger families and a few boxes with the small families or material that you are unable to identify to family. You will need the help of specialists with some of this material.

It is best to use an elbow or L-cork (Fig. 17) to hold specimens for study with a hand lens or a low power dissecting microscope. A good hand lens is suitable for observing the features used in the keys that follow. Eventually, the advanced collector will find it necessary to use a steroscopic dissecting microscope. These usually magnify up to 60X, but for some of the very small beetles it may be necessary to magnify to 120X or even higher.

A collection of beetles should be arranged according to a standard list of families. As explained previously, the list used in this book is the one generally accepted in the

Figure 17 The L-cork with plasticine; this is used to hold a specimen in any position for observation under a microscope. The head of the pin may be inserted into the plasticine for viewing a specimen from the ventral side.

United States. The genera and species within the families may follow the standard catalog list (best way) or they may be arranged alphabetically by genus under the family and alphabetically by species under the genus. In this way a species may be quickly located in a large collection. Some specialists will wish to arrange their material according to their ideas of the "relationship of the species", but this is subject to rapidly changing lists; so this method is not practical to use unless the tray system for housing specimens is followed.

Storage

Collections should be stored in standard size insect boxes with a suitable pinning bottom, usually polyethylene foam. These measure 8⅞ x 12⅞ x 2½ inches. Various grades of boxes are available commercially. Some are made of redwood, others of basswood, and certain economically priced boxes are made of chip board. The top is hinged, fits tightly to the bottom, and is held in place with hooks. This helps protect the contents from damage any skin beetles (Dermestidae). By using a standard sized box, the boxes are easily stored in a cabinet and may be moved along as the collection grows without the need to repin the entire collection.

Another storage system uses glass top drawers, also of a standard size, that slide into a cabinet. These cabinets are airtight to retain the fumigant placed in the drawer to protect the specimens from being damaged by museum pests. The drawers may have a pinning bottom or more usually, a masonite one. Unit trays of various sizes (Fig. 18) fill the drawers and insects are placed within these trays. Specimens of a single species occupy a single tray. Users of this more expensive storage system need not repin the contents of a tray unless new specimens are added and a larger tray is needed. As additional species are added to the collection additional trays are inserted in the proper place and the other trays moved along the drawer to the next drawer, etc. Three sizes and styles of drawers are available: Cornell University, U. S. National Museum, and California Academy of Science. Each of these is available commercially, with trays that fit each size of drawer. Carefully plan your collection for future expansion before you decide on the style of storage you wish to use. The best thing to do is to visit a major museum and discuss the matter with one of the curators of

Figure 18 Unit trays of various sizes are used in drawers; each tray holds all of the specimens of one species. As additional species are added to the collection trays are moved to make room for them thus eliminating the need to repin specimens.

the insect collection. Study the catalogs and price lists of several of the biological supply companies to see what is available. If the offerings of the various supply companies seem too expensive, then improvise by making your own receptacles. Boxes and drawers can be made at home very practically and economically.

Use of the Pictured Key

The successful use of any identification key is dependent on the user's knowledge of the characteristics featured in the key, the correct interpretation of the features as found on the specimen at hand, and the condition of the specimen being examined. The present key is simplified by the use of many illustrations in the text. The only word of caution is that it is necessary for the user to make certain that the specimen fits all of the conditions of the key couplet. If only part fits, unless there is an "or" separating sections, be prepared to "back track" in the key. Try both parts of the couplet until you are sure that one part fits better than the other. The numbers in parenthesis show the previous couplet number. This makes it easier to back-track.

For those who have not previously used an identification key, these brief instructions will tell how. With the specimen in place on an elbow cork, either under a microscope or hand lens, read *both* parts of the first couplet (1a and 1b). With the characters in mind examine the specimen. Decide which part of the couplet best describes the specimen. At first it will be necessary to check back with the key and examine one feature at a time. When a decision is reached, proceed to the couplet number indicated at the end of the line that fits best. Repeat the same procedure, always going to the couplet number indicated. Finally you will reach a name. In this book there is a separate key to the families of beetles. Each beetle family is numbered. As soon as you find the name of the family of your specimens, turn to the family section and repeat the keying operation until you find the generic, and, if given, the specific name. Always read the notes with the name. The distribution statement, the sizes range, along with any further descriptive information are all helpful in making the correct identification.

Pictured-Key to the Families of Beetles

1a Mouthparts reduced, front of head usually prolonged into a slender snout (beak); gular sutures (Fig. 19) on under side of head fused into one, or absent; tarsi 5-segmented, 4th very small, usually hidden in base of bilobed 3rd; prosternal sutures absent. (Snout Beetles, Weevils, Engraver Beetles, etc.) 131

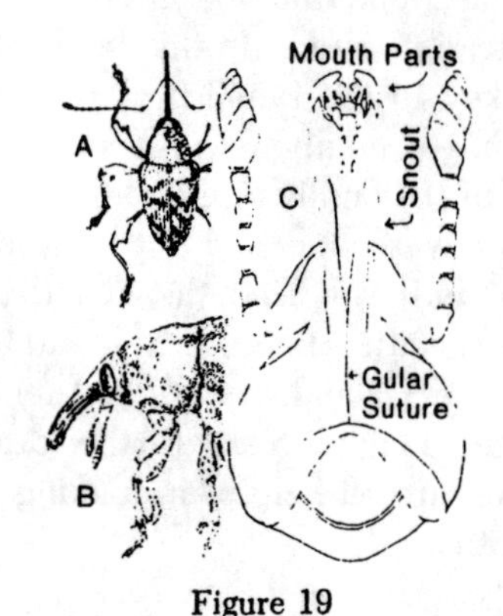

Figure 19

Figure 19 A, dorsal view of a snout beetle; B, side view; C, ventral view of head and snout showing gular sutures, mouth parts and antennae.

1b Head usually not prolonged into a narrow cylindrical snout (beak), but if so, the two gular sutures are distinct at both ends and not united; prosternal sutures distinct (Fig. 20). 2

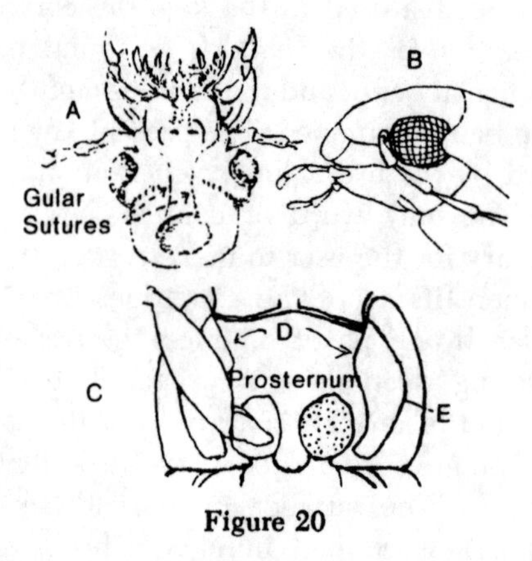

Figure 20

Figure 20 A, ventral view of head showing the two distinct gular sutures, mouth parts, etc.; B, side view of head; C, ventral view of prothorax showing prosternal suture (D).

2a (1) Hind tarsi with the same number of segments as the fore tarsi (Fig. 21). 3

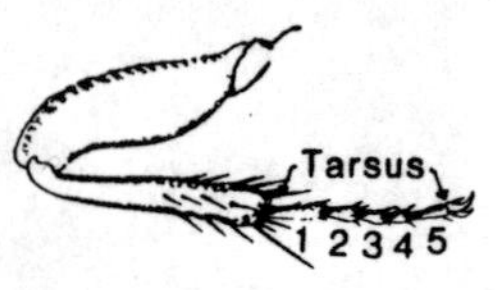

Figure 21

2b Hind tarsi with only 4 segments; fore tarsi with 5 segments (heteromerous beetles). 96

3a (2) All tarsi with five segments; (if 4th segment is very small and nearly hidden by 3rd (Fig. 22A, see 3B). 5

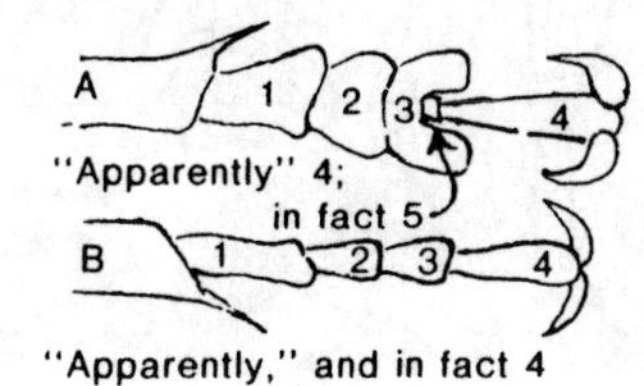

Figure 22

3b All tarsi with less than 5 more or less equal segments, (one segment may be elongate and the others nearly equal, or 4th very small and nearly hidden by 3rd) (Fig. 22A and B). 4

4a (3) All tarsi with 4 segments, or if 5, 4th is nearly hidden by bilobed 3rd (Fig. 22). 116

4b All tarsi with 3 segments (Fig. 23). 87

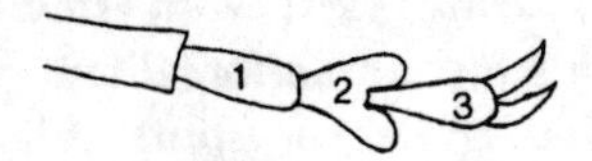

Figure 23

5a (3) First ventral segment of abdomen completely divided by hind coxal cavities (Fig. 24, C); outer lobe of galea of maxilla (Fig. 24, B) palpiform; antennae almost always filiform (Fig. 24, A) Suborder *Adephaga.* 9

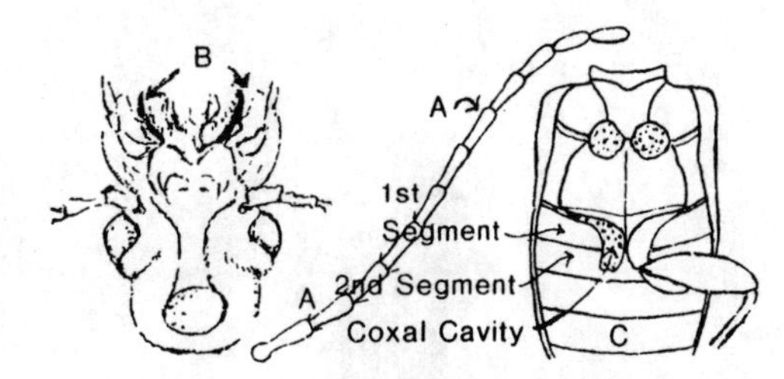

Figure 24

5b First abdominal segment not completely divided by hind coxal cavities (Fig. 25); outer lobe of galea of maxilla not palpiform. 6

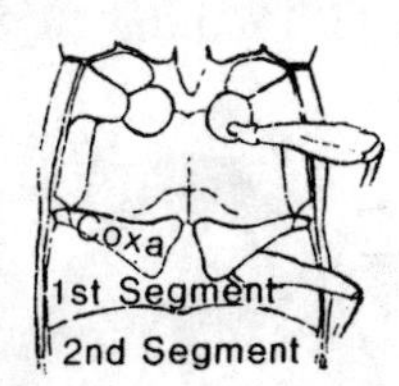

Figure 25

6a (5) Pronotum with notopleural sutures (Fig. 20, E); species either very small (not over 1.5 mm) with clavate or capitate antennae (Fig. 12) or if large (over 6 mm), antennae filiform. 7

6b Pronotum without notopleural sutures; antennae of various shapes. Suborder *Polyphaga* (in part). 17

7a (6) Minute beetles, less than 1.5 mm, with clavate or capitate antennae. Suborder *Myxophaga.* 8

7b Antennae filiform; elongate flattened beetles with rows of closely placed square punctures on the elytra. Fig. 26. Suborder *Archostemata.* ....
.... (p. 61) Family 1. CUPESIDAE

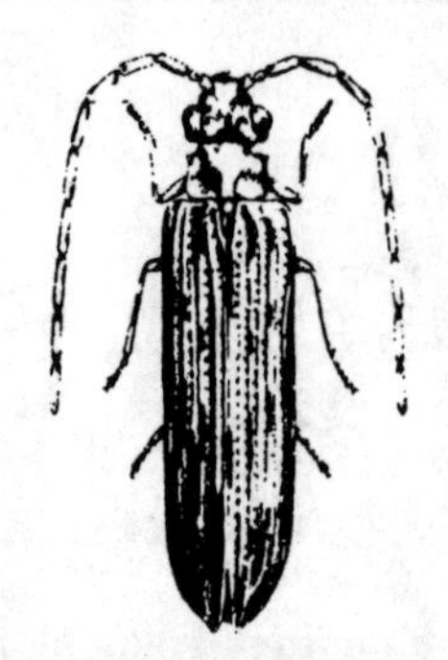

Figure 26

8a (7) Abdomen with 3 visible sternites; hind coxae very large; elytra entire. Fig. 26a. .... (p. 117) Family 10. SPHAERIDAE

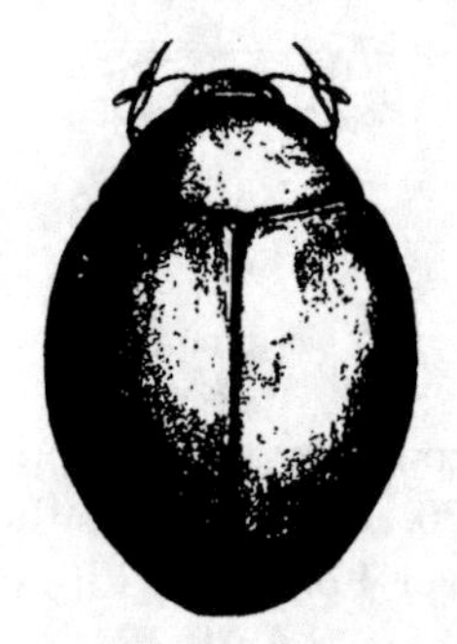

Figure 26a

8b Abdomen with 6-7 visible sternites; hind coxae smaller, well-separated; elytra truncate with at least 3 segments exposed. Fig. 26b. .... (p. 117) Family 11. HYDROSCAPHIDAE

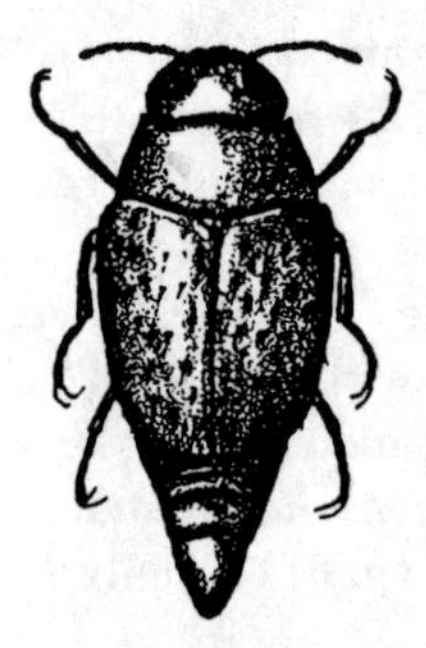

Figure 26b

9a (5) Eyes divided, one pair on the upper surface, one on the lower surface of the head (Fig. 27 B); antennae very short and irregular (auriculate); legs for swimming (Fig. 27C). (Whirligig Beetles). .... (p. 115) Family 9. GYRINIDAE

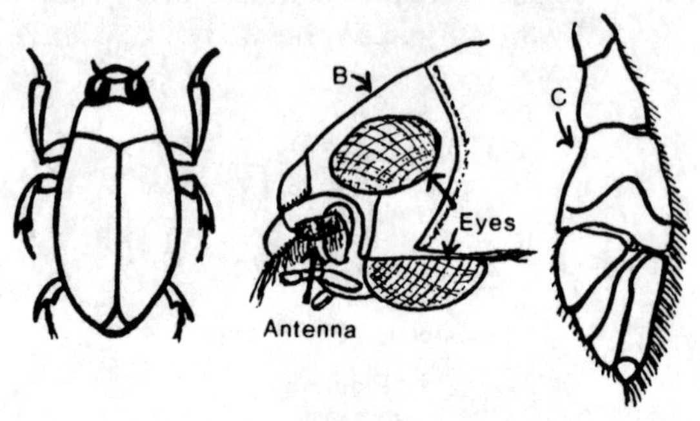

Figure 27

9b (9) Eyes not divided; antennae filiform or nearly so. 10

10a (9) Hind legs with fringes of long hair and a long spur; aquatic beetles. 11

10b (10) Hind legs without fringes of long hairs. 12

11a (10) Hind tarsi with single, straight, thick claw; anterior tibiae without recurved spines or spurs. Fig. 28. (The predaceous diving beetles). .... (p. 108) Family 7. DYTISCIDAE

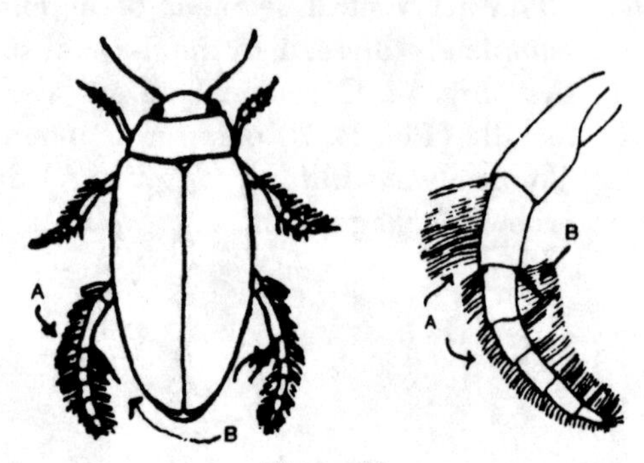

Figure 28

11b Hind tarsi with 2 slender, curved claws of equal length; anterior tibiae with strong, recurved spine at outer apical angle. Fig. 29. .... (p. 114) Family 8. NOTERIDAE

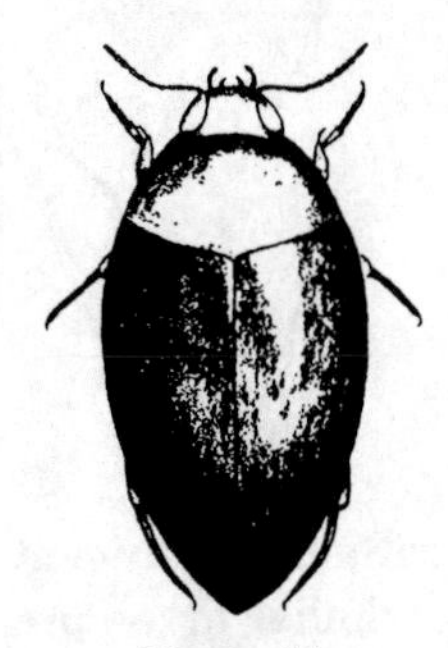

Figure 29

12a (10) Small oval water beetles; antennae with 10 segments, entirely smooth; hind coxae forming large plates covering the base of the abdomen. Fig. 30. (Crawling Water Beetles) .... (p. 106) Family 6. HALIPLIDAE

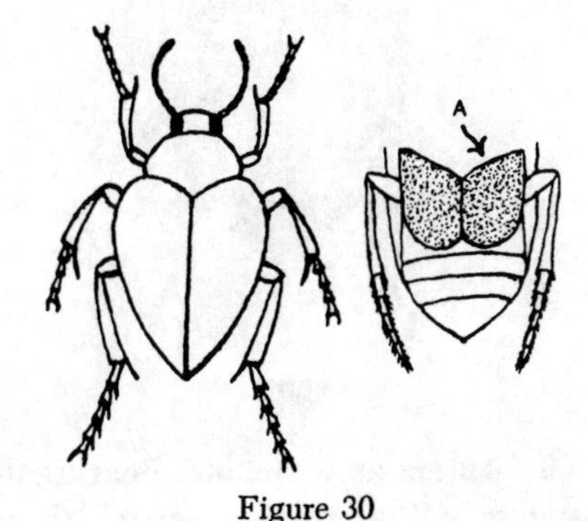

Figure 30

12b Terrestial; antennae with the 6 apical segments pubescent or moniliform. 13

13a (12) Antennae moniliform; elongate pronotum deeply grooved. (Fig. 32). .... (p. 62) Family 2. RHYSODIDAE

Figure 32

13b Antennae with at least the 6 apical segments pubescent (Fig. 31); hind coxae not greatly enlarged. 14

Figure 31

14a (13) Head including the eyes almost always wider than the pronotum (Fig. 33); antennae arising above base of mandible (Fig. 33) (Tiger Beetles). .... (p. 62) Family 3. CICINDELIDAE

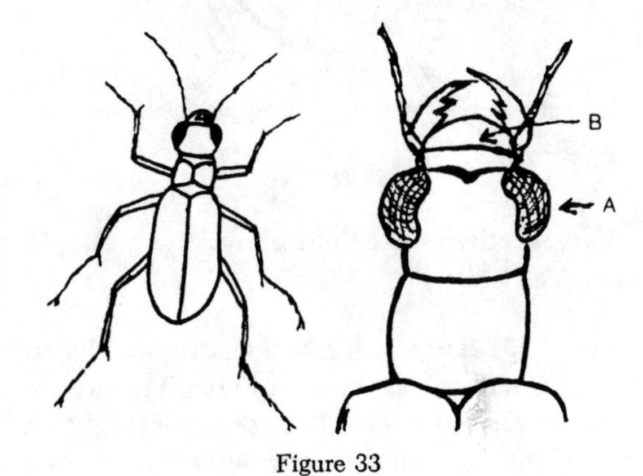

Figure 33

14b **Head almost always narrower than the pronotum; inner lobe of maxilla without a movable hook at its end; clypeus not extending beyond the base of the antennae; antennae arising between the eye and the base of the mandible. 15**

15a **(14) Beetles of round convex form; less than 8 mm in length; scutellum hidden; prosternum scoop-shaped, entirely covering the metasternum. (Fig. 34) Genus *Omophron*. .... (p. 70) Family 4. CARABIDAE**

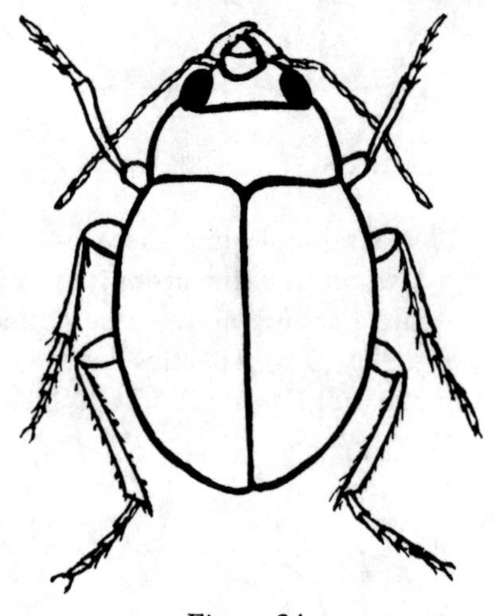

Figure 34

15b **Beetles unlike those above 16**

16a **(15) Body with prominent sensory hairs; usually elongate in form; small to large in size; often dark but sometimes light or with metallic colors; usually fast running. Fig. 35. .... (p. 70) Ground Beetles. Family 4. CARABIDAE**

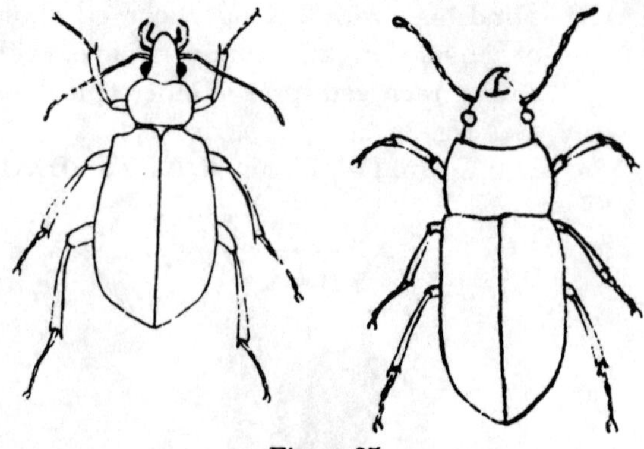

Figure 35

16b **Body without prominent sensory hairs; aquatic; similar to the preceding. Fig. 36. .... (p. 106) Family 5. AMPHIZOIDAE**

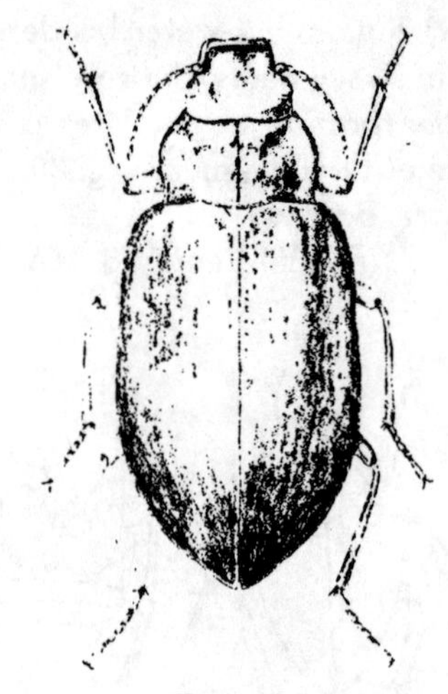

Figure 36

17a **(6) Antennae lamellate, bearing 3-7 segments at the end each of which extends to one side into a flattened plate-like structure (Fig. 37a); legs often fitted for digging. (Lamellicornia, Superfamily SCARABAEOIDEA). 18**

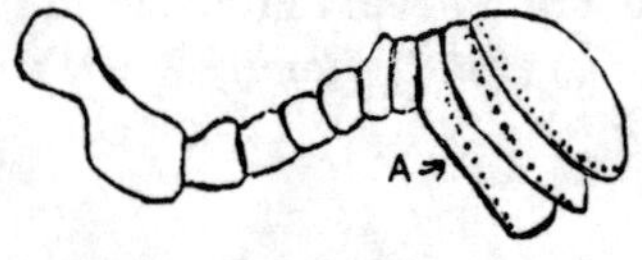

Figure 37

17b Antennae not lamellate. 21

18a (17) Plates composing antennal club flattened and capable of close apposition (Fig. 38). 19

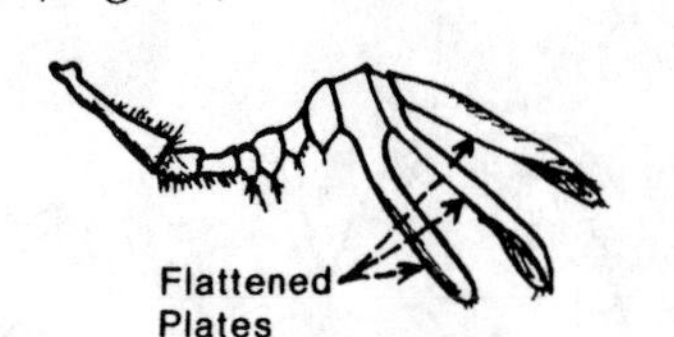

Figure 38

18b Plates of antennal club not capable of close apposition, usually but slightly flattened (Fig. 39). 20

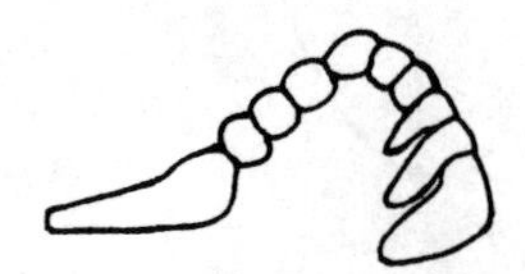

Figure 39

19a (18) Abdomen with 5 visible ventral segments; epimera of the mesothorax not reaching the coxae (Fig. 40); elytra usually covered with tubercles. (The Skin Beetles) TROGINAE. .... (p. 169) Family 30. SCARABAEIDAE

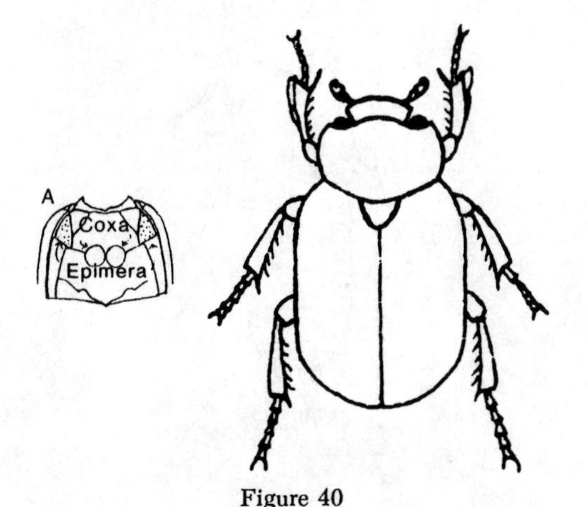

Figure 40

19b Abdomen with 6 visible segments, or if only 5, then the epimera of the metathorax reaching the coxae (Fig. 41). (p. 169). The Scarabaeids; The Dung Beetles, etc. .. Family 30. SCARABAEIDAE

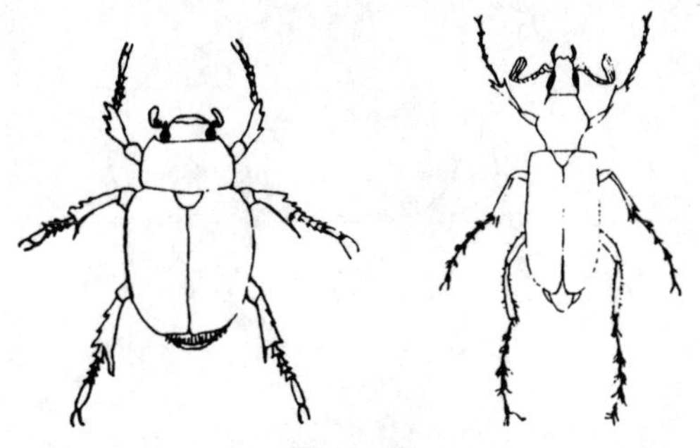

Figure 41

20a (18) Antennae not elbowed; mentum deeply notched, the ligula filling the notch. (Fig. 42). .... (p. 169) Family 29. PASSALIDAE

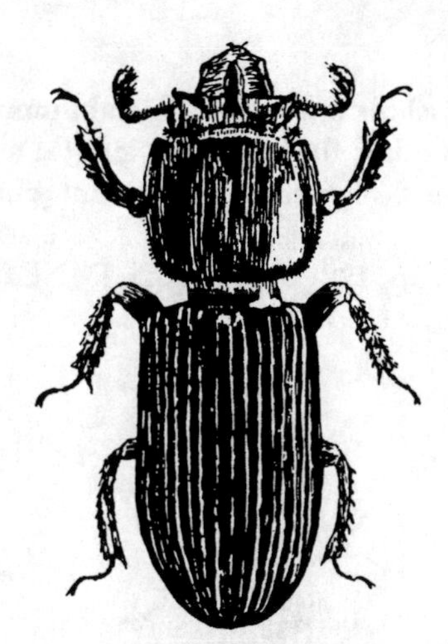

Figure 42

20b Antennae nearly always geniculate (elbowed); mentum entire. (Fig. 43) (The Stag Beetles). .... (p. 167) Family 28. LUCANIDAE

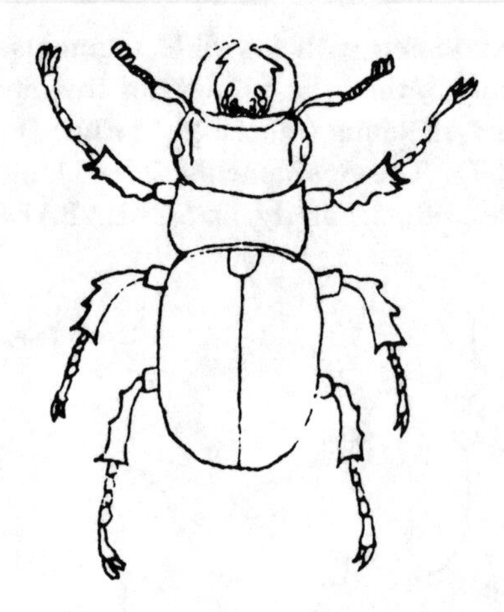

Figure 43

21a **(17) Antennae with segment 6 modified, cup-shaped. 22**

21b **Antennae various, but never as above. 23**

22a **(21a) Six or seven visible abdominal sternites; less than 2 mm in length; antennae with 5-segmented pubescent club. Fig. 45. .... (p. 122) Family 13. LIMNEBIIDAE**

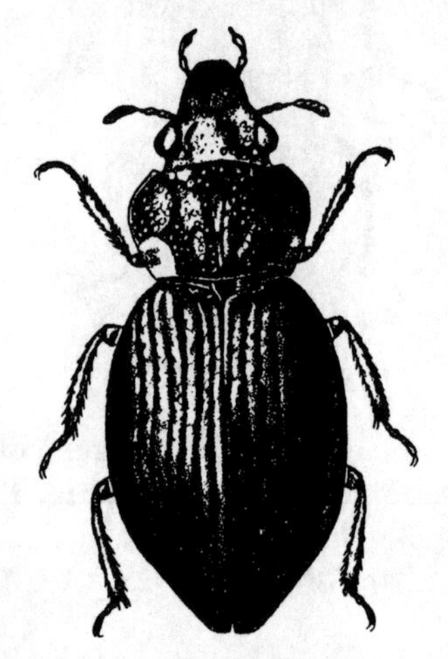

Figure 45

22b **Five visible abdominal sternites; antennae with 3 pubescent segments beyond the cupule. Fig. 44. (The Water Scavenger Beetles). .... (p. 108) Family 12. HYDROPHILIDAE**

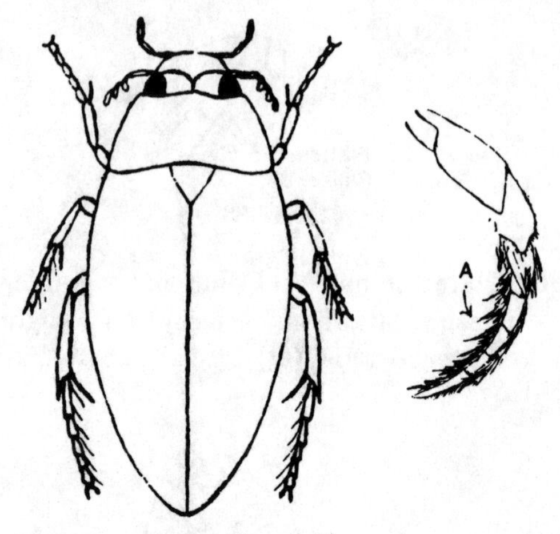

Figure 44

23a **(21) Elytra (Fig. 46) covering less than half the length of the abdomen which is exposed and hardened; wings folded under short elytra when not in use. 24**

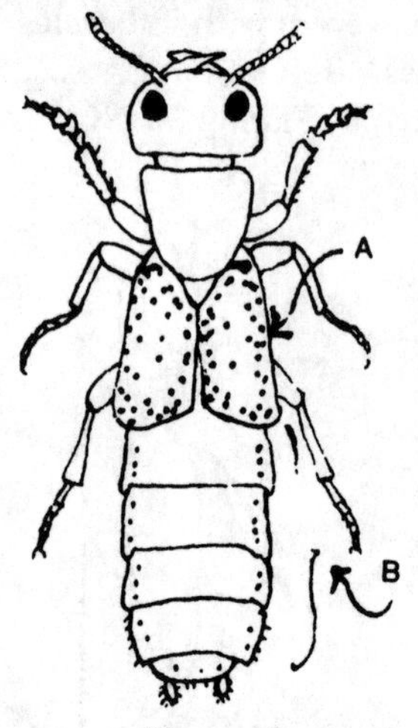

Figure 46

23b Elytra covering all or at least more than half of the abdomen or if short, the wings are not folded at rest or are absent; upper part of some of the abdominal segments membranous. 34

24a (23) Abdomen not flexible, very much wider than the prothorax, swollen, oval; usually 3-6 visible sterna; antennae often with less than 11 segments; tarsi 3-segmented; small or minute (0.5-5.5 mm, avg. 1.5 mm); robust species. Fig. 47, 48. ...... (p. 156) Family 25. PSELAPHIDAE

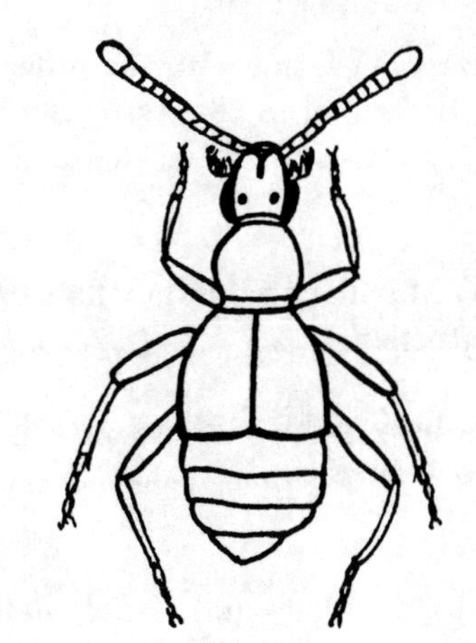

Figure 47

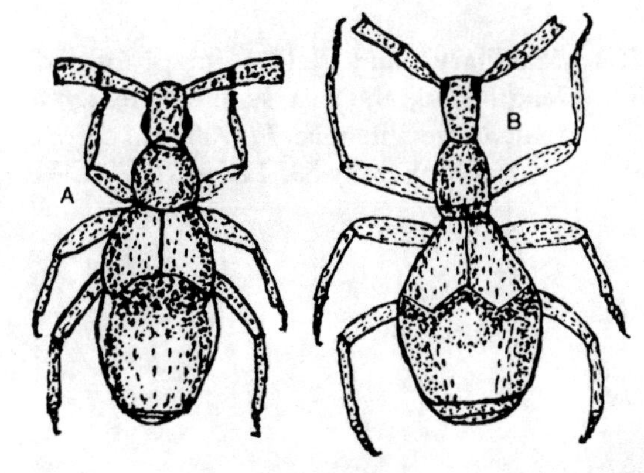

Figure 48

24b Abdomen flexible, not enlarged apically, parallel or tapering with 5-8 freely movable sterna; antennae of variable number of segments, usually 11; tarsi with many combinations of segments; usually small species, but occasionally large, slender species. 25

25a (24) Abdomen with 5 visible sterna. .. 26

25b Abdomen with 6-8 visible sterna. 28

26a (25) Inner wings present. 27

26b Inner wings absent; elytra short; extremely flat beetles (Inopeplinae). .... (p. 235) Family 72. CUCUJIDAE

27a (26) Three abdominal terga exposed, heavily sclerotized; antennae distinctly capitate. Fig. 49. .... (p. 231) Family 70. NITUDULIDAE

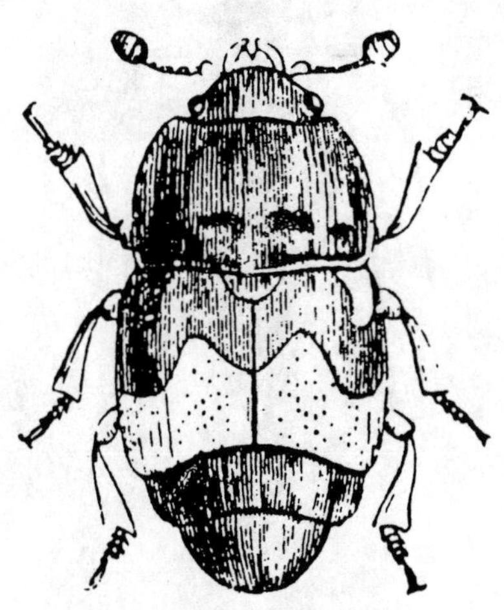

Figure 49

27b Elytra reduced, exposing entire abdomen; antennae pectinate; some resembling flies. Fig. 50. .... (p. 234) Family 71. RHIPHORIDAE

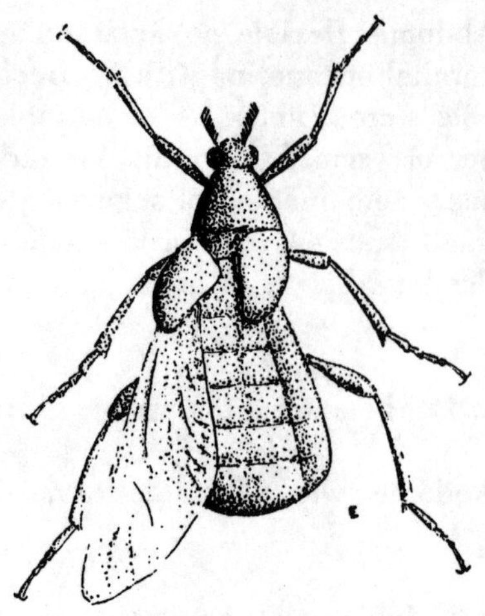
Figure 50

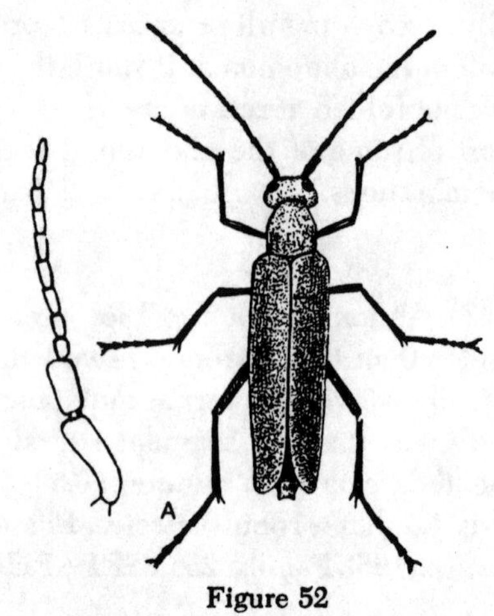

Figure 52

28a (25) Tarsi 5-5-4; abdomen soft; head deflexed, much broader than pronotum, latter narrowed in front. Figs. 51 and 52. (The Blister Beetles) .... (p. 283) Family 101. MELOIDAE

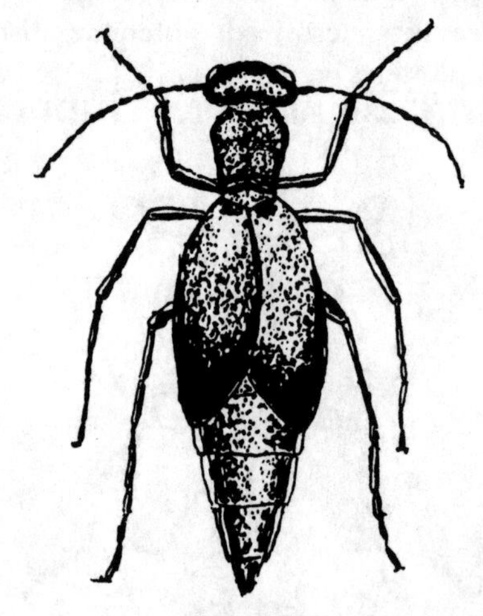
Figure 51

28b If tarsi 5-5-4, not with the other characters described in 28a; tarsi usually 5-5-5. 29

29a (28) Maxillary palpi flabellate or greatly lengthened. 30

29b Maxillary palpi without greatly lengthened segments. 31

30a (29) Maxillary palpi flabellate; hind wings with series of radiating veins. .. (p. 231) Family 69. LYMEXYLONIDAE

30b Maxillary and labial palpi greatly lengthened, the last segment nearly as long as the antennae. Fig. 53. .... (p. 206) Family 53. TELEGEUSIDAE

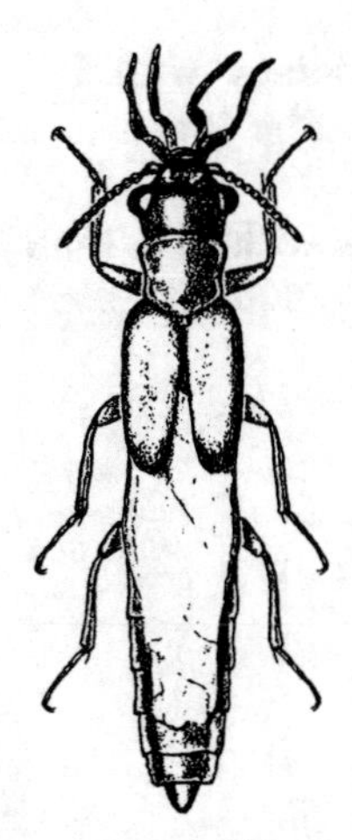

Figure 53

31a (29) Antennae reduced to 3 segments; eyes absent; flattened; beaver ectoparasites (*Platypsyllus*). Fig. 56. .... (p. 124) Family 18. LEPTINIDAE

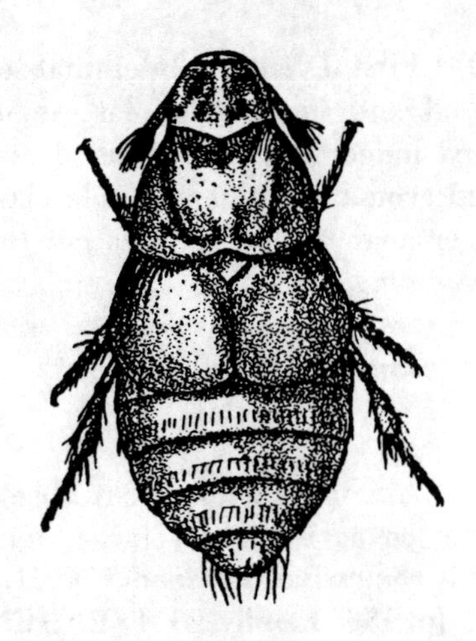

Figure 56

31b Not the above combination of characters. 32

32a (31) Antennae inserted at sides of frons in front of eyes, or, if inserted between the eyes, elytra exposing all abdominal terga. Figs. 54 and 55. .... (p. 134) Family 24. STAPHYLINIDAE*

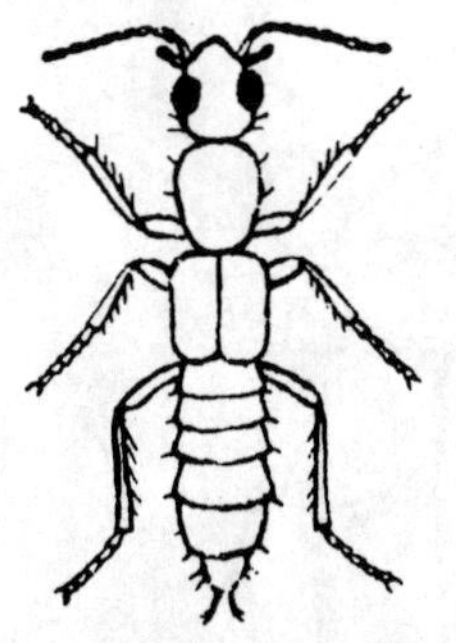

Figure 54

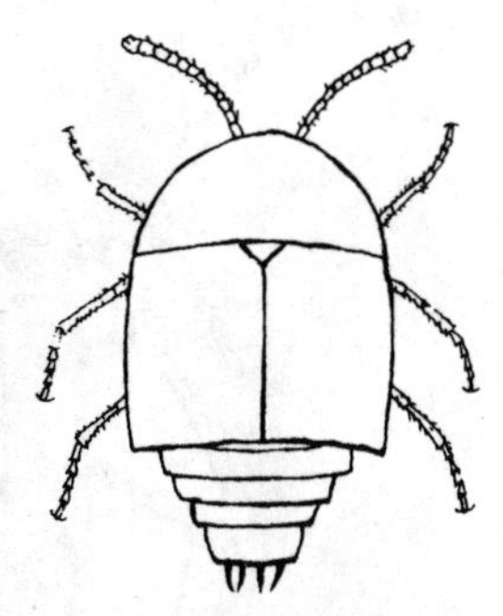

Figure 55

32b Antennae inserted on upper part of frons or at base of its anterior lobe. 33

33a (32) Antennae 12-segmented, flabellate. Fig. 57. .... (p. 206) Family 54. PHENGODIDAE

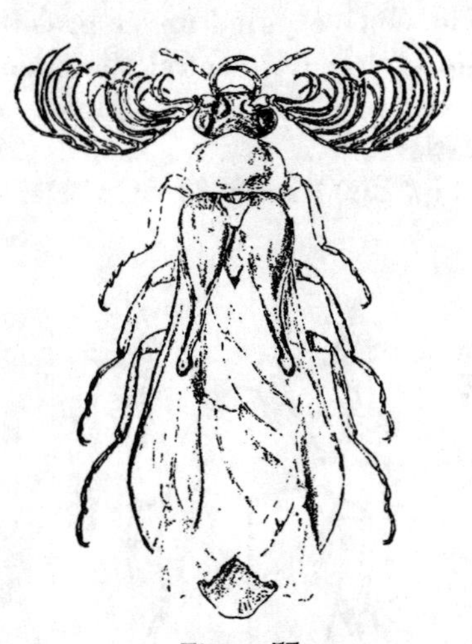

Figure 57

33b Antennae 11-segmented, filiform, serrate, or flabellate. Fig. 58. .... (p. 209) Family 56. CANTHARIDAE

*Family 105 Stylopidae will key out here. See p. 293.

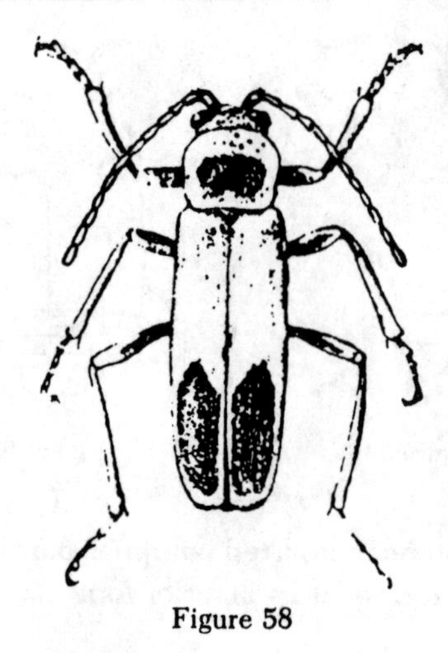

Figure 58

34a (23) Small, oval, convex, very shiny beetles with conical tipped abdomen (Fig. 59a) exposed under broadly truncate elytra; 6-7 ventral abdominal segments. Fig. 59. (The Shining Fungus Beetles). .... (p. 132) Family 23. SCAPHIDIIDAE

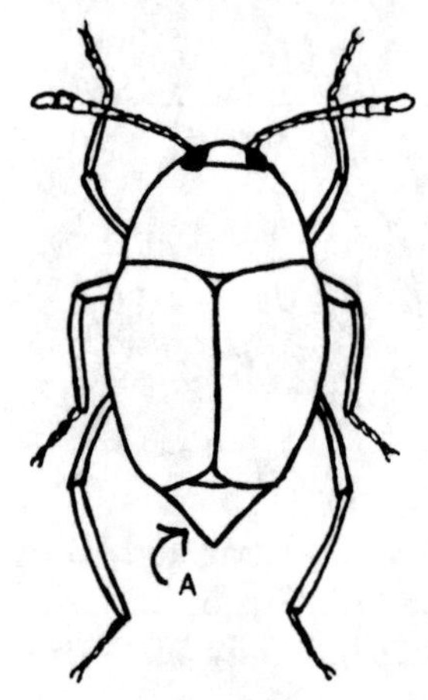

Figure 59

34b Not as in 34a. 35

35a (34) Abdomen with 7-8 visible ventral segments. Fig. 60a. 50

35b Abdomen with less than 7 ventral segments. Fig. 60b. 36

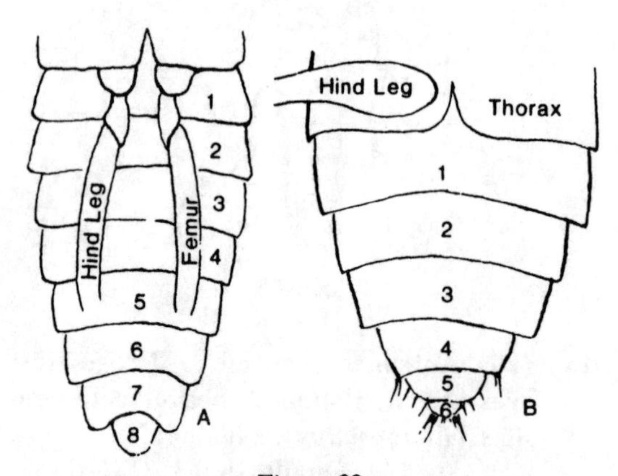

Figure 60

36a (35) First 3 ventral abdominal segments fused and immovable; last segment of tarsi longer than the other 4 combined and armed with long, simple claws; live in or near water, but legs not fitted for swimming. 37

36b Not as in 36a above. 38

37a (36) Abdomen with 6 ventral segments; front coxae with very large trochantin; body somewhat flattened. Fig. 61. .... (p. 156) Family 25. PSEPHENIDAE

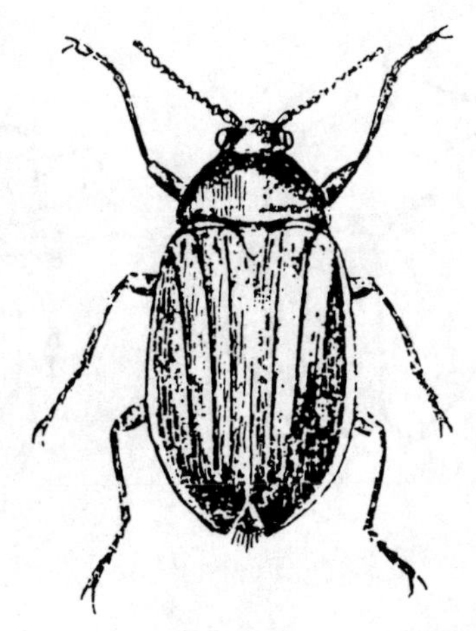

Figure 61

37b Abdomen with 5 ventral segments; front coxae rounded, without trochantin; body convex and lightly pubescent. Fig. 62. .... (p. 189) Family 45. ELMIDAE

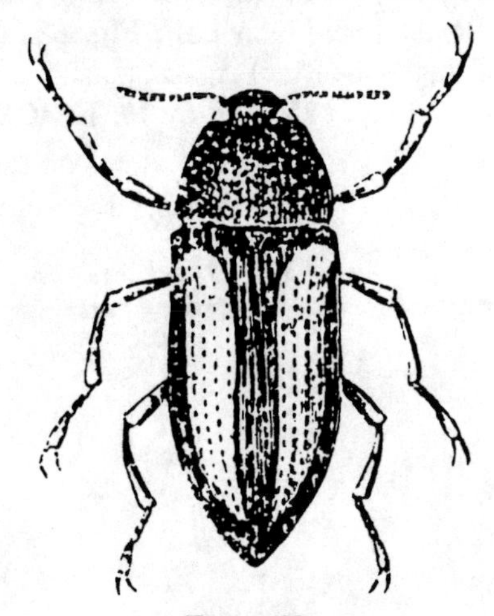

Figure 62

38a (36) Abdomen with 6 ventral segments. 39

38b Abdomen with 5 ventral segments. 53

39a (38) Small (2-5 mm) flattened beetles with only rudimentary eyes or none; live in nests of small mammals, birds, and bumblebees. Fig. 63. .... (p. 124) Family 18. LEPTINIDAE

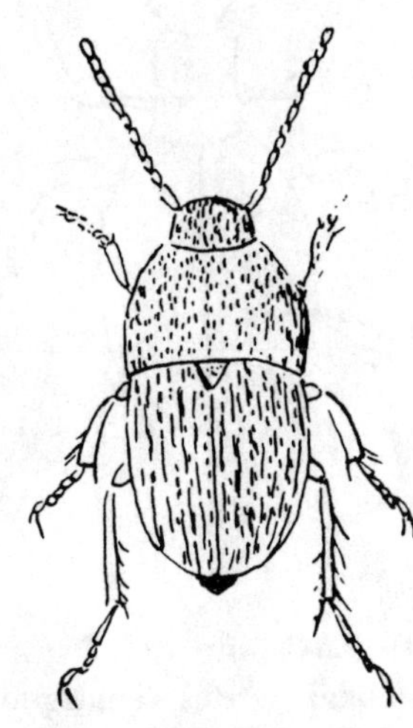

Figure 63

39b Not as in 39a. 40

40a (35) Less than 3 mm in length. 41

40b Mostly over 3 mm in length, if smaller, without enlarged coxal plates; broad oval in shape. 42

41a (40) Length less than 1 mm to about 3 mm; convex, shiny, brownish or blackish beetles; usually covered with erect hairs; eyes coarsely granulate. Fig. 64. (Ant-like Stone Beetles) .... (p. 126) Family 20. SCYDMAENIDAE

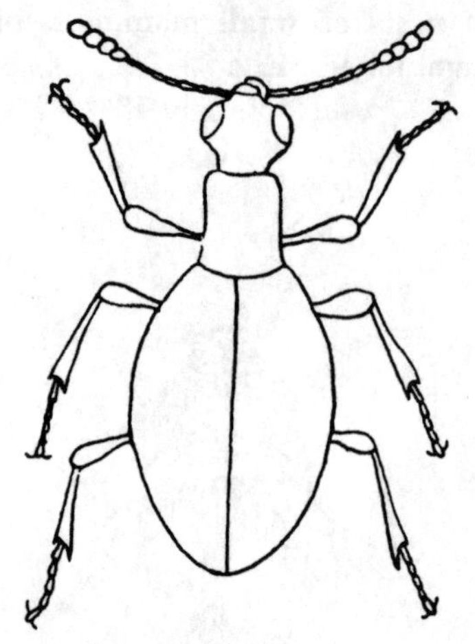

Figure 64

41b Length 2.5-3 mm; dull brown or black; hind coxal plates enlarged, covering most of the first abdominal segment. Fig. 65. .... (p. 182) Family 31. EUCINETIDAE

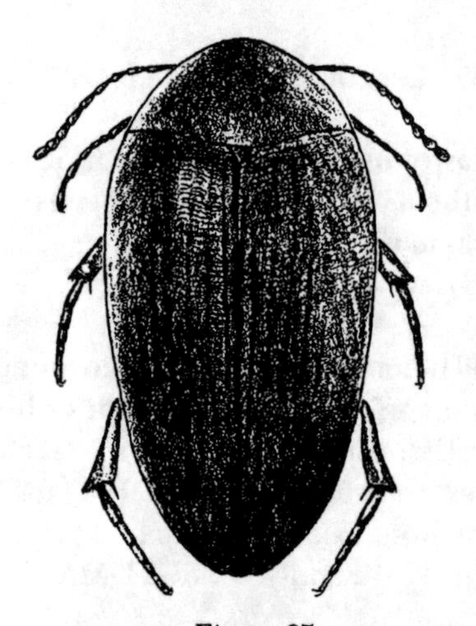

Figure 65

42a (40) Mostly large beetles, usually over 12 mm long, either broadly flattened (Fig. 66a) or heavy with elytra short, exposing 2 or 3 segments of abdomen (Fig. 66b) (Carrion Beetles). Fig. 66. .... (p. 127) Family 21. SILPHIDAE

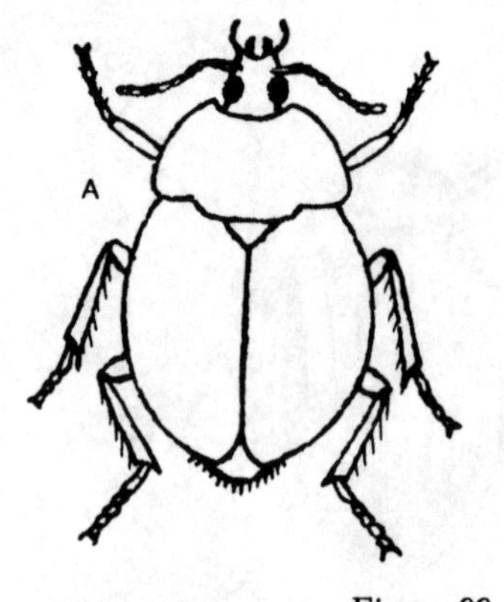

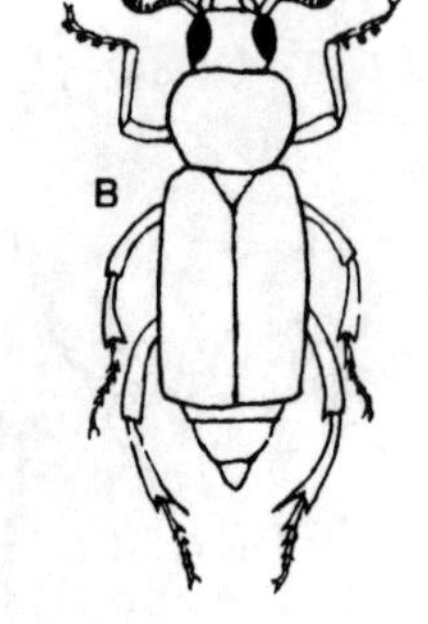

Figure 66

42b Smaller beetles, size less than 12 mm. 43

43a (42) Elytra embracing the sides; smooth and shiny; head and appendages capable of being retracted; head and thorax deflexed forming a ball. Fig. 68. (Round Fungus Beetles) .... (p. 125) Family 19. LEIODIDAE

Figure 68

43b Not possessing the above characters. .. 44

44a (43) Antennae clavate with segment 8 much smaller than 7 or 9; pronotum and elytra usually with cross striations. Fig. 67. (Small Carrion Beetles). .... (p. 131) Family 22. LEPTODIRIDAE

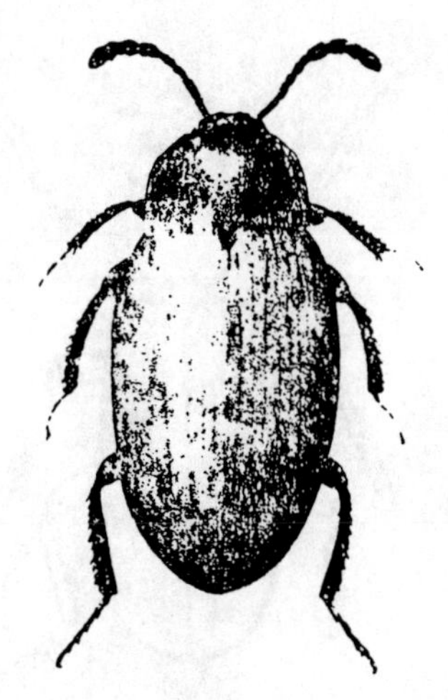

Figure 67

44b Not possessing the above characters. Usually under 10 mm in length; body may be cylindrical, elongate, or broad and oval, or soft-bodied; if associated with carrion, the front coxae are long with distinct trochantins. Fig. 69a. 45

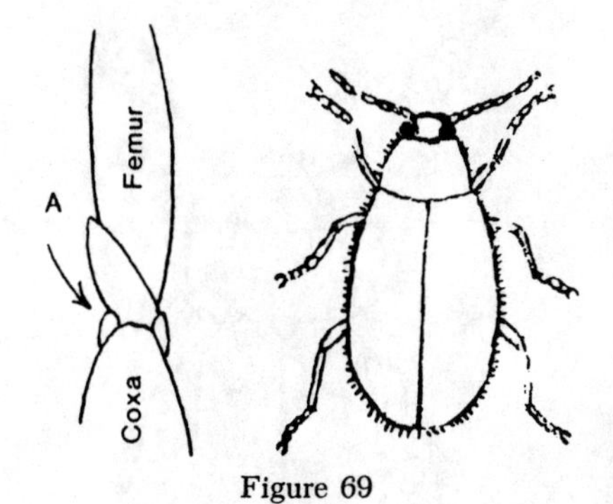

Figure 69

45a (44) Hind coxae conical. 46

45b Hind coxae flat; covered by femora when at rest. 49

46a (45) Front coxae long with distinct trochantin (Fig. 69a); soft-bodied, rarely associated with carrion. Fig. 69. (Soft-winged Flower Beetles). ....
.... (p. 229) Family 68. MELYRIDAE

46b Front coxae without visible trochantin. ....
.... 47

47a (46) Two ocelli present on tip of head between the eyes; maxillary palpi simple; body more or less ovate. *Brathinus*, 3 spp. Fig. 70. ....
.. (p. 134) Family 24. STAPHYLINIDAE

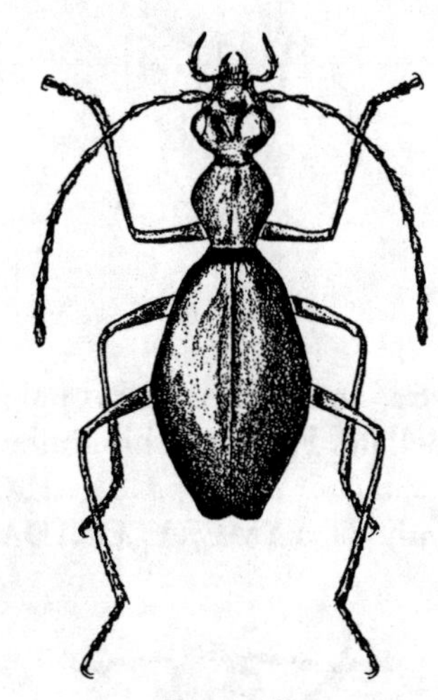

Figure 70

47b Ocelli absent; maxillary palpi may be flabellate; body elongate. 48

48a (47) Elytra shortened, exposing several segments; length about 2 mm. Fig. 71. (p. 213) Family 58. MICROMALTHIDAE

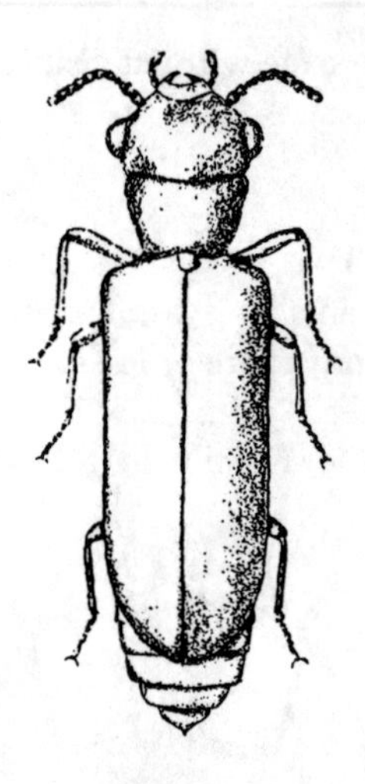

Figure 71

48b Elytra entire; maxillary palpi of male flabellate. Fig. 72. (Ship-timber beetles). .... (p. 231) Family 69. LYMEXYLONIDAE

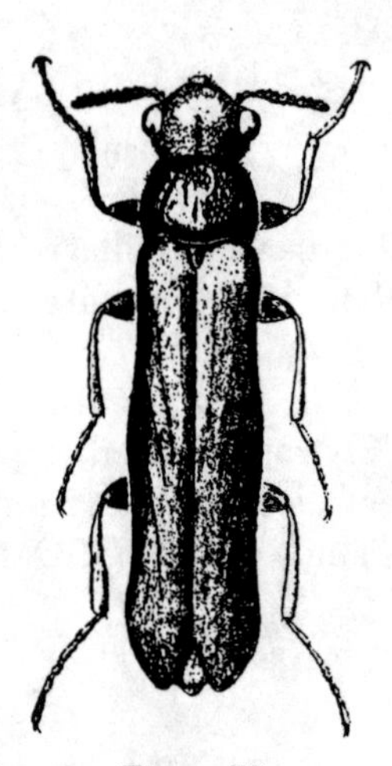

Figure 72

49a (45) Fourth tarsal segment equal to the others in length; disk of thorax not separated from the sides. Fig. 73. (Checkered Beetles). .... (p. 223) Family 67. CLERIDAE

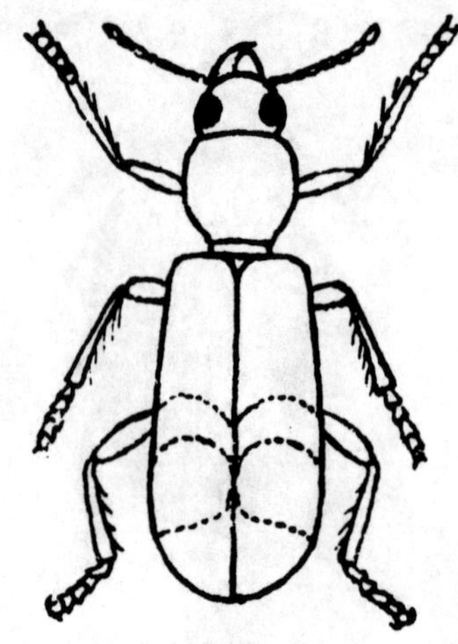

Figure 73

49b Fourth tarsal segment very small, scarcely, if at all, visible from above; disk of thorax separated from the sides by an elevated marginal line. Fig. 74. (Subfamily Corynetinae) .... (p. 223) Family 67. CLERIDAE

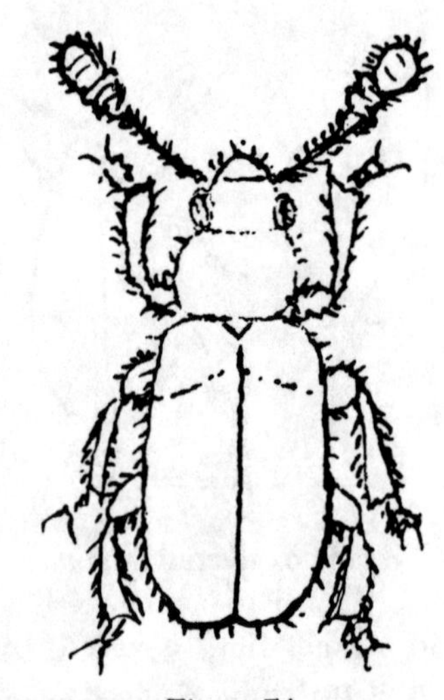

Figure 74

50a (35) Middle coxae separated from each other; epipleura absent. Fig. 75. (Net-winged beetles). .... (p. 211) Family 57. LYCIDAE

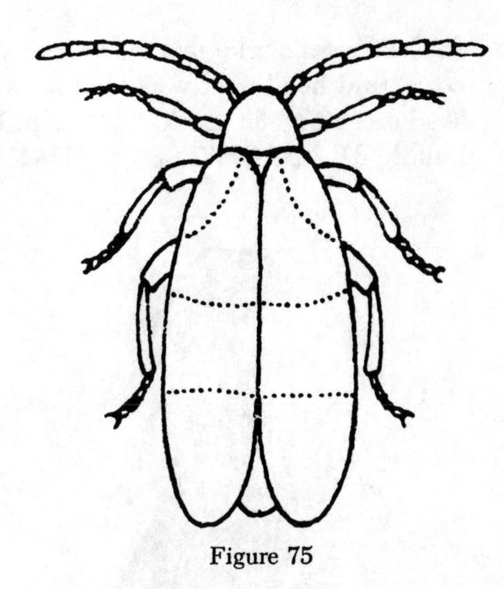

Figure 75

50b Middle coxae touching, distinct. 51

51a (50) Episternum of metathorax sinuate with an S curve on inner side (Fig. 76a); head, if at all covered, less than half covered by pronotum. Fig. 76. (Soldier Beetles). ....
.... (p. 209) Family 56. CANTHARIDAE

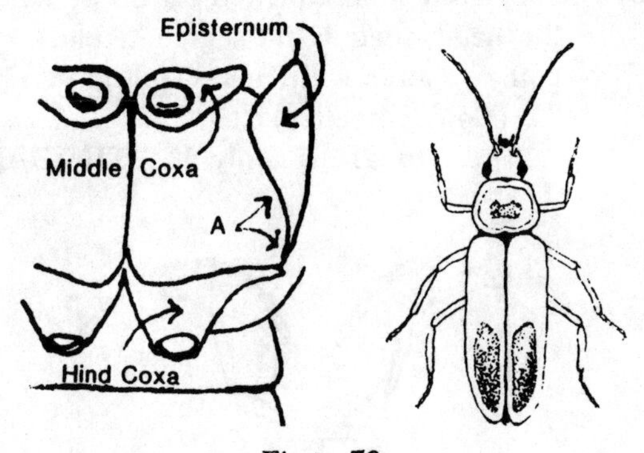

Figure 76

51b Episternum of metathorax not sinuate on inner side. (Fig. 77b) 52

52a (51) Head more or less completely covered by the pronotum; antennae usually close at base. Fig. 77. (Firefly Beetles) ..
.... (p. 207) Family 55. LAMPYRIDAE

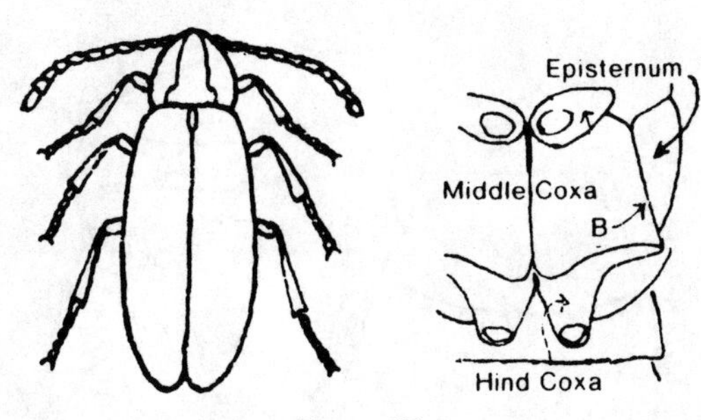

Figure 77

52b Head not at all covered by thorax; antennae well separated at their base. Fig. 78. (Glowworms) ....
.... (p. 206) Family 54. PHENGODIDAE

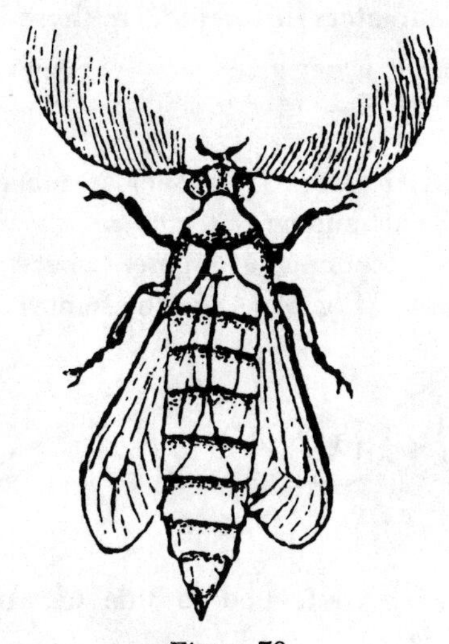

Figure 78

53a (38) Antennae both elbowed and club-shaped (Fig. 79a); hard, usually small, black, red and black, brown, or dark green beetles; elytra square-cut (truncate) at apex (Fig. 79b) exposing 2 segments of the abdomen (pygidium). Fig. 79 (Clown Beetles). ....
.... (p. 161) Family 26. HISTERIDAE

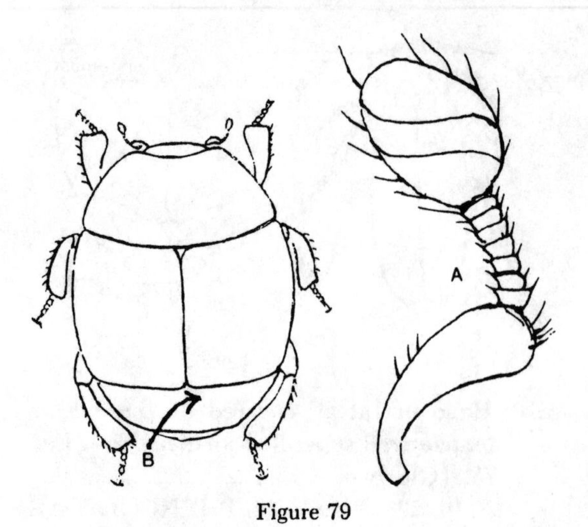

Figure 79

53b Characters different from those in 53a. 54

54a (53) Femora (Fig. 80f) attached to end of trochanter (Fig. 80t) or very near the end forming a segment between the coxae (Fig. 80c) and the femora. 55

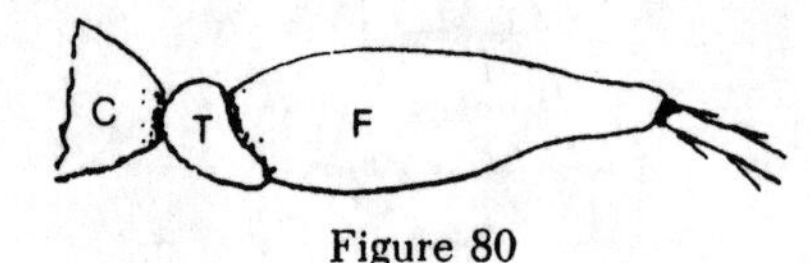

Figure 80

54b Femora attached to side of trochanter (Fig. 81). 60

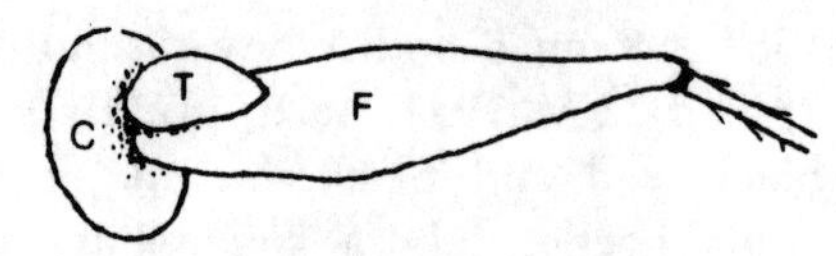

Figure 81

55a (54) First segment of tarsus shorter than second. 56

55b First segment of tarsus equal to or longer than the second. 58

56a (55) Prosternum large, area between the coxae and head nearly as long as wide; head erect. Fig. 82. .... (p. 185) Family 37. BRACHYPSECTRIDAE

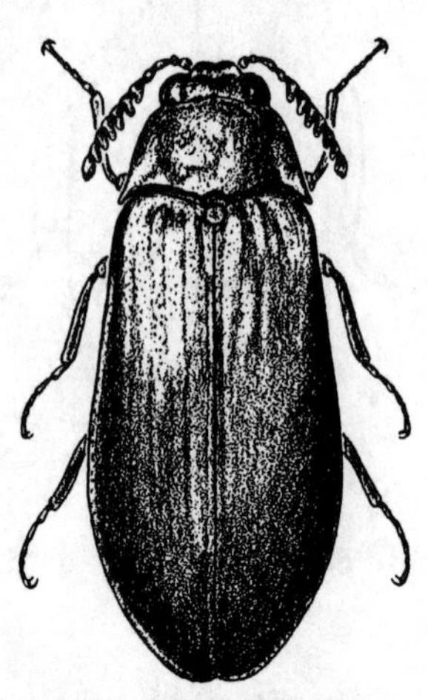
Figure 82

56b Prosternum short, head reaching coxae when in deflexed position. 57

57a (56) Antennae attached on the front of the head, their basal segments close together; thorax without side margins. Fig. 83. (Spider Beetles) .... (p. 216) Family 62. PTINIDAE

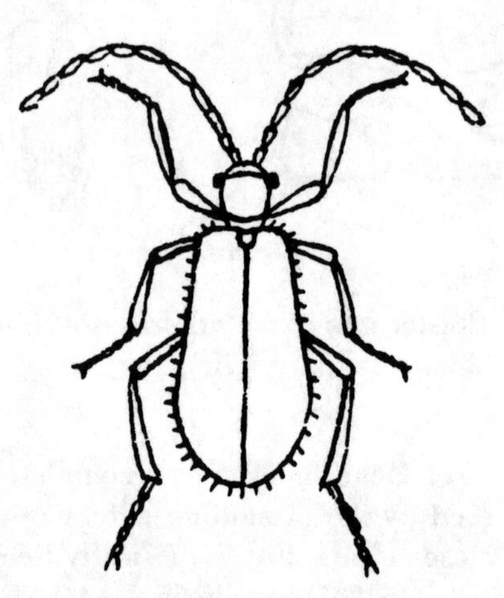
Figure 83

57b Antennae attached in front of eyes on side of head, their basal segments usually more widely separated. Fig. 84. (Death Watch Beetles) .... (p. 217) Family 63. ANOBIIDAE

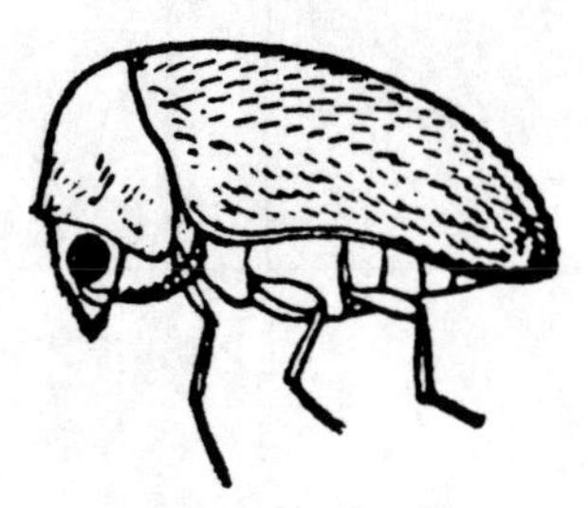

Figure 84

58a (55) First tarsal segment as long as second to fourth combined; hind trochanters nearly as long as the femora. (Fig. 85) .. .... (p. 204) Family 50. CEROPHYTIDAE

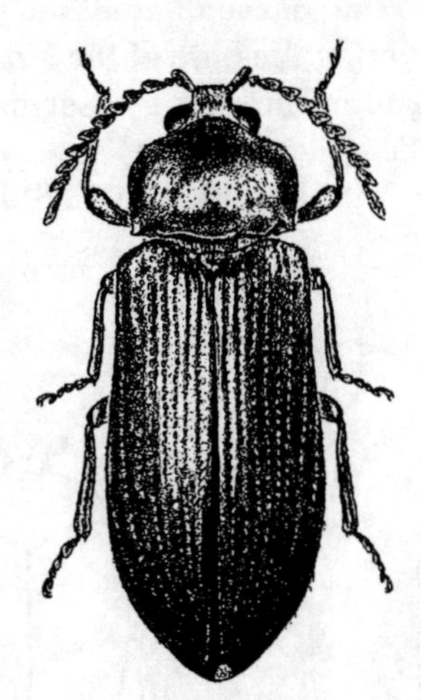

Figure 85

58b First tarsal segment no longer than second and third combined; trochanters not greatly elongated. 59

59a (58) First ventral segment of abdomen not longer than the others; antennae with club of 3-4 segments; black or brown cylindrical beetles with the head usually directed downward and hidden by the pronotum. Fig. 86. (Horn Powder-post Beetles) .... (p. 000) Family 64. BOSTRICHIDAE

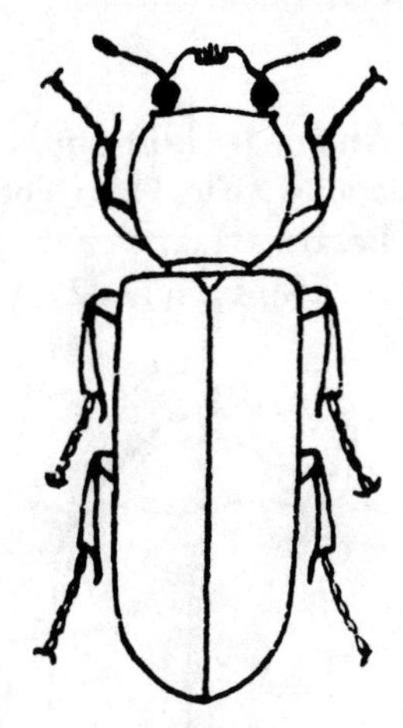

Figure 86

59b First ventral segment of abdomen longer than the others; antennal club with only 2 segments serrate; slender, brown or black; head distinct. Fig. 87. (Horn Beetles) .... (p. 221) Family 65. LYCTIDAE

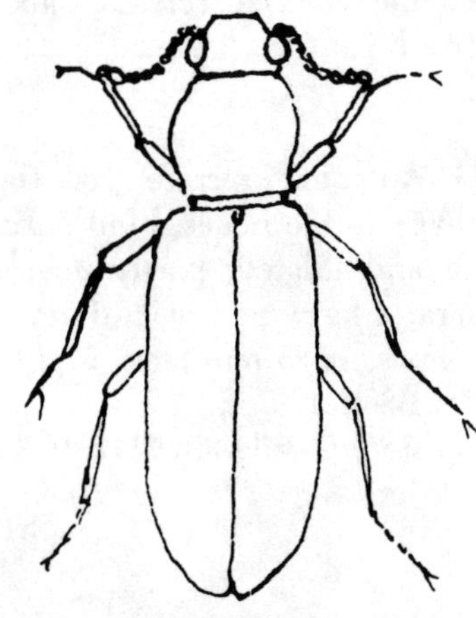

Figure 87

60a (54) Front coxae conical; projecting prominently from the coxal cavity. 61

60b Front coxae globular or transverse, usually projecting only slightly from the coxal cavity. 64

61a (60) Antennae filiform; mandibles inconspicuous. (Fig. 92) (The Soft-bodied Plant Beetles. .... (p. 183) Family 32. HELODIDAE

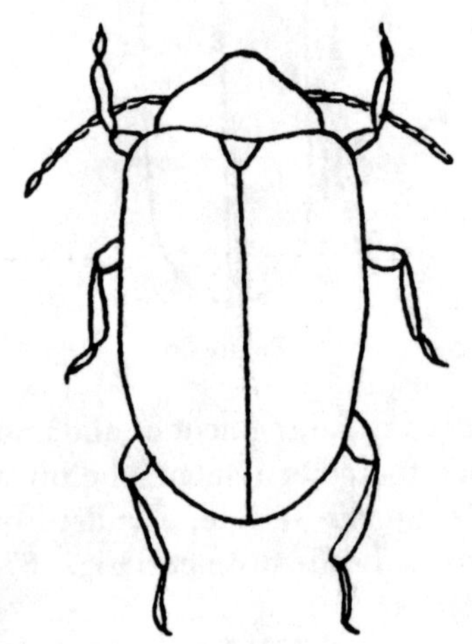

Figure 92

61b Antennae modified, serrate, flabellate, or clubbed. 62

62a (61) Antennae serrate (or flabellate, fan-like) in the males; hind coxae transverse and dilated, partly covering the femora; a hairy pad or cushion between the claws; 10-25 mm long. Fig. 88. (The Cedar Beetles) .... (p. 186) Family 39. RHIPICERIDAE

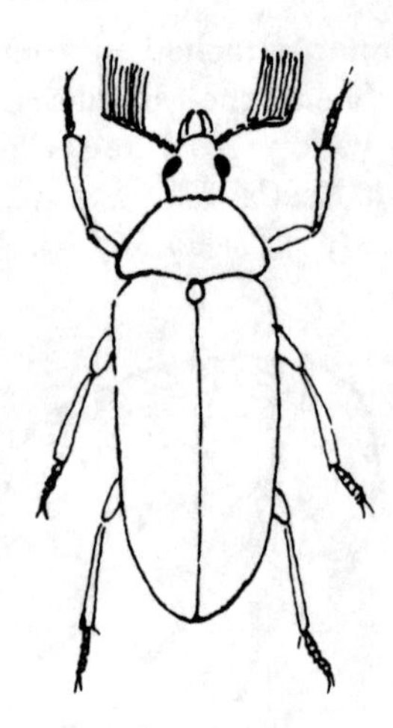

Figure 88

62b Antennae with the last 3 segments clubbed. 63

63a (62) Hind coxae dilated into plates partly covering the base of the femora; antennae ending in a large 3-segmented club. Fig. 89. (Skin Beetles) .... (p. 215) Family 61. DERMESTIDAE

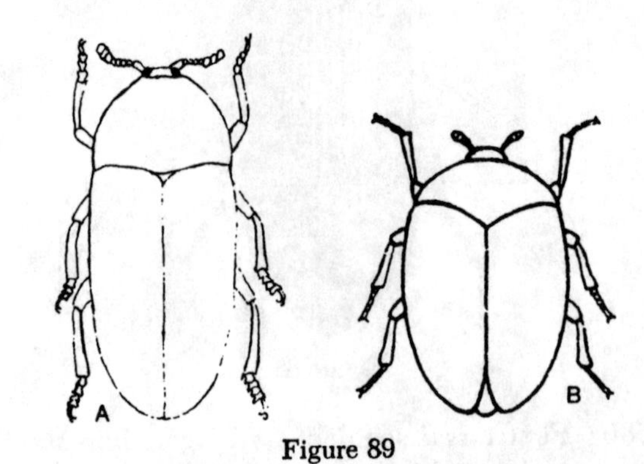

Figure 89

63b Hind coxae flat, not dilated into plates; 4th segment of tarsi equal to each other. Fig. 73. (Checkered Beetles) .... (p. 223) Family 67. CLERIDAE

64a (60) Front coxae transverse; hind coxae flat. 65

64b Front coxae globular or oval. 78

65a (64) Elytra covering entire length of abdomen, never truncate. 67

65b Elytra usually truncate. 66

66a (65) Tarsi more or less dilated, first segment not short. Fig. 90. (Sap-feeding Beetles) ....
...... (p. 231) Family 70. NITUDULIDAE

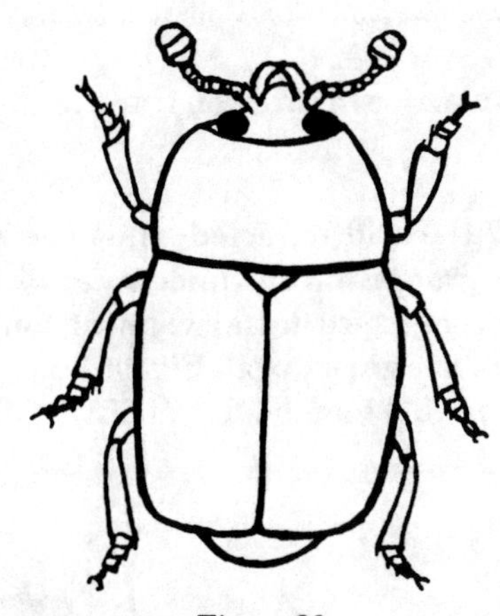

Figure 90

66b Tarsi not dilated; elytra at most exposing one abdominal segment; first tarsal segment as long as 2 and 3 combined; antennae with a 3-segmented club, not elbowed. Fig. 91. ....
...... (p. 166) Family 27. SPHAERITIDAE

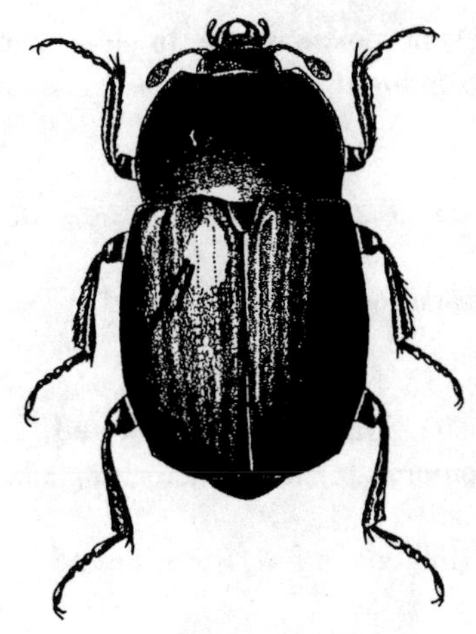

Figure 91

67a (65) Body covering thin and soft; antennae usually serrate, not clavate; mandibles prominent. Fig. 93. ....
.... (p. 184) Family 34. DASCILLIDAE

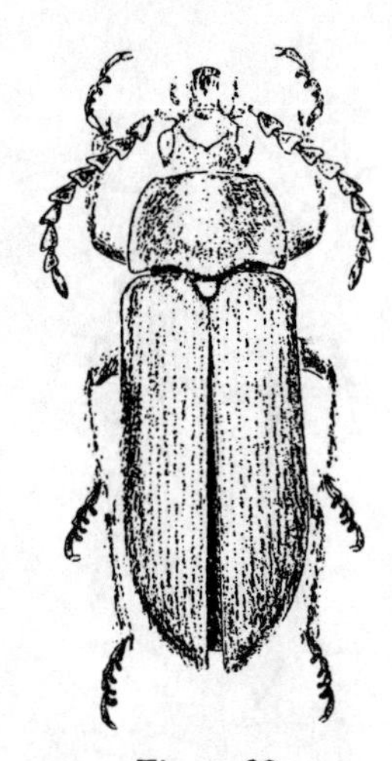

Figure 93

67b Body normally hard as with most beetles; antennae with usually small club that may be serrate or pectinate. 68

68a (67) Front coxae transverse, more or less cylindrical. 69

68b Front coxae more or less conical and prominent. 76

69a (68) Tarsi slender. 77

69b Tarsi more or less dilated. 70

70a (69) Hind coxae grooved to receive femora. 72

70b Hind coxae flat, not grooved. 71

71a (70) Antennae serrate or pectinate; 3rd. tarsal segment 3 with large lobe. Fig. 94. .... (p. 186) Family 38. PTILODACTYLIDAE

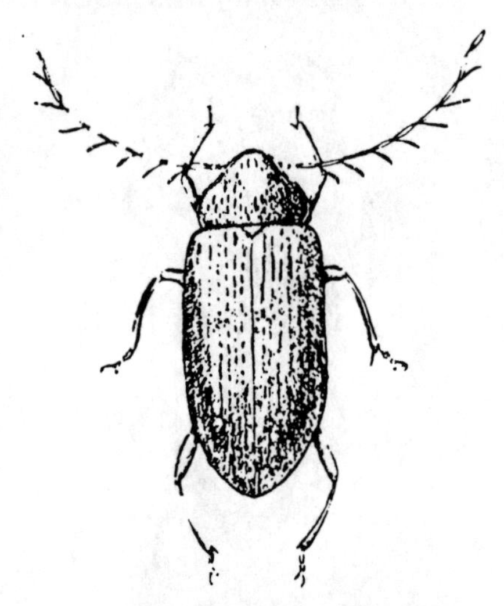

Figure 94

71b Antennae with 3-segmented club; 2nd and 3rd tarsal segments lobed; body with dense, long pubescence. Fig. 95. .... (p. 256) Family 84. BYTURIDAE

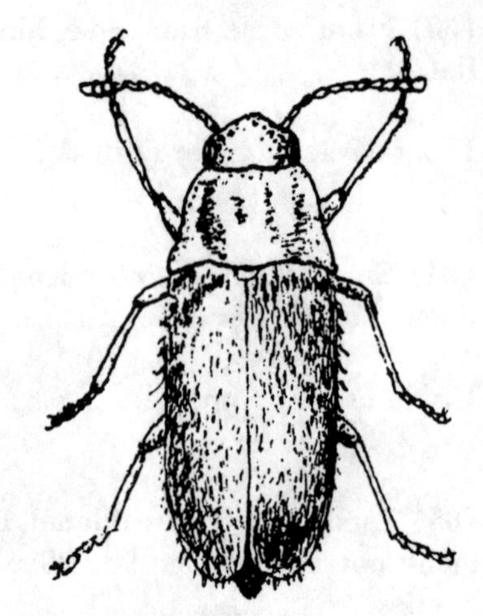

Figure 95

72a (70) Antennae inserted at sides of head. 74

72b Antennae inserted on frons. 73

73a (72) Head retracted; antennae received in grooves on the undersides of the pronotum; third tarsal segment lobed; thorax margined, oval. Fig. 96. .... (p. 187) Family 41. CHELONARIIDAE

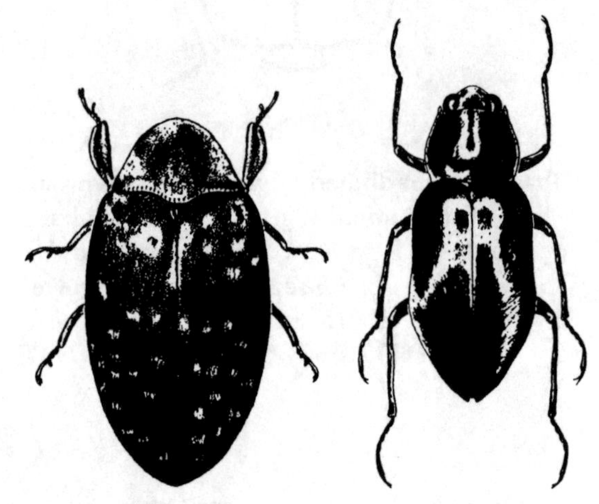

Figure 96 Figure 96a

73b Head not retracted; antennae short, compact, received under a frontal ridge; tarsi slender; body covered with more or less fine hairs; aquatic. Fig. 96a. .... (p. 189) Family 44. DRYOPIDAE

74a (72) Head prominent; mentum large, elongate and sub-elliptical; tarsi not lobed. Fig. 97. .... (p. 214) Family 60. NOSODENDRIDAE

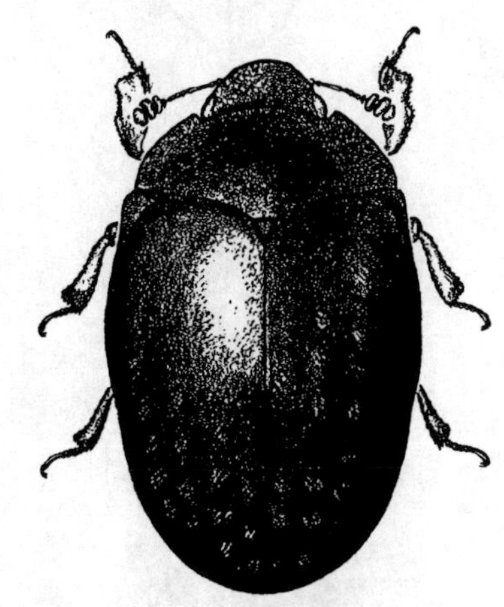

Figure 97

74b Head retracted; mentum small and quadrate. 75

75a (74) Clypeus not distinct from frons. Fig. 98. .... (p. 184) Family 35. BYRRHIDAE

Figure 98

75b Clypeus separated from frons by fine epistomal sulcus. Fig. 99. .... (p. 188) Family 43. LIMNICHIDAE

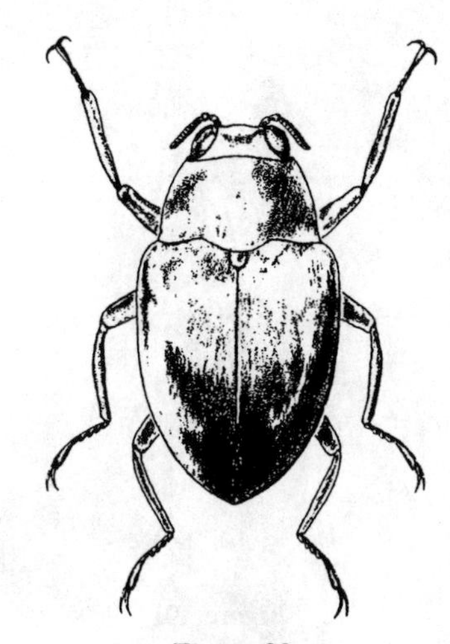

Figure 99

76a (68) Tarsal segments 1 to 4 lobed beneath. Fig. 100. .... (p. 186) Family 39. RHIPICERIDAE

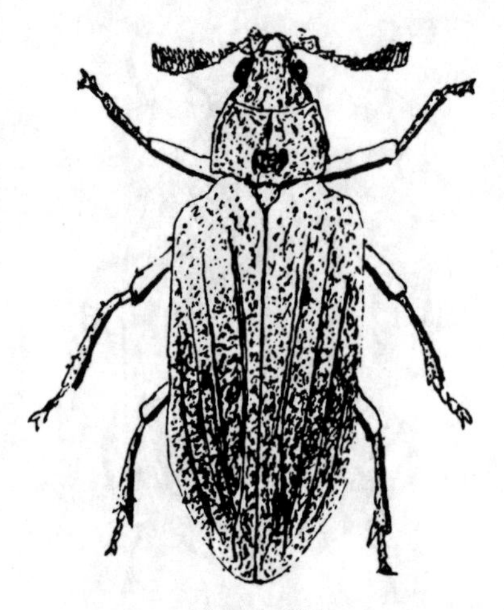

Figure 100

76b Tarsal segments not lobed beneath. Fig. 101. .... (p. 187) Family 40. CALLIRHIPIDAE

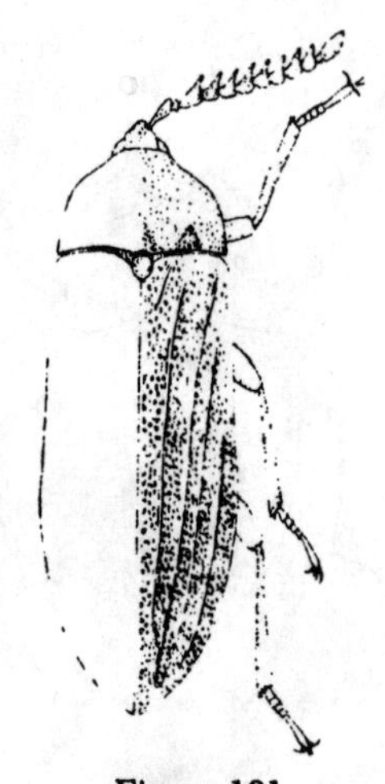

Figure 101

77a (69) Tarsi slender, first segment short, small to medium beetles living in grains or under bark, cylindrical or flat. Figs. 102 and 103. (Bark-gnawing Beetles) ...... Beetles) .... (p. 223) Family 66. TROGOSSITIDAE

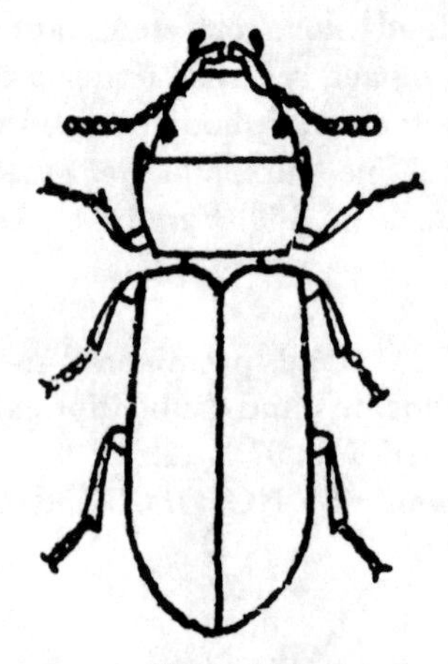

Figure 102

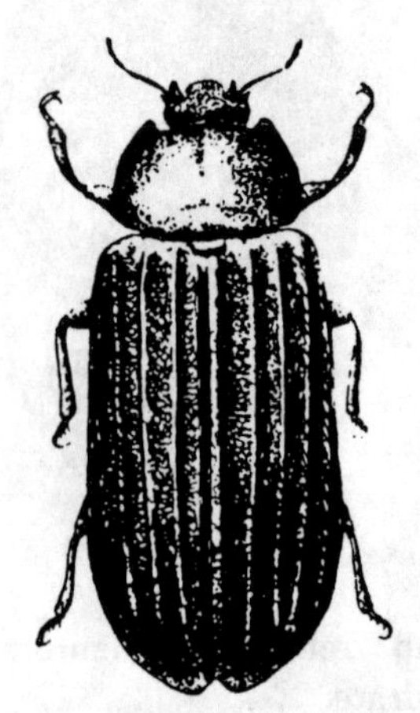

Figure 103

77b Last segment of tarsi about as long as the other 4 combined; hind coxae extending to margin of andomen. (Tooth-necked Fungus Beetles). Fig. 104. .... (p. 214) Family 59. DERODONTIDAE

Figure 104

78a (64) Prosternum with a spine that fits into a groove in the metasternum. Fig. 105. 79

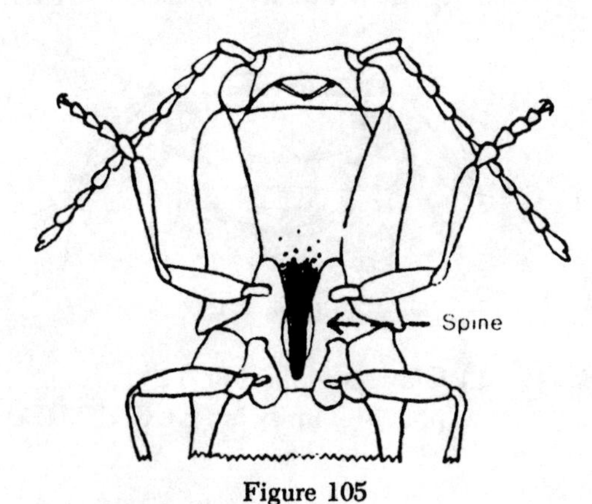

Figure 105

78b Without such spine. 84

79a (78) First and second abdominal segments fused; prothorax closely joined to mesothorax. Fig. 106. (Metallic Woodborers) ....
.... (p. 190) Family 46. BUPRESTIDAE

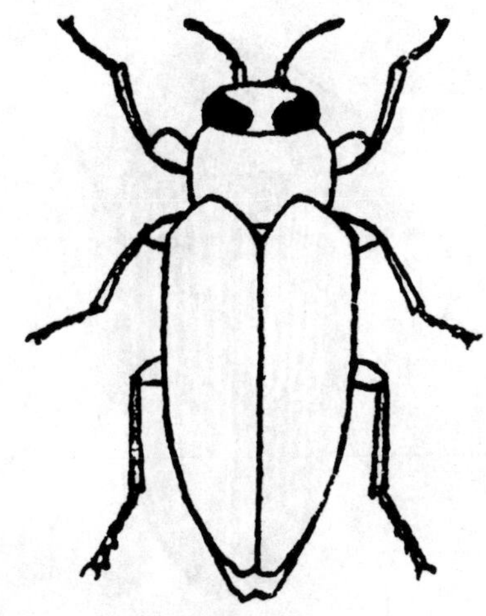

Figure 106

79b Ventral segments not fused. 80

80a (79) Prothorax loosely joined to mesothorax and moving freely on it; its hind angles prolonged into a tooth. Fig. 108. 81

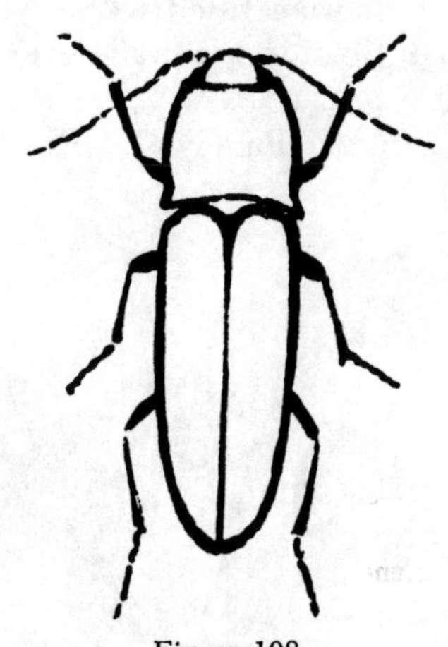

Figure 108

80b Prothorax not movable but firmly fixed to the mesothorax; length not over 4 mm. Fig. 107. (False Click Beetles) ....
.... (p. 203) Family 49. THROSCIDAE

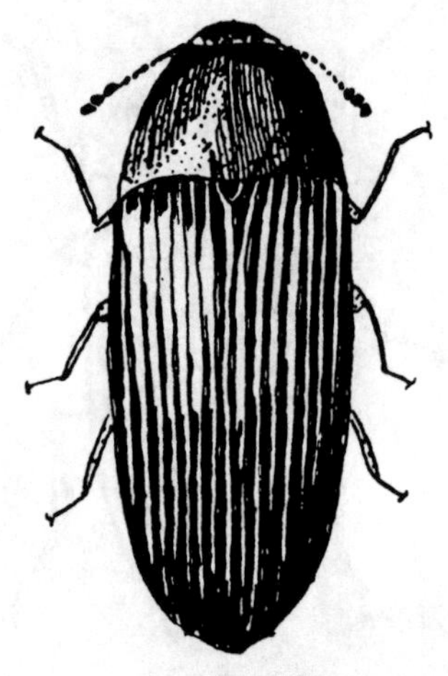

Figure 107

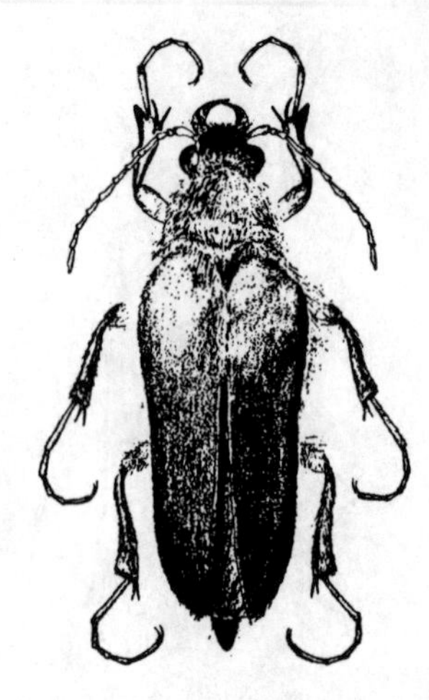

Figure 110

81a (80) Prosternum with lobe in front and prolonged behind into a process that fits into a notch in the mesosternum. 82

81b Prosternum without a lobe or process; labium hidden; size usually small. 83

82a (81) Labrum fused with clypeus (Fig. 109); tibial spurs well-developed. Fig. 110. .... (p. 195) Family 47. **CEBRIONIDAE**

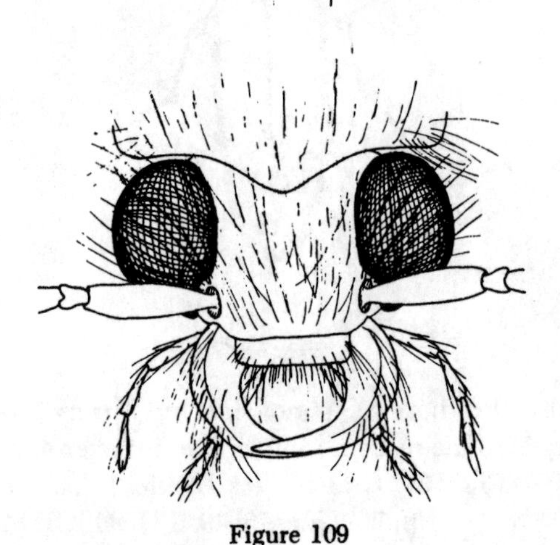

Figure 109

82b Labrum free (Fig. 111); tibia spurs very weak; beetles able to click. (Click Beetles) Fig. 108. .... (p. 195) Family 48. **ELATERIDAE**

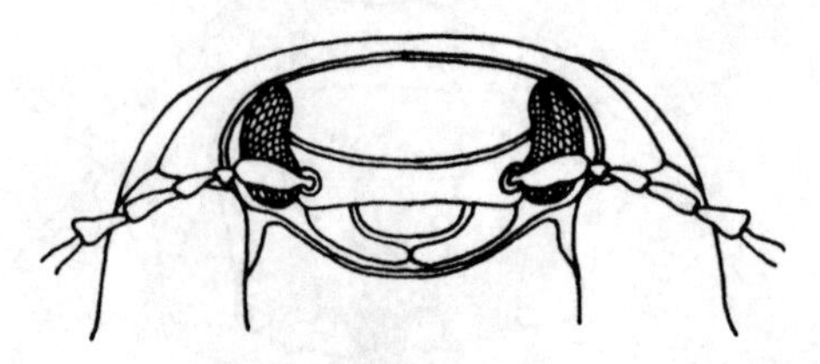

Figure 111

83a (81) Claws simple. Fig. 113. .... (p. 205) Family 52. **EUCNEMIDAE**

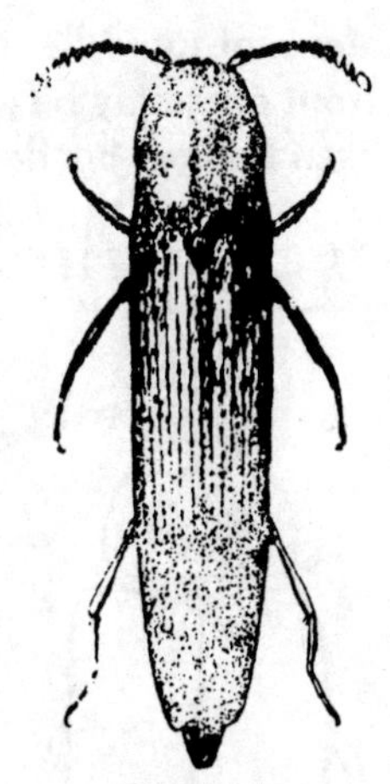

Figure 113

83b Claws pectinate. (Fig. 112) ....
.... (p. 204) Family 51. PEROTHOPIDAE

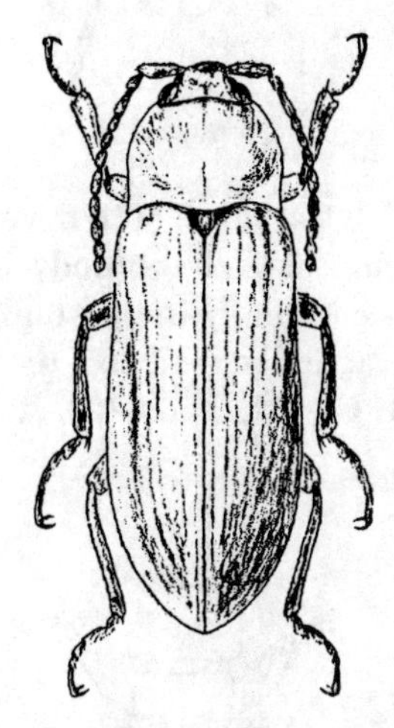

Figure 112

84a (78) Prosternum not prolonged behind; small (2-3 mm), oval coarsely punctate species with the 4th tarsal segment short, 3rd lobed. Fig. 114. ....
.... (p. 255) Family 83. BIPHYLLIDAE

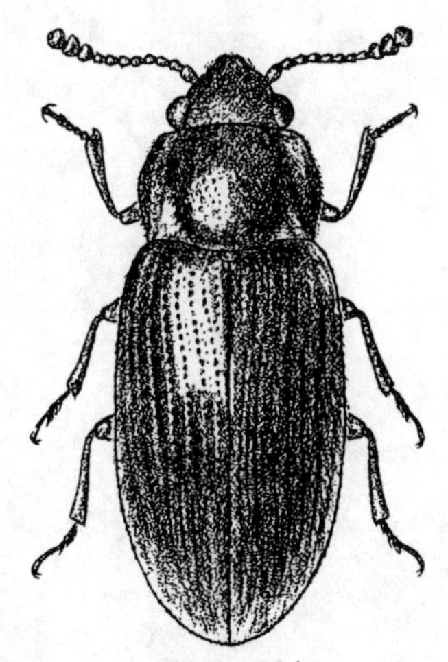

Figure 114

84b Prosternum prolonged behind, meeting mesosternum. Fig. 115. 85

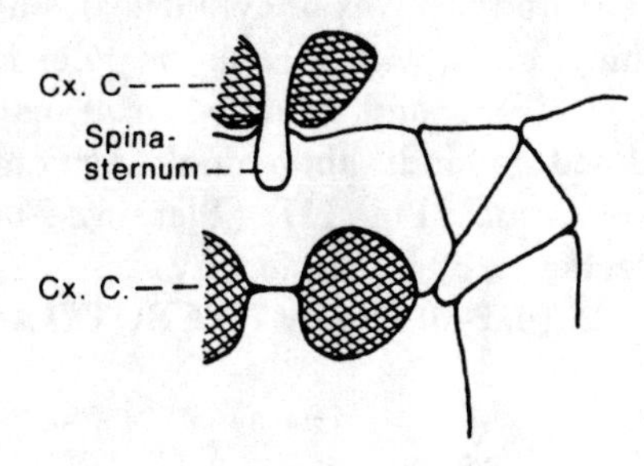

Figure 115

85a (84) Body flattened, middle coxal cavities open behind; abdominal segments equal. (Flat Bark Beetles) Fig. 116. ....
.... (p. 235) Family 72. CUCUJIDAE

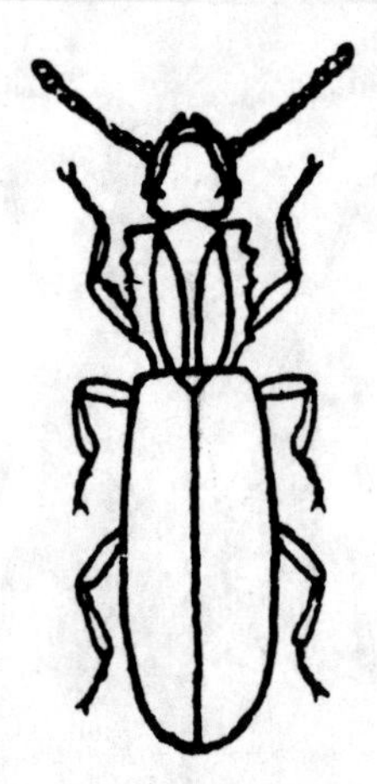

Figure 116

85b Not having the above characteristics. 86

86a (85) Body convex or cylindrical, smooth, shiny; black with orange, or paler markings; front and middle coxal cavities closed behind; abdominal segments 5, each equal. Fig. 117. (Pleasing Fungus Beetles) .... (p. 240) Family 76. EROTYLIDAE

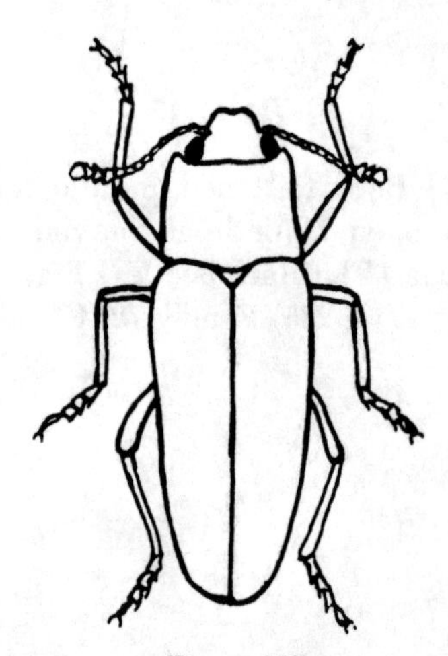

Figure 117

86b Body convex, elongate oval, densely hairy; 1st ventral segment longer than the others; front coxae oval, separated by prosternum; middle coxal cavities closed, front ones may be open or closed. (The Silken Fungus Beetles) Fig. 118. .... (p. 238) Family 74. CRYPTOPHAGIDAE

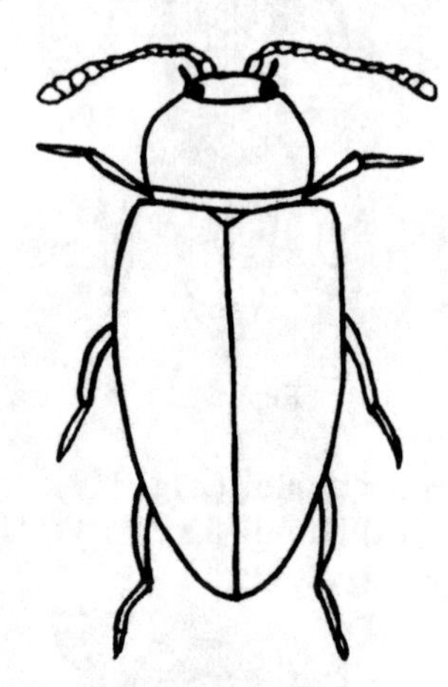

Figure 118

87a (4) Elytra short (often very short and covering half of the body or much less, but occasionally in the staphylinids nearly or actually reaching the tip of abdomen). Figs. 119 and 120. 88

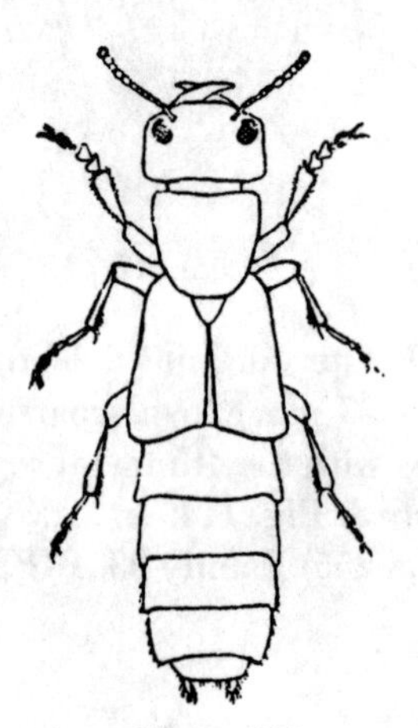

Figure 119

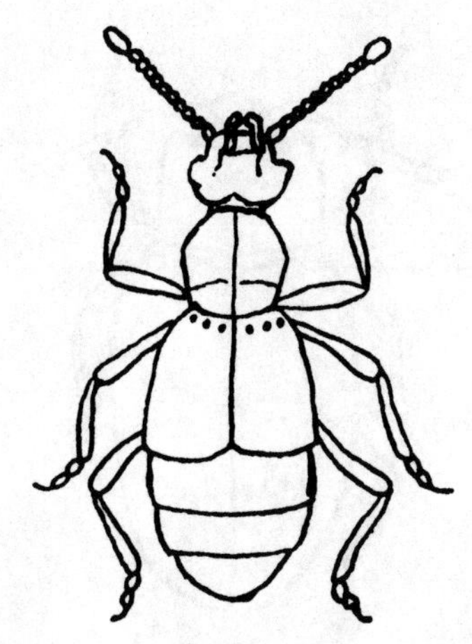

Figure 120

87b Elytra entire, or if shortened, at most exposing only one abdominal segment; body shaped differently from those shown in Figs. 119 and 120. 89

88a (87) Abdomen flexible with 7 or 8 (rarely 6) segments visible below. (Rove Beetles). Fig. 119. ....
.. (p. 134) Family 24. STAPHYLINIDAE

88b Abdomen rigid; elytra short, covering about half of the body; antennae usually clubbed, occasionally bead-like; tarsi sometimes with only 2 segments; less than 4 mm in length. (Ant-loving Beetles). Fig. 120. ....
.... (p. 156) Family 25. PSELAPHIDAE

89a (87) Small (0.25-5 mm in length, usually under 2 mm); wings with fringe of long hairs. 90

89b Usually over 3 mm in length, but some under 3 mm; wings not fringed, or wings absent. 91

90a (89) Tarsi actually with 4 segments, 3rd very small hidden within a notch in the 2nd; wings broad. Fig. 121. ....
.... (p. 243) Family 79. CORYLOPHIDAE

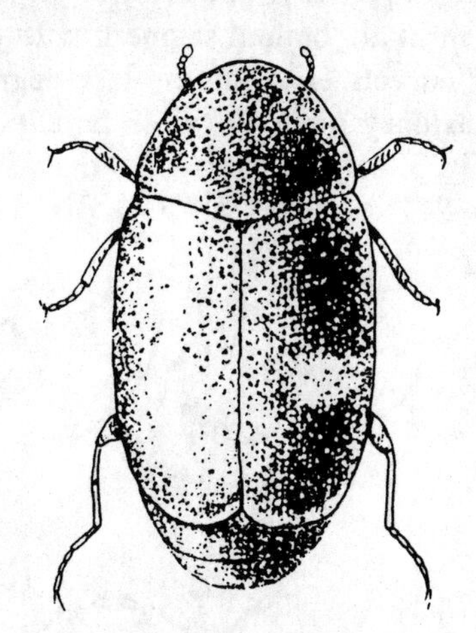

Figure 121

90b Tarsi with only 3 segments; wings narrow; fringe longer than in the preceding. Fig. 122. (Feather-winged Beetles). ....
.... (p. 123) Family 15. PTILIDAE

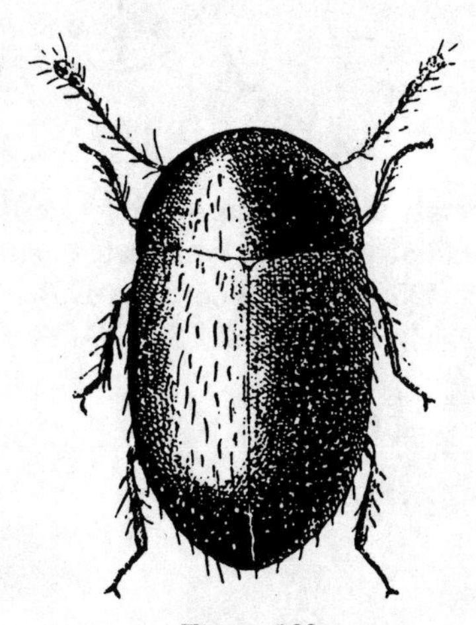

Figure 122

91a (89) Tarsal segments actually 4, the 3rd hidden, very small, not easily seen. 92

91b With only 3 tarsal segments. 93

92a (91) Tarsal claws toothed or appendiculate (Fig. 123); epimeron triangular; 1st ventral abdominal segment with distinctly curved coxal lines; last segment of maxillary palpus oval or hatchet-shaped (Fig. 124a). (Ladybird Beetles). Fig. 124. (p. 244) Family 80. COCCINELLIDAE

Figure 123

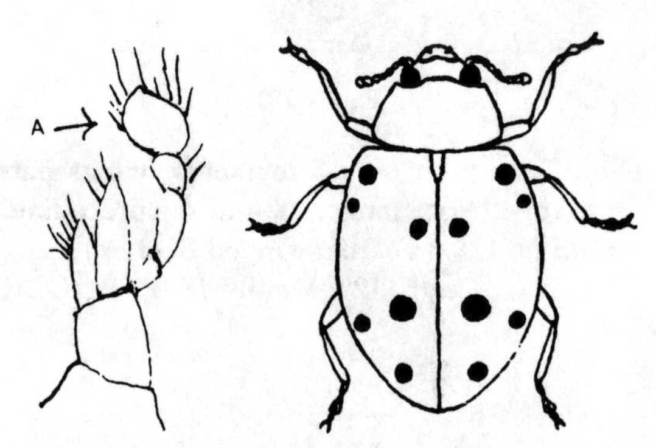

Figure 124

92b Tarsal claws simple; 1st ventral abdominal segment without coxal lines. Fig. 125. (Handsome Fungus Beetles). .. (p. 254) Family 81. ENDOMYCHIDAE

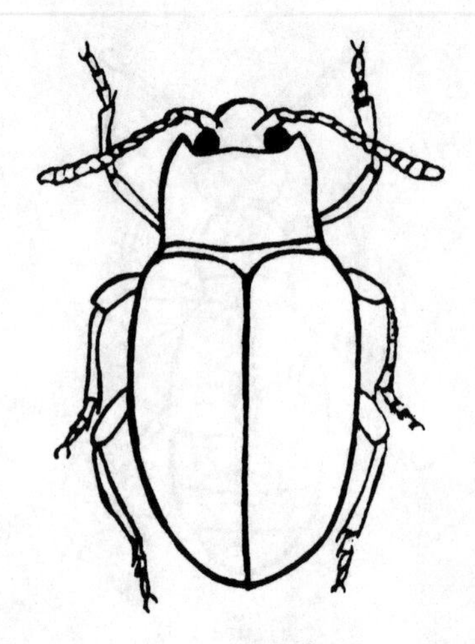

Figure 125

93a (91) Horse-shoe shaped; wingless. Fig. 126. .... (p. 123) Family 16. LIMULODIDAE

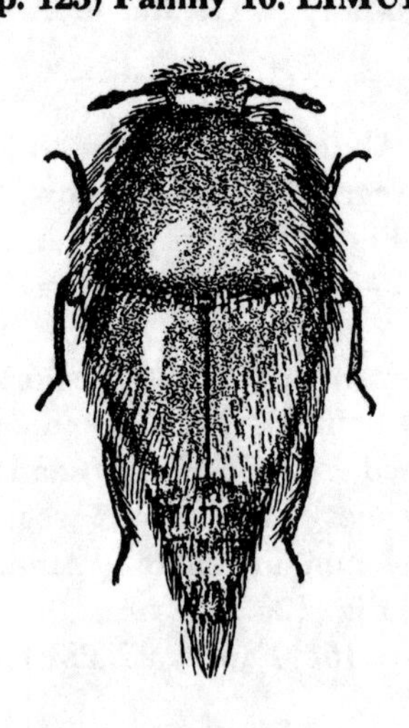

Figure 126

93b Shaped otherwise; wings present. 94

94a (93) Elytra truncate, exposing the last abdominal segment; 1st and 5th abdominal segments longer than the others; body somewhat depressed. Fig. 127. .... (p. 256) Family 85. MONOMMATIDAE

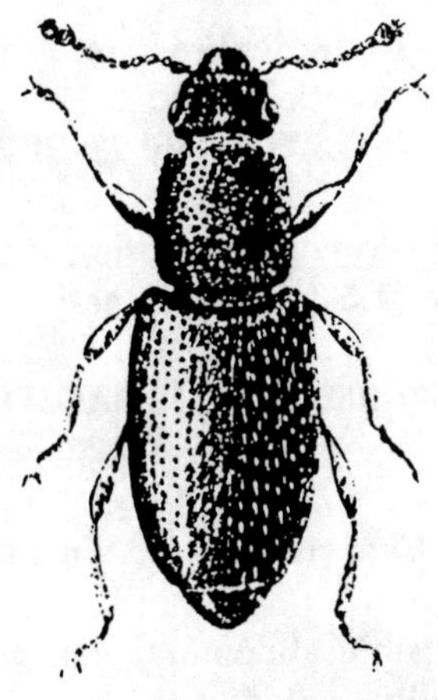

Figure 127

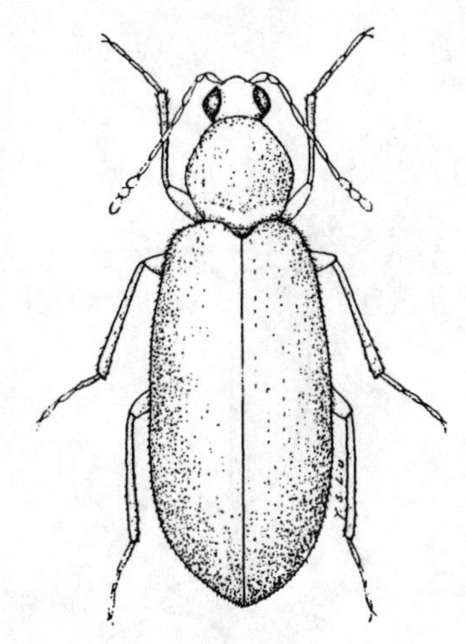

Figure 128a

94b Elytra not truncate, entirely covering the abdomen; ventral segments about equal in length. 95

95a (94) Procoxae open behind (Fig. 128-3); antennae with 2 basal segments enlarged, rest small, forming a loose 3-segmented club; length 2 mm (Fig. 310a). .... (p. 124) Family 17. DASYCERIDAE

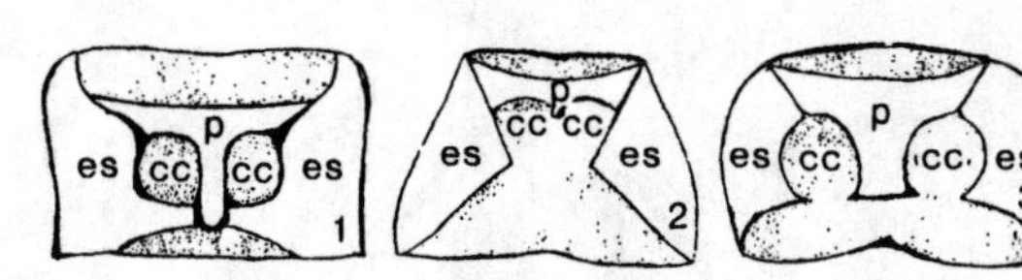

Figure 128

95b Procoxae closed behind. (Minute Brown Scavenger Beetles) (Fig. 128a) .... (p. 254) Family 82. LATHRIDIIDAE

96a (2) Front coxal cavities closed behind (Fig. 128-1); abdomen with 5 ventral segments partly fused together in some families. 97

96b Front coxal cavities open behind Figs. 128-2 and 128-3. 104

97a (96) Tarsal claws comb-like (pectinate) (Fig. 129a); abdominal sterna free; bodily usually elongate and rather soft. (The Comb-clawed Bark Beetles) Fig. 129. .... (p. 269) Family 88. ALLECULIDAE

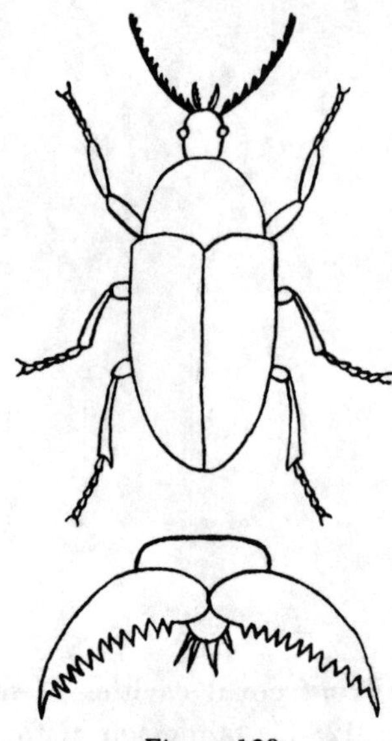

Figure 129

97b Tarsal claws simple. 98

98a (97) Abdominal sterna all freely movable, i.e. membrane visible between at least the middle of the suture. 99

98b First 2 to 4 visible abdominal sterna more closely united more or less fused or immovable, i.e. no membrane visible between segments, a single suture. 102

99a (98) Antennae 11-segmented. 100

99b Antennae 10-segmented. 101

100a (99) Prothorax quadrate, not wider than head; narrow-bodied beetles. .... (p. 275) Family 95. OTHNIIDAE

100b Pronotum greatly expanded at sides, much wider than head; antennae clavate. .... (p. 231) Family 70. NITIDULIDAE

101a (99) Elytra entire; small (1.5-3 mm) convex beetles. .... (p. 237) Family 73. SPHINDIDAE

101b Elytra truncate, exposing the pygidium; small (1.5-3 mm); sometimes flattened. .... (p. 282) Family 100. RHIZOPHAGIDAE

102a (98) Five visible abdominal sterna. .. 103

102b Six visible abdominal sterna, first 2 immovably united. .... (p. 271) Family 90. SALPINGIDAE

103a (102) Next to last tarsal segment spongy beneath. (Long jointed Beetles). Fig. 130. .... (p. 270) Family 89. LAGRIIDAE

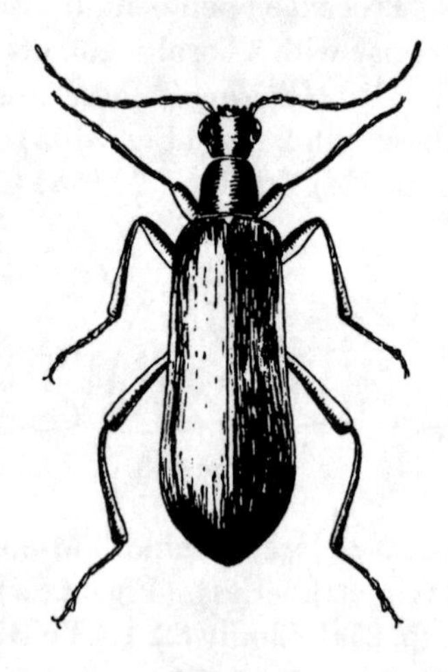

Figure 130

103b Next to last tarsal segment not spongy beneath. (Darkling Beetles). Figs. 131a and b. .... (p. 257) Family 87. TENEBRIONIDAE

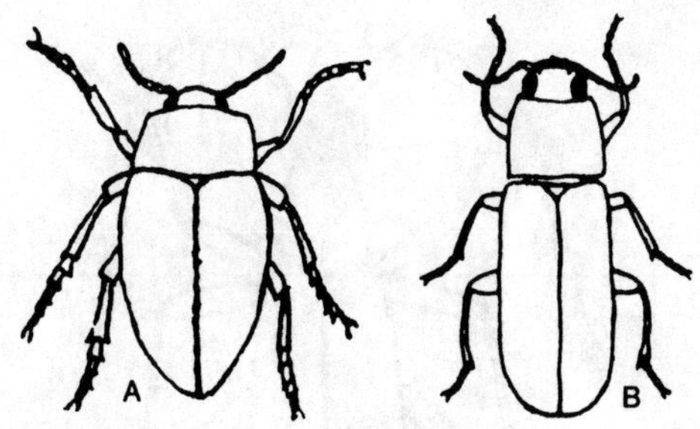

Figure 131

104a (96) Head normal, neither suddenly constricted at base or tapered behind. 105

104b Head either strongly constricted behind or tapered. 107

105a (104) Middle coxae large; body modernately long and narrow. (Pollen feeding Beetles) Fig. 133. .... (p. 276) Family 97. OEDEMERIDAE

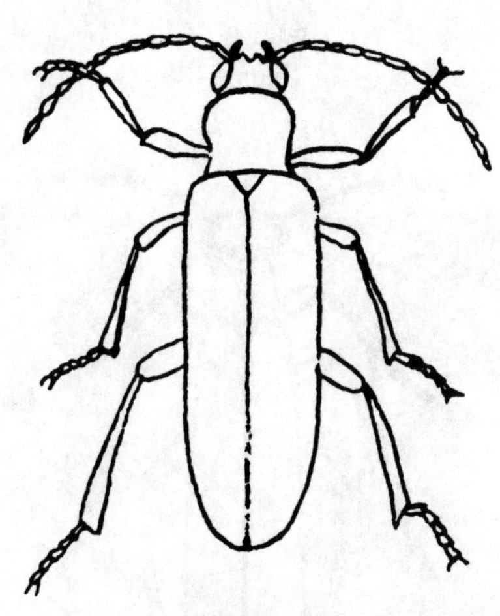
Figure 133

105b Middle coxae not prominent. 106

106a (105) Mesosternum long; epimera of metathorax visible. (False Darkling Beetles). Fig. 134. .... (p. 277) Family 98. MELANDRYIDAE

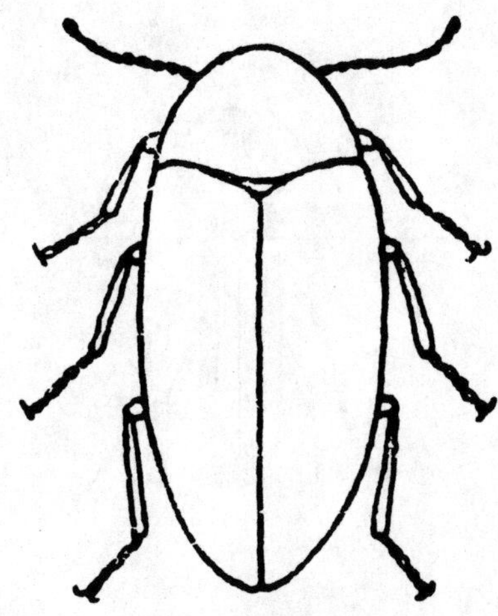
Figure 134

106b Mesosternum quadrate; epimera of metathorax covered. (Flat Bark Beetles). Fig. 135. .... (p. 235) Family 72. CUCUJIDAE

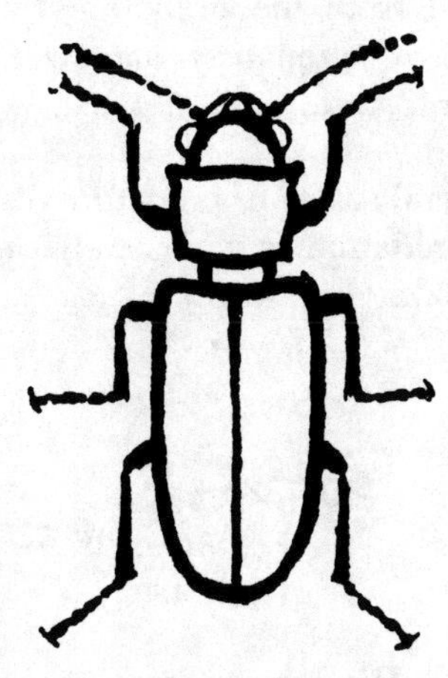
Figure 135

107a (104) Head tapered behind; claws with fleshy lobe, usually pectinate. Fig. 132. .. .. (p. 257) Family 86. CEPHALOIDAE

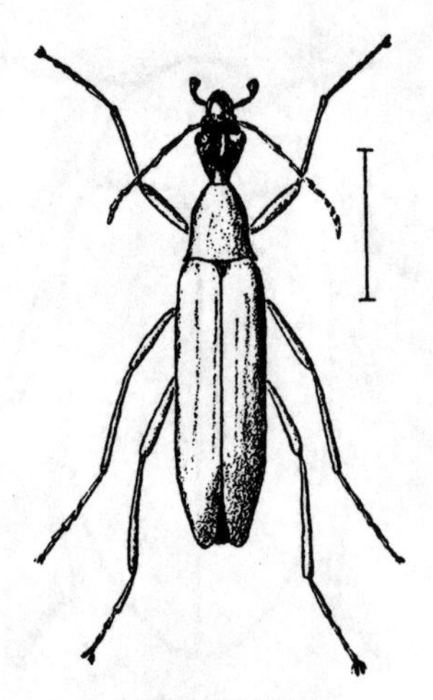

Figure 132

107b Head strongly constricted at base, abruptly narrowed behind. 108

108a (107) Side pieces of prothorax not separated from the pronotum by a suture; base of pronotum narrower than the elytra. 109

108b Lateral suture of pronotum distinct; base of pronotum as wide as elytra; antennae filiform (Fig. 136). 114

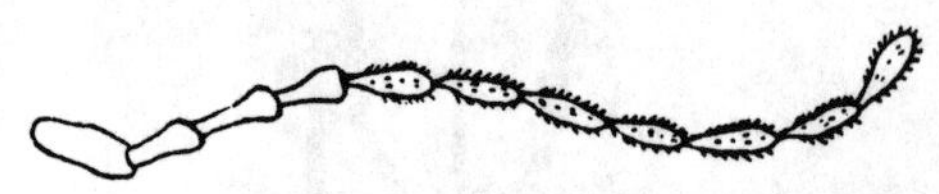

Figure 136

109a (108) Hind coxae large and prominent. 110

109b Hind coxae at most only slightly prominent. 111

110a (109) Tarsal claws simple; head horizontal in position. (Fire-colored Beetles). Fig. 137. ....
.. (p. 274) Family 94. PYROCHROIDAE

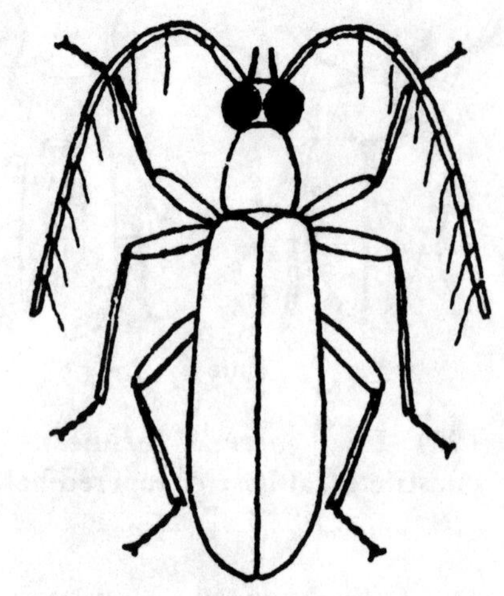

Figure 137

110b Tarsal claws toothed or cleft; front vertical. (Blister Beetles). Figs. 138, 139. ....
.... (p. 283) Family 101. MELOIDAE

Figure 138

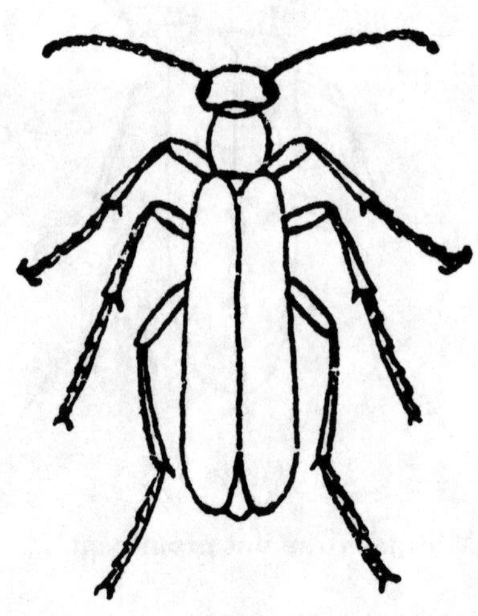

Figure 139

111a (109) Anterior coxae globular, not prominent. (Flat Bark Beetles). Fig. 135. ....
.... (p. 235) Family 72. CUCUJIDAE

111b Anterior coxae conical, prominent. .. 112

112a (111) Eyes usually with a notch in their edge (emarginate). 113

112b Eyes entire, elliptical; hind coxae usually well-separated. (Ant-like Flower Beetles). Fig. 142. .... (p. 288) Family 102. ANTHICIDAE

Figure 142

113a (112) Eyes usually with a slight notch on their edge, finely granulate; hind coxae touching or very close. (False Antloving Flower Beetles). Fig. 140. .... (p. 291) Family 103. PEDILIDAE

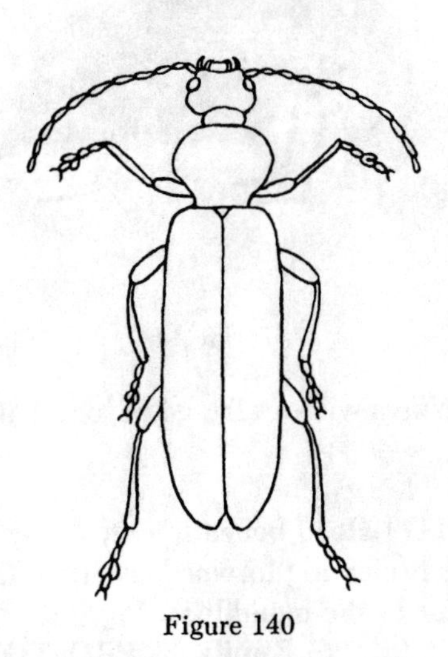

Figure 140

113b Eyes usually deeply emarginate, coarsely granulate; hind coxae contiguous or not; tarsi apparently 4-4-3 because the next to last is extremely small. (Ant-like Leaf Beetles) Fig. 141. .... (p. 292) Family 104. EUGLENIDAE

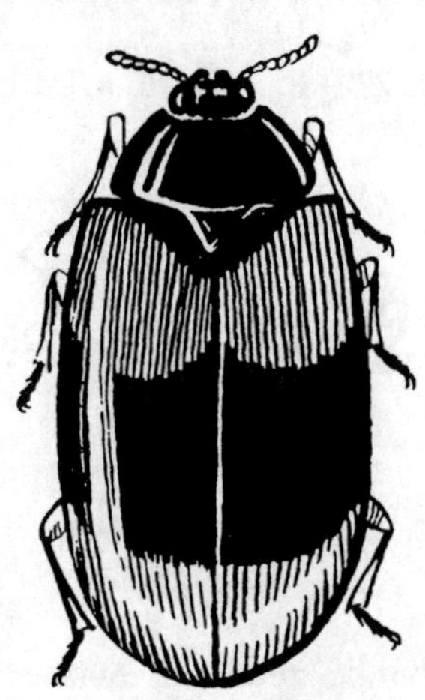

Figure 141

114a (108) Hind coxae not plate-like, not covering the femora; (False Darkling Bark Beetles). (Fig. 143). .... (p. 277) Family 98. MELANDRYIDAE

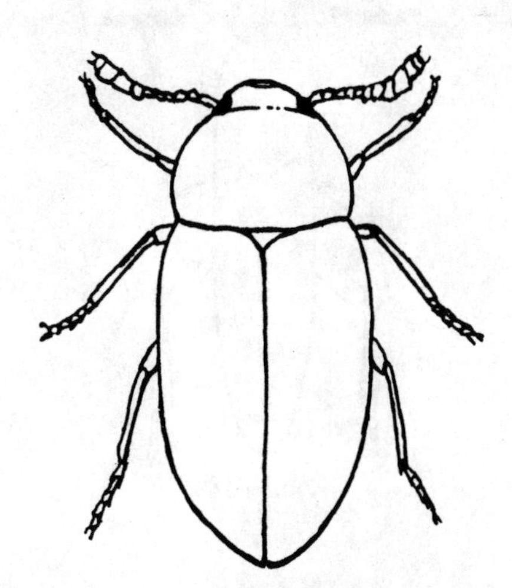

Figure 143

114b Hind coxae plate-like, covering part of femora; abdomen usually pointed. Fig. 144a. 115

115a (114) Prothorax with a sharp marginal ridge at sides. (Tumbling Flower Beetles). Fig. 144. ....
.... (p. 279) Family 99. MORDELLIDAE

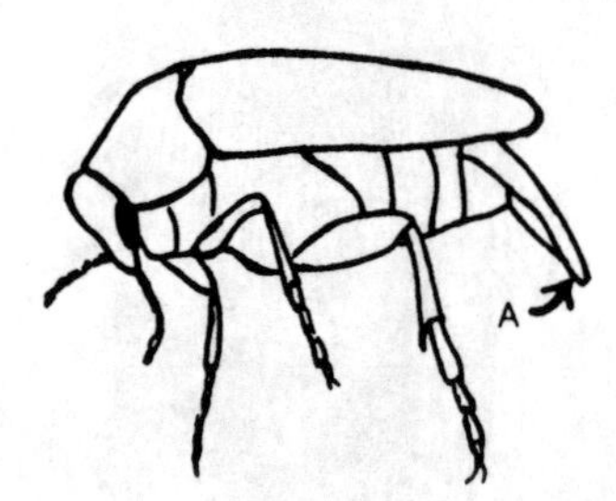

Figure 144

115b Prothorax rounded, without a sharp ridge laterally. (Wedge-shaped Beetles). Fig. 145. ....
(p. 282) Family 100. RHIPIPHORIDAE

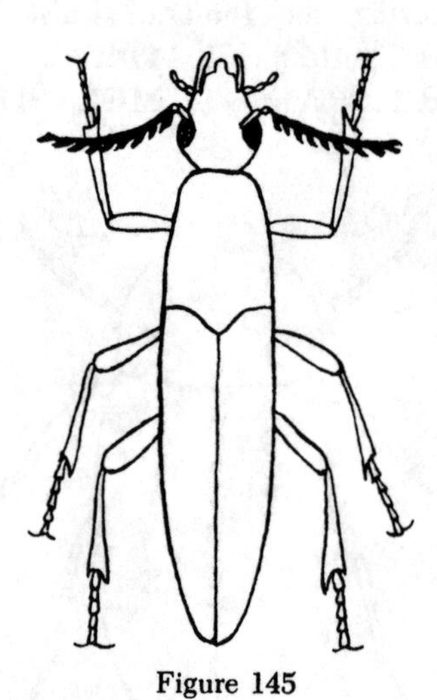

Figure 145

116a (4) Tarsi each with 5 segments, the 4th, very small hidden between the lobes of the 3rd. Fig. 146. 128

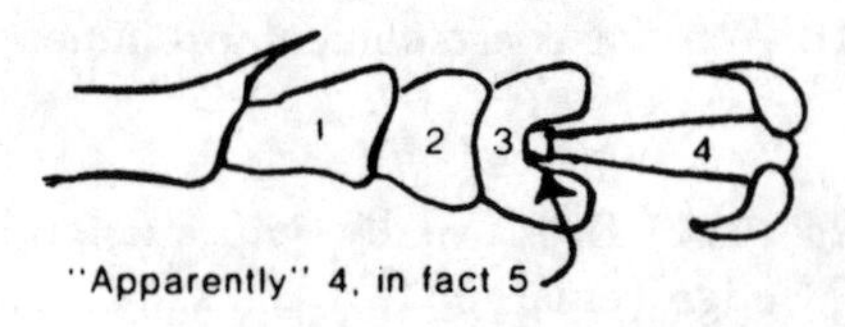

Figure 146

116b Tarsi each with 4 segments. Fig. 147. ......
.... 117

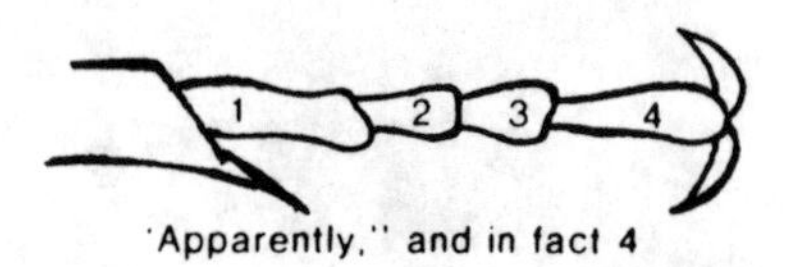

Figure 147

117a (116) Wings fringed with long hairs usually extending beyond elytra even when elytra are closed; hind coxae with plates partly covering femora; length about 1 mm. Fig. 148. ....
.... (p. 183) Family 33. CLAMBIDAE

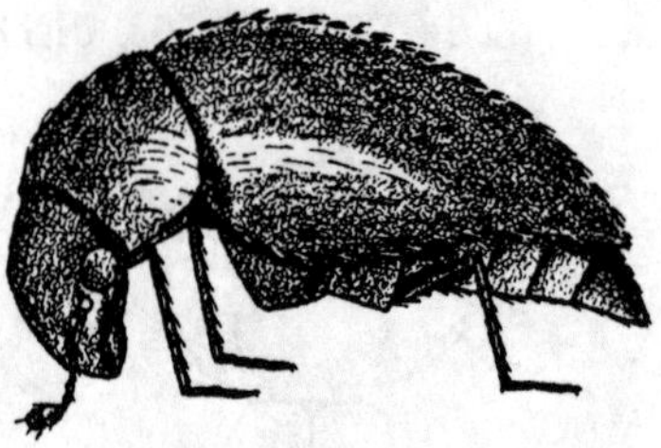

Figure 148

117b Wings without fringe of long hairs. .. 118

118a (117) Head beneath with 2 long processes projecting forward nearly as far as the tips of the mandibles. Fig. 149. ....
...... (p. 275) Family 96. PROSTOMIDAE

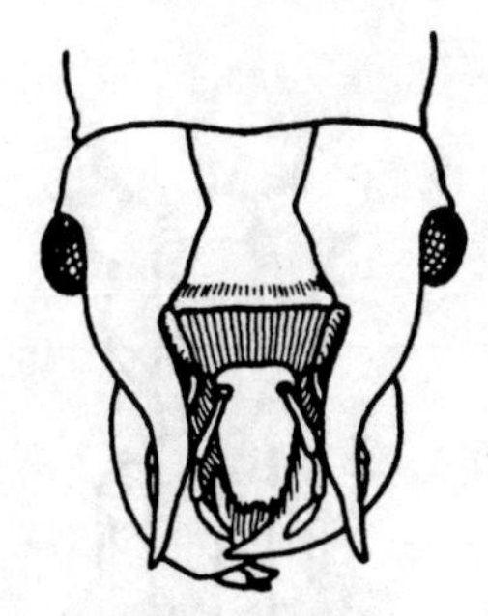

Figure 149

118b Head without processes beneath. 119

119a (118) First 4 abdominal segments fused, only the 5th freely movable or all 5 fused. 120

119b Ventral abdominal segments not fused together, membrane visible and segments movable. 124

120a (119) Tibiae dilated and armed with rows of spines for digging; antennae short. (Variegated Mud-loving Beetles). Fig. 150. ....
(p. 188) Family 42. HETEROCERIDAE

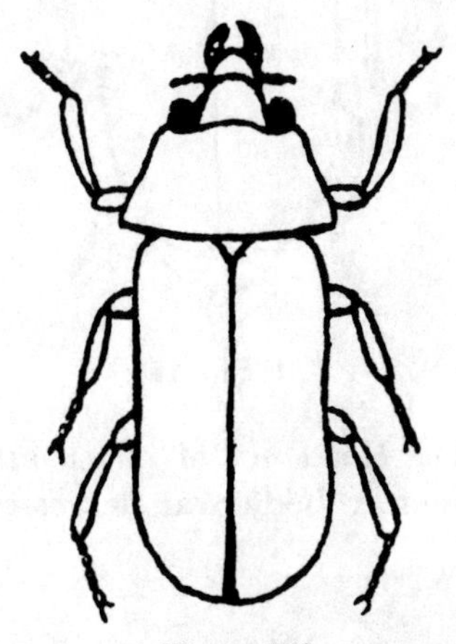

Figure 150

120b Tibiae and antennae not as above. 121

121a (120) Head covered by pronotum; abdomen very small; thoracic sterna much larger; length 1.5-3 mm. (Minute Mud-loving Beetles) Fig. 151. ....
...... (p. 122) Family 14. CEORYSSIDAE

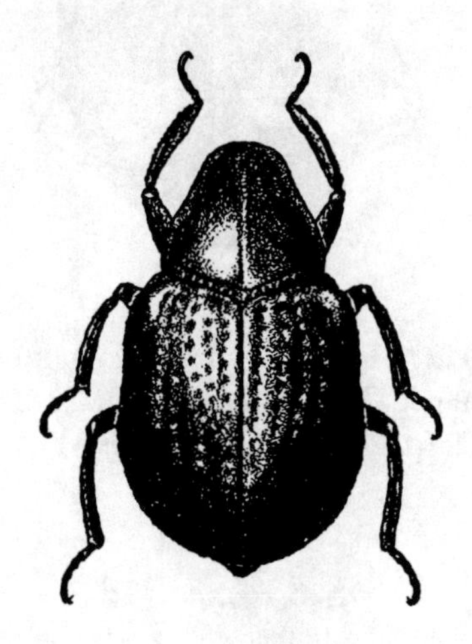

Figure 151

121b Head usually visible from above, or if covered, abdomen and thoracic sterna otherwise. 122

122a (121) Antennae inserted under frontal ridge. 123

122b Antennae inserted on front. (Mycetaeinae). Fig. 152. ....
.. (p. 254) Family 81. ENDOMYCHIDAE

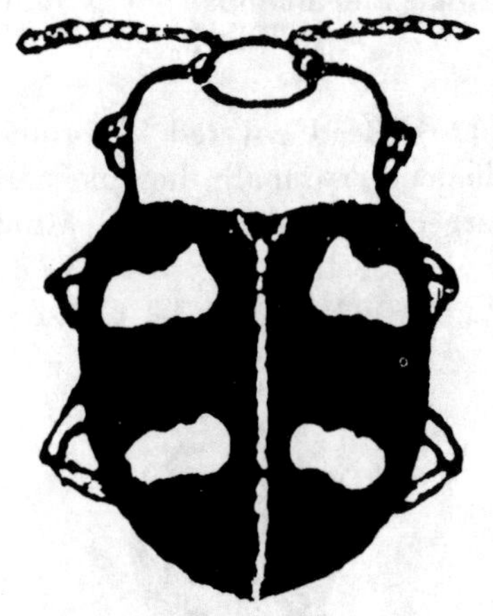

Figure 152

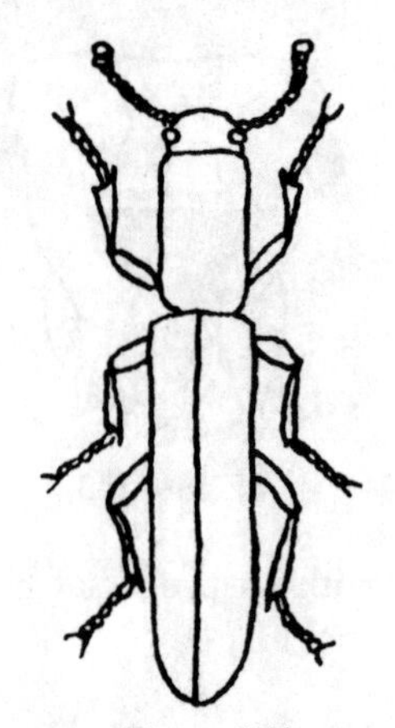

Figure 154

123a (122) Last segment of palpi needle-shaped. Fig. 153. ....
...... (p. 243) Family 78. CERYLONIDAE

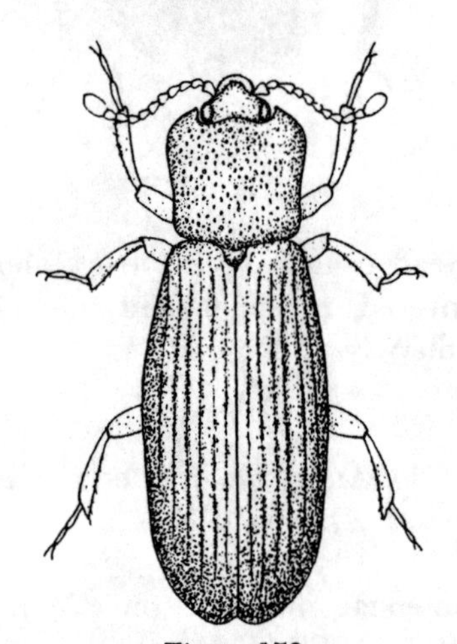

Figure 153

123b Last segment of palpi not needle-shaped. (Cylindrical Bark Beetles). Fig. 154. ......
.... (p. 273) Family 93. COLYDIIDAE

124a (119) Front coxae globose. 125

124b Front coxae oval, separated; small, depressed beetles. (Hairy Fungus Beetles). Fig. 155. .... (p. 272) Family 91. MYCETOPHAGIDAE

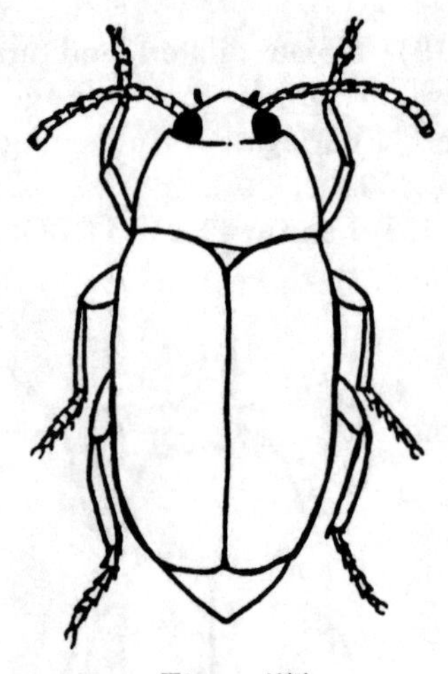

Figure 155

125a (124) Head not at all covered by the prothorax; body oval, depressed. 126

125b Head covered at least in part by the prothorax; last segment of tarsi very long; body cylindrical. (Minute Tree Fungus Beetles). Fig. 156. .... (p. 273) Family 92. CIIDAE

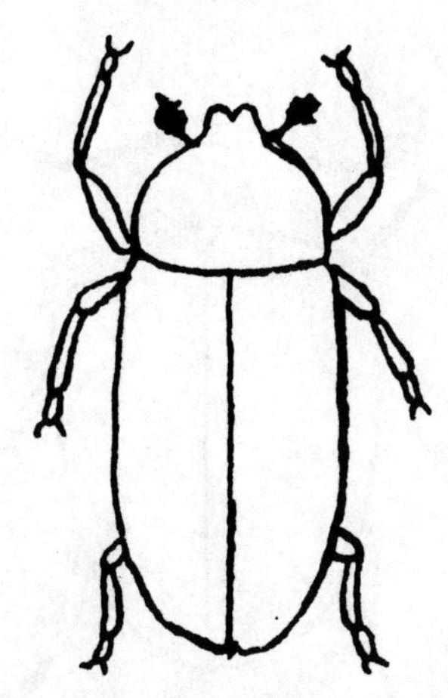

Figure 156

126a (125) Tarsi slender, 3rd segment distinct, but shorter than the 2nd. (Handsome Fungus Beetles) Fig. 157. .... (p. 254) Family 81. ENDOMYCHIDAE

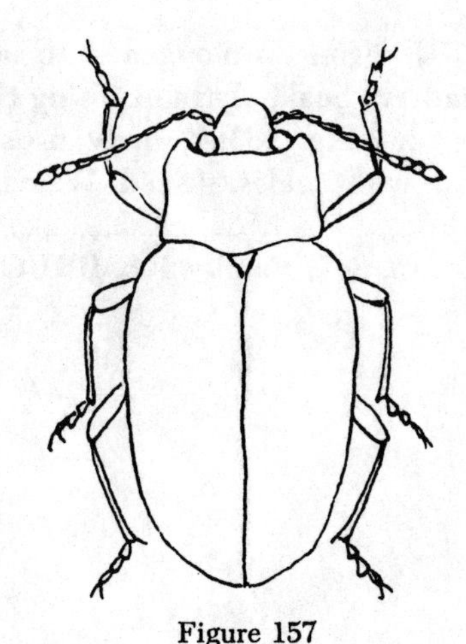

Figure 157

126b Tarsi more or less dilated and spongy beneath. 127

127a (126) Form cylindrical, very elongate; front coxal cavities open behind; antennal club 5 or 6-segmented. (Lizard Beetles). Fig. 158. .... (p. 239) Family 75. LANGURIIDAE

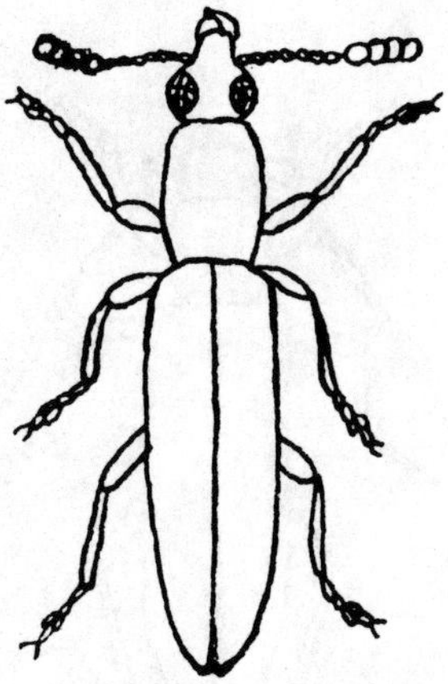

Figure 158

127b Body more ovate, not long and cylindrical; front coxal cavities closed behind; club of 3 or 4 segments. (Pleasing Fungus Beetles). Fig. 159. .... (p. 240) Family 76. EROTYLIDAE

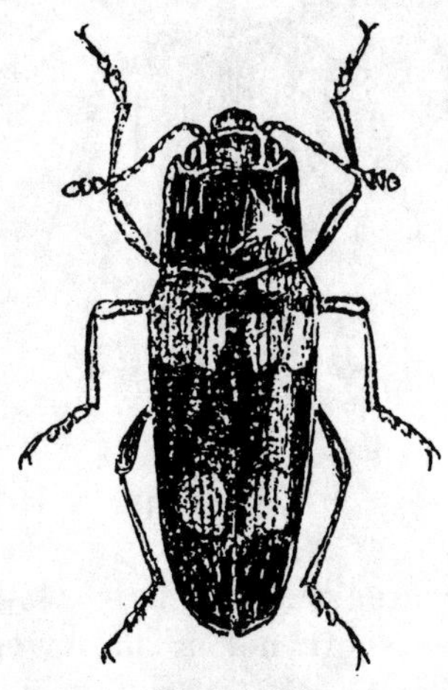

Figure 159

128a (116) Body elongate, or if ovate, somewhat depressed; antennae almost always long, often as long as or much longer than the body; base of antennae usually partly surrounded by the eyes, (Long-horned Beetles). Figs. 160, 161. .... (p. 293) Family 106. CERAMBYCIDAE

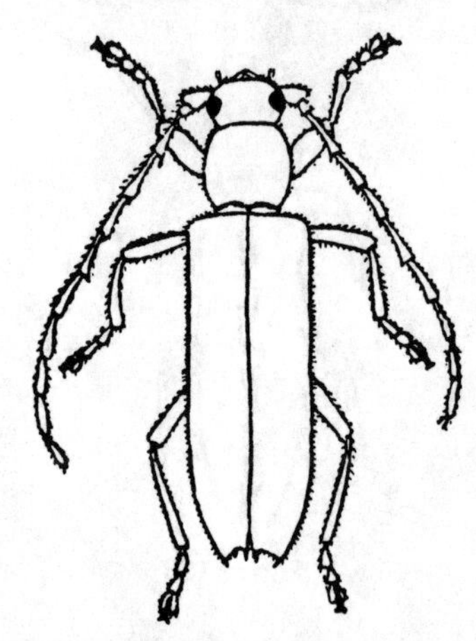

Figure 160

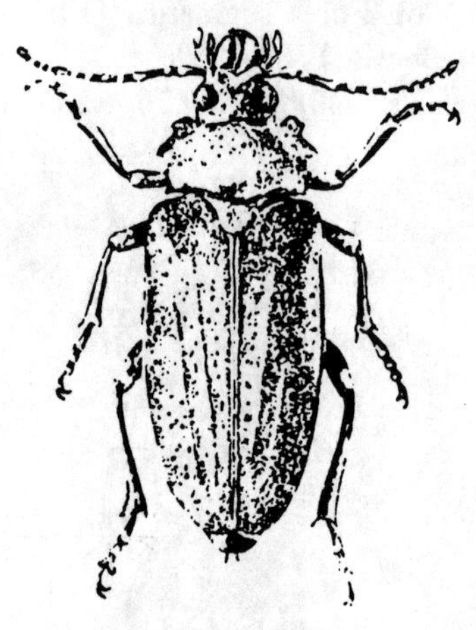

Figure 161

128b Body usually short, more or less oval; antennae short, not at all surrounded by the eyes. 129

129a (128) Small, less than 3 mm, shining, very compact, rounded convex beetles; elytra wholly covering the abdomen; last 3 antennal segments forming a club. (Shining Flower Beetles). Fig. 162. .... (p. 241) Family 77. PHALACRIDAE

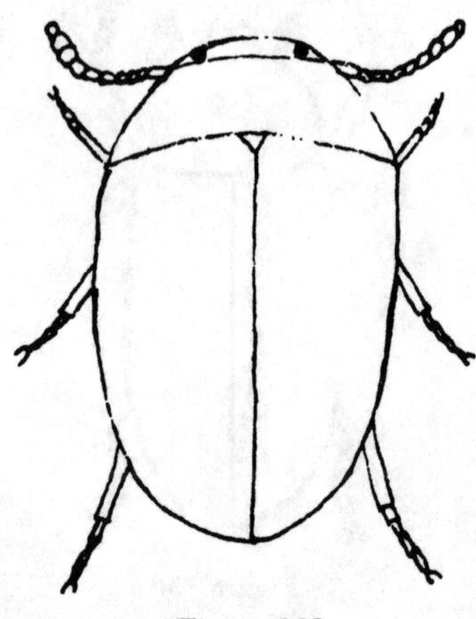

Figure 162

129b Usually over 3 mm in length, often much larger; antennae not clubbed, but may be clavate. 130

130a (129) Front prolonged into a broad quadrate beak; elytra exposing tip of abdomen (Fig. 163a); body usually covered with scales. (Seed Weevils) Fig. 163. .... (p. 347) Family 108. BRUCHIDAE

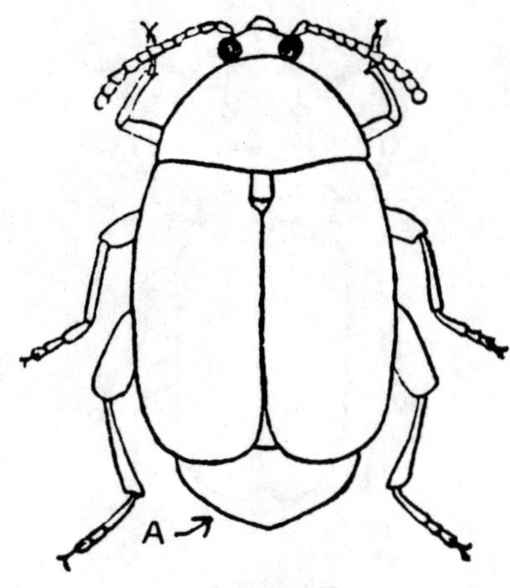

Figure 163

Figure 165

130b Front not prolonged into a beak; abdomen usually wholly covered by the elytra. (Leaf Beetles). Fig. 164. .... (p. 316) Family 107. CHRYSOMELIDAE

132b First segment of hind tarsus shorter than the next 3 combined. Fig. 166. (Bark Beetles). .... (p. 352) Family 112. SCOLYTIDAE

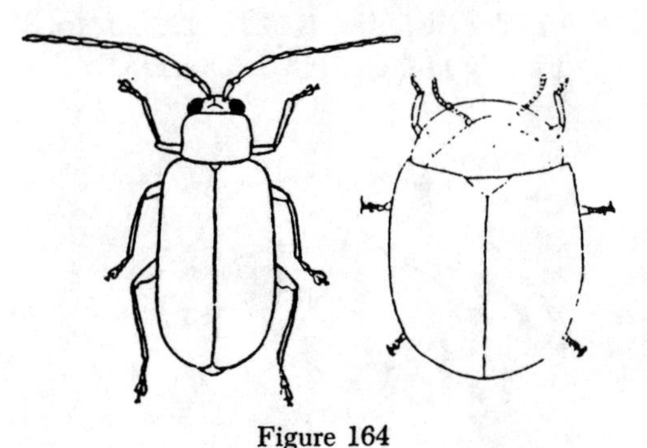

Figure 164

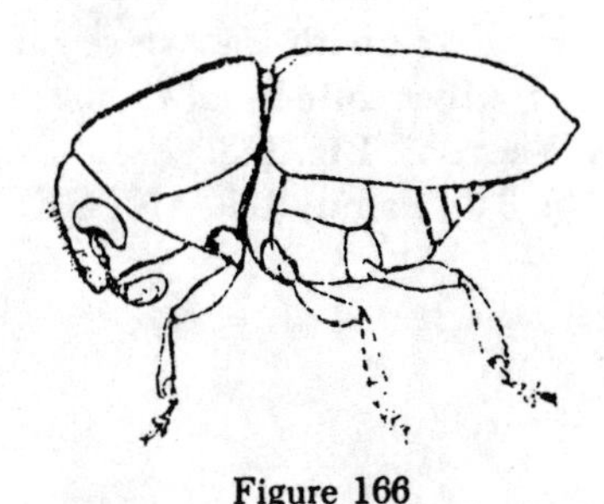

Figure 166

131a (1) Beak absent or very short and broad; antennae short and always elbowed; tibiae usually with teeth. 132

131b Beak usually longer than broad; tibiae without teeth on outer edge. 133

132a (131) First segment of front tarsi longer than the next 3 segments combined. (Ambrosia Beetles). Fig. 165. .... (p. 352) Family 111. PLATYPODIDAE

133a (131) Antennae without a distinct club; not elbowed; body long, slender, usually cylindrical. (Straight-snouted Weevils). Fig. 167. .... (p. 351) Family 110. BRENTIDAE.

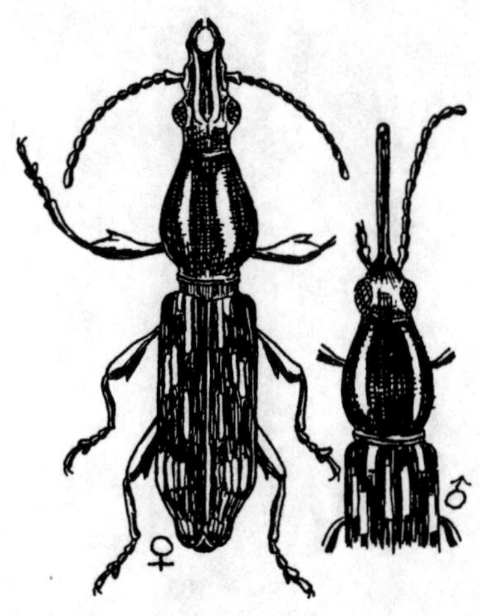

Figure 167

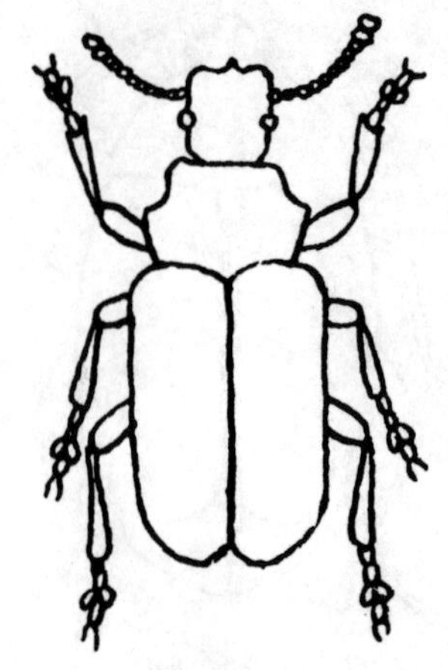

Figure 168

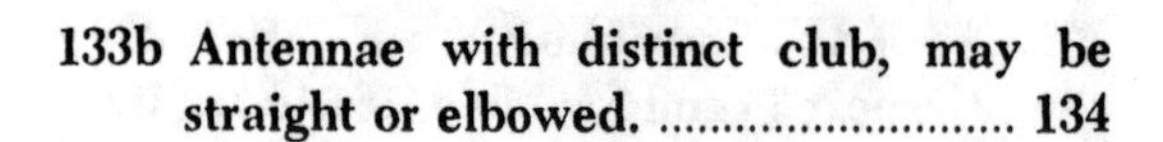

133b Antennae with distinct club, may be straight or elbowed. 134

134a (133) Palpi flexible; antennal club rarely compact; beak always short and broad; thorax with transverse raised line that is either ante-basal or basal. (Fungus Weevils). Fig. 168. .... (p. 349) Family 109. ANTHRIBIDAE

134b Palpi rigid and labrum absent except rarely, then always with a long, well-developed beak; antennal club usually compact; beak variable in length, often long and curved downward. (Snout Beetles or Weevils). Fig. 169. .... (p. 358) Family 113. CURCULIONIDAE

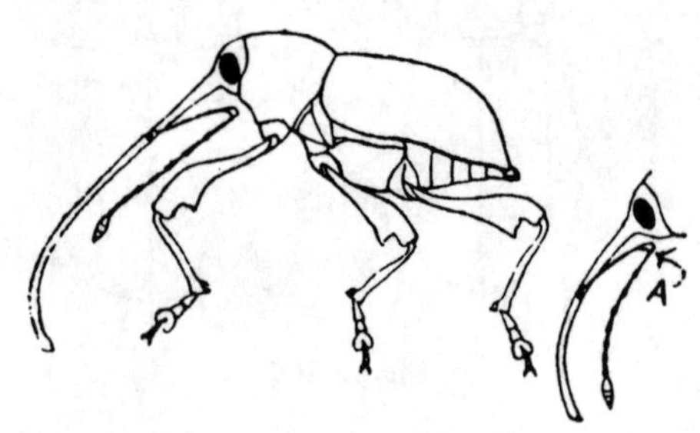

Figure 169

Pictured-Key to the Genera and Species of Some Common Beetles

SUBORDER ARCHOSTEMATA

Only one family is assigned to this ancient suborder. These beetles are thought to be among the most primitive known.

FAMILY 1. CUPEDIDAE
The Reticulated Beetles

These beetles are usually rare in collections, but probably are fairly common once one learns where to look for them. They are yellowish-brown to black, sometimes grayish, in color and vary between 7 to 25 mm in length. Their distinctive sculpturing quickly separates them from other beetles of a similar shape.

The two genera are separated in the following key. *Cupes* spp. are wood-borers as larvae and are found under bark as adults. Little is known about *Priacma serrata* (LeConte) our only species of the genus. It has been taken by beating dead branches of white fir and under the loose bark of the stumps of coniferous trees.

1a Antennae short, scarcely half as long as body, separated at base by at least 4 times width of first segment. .... *Priacma*

P. serrata (LeConte), Western United States, from Montana to Washington and south in mountains to southern California.

1b Antennae long, much longer than half length of body, separated by no more than width of first antennal segment. .... *Cupes*

This genus contains 4 species, 3 of which are eastern and 1 western.

C. concolor Westwood (Fig. 26) is probably the most common species of the genus. It is a yellowish-brown color with mottled elytra. Specimens have been taken throughout eastern United States, east of the Mississippi. The larvae occur in solid but decaying oak.

Eight families, 5 of which are aquatic, fall in this suborder. They have well developed mouth parts and live an active predacious or carnivorous life. The ventral abdominal segments count one more along the sides than at the middle, a character which makes identification of the group more positive. If the antennae of any specimens are filiform, the first ventral segment should always be examined to see if it belongs to this suborder.

FAMILY 2. RHYSODIDAE
The Wrinkled Bark Beetles

This is a small family of relatively rare beetles, varying in length from 5 to 8 mm; color brownish to black. The two genera are separated in the following descriptions of the more common species. All species are found under the bark of decaying trees.

Figure 170

Rhysodes americanus Lap. (Fig. 170). Length: 6-8 mm. Dark reddish-brown, shining. Deep furrows on thorax; elytra striae with large punctures. Male with distinct tooth on front femur. Found under bark of logs and trees of beech, oak, and elm. Distribution: Northeastern United States and southeastern Canada.

Clinidium sculptile Newn. is similar to the preceeding in shape and size, and color, but has grooved elytra without punctures. Distribution: Northeastern United States from the Catskill Mts. south to North Carolina.

FAMILY 3. CICINDELIDAE
Tiger Beetles

Medium sized, often brilliant colored, swift moving land beetles. Both larvae and adults feed on insects which they pounce upon to catch. The larvae await their prey at the doorway of the tunnels in which they live. Members of the family are distributed all over the world but are at their best in the tropics. About 2,000 species are known.

They are favorites with collectors. When flushed they fly some distance and alight facing their pursuer. Some species are very difficult to catch.

Both the markings and habits could suggest the name tiger beetles. Many brilliant metallic and iridescent colors with white or ivory markings are common. They are mostly of medium size and have long slender legs.

The long, sharply-pointed, overlapping mandibles are characteristic. The filiform antennae bear 11 segments. Many species are strong flyers, but the members of the western genus *Omus* do not possess membranous wings and are flightless. The larvae await at burrows to grasp passing prey. A peculiar hooked-hump located about midway of the dorsal abdomen, enables the larva to hold itself within its burrow while subduing a struggling insect.

1a **Third coxae touching each other (a); eyes large, prominent. Fig. 171. .... 5**

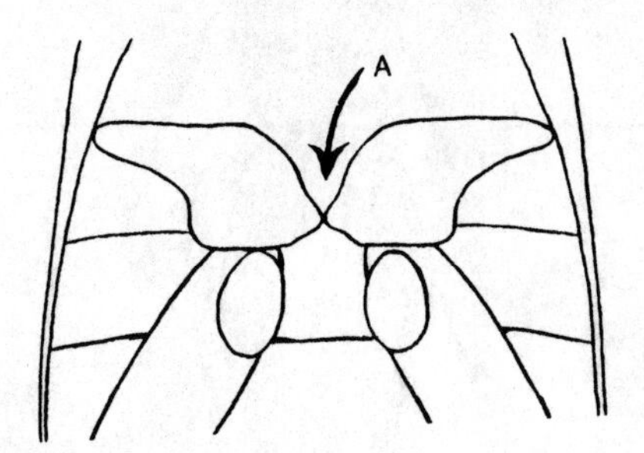

Figure 171

1b **Third coxae separated (b); eyes small. Fig. 172. .... 2**

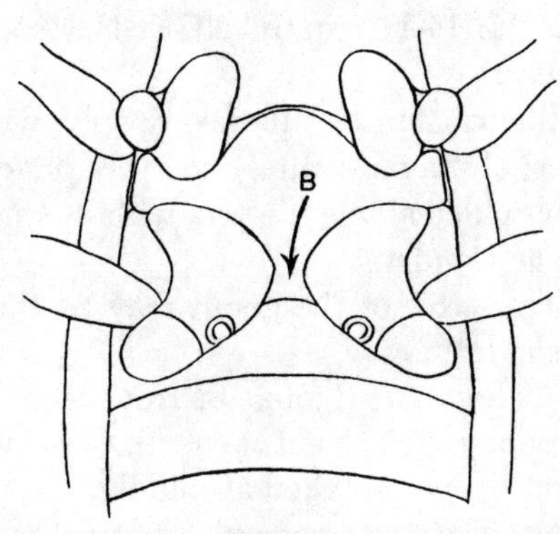

Figure 172

2a **(1) One inch or more in length. Sides of elytra widely inflexed. Fig. 173. .... *Amblycheila cylindriformis* Say**

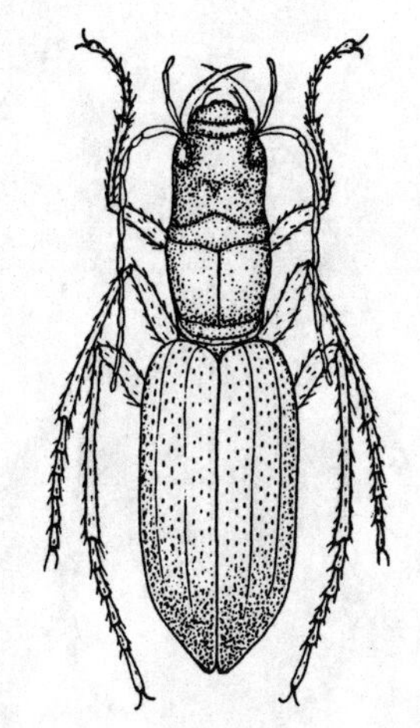

Figure 173

LENGTH: 30-38 mm. RANGE: Kan., Colo., Ariz., Okla., N. M., Tex.

Brown or blackish with pale brown elytra; head large, eyes small; elytra with three ridges; legs long and heavy; tarsi short; wingless.

They are nocturnal in habit and live in holes in clay banks. When moving the body is raised high and the antennae kept in continuous motion.

Another species, *A.** *baroni* Riv. is known from Arizona. It is considerably smaller than *cylindriformis*.

2b **Less than one inch in length. Sides of elytra narrowly inflexed. Flightless beetles with elytra fused at the suture. Usually black. Genus *Omus*. 3**

3a **(2) Elytra with distinct pits (fovea). Usually more than 2/3 inch in length. Fig. 174. .... *Omus dejeani* Reiche**

*This "*A*" refers to *Amblychella*. It is often customary to thus abbreviate the generic name when it appears a second or more times in close association with the place where it is spelled out in full.

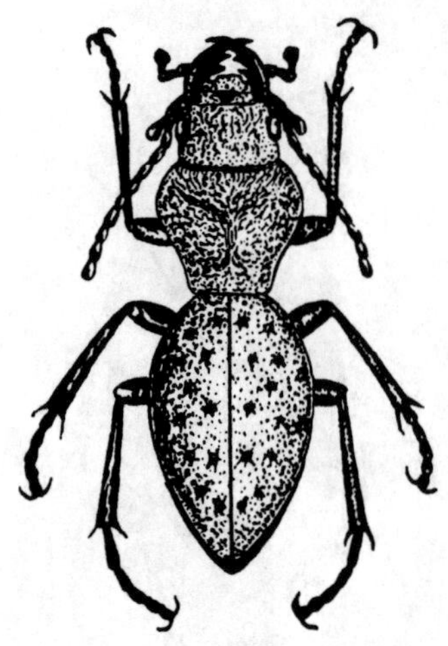

Figure 174

LENGTH: 15-21 mm. RANGE: B. C., Wash., Ore., Mont.

Black. Elytra with conspicuous pits; side margins of thorax apparent at front only; labrum with two notches; head and thorax much wrinkled.

The members of this genus hide by day and hunt at night. Trapping by placing finely chopped meat under a board is said to yield good results.

3b Elytra without distinct pits. Usually less than 2/3 inch long. 4

4a (3) Thorax with deep wrinkles; shining. Fig. 175. .... *Omus californicus* Esch.

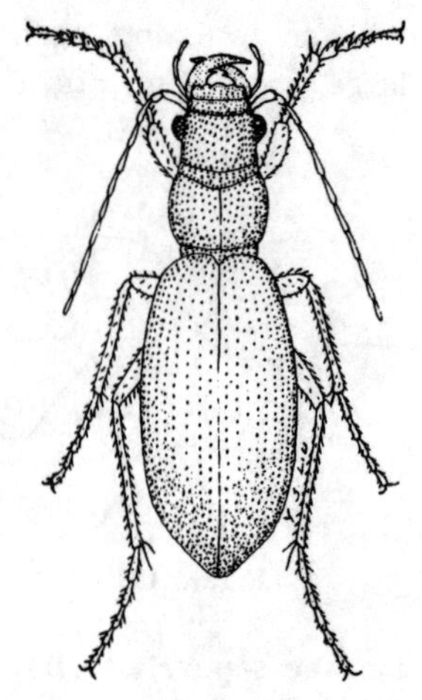

Figure 175

LENGTH: 14-17 mm. RANGE: Northern Cal. near coast.

Black, shining; thorax deeply wrinkled all over; elytra moderately coarsely punctured; head wrinkled all over; elytra widest at middle; antennae slender.

The sexes of the genus may be thus distinguished.

Males: Three tarsal joints of front legs dilated and spongy pubescent beneath, last ventral segment deeply notched at middle.

Females: Front tarsi simple; last ventral segment without notch, oval at tip.

O. edwardsii Cr. is similar in size but with shallower wrinkles on the thorax. It has been found in Eldorado, Placer and Sierra Counties in California.

O. californicus lecontei Horn, measuring 15-16 mm, is slender with its greatest width in front of the middle. The labrum is truncate while in *O. intermedius* Leng (17 mm), it is bisinuate. Both species are found in central and west-central California.

4b Thorax entirely smooth without any wrinkles; elytra usually smooth and dull. Fig. 176. .... *Omus laevis* Horn

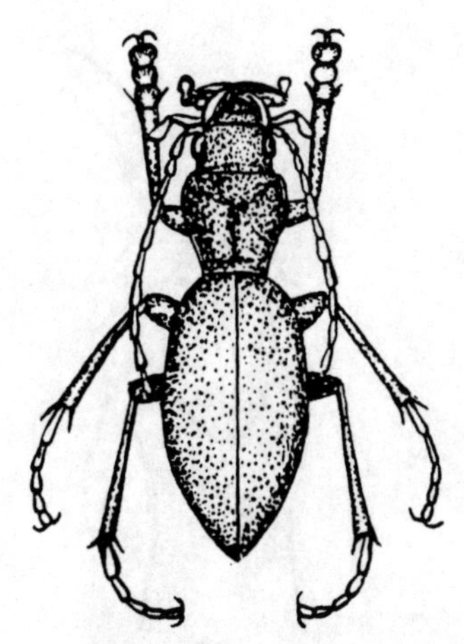

Figure 176

LENGTH: 13-16 mm. RANGE: Calif.

Dull black; antennae stout; sides of thorax somewhat sinuate, elytra widest at middle.

5a (1) Body, heavy, cylindrical; third segment of maxillary palpi longer than the fourth. (See fig. 178a) Fig. 177.
(a) *Megacephala virginica* L.

LENGTH: 20-24 mm. RANGE: Pa. to Fla. and west to Neb. and Tex.

Dark golden green, last ventral segments and antennae rusty brown, lateral margins of thorax and elytra metallic green, middle black; no lunule (moon-shaped mark) on elytra.

(b) *M. carolina* L.

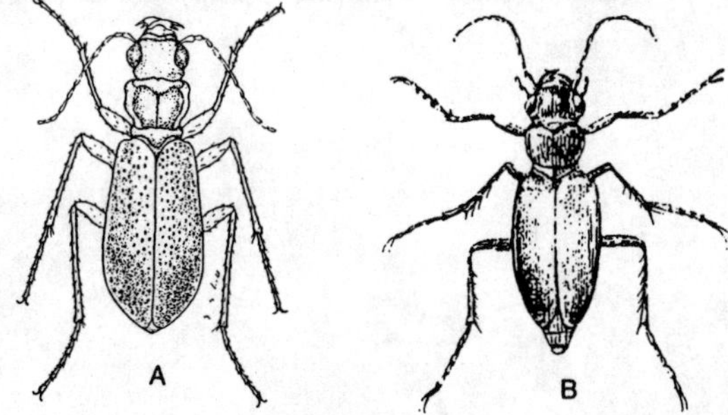

Figure 177

LENGTH: 20 mm. RANGE: entirely across southern U. S.

Light golden green, elytra purplish at middle with a lunule at the apex.

These two are the only species of this genus in our region. They hide by day and are active at night often attracted by lights.

5b Elytra usually somewhat flattened; third segment of maxillary palpi shorter than the fourth. Fig. 178b. Genus *Cicindela* 6

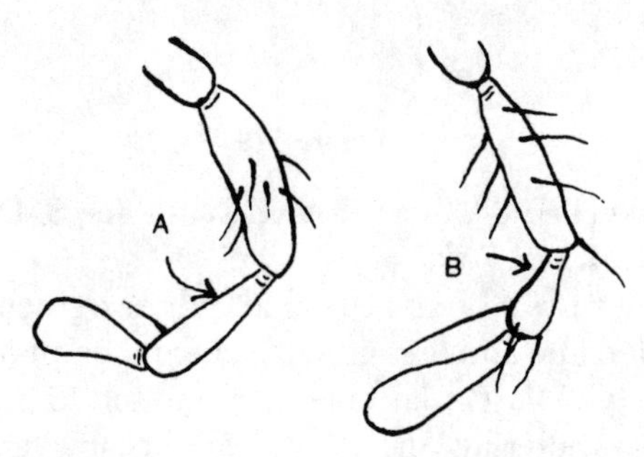

Figure 178

Among the members of this genus are some of our most beautiful beetles. They are often brilliantly colored and live such an active life that it offers rare sport to catch them.

6a (5) Humeral angles distinct; with flying wings. 7

6b Humeral angles rounded; small flightless species. Fig. 179. .... *Cicindela celeripes* Lec.

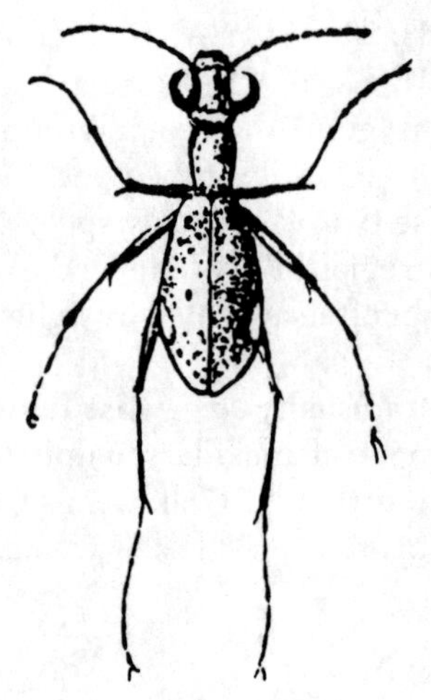

Figure 179

LENGTH: 6-8½ mm. RANGE: Ill., Ia., S. Dak., Neb., Kans., Ark.

Brown-bronze, head and thorax greenish; undersides and legs green; tibia and palpi pale; eyes very large, labrum with one tooth; elytra with scattering short hairs and coarse green punctures, the markings arranged as shown in the picture are rather indistinct and often wanting in part.

This is one of our smallest tiger beetles. They cannot fly but run "like a deer" as the species name indicates. It seems to prefer prairie hillsides.

Prothyma leptalis Bates, found in Mexico, is only 6 mm in length. This genus as well as several other genera of tiger beetles have no representatives within our region.

7a (6) Thorax cylindrical with no margin. 8

7b Thorax with narrow margin, abdomen not hairy, elytra much flattened. Fig. 180. .... *Cicindela unipunctata* Fab.

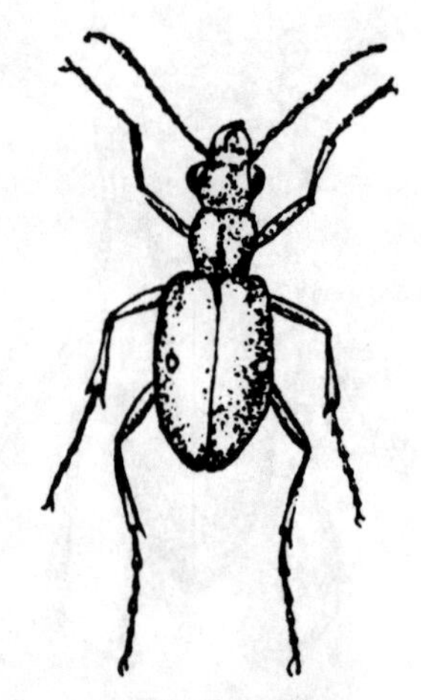

Figure 180

LENGTH: 16-18 mm. RANGE: N. Y. and N. J. west to Ia. and Mo. and south to Ga. and N. C.

Dull brown, dark blue beneath; the flattened elytra, rough and with green pits and punctures. A small triangular white spot near the margin of each elytron is its only mark.

This species lives along shaded woods paths and hides instead of flying when disturbed.

8a (7) Pubescence beneath either erect or absent; outer margin of elytra of female not angulate. 9

8b Pubescence on under parts prostrate; outer margin of female angulated near apex (a); elytra distinctly bronzed. Fig. 181. .... *Cicindela cuprascens* Lec.

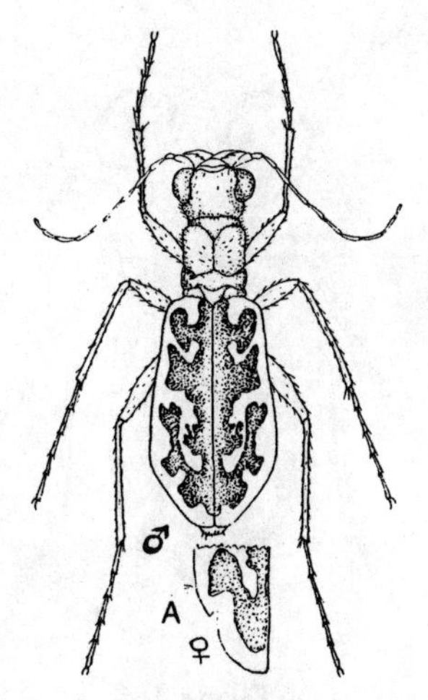

Figure 181

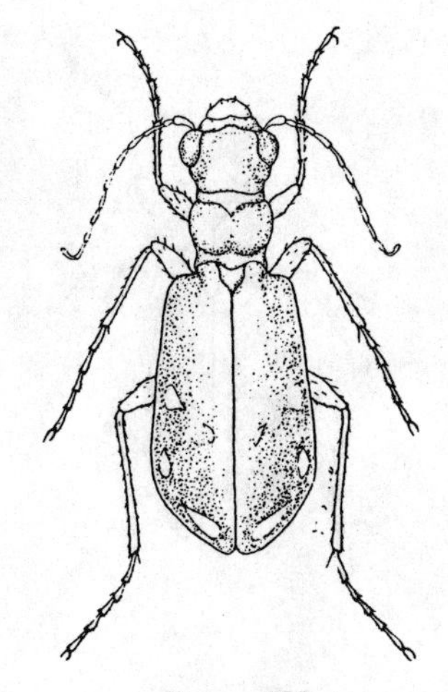

Figure 182

LENGTH: 12-14 mm. RANGE: Eastern half of U. S.

Coppery or greenish bronzed above with complete markings of ivory as pictured. Thorax squarish flattened. Thinly clothed below with decumbent white hair.

They frequent sandbars along water courses but are usually comparatively rare. Two varieties *macra* and *puritana* have been named.

9a (8) Thorax narrowed behind; markings almost always reduced to dots or spots. 10

9b Thorax not noticably narrowed behind; markings largely connected and entering the interior of wing cover. 11

10a (9) (a, b, c) Bright green or bluish-purple above with 2 to 8 white dots, rarely none. Not hairy beneath. Elytra punctate. Fig. 182. .... *Cicindela sexguttata* Fab.

LENGTH: 10-14 mm. RANGE: much of the U. S. east of the Rockies.

Bright green above, often with a bluish or purplish reflection. Each elytron may have from one to five white dots or sometimes has no markings at all.

Several varieties have been named, *violacea* Fab. having a rich violet color and is usually without dots though not always so.

This highly attractive species is fairly common along sun-lit paths where green vegetation grows.

10b Black, dark bronze or green-bronze above with rather indistinct markings. Fig. 183. .... *Cicindela punctulata* Oliv.

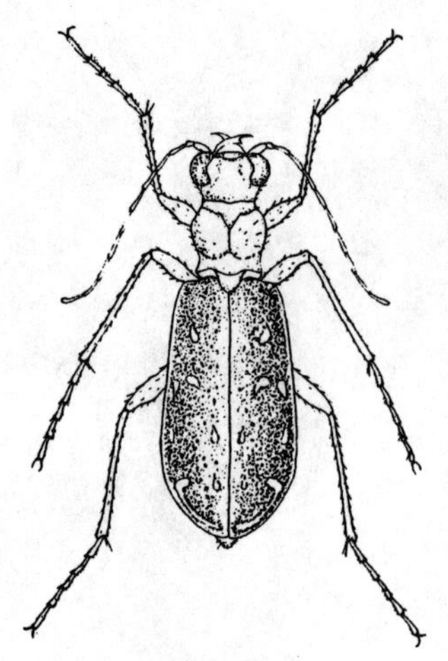

Figure 183

LENGTH: 10-13 mm. RANGE: much of the U. S. and Can. except the extreme west and north.

Form slender, legs long; black dark bronze or greenish-bronze above; greenish-blue beneath. The markings of bands or small dots are usually indistinct and sometimes wanting.

It is a very common species, seems to like dusty paths, often flies to light and in summer may frequently be seen on city streets.

10c Elytra practically without punctures. Color variable leading to several varieties; var. *lecontei* here pictured, coppery-bronze or greenish-bronze. Fig. 184.
.... *Cicindela scutellaris* Say

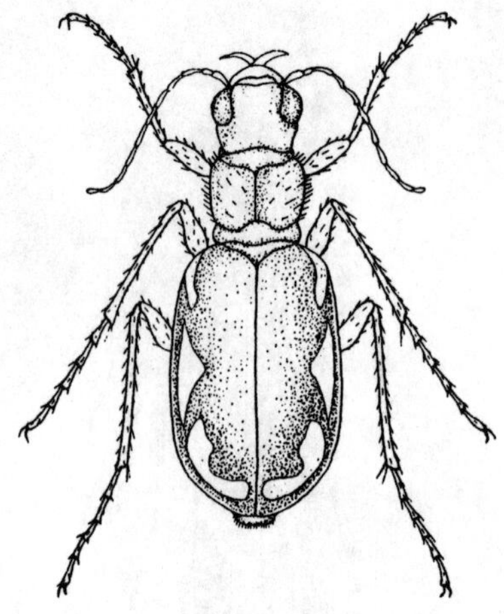

Figure 184

LENGTH: 11-13 mm. RANGE: East of the Rockies.

Head and thorax green or sometimes blue; elytra coppery or greenish with marginal dots distinct or connected or unmarked. Some varieties are—

Var. *unicolor* Dej.—Unmarked green or blue.

Var. *modesta* Dej.—Black with apical lunule and one or two marginal dots.

Var. *nigrior* Schaupp—Wholly black.

Var. *rugifrons* Dej—Green or blue with apical lunule and one or two marginal dots.

11a (9) Coppery-red (sometimes brown or green shaded above) large. Fig. 185. ......
.... *Cicindela formosa* Say

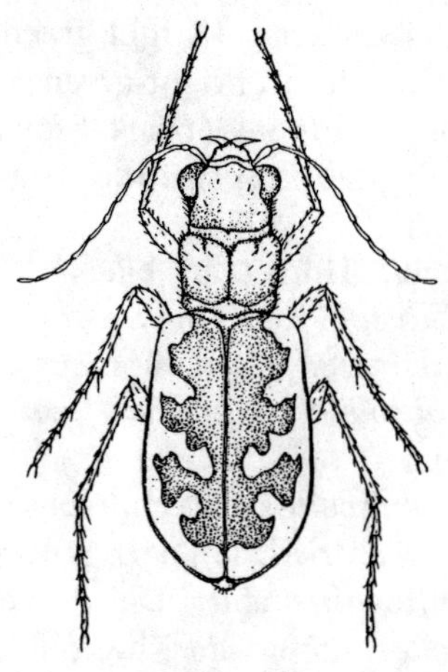

Figure 185

LENGTH: 16-18 mm. RANGE: East of the Rockies to the Missouri river.

Red cupreous; under parts metallic blue or green, thorax broader than long, much hair on head, sides of thorax, the abdomen, trochanters and femora; labrum with three teeth.

This species is found near sand dunes and blow-outs.

Var. *generosa* Dej. Dull red, green-bronze or brown-bronze above, green beneath; markings variable but when complete, as pictured. It ranges from Colo. and Manitoba east to the coast.

Var. *manitoba* Leng has heavy white markings.

11b Brown-bronze above; smaller. Fig. 186.
.... *Cicindela repanda* Dej.

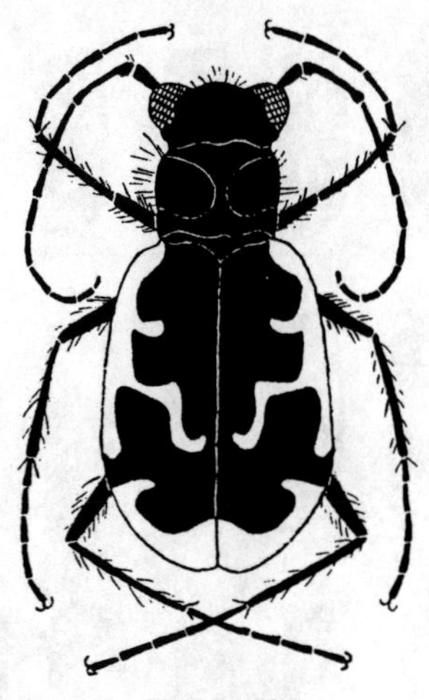

Figure 186

LENGTH: 12-13 mm. RANGE: U. S. and Canada east of the Rockies.

Brown-bronze usually with a strong greenish or coppery reflection above, underparts green; labrum short, one toothed. Underparts rather thinly clothed with long white hairs. This is a very common and widely distributed species. It is easily confused with *Cicindela duodecimguttata* Dej. which is similar in color and range. It is more flattened than *repanda* and usually has incomplete markings.

Other Tiger Beetles

Only a comparatively few species of any family, large or small, can be included in a book of this size but since the Tiger Beetles are such strong favorites with collectors we are reproducing from Charles W. Leng's "Revision of the Cicindelidae," some drawings of the left wing cover of a number of species. These figures are all drawn to the same scale and are but slightly larger than natural size. In variable forms the most completely marked pattern is shown.

The scientific names follow the numbers. All belong to the genus *Cicindela*. The chief color of the elytra is given. Fig. 187.

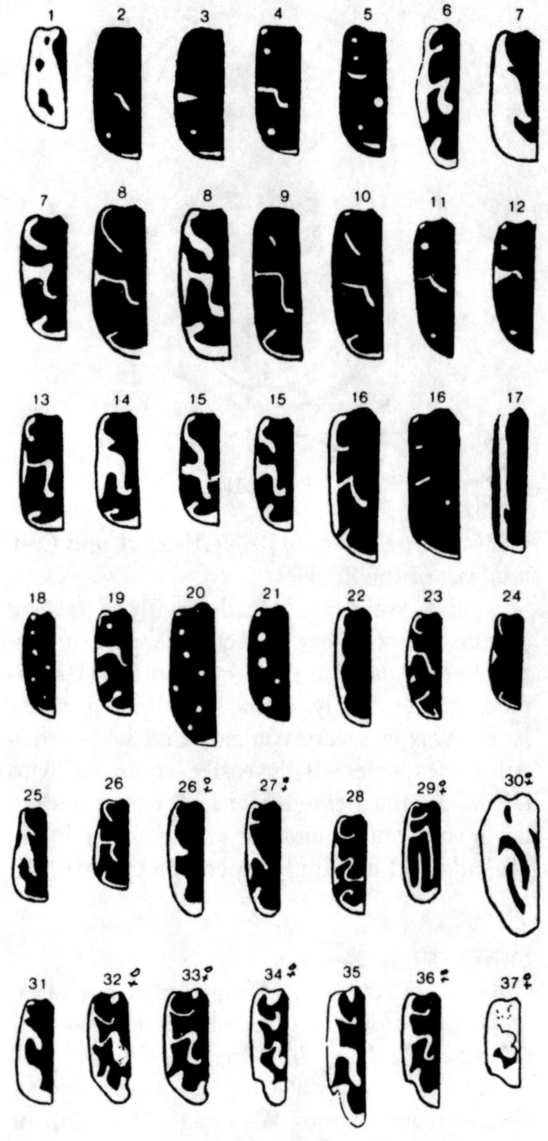

Figure 187

1. *C. limbata* Say (green)
2. *C. purpurea* Oliv. (cupreous)
3. *C. limbalis transversa* Leng (cupreous)
4. *C. limbalis* Klug (cupreous)
5. *C. duodecimguttata* Dej. (brown-bronze)
6. *C. hirticollis* Say (bronze-brown)
7. *C. latesignata* Lec. (brownish-black)
8. C. *tranquebarica* Hbst. (black)
9. *C. t. sierra* Leng (bright green)
10. *C. vibex Horn* (bright green)
11. *C. longilabris* Say (black, brown or green)
12. *C. longilabris perviridis* Schp. (bright green)
13. *C. senilis* Horn (dull black)
14. *C. willistoni* Lec. (brown or green bronze)
15. *C. fulgida* Say (red-cupreous or green)
16. *C. obsoleta vulturina* Lec. (black or green)
17. *C. lemniscata* Lec. (cupreous)
18. *C. rufiventris* Dej. (dark brown)
19. *C. rufiventris hentzi* Dej. (black)
20. *C. sedecimpunctata* Klug (black)
21. *C. ocellata rectilatera* Chd. (dark brown)
22. *C. marginipennis* Dej. (olive or brown)
23. *C. haemorrhagica* Lec. (greenish black)
24. *C. pusilla terricola* Say (dull black)
25. *C. p. cinctipennis* Lec. (brown, green, blue or black)
26. *C. circumpicta* Laf. (green, blue, brown or black)
27. *C. californica* Men. (dark bronze)
28. *C. trifasciata tortuosa* Lec. (brown)
29. *C. gabbi* Horn (olive bronze)
30. *C. dorsalis* Say (bronze)
31. *C. pamphila* Lec. (dull olive)
32. *C. hamata* Aud. and Br. (brown or green bronze)
33. *C. marginata* Fab. (brown or green bronze)
34. *C. blanda* Dej. (green bronze)
35. *C. cuprascens* Lec. (cupreous or greenish bronze)
36. *C. sperata* Lec. (brown to bright green)
37. *C. lepida* Dej. (green or bronze)

FAMILY 4. CARABIDAE
The Ground Beetles

The members of this large family live mostly in or on the ground and feed upon other insects or small forms of life. They are most active at night. In warmer climates many of the species live in trees. Members of the family vary considerably in size, shape, and color. As a rule they are dull and blackish in color although some species are brilliantly colored or metallic. Usually ground beetles have long slender legs and antennae composed of 11 segments. Members of this family are found on all the continents, over 20,000 species being known. Although there is this variety among species, the collector soon learns to recognize members of this family by sight, in many cases down to species. Because of the predacious habits of the family, this is a most beneficial family.

1a **Scutellum normally exposed; pronotum not scooped-shaped covering the entire mesosternum; usually more or less elongate beetles. 2**

1b **Scutellum not visible; prosternum scoop-shaped completely covering the mesosternum; body oval. Members of this genus live in burrows along water and may be flushed out by splashing over the bordering mud or sand. 11 species in U.S., six listed below. *Genus Omophron.***

(a) Thorax with only side margins pale; elytra with 15 striae. Fig. 188. .... *Omophron americanum* Dej.

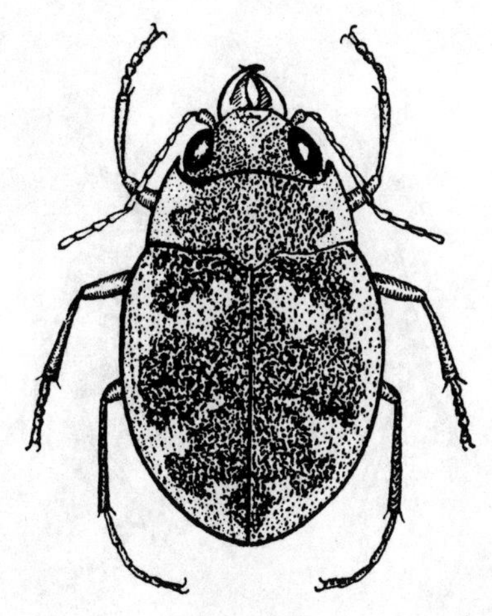

Figure 188

LENGTH: 6-7 mm. RANGE: Northeastern and central United States and Canada.

Bronzed or greenish-black; head largely green, thorax and elytra with narrow, pale margins; Pronotum striae with fine punctures; the intervals convex.

O. robustum Horn, with about the same size and range as *americanum*, is pale brownish-yellow with broken green cross markings. It may be recognized by its 14 shallow striae.

(b) Thorax with a somewhat square green spot in center wholly surrounded by wide yellowish margins. Fig. 189. .... *Omophron tessellatum* Say

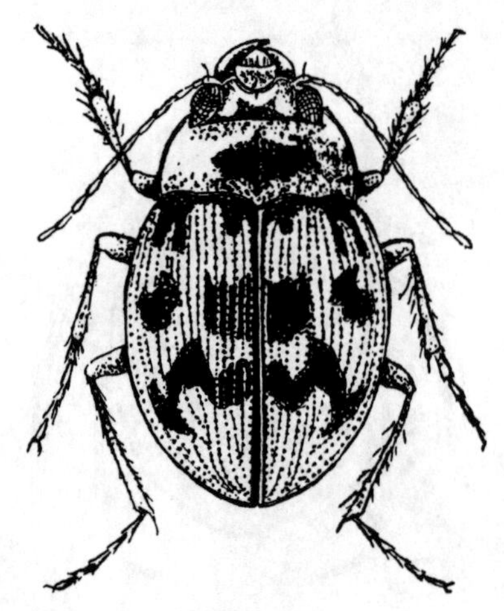

Figure 189

LENGTH: 6-7 mm. RANGE: Middle and eastern United States and Canada.

Pale brownish-yellow. Head with a green band across the base; thorax with center green spot; cross markings on elytra metallic-green. Elytra with 15 striae with close fine punctures.

O. ovale Horn is found in the Pacific States while *O. grossum* Csy. is known from the south-central states.

(c) More broadly oval; elytral striae plain only at base. Fig. 190. .... *Omophron labiatum* (Fab.)

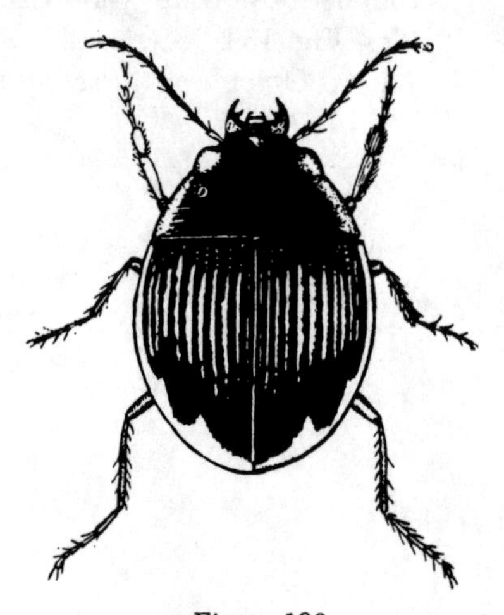

Figure 190

LENGTH: 6 mm. RANGE: east of the Rockies.

Nearly black with pale side margins. The punctures are only in the basal end of the striae. (After Hart)

O. *nitidum* Lec. 5-6 mm long and ranging through central U. S. is a shining metallic green, pale at sides.

2a (1) Middle coxal cavities entirely enclosed by the sterna alone, the epimeron not touching the coxa on its outer side; one or more bristly hairs above each eye. Fig. 191. Sub-family HARPALINAE. 11a

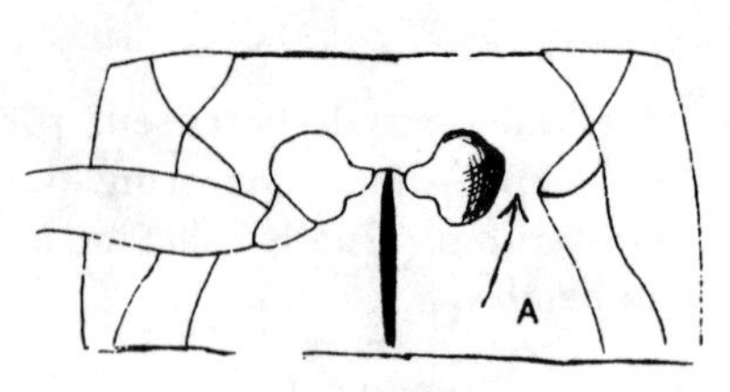

Figure 191

This is the major group of ground beetles, representing around 85% of the entire family.

2b The epimeron aiding the sterna to enclose the middle coxal cavities. Fig. 192. Sub-family CARABINAE. 3

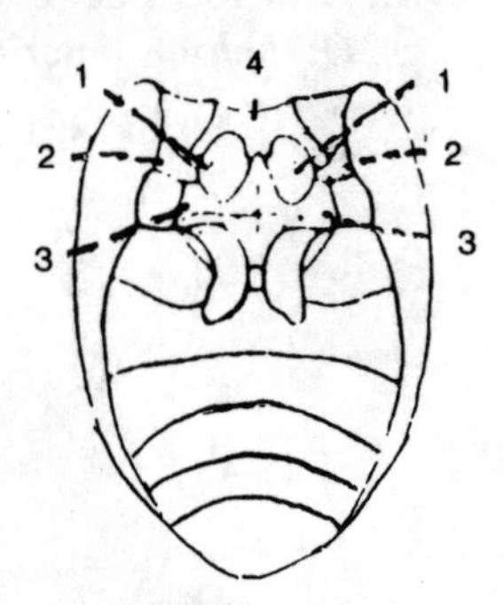

Figure 192

This much-smaller sub-family includes some of the largest and most attractively marked-and-colored members of the family.

A large percentage of the beetles taken from under stones, logs, and debris, especially in the spring, belong to this family. A good share of the beetles flying to lights are often carabids. A few species have a partial diet of seeds and fruit but their food habits on the whole make them friends of the gardner.

3a **(2) Front coxal cavities open behind. Fig. 193a. 4**

3b **Front coxal cavities closed behind. Fig. 193b. 8**

4a **(3) Hind coxae not separated; labrum not forked. 5**

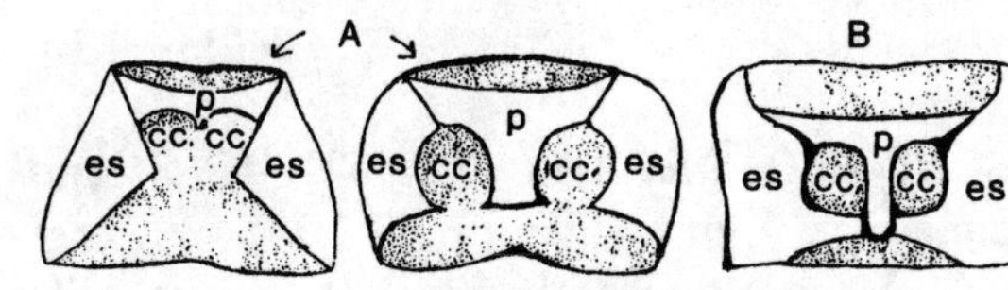

Figure 193

4b **Hind coxae separated; labrum deeply forked, antennae with 4 basal joints glabrous, thorax nearly as wide as elytra, the sides elevated. Fig. 194.**

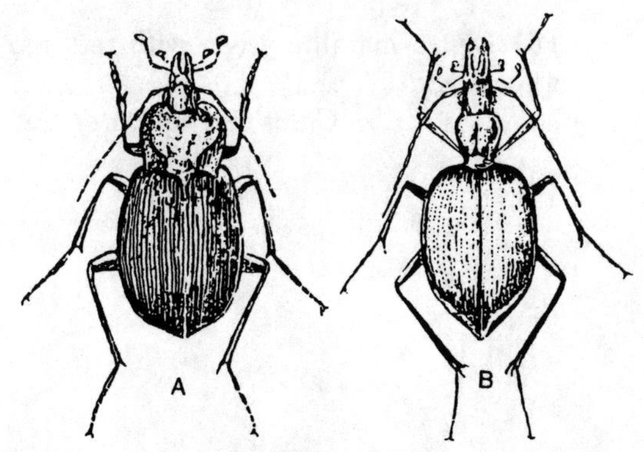

Figure 194

(a) ***Scaphinotus elevatus*** **(Fab.)**

LENGTH: 18-20 mm. RANGE: middle and eastern U. S., not common.

Violaceous or cupreous. Antennae long and slender. Thorax with hind angles prolonged over the elytra, margins of both thorax and elytra reflexed. This species which runs to several varieties is one of a group of snail eating beetles which usually live in moist woods and are found under logs.

(b) ***Scaphinotus andrewsi*** **Harr.**

LENGTH: 19-22 mm. RANGE: from Ind. eastward.

Blackish beneath; violet with iridescence above. Disk of thorax nearly smoth with punctures at the sides.

Other species are *S. interruptus* (Men.) 16-18 mm long. Black shining with a purplish metallic sheen; found in Cal. and Or.

Sphaeroderus nitidicollis Chev. 20-22 mm in length, bronzy black with iridescent sheen is found in Canada and our northeastern states and *Scaphinotus fissicolis* (Lec.) with a length of 10-13 mm, deep violaceous ranges in East and South. This latter species has but 2 basal antennal joints glabrous.

5a **(4) Third segment of antennae compressed. Genus *Calosoma*. 6**

5b **Third segment of antennae cylindrical. Fig. 195. .... *Carabus vinctus* Web.**

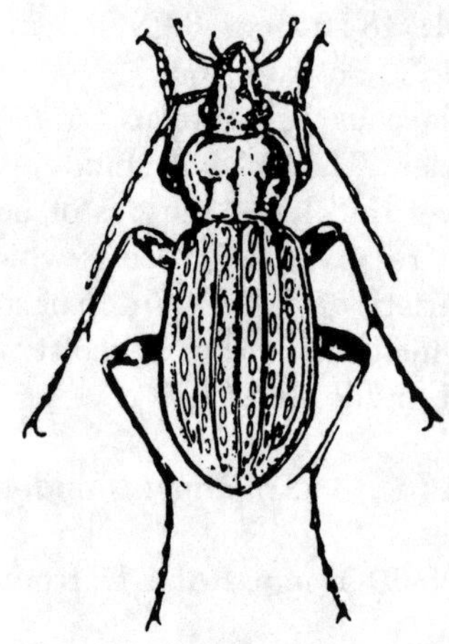

Figure 195

LENGTH: 25-30 mm. RANGE: U. S., east of the Mississippi River.

Dull black, bronzed, thorax tinged with greenish at borders, without punctures though finely rugose; elytral striae with irregular granulate punctures.

C. sylvosus Say 27-30 mm. East and South. Black, margins of thorax and elytra blue; 3 rows of pits on each elytron.

6a (5) Elytra without rows of metallic spots. 7

6b Elytra black, each with three rows of golden spots. Fig. 196.

(a) *Calosoma calida* Fab.

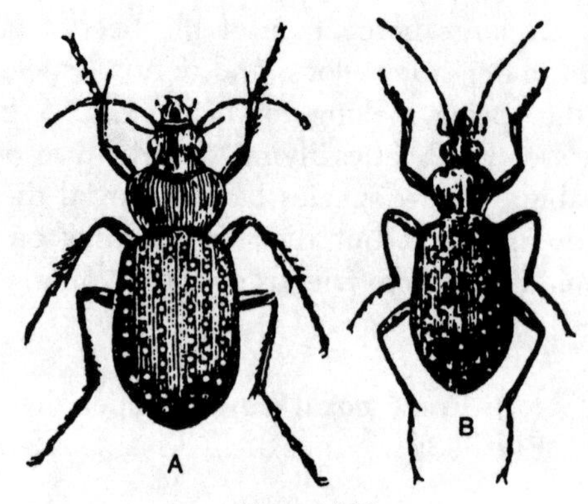

Figure 196

LENGTH: 20-25 mm. RANGE: U. S.

Heavy. Black above and below; elytra with three rows of reddish or coppery pits. Elytral striae deep with broad intervals.

(b) *C. frigidum* Kby.

LENGTH: 22-24 mm. RANGE: U. S. and Canada.

More slender than *calida;* elytra with green punctures and greenish margin, ranges through our northeast.

The members of this genus are highly useful in killing caterpillars and grubs, often climbing trees to find them.

7a (6) Elytra metallic green with red margins. Fig. 197. .... *Calosoma scrutator* Fab.

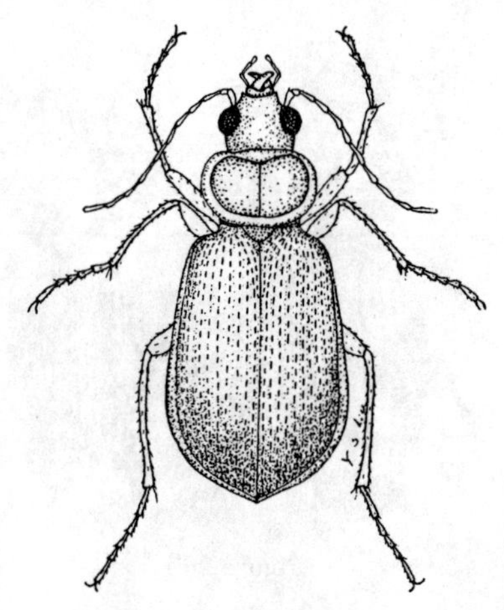

Figure 197

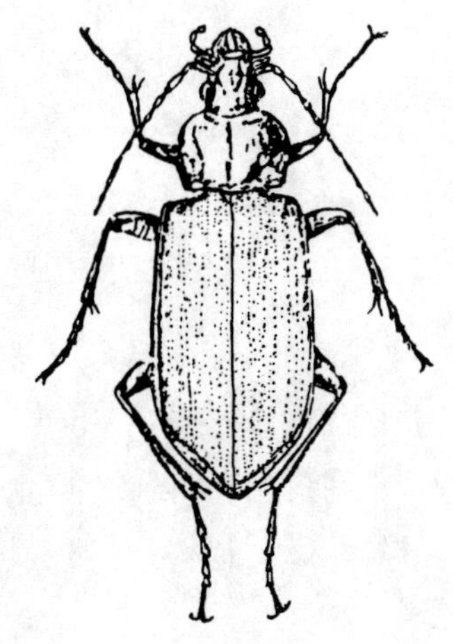

Figure 198

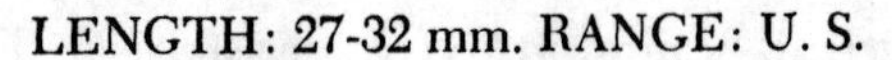

LENGTH: 27-32 mm. RANGE: U. S.

Metallic and highly iridescent throughout. Disk of thorax blue or dark purple, the margins golden or red; legs blue; abdomen green and red. Male distinguished by a dense brush of hairs on inner surface of curved middle tibiae.

C. willcoxi Lec. is very much like *scrutator* but only 17 to 20 mm long.

C. sycophanta L. (182) with a length of 25-30 mm is an European species which has been introduced in our East to destroy the Gipsy Moth. Its thorax is blue and the elytra brighter green than in *scrutator,* the elytral margins are also green.

7b **Elytra black with blue border. Fig. 198. .... *Calosoma externum* Say**

LENGTH: 28-32 mm. RANGE: east of the Rockies.

Black, somewhat dull; thorax and elytra with blue side margins. Elytra parallel for ¾ of length, striae distinctly punctured.

C. marginale Csy. is a plain shining black species about the size and shape of *scrutator*... RANGE: South central states.

C. peregrinator Guer. is plain dull black. It belongs to the whole Southwest.

8a **(3) Base of antennae covered by a frontal plate; body pedunculate (thorax and elytra joined by stem like part); scrutellum not visible. Tribe SCARITINI. 9**

8b **Base of antennae not covered; base of elytra joining closely with thorax; scrutellum visible. Fig. 199. .... *Elaphrus ruscarius* Say**

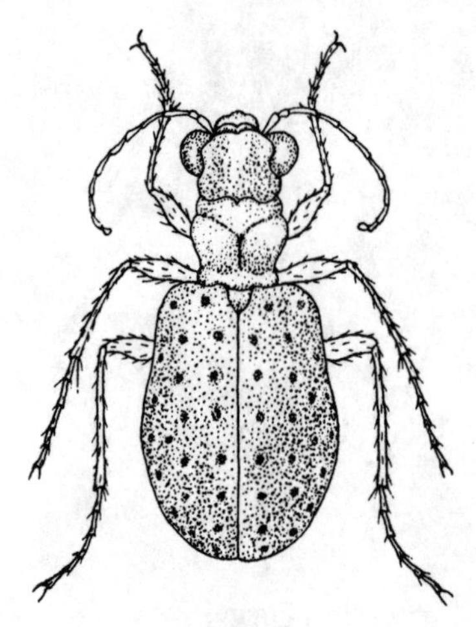

Figure 199

LENGTH: 8-10 mm. RANGE: northern half of U. S. and parts of Canada.

Dull drab with metallic sheen above; metallic green beneath; large impressed pits on elytra dull purplish, legs reddish brown.

The members of this genus of which there are several similar species may be found running on mud or sand flats on sunny days. They are sometimes mistaken for Tiger Beetles which they somewhat resemble in shape and habits. The head including the eyes is wider than the thorax.

9a (8) 15 or more mm in length; basal joint of antenna long; one bristle-bearing puncture above the eye and one at hind angle of thorax. 10

9b Less than 10 mm long; basal joint of antennae short; two bristle-bearing punctures above each eye and two at hind angles of thorax. Fig. 200.

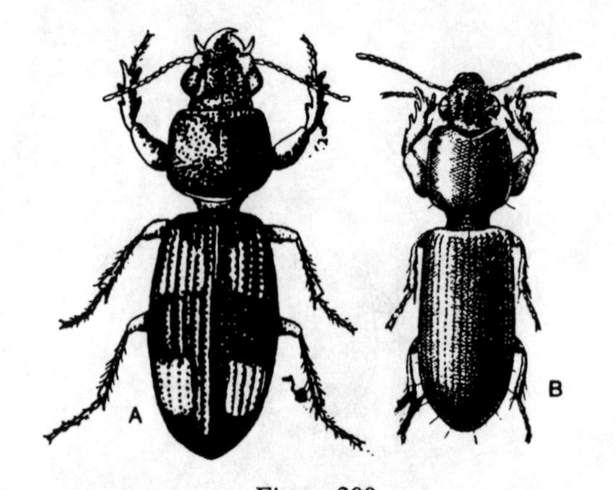

Figure 200

(a) *Clivinia bipustulata* (Fab.)

LENGTH: 6-8 mm. RANGE: Eastern and southern U. S.

Black, with two more-or-less-obscure reddish spots near base of elytra and another pair near apex; elytral striae deeply punctate; legs and antennae reddish brown. Often taken abundantly at lights.

(b) *C. impressifrons* Lec.

LENGTH: 6-6.5 mm. RANGE: Eastern half of U. S.

A similarly shaped uniformly reddish-brown beetle, sometimes eats sprouting grains of corn and thus becomes a "black sheep" in a, for-the-most-part, helpful family.

10a (9) Large broad beetles with hind angles of thorax distinct. Fig. 201. .... *Pasimachus punctulatus* Hald.

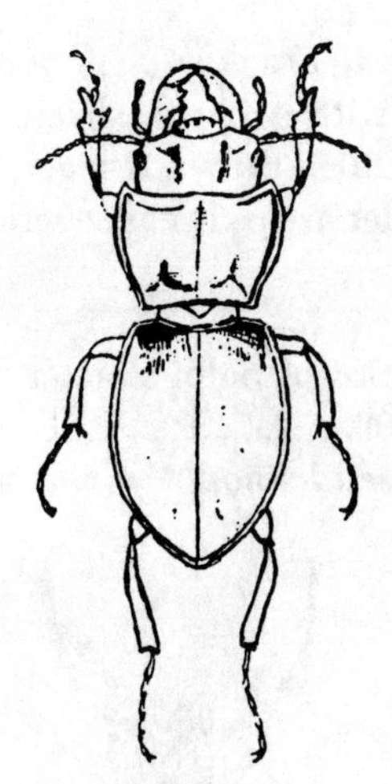

Figure 201

LENGTH: 27-30 mm. RANGE: Central and Southern States.

Black with blue margins; elytra usually with pairs of punctures; mandibles deeply and coarsely striate.

P. sublaevis Bov. 21-28 mm, ranging east of the Mississippi, is colored like *punctulatus* but differs in having the elytra striate, and the tips of the elytra rounded instead of angulate.

These beetles are found under stones etc. in open woods, pastures and along cultivated fields where they feed on caterpillars and other larvae.

10b Medium sized narrow beetles, without angles at back of thorax. Fig. 202. .... *Scarites subterraneus* Fab.

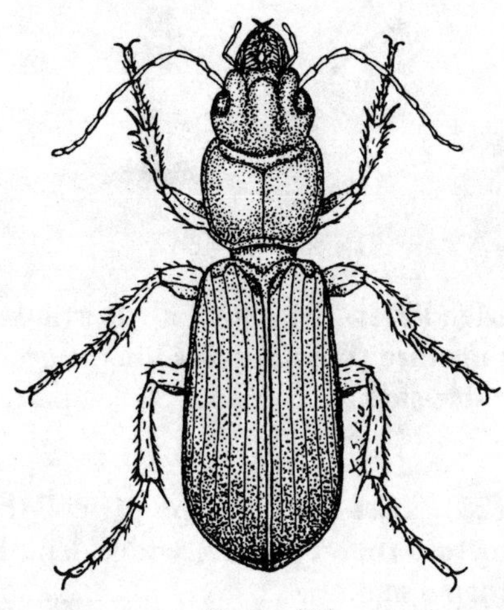

Figure 202

LENGTH: 15-30 mm. RANGE: Central and Eastern U. S. Abundant.

Black, shining. Elytra with striae but without punctures.

These beetles are found under stones etc. near cultivated fields and are highly beneficial in killing other insects.

11a (2) Head with two punctures above the eye, with a single heavy bristle in each. Fig. 203A. 12

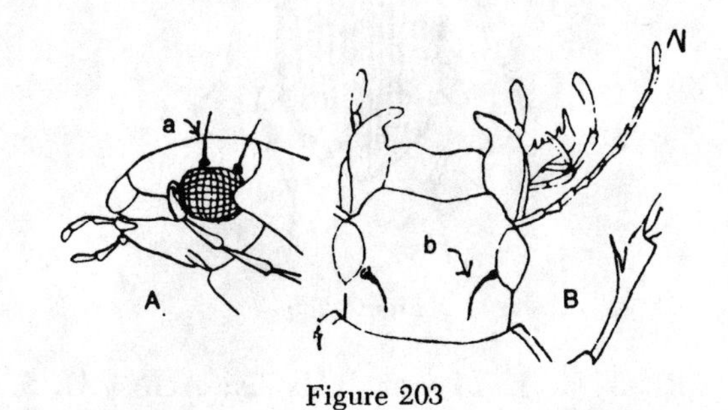

Figure 203

11b Head with but one bristle-bearing puncture above the eye. Fig. 203B. 55

12a (11) Mandibles with a bristle-bearing puncture (a) in the groove on the outer side. Fig. 204. 13

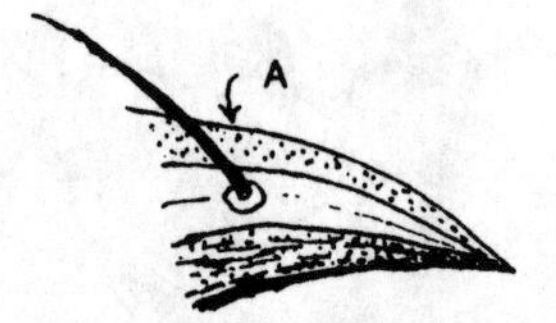

Figure 204

12b Mandibles without a bristle-bearing puncture (see a, fig. 204) in the groove on outer side. 19

13a (12) Last joint of palpi awl-shaped; mesosternal epimera wide; length less than 8 mm. 14

13b Last joint of palpi elongate, obtuse at tip or subcylindrical; mesosternal epimera narrow. Fig. 205. .... *Patrobus longicornus* (Say)

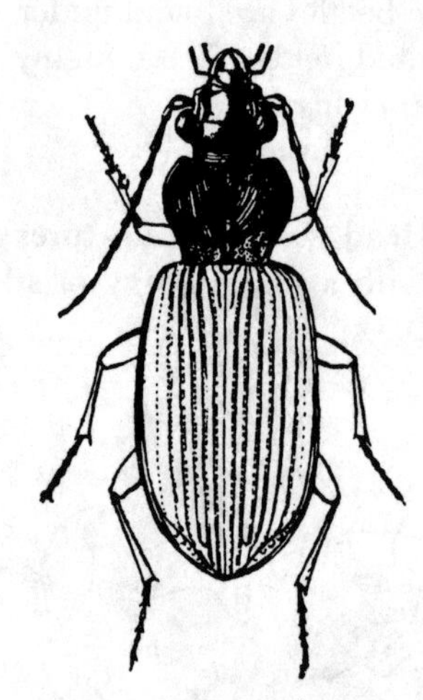

Figure 205

LENGTH: 12-14 mm. RANGE: Across U. S. and Canada.

Black above; under parts piceous; antennae reddish brown; legs paler. Thorax slightly broader than long, a median impressed line and a front transverse impression deep, punctured. Striae of elytra deep and punctured.

The above genus includes a number of species, most of which range north of the United States. *Trechus chalybeus* Dej. 5 mm long, black with a bluish sheen and antennae and legs reddish-brown, ranges widely, especially in colder areas. It has several named varieties.

13c Last joint of palpi slender, without eyes. Fig. 206. .... *Pseudoanophthalmus tenuis* (Horn)

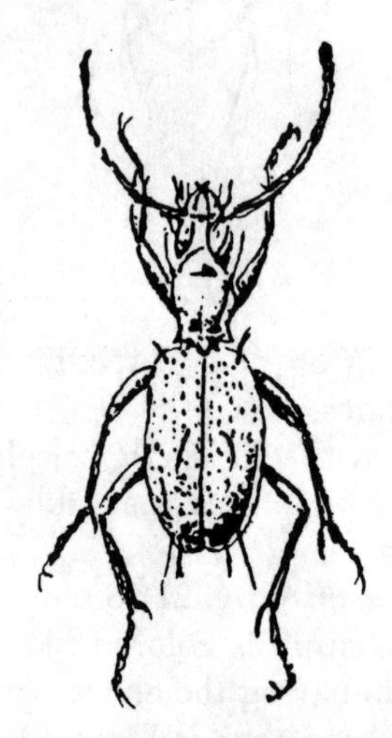

Figure 206

LENGTH: 4.5-6 mm. RANGE: Indiana.

Shining. Pale brownish-yellow. This beetle is one of several species of eyeless ground beetles found in the remote parts of caves. More than 50 species are known from Ky., Ind. and Va.

14a (13) Front tibia not obliquely truncate at apical end; sutural striae of elytra not recurving at tip of elytra. Genus *Bembidion*. 16

14b Front tibia obliquely truncate at apex; sutural striae recurving at tip of elytra. Very small beetles. 15

15a **(14) Thorax with broad translucent margins. Fig. 207.** .... ***Mioptachys flavicauda*** **(Say)**

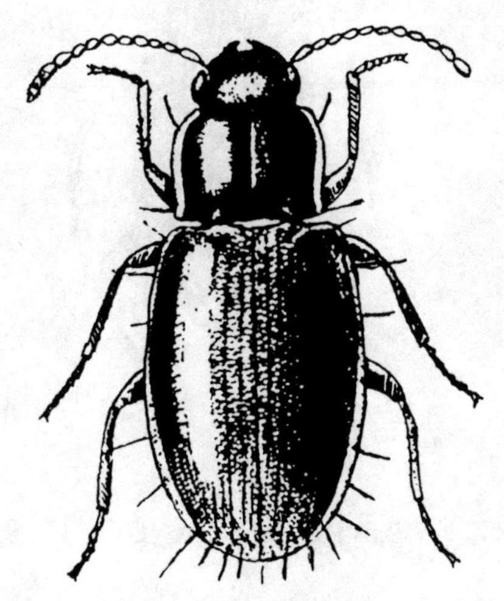

Figure 207

LENGTH: 1.5-1.8 mm. RANGE: U. S. and Can.

Dark piceous to black with ⅓ of elytra yellowish at tips. Antennae and legs reddish yellow. Frequently found under bark of trees.

Tachyta nana (Gyll.) differs from the above in being larger (2.2-3 mm), jet black throughout, and in having a narrow opaque margin on the thorax. It is common and apparently widely distributed.

15b **Margins of thorax narrow and not translucent; recurved arm of sutural stria at tip of elytra, short. Fig. 208.** .... ***Elaphropus incurvus*** **(Say)**

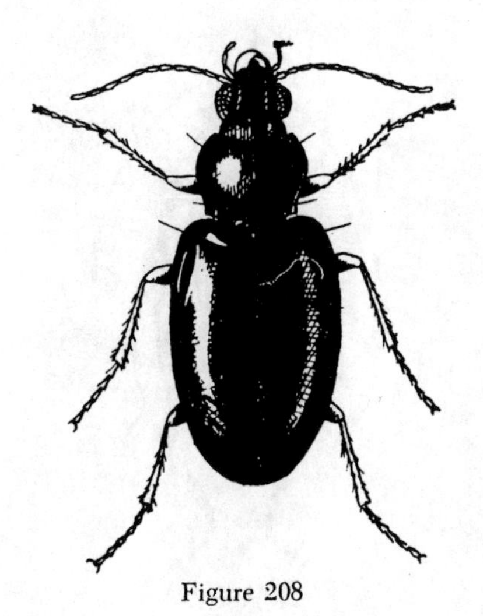

Figure 208

LENGTH: 1.7-2.5 mm. RANGE: U. S. and Can. Common.

Shining. Dark reddish brown to nearly black. Antennae brownish except basal joints which with legs are dull yellow. Indistinct stripe on elytra pale yellow. Found in open woods and in ants' nests.

Paratachys scitulus (LeC.) found in rubbish near water and on mud flats is from 2.5 to 3 mm long, dull reddish yellow, the head darker. The elytra have an indistinct darker crossband behind their middle. It is widely distributed.

16a **(14) Dorsal punctures (2) on the third striae of elytra.** **17**

16b **Dorsal punctures (2 or more) on the third interval of elytra. Fig. 209.** .... ***Bembidion inaequale*** **Say**

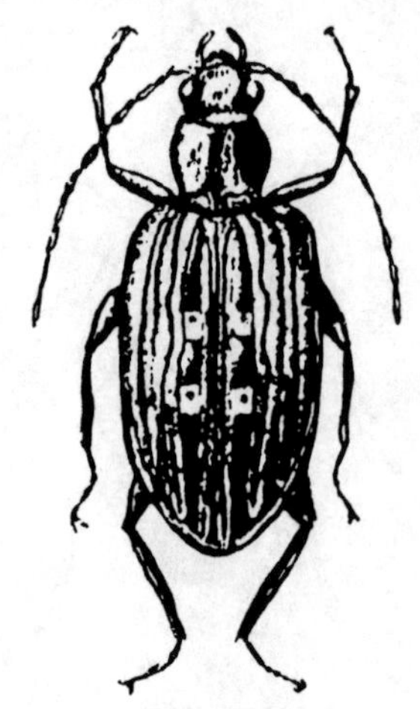

Figure 209

LENGTH: 4.5-5.5 mm. RANGE: No. U. S. and So. Canada.

Bronzed, shining; antennae piceous, with pale red basal joints; legs greenish; tibia and base of femora dull yellow; third and sixth intervals irregular.

The genus *Bembidion* includes almost 400 species in our region, many of which are attractively marked.

17a (16) Thorax not appreciably narrower at base than at apex. 18

17b Base of thorax narrower than its apex, very small. Common. Fig. 210. .... *Bembidion versicolor* (Lec.)

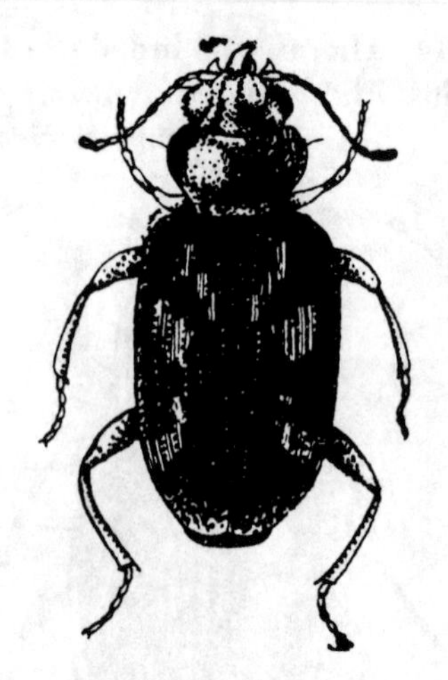

Figure 210

LENGTH: 2.5-3.2 mm. RANGE: U. S. and Can.

Conxev. Head and thorax bronzed greenish-black; elytra dull yellow marked in black or near black as shown. Striae deep and punctate at basal half of elytra.

18a (17) Brownish or black, marked with dull yellow as pictured. Fig. 211.

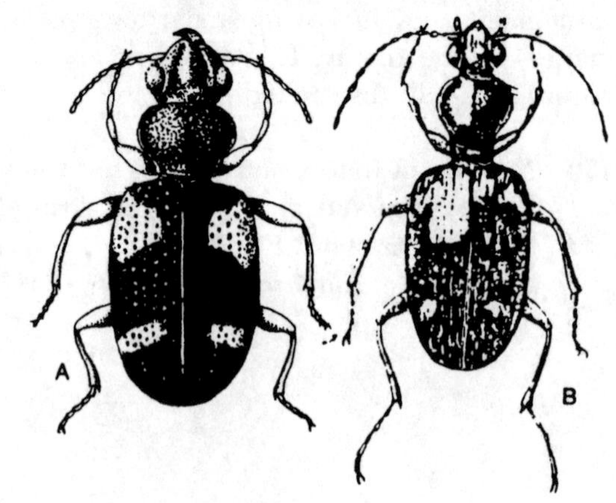

Figure 211

(a) *Bembidion quadrimaculatum* (L.)

LENGTH: 2.7-3.7 mm. RANGE: U. S. and Canada.

Head and thorax shining dark bronzed; legs and basal part of antennae dull yellow.

(b) *Bembidion pedicellatum* Lec.

LENGTH: 3-3.7 mm. RANGE: E. half of U. S.

It is closely similar in color and markings to *quadrimaculatum*. The elytral spots are smaller and less distinct.

18b Elytra dull yellow with black markings as pictured. Fig. 212. .... *Bembidion patruele* Dej.

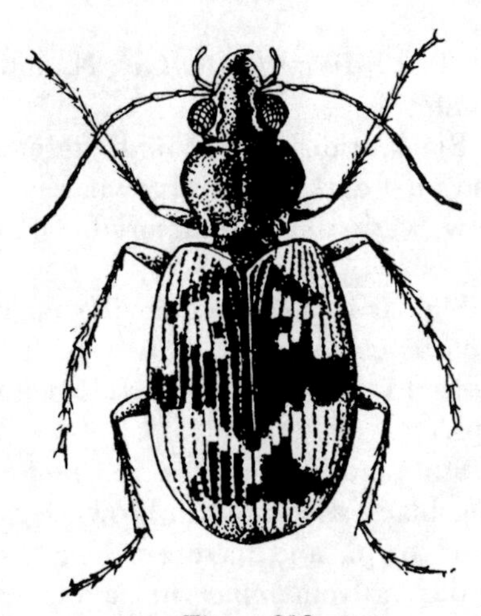

Figure 212

LENGTH: 3.5-4.7 mm. RANGE: U. S. except Southwest.

Head and thorax dark bronzed elytra dull yellow with black markings; legs reddish brown. Often taken at lights. Common.

B. variegatum Say, is quite similar except that the elytra are black with a few dull yellow markings.

B. americanum Dej. 5-6 mm long and colored a uniform blackish bronze in a very common species of wide distribution.

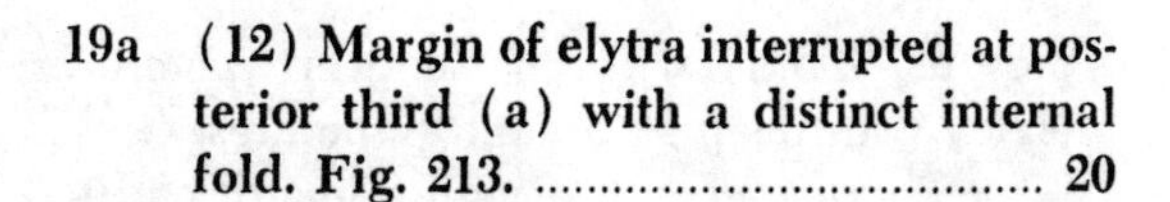

19a (12) Margin of elytra interrupted at posterior third (a) with a distinct internal fold. Fig. 213. 20

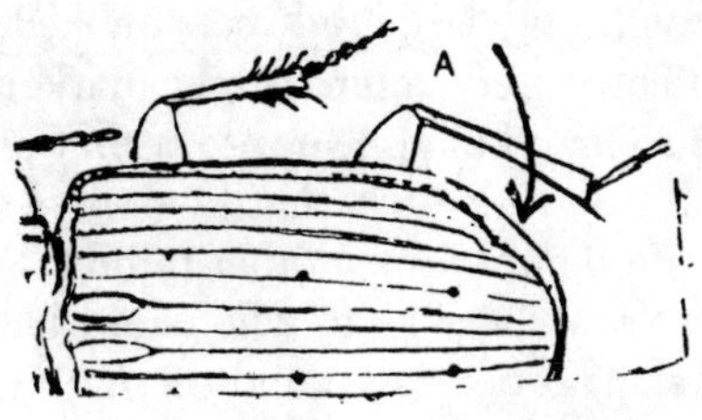

Figure 213

19b Margin of elytra without an internal fold; not interrupted. 33

20a (19) Head not constricted behind the eyes. 21

20b Head constricted behind the eyes then enlarged into a roll around the neck. Reddish-brown marked with black. Fig. 214. .... *Panagaeus fasciatus* Say

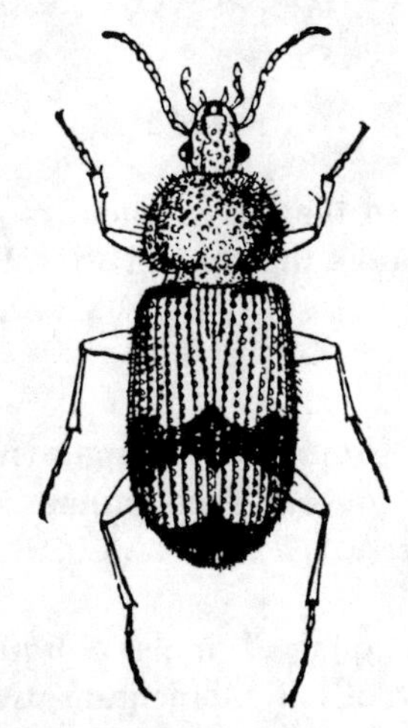

Figure 214

LENGTH: 7-9 mm. RANGE: East of the Rockies. Scarce.

Head, thorax, and elytra reddish brown, the elytra marked with black as pictured; legs piceous; thorax with coarse deep punctures; elytra with deep striae and punctures.

P. crucigerus Say, with a length of 10-12 mm and much the same range as *fasciatus* is black with two large red spots on each elytron.

These very interestingly marked beetles differ so much in appearance from the average run of ground beetles that beginners often try to locate them in some other family. We had a student some years ago who could bring them in on short order. He found them under lumps of coal along the railroad tracks.

21a (20) Next to the last segment of labial palpi with but two bristles. Fig. 215a. 22

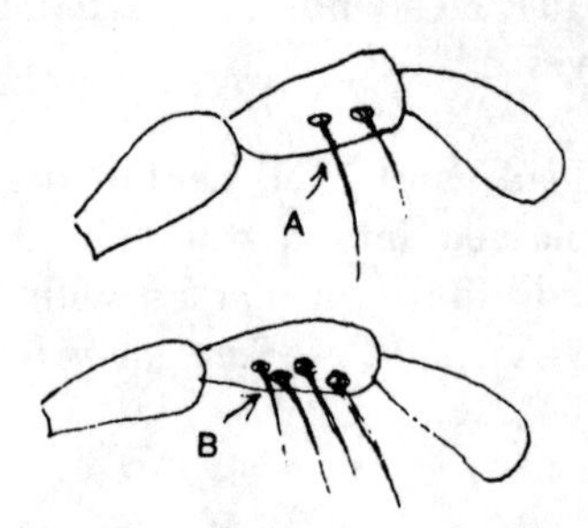

Figure 215

21b Next to the last segment of labial palpi with more than two bristles. Fig. 215b. 30

22a (21) Front tarsi of male normally dilated; tooth of submentum usually emarginate. 23

22b Front tarsi of male obliquely dilated (a); tooth of submentum usually entire, pointed; dorsal punctures one. Fig. 216. .... *Loxandrus minor* (Chaud.)

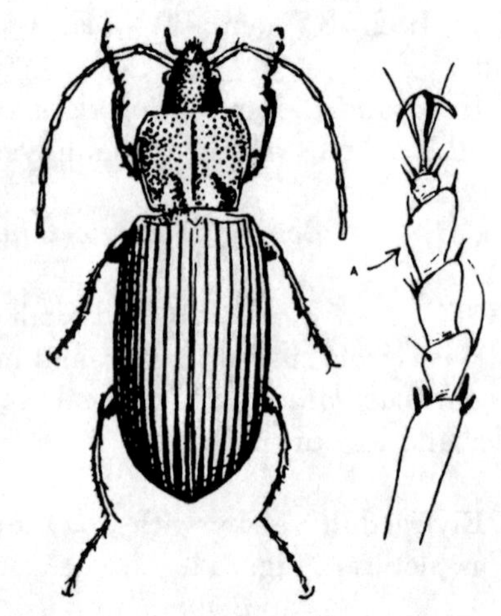

Figure 216

LENGTH: 9-10 mm. RANGE: N. and S. Central States.

Black, iridescent, shining; antennae, tibiae and tarsi dark reddish-brown. Elytral striae shallow, very finely punctured. Often abundant.

Similar in color but smaller (5.5-7 mm), *Loxandrus agilis* (Dej.), lives in Texas and Florida. It is found under bark of oak logs and stumps.

Still another species, *L. rectus* (Say) shining black with reddish-brown legs, antennae and palpi, and having a length of 11-13 mm., differs from *minor* in having its thorax one-half wider than long.

23a (22) With two or more dorsal punctures on each elytron. Fig. 217a. 25

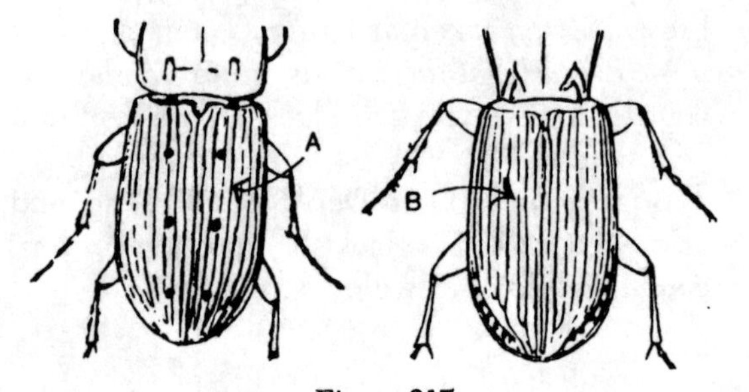

Figure 217

23b **With but one or not any dorsal punctures. Fig. 217b. 24**

24a **(23) With no dorsal punctures; head very large. Fig. 218. ....**
.... *Pterostichus rostratus* Newn.

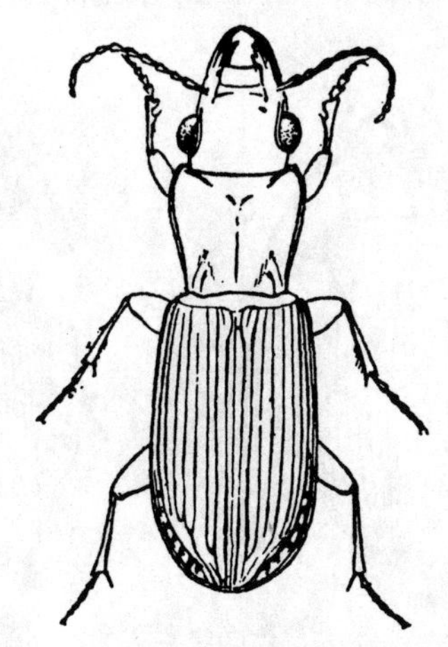

Figure 218

LENGTH: 14-16 mm. RANGE: N. C. to Ga.

Black, thorax quadrate with broad, basal impressions. *Pterostichus adoxus* (Say) with a length of 13-15 mm and ranging through much of Eastern U. S. and Can., is like *rostrata* in having no dorsal puncture; the head is smaller and the thorax much narrowed behind. This genus contains close to 200 species in the U. S. and Canada.

24b **One dorsal puncture on each elytron, placed on the third interval behind the middle; but one basal impression on thorax. Fig. 219. .. *Pterostichus honestus* Say**

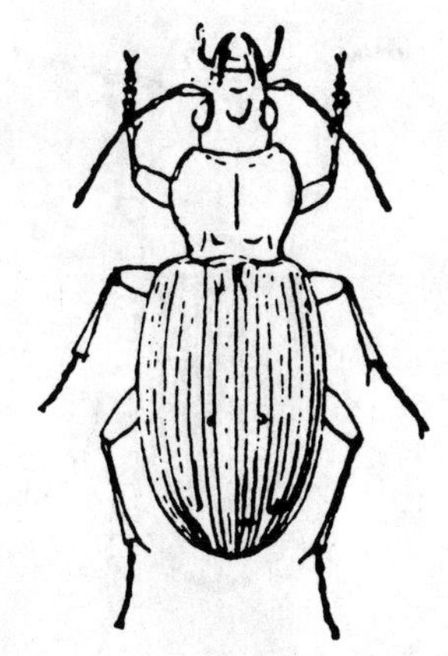

Figure 219

LENGTH: 8 mm. RANGE N.E. U. S.

Black; antennae and legs reddish-brown; basal impressions of thorax narrow and deep, the space between them with a few coarse punctures. Elytra quite convex, the striae deep and without punctures.

Evarthrus obsoletus Say about 2 mm longer than *P. honestus* and of similar color and range. The hind angles of the thorax are rounded instead of being rectangular as in *honestus.*

25a **(23) Usually 10 or more mm in length, last segment of palpi cylindrical with truncate tip. 26**

25b **Not over 8.5 mm; last segment of palpi elongate oval, scarcely truncate. Fig. 220. .... *Pterostichus leconteianus* Lut.**

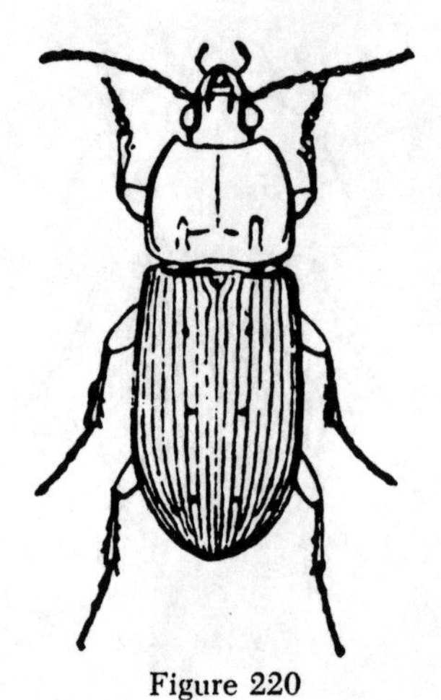

Figure 220

LENGTH: 8-8.5 mm. RANGE: Northern and Eastern U. S. and Can.

Convex, black, much shining; antennae and legs reddish-brown. Elytral striae without punctures. It lives near water and is often very abundant.

Pterostichus femoralis Kby. is a similar species. It is black but not so shining. Its length is 7.5 mm and is common in its range throughout our Middle States and Southern Canada.

26a (25) Side pieces of metathorax less than twice as long as their width at base. Fig. 221A. 27

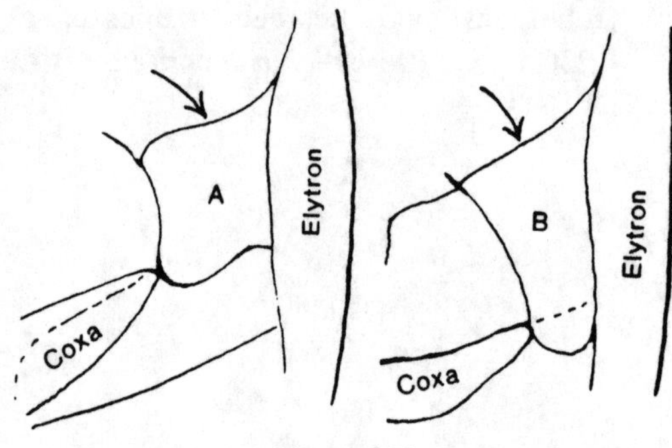

Figure 221

26b Side pieces of metathorax long, narrowed from base to apex, their length being more than twice the width at base. Fig. 221B. 28

27a (26) Thorax narrower at base than apex; hind angles of thorax carinate. Fig. 222.

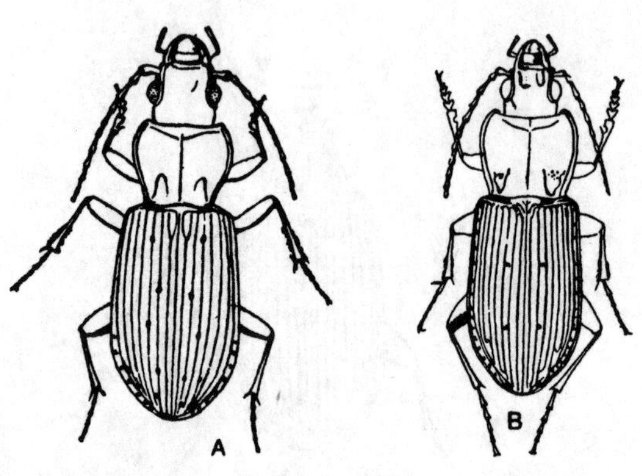

Figure 222

(a) ***Pterostichus adstrictus* (Esch.)**

LENGTH: 11 mm. RANGE: Can. and Northern U. S.

Black, sides of prothorax broadly depressed, hind angles obtuse, dorsal punctures five or six.

(b) ***Pterostichus coracinus* Newn.**

LENGTH: 15-18 mm. RANGE: E. U. S. and Can. Often abundant.

Black, shining, basal impressions of thorax deep. Striae of elytra deep but without punctures.

Another very common black species is *P. stygicus* Say measuring 14-16 mm. The basal impressions have a tooth-like projection which distinguishes it from *coracina.*

Still another closely similar to the two above species, *P. relictus* (Newn.), differs in that the hind angles of the thorax are not carinate and that the elytral striae are deeper with their intervals being narrower and more convex. It measures 16-17 mm and ranges from Indiana east and north.

27b Thorax broader at base than apex; dorsal punctures two. Fig. 223. .... *Abacidus sculptus* (Lec.)

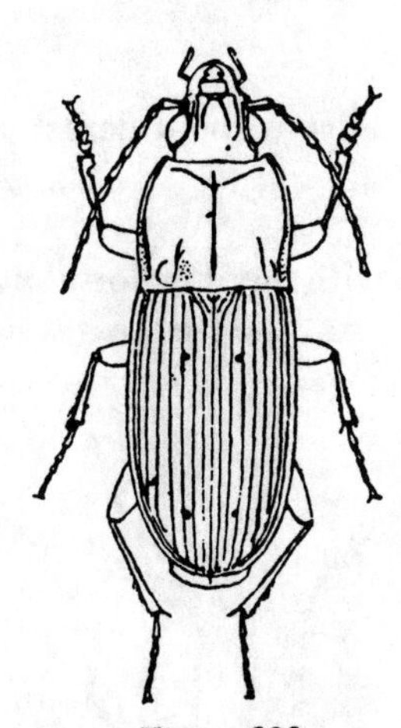

Figure 223

LENGTH: 13-15 mm. RANGE: N. Y. to S. Carolina.

Black. Thorax about as wide as elytra with deeply impressed mid-line; dorsal punctures on the third striae.

A. permundus (Say). Length 12-14 mm is a similar species differing in that it has three dorsal punctures. It is black or purplish and often displays an iridescent sheen. It is frequently common in the North and South Central States.

28a (26) Basal joints of antennae smooth without a ridge. Color black or purplish. 29

28b Three basal joints of antennae bearing a ridge, upper parts green or bluish, dull. Fig. 224. .... *Pterostichus lucublandus* (Say)

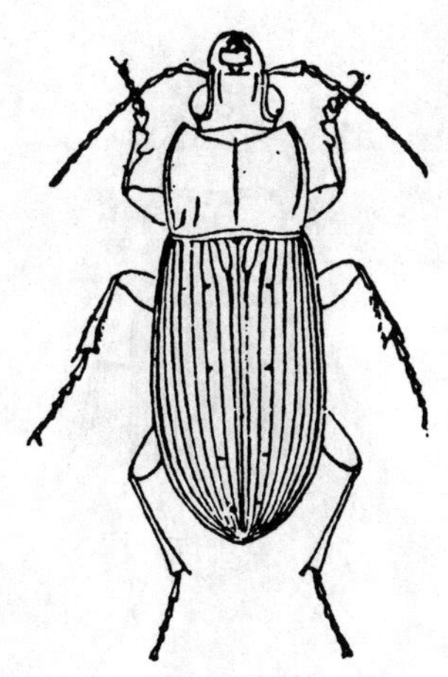

Figure 224

LENGTH: 10-14 mm. RANGE: U. S. and Can. Common.

Color highly variable, usually green or bluish; dorsal punctures 4 located on the third interval. Legs usually reddish; under surface punctured.

P. chalcites Say. length 10-13 mm, common throughout the eastern half of U. S. Is green or bronzed but highly polished. It has but two dorsal punctures. The legs are black and the undersurface is not punctured.

29a (28) Basal impressions wide and deep; thorax with prominent margins; striae deep. Fig. 225. .... *Pterostichus tartaricus* Say

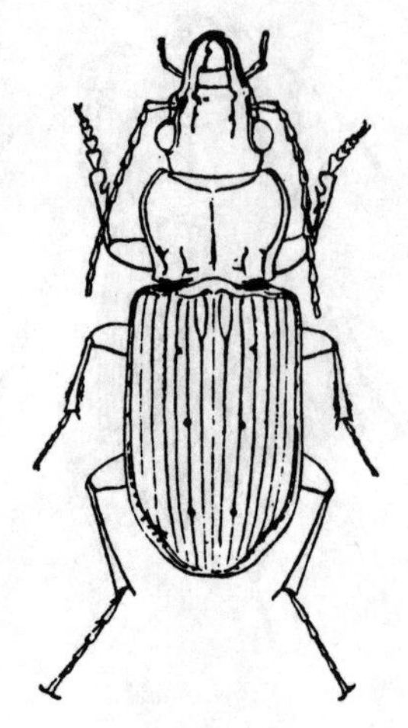

Figure 225

LENGTH: 16-20 mm. RANGE: Eastern U. S. Common.

Depressed, black, shining. Elytral striae indistinctly punctate.

P. scrutator (Lec.), differs in being more highly polished and the striae less prominent, becoming very faint near apex of elytra. Its length is 15-16 mm and range the North Central States.

29b Basal impressions linear; marginal lines of thorax light. Fig. 226. .... *Pterostichus mutus* Say

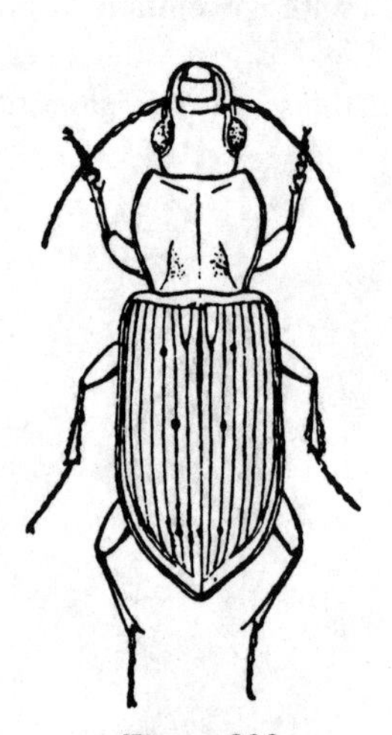

Figure 226

LENGTH: 10-13 mm. RANGE: N. and E. U. S. and Canada.

Subdepressed. Black, shining. Striae of elytra deep with fine punctures, intervals subconvex.

P. ohionis Csiki is closely related. It is purplish instead of black. It is 13-14 mm long and is found in the North Central States though usually not common.

30a (21) Elytra without dorsal punctures. 31

30b Elytra with but one dorsal puncture. Fig. 227. .... *Evarthrus sigillatus* (Say)

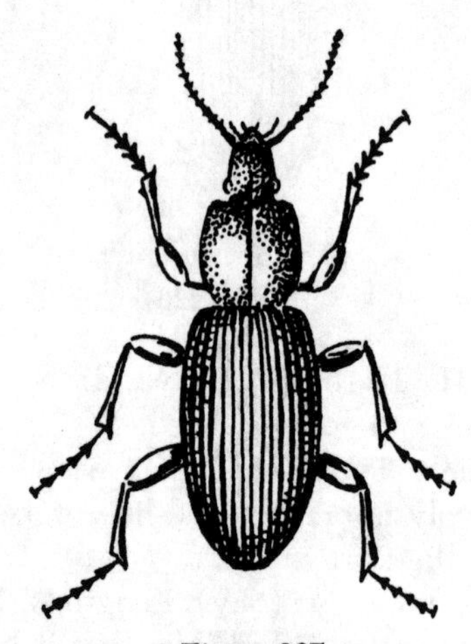

Figure 227

LENGTH: 15-17 mm. RANGE: United States east of the Rockies.

Black; thorax as long as broad, but slightly narrowed behind; elytral striae deep with fine punctures; intervals convex.

E. sodalis (LeC.) is similar with the thorax much narrowed behind and the elytral striae fine. It is recorded for Eastern Canada and U. S. east of the Rockies.

31a (30) Thorax broadest at its base, gradually narrowing to apex. 32

31b Thorax broader in front than at base; often somewhat heart-shaped. Fig. 228. .... *Amara avida* Say

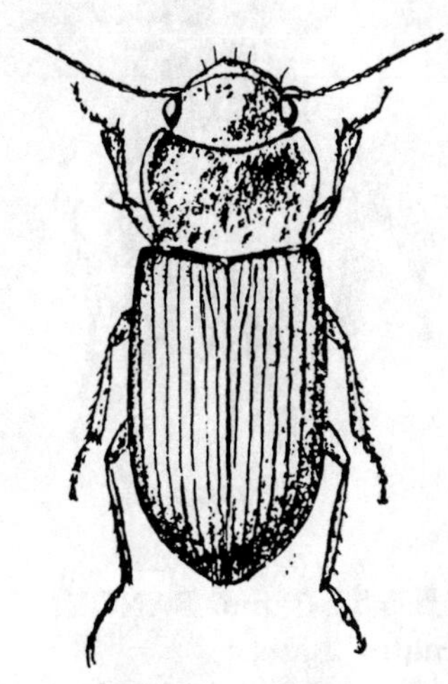

Figure 228

LENGTH: 8-10 mm. RANGE: B. C. to Atlantic.

Convex. Black, shining; antennae and legs reddish-brown; thoracic disk smooth at middle, densely punctate at base and sparingly so at apex; elytral striae deep.

A closely related species, *A. latior* Kby., but slightly larger often shows a bronze reflexion from its black surface. It ranges from Northern Illinois east to the Atlantic and Canada.

32a (31) Hind tibia of males pubescent on inner side. Apical spur of front tibia simple. Fig. 229. .... *Amara impuncticollis* Say

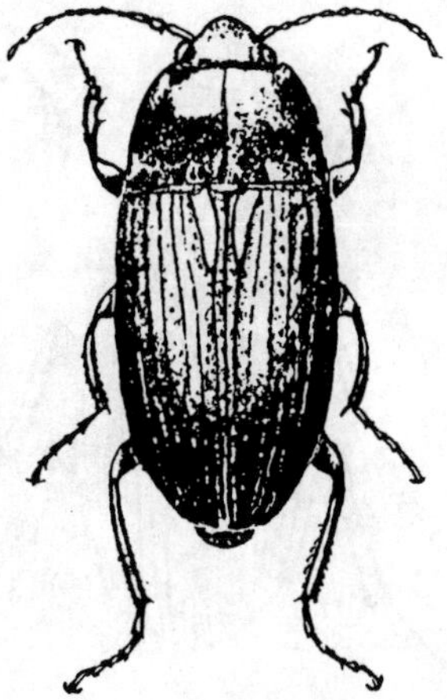

Figure 229

LENGTH: 7-9 mm. RANGE: U. S. and Canada. Abundant.

Dark bronze sometimes with a greenish tinge. Elytral striae shallow not punctured, but with a single large puncture where the scutellar stria joins the second stria. Flies freely to lights.

A. cupreolata Putz. is a smaller (6-7) bronzed or purplish-black species, quite oval and convex, which is very common over a wide range of the U. S. and Canada. It is readily distinguished from *impuncticollis* by the lack of the large puncture at the end of the scutellar stria.

32b Hind tibia of males not distinctly pubescent on inner side; antennae and legs reddish-brown. Fig. 230. .. *Amara obesa* Say

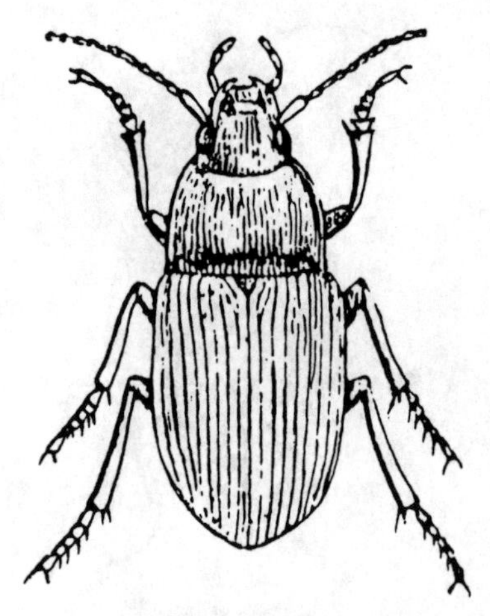

Figure 230

LENGTH: 9-12 mm. RANGE: Wide area of the U. S. and Canada.

Somewhat depressed. Black, shining, elytra dull in females; antennae and legs reddish-brown; elytral striae deeper at apex, with very fine punctures.

Amara rubrica Hald, found from Quebec to S. Carolina west to S. Dakota, is reddish to chestnut-brown and but 6-7 mm long. *A. musculis* (Say) is a darker reddish-brown and still smaller (5 mm). It seems to range over much of U. S. and Canada. This genus is large with about 250 N. American species.

33a (19) Front of head short, the labrum impressed. 34

33b Front of head normal. 38

34a (33) Three basal joints of antennae entirely glabrous. Elytra without dorsal punctures. Genus *Dicaelus*. 35

34b But two basal joints of antennae glabrous; each elytron with two dorsal punctures. Fig. 231. .... *Badister pulchellus* Lec.

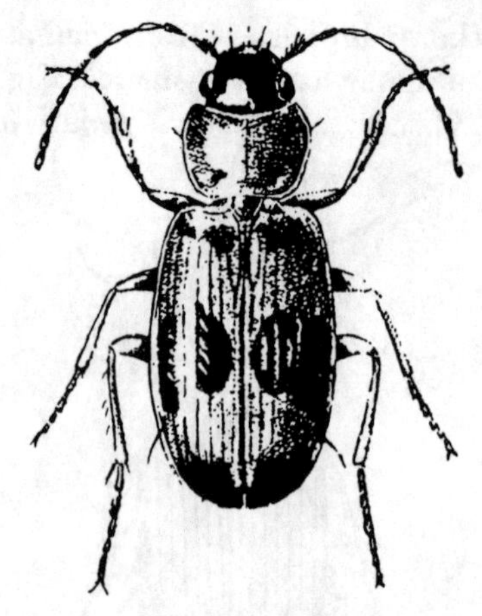

Figure 231

LENGTH: 5.5-6.5 mm. RANGE: Central and Eastern States. Rare.

Head and markings on elytron black; antennae dusky with first joint yellow.

B. maculatus Lec. is similar in size and structure to ***pulchellus***. The color is black except that the basal third of the elytra and a spot near the apex are orange. It is reported from Pa. and Ind. to Fla.

35a (34) Intervals (spaces between striae) on elytra regular and continuous. 36

35b Intervals on elytra broken by numerous large punctures. Fig. 232. .... *Dicaelus sculptilis* Say

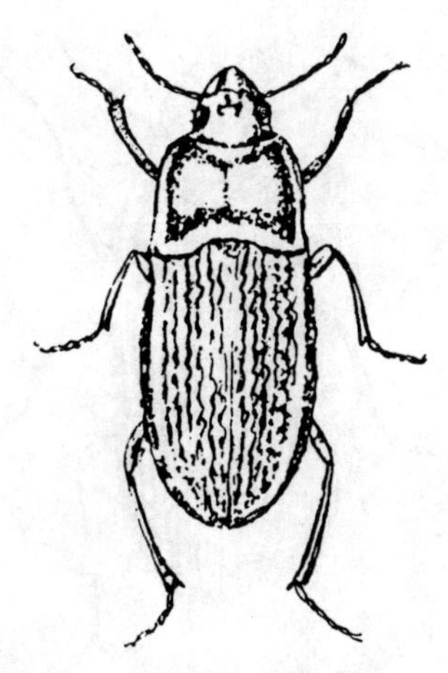

Figure 232

LENGTH: 17-19 mm. RANGE: U. S., east of the Rockies.

Subconvex. Black, shining; antennae paler at tip. In low damp woods, under logs and debris. The members of this genus are our most attractive ground beetles. They are also of high economic importance in destroying harmful insects. If it were not for the aid given by the many useful insects in keeping the harmful species in check, men could not begin to compete with the insects.

36a (35) Elytra black, the intervals equal in width and shape. (See *D. furvus* under 37a). 37

36b Elytra purplish or brassy. Fig. 233. .... *Dicaelus purpuratus* Bon.

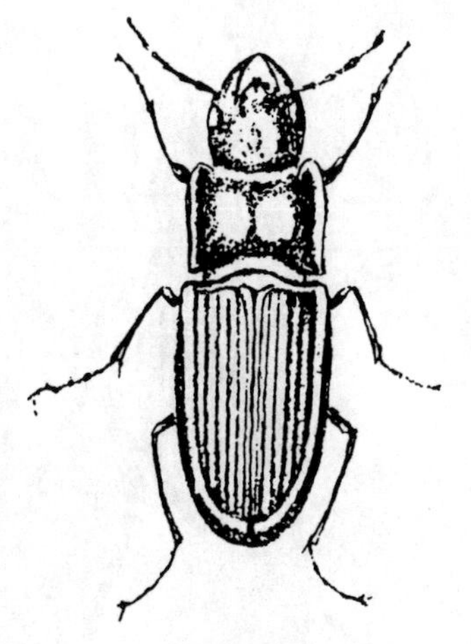

Figure 233

LENGTH: 20-25 mm. RANGE: Central and Eastern States. Sometimes fairly common.

Purplish without brassy tinge, legs black, elytra with deep striae and very convex intervals. It is a grandly decorated beetle.

D. spendidus Say is of similar size and *tus.* The upper parts are purplish to rose violet shape but even more beautiful than *purpura-* and all adorned with a strong brassy iridescent sheen. It is not rare but scarce enough to give the collector a real thrill when he finds one. It belongs to the West Central States.

37a (36) Two bristle-bearing punctures on margin of thorax. Fig. 234. .... *Dicaelus elongatus* Bon.

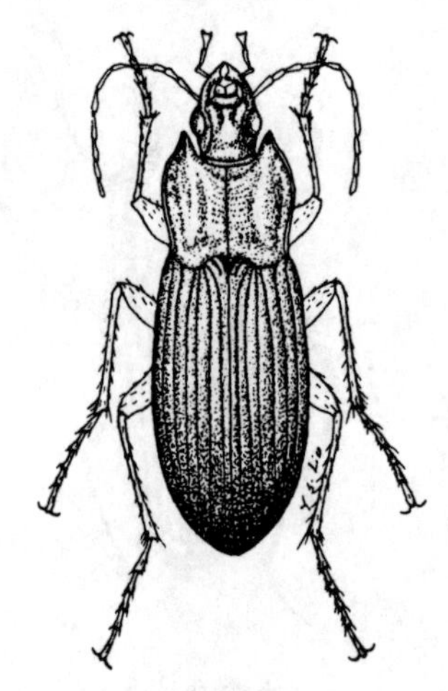

Figure 234

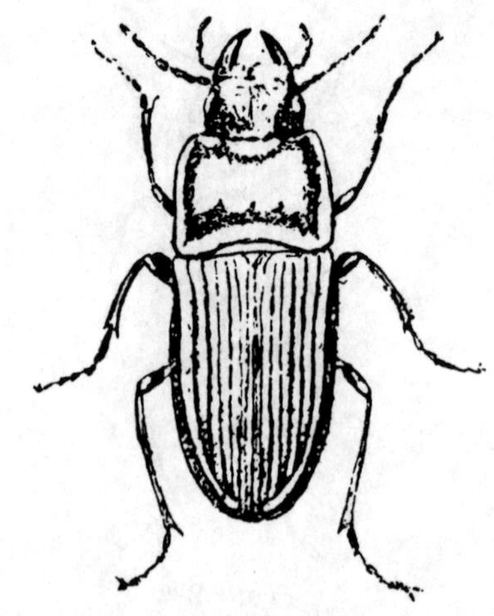

Figure 235

LENGTH: 15-18 mm. RANGE: Central and Eastern States and Canada.

Elongate. Black, shining thorax with deep median line and 2 bristle-bearing punctures at side in front of the middle; humeral carina reaching the middle.

D. furvus Dej. is readily told from the other black species by the wide and narrow intervals alternating on the elytra. The humeral carinae almost reach the apex. It is rather dull, 15-16 mm long and ranges in the North Central States.

37b But one bristle-bearing puncture on the margin of the thorax. Fig. 235. .... *Dicaelus dilatatus* Say

LENGTH: 20-25 mm. RANGE: Central and Eastern States.

Black, dull. Elytral striae deep and faintly punctured near tip. Humeral carina reaching 2/3 to tip of elytra.

D. ovalis Lec. is something like a small edition of *dilatatus.* Its length is 15-16 mm and ranges through much of the Central and Eastern States. It is also much like *furvus* except for the alternating intervals.

38a (33) Next to last segment of labial palpi with but two bristly hairs. 39

38b Next to last segment of labial palpi longer than the last segment and bearing several bristly hairs; first joint of antennae elongate. Fig. 236. .... *Galerita janus* Fab.

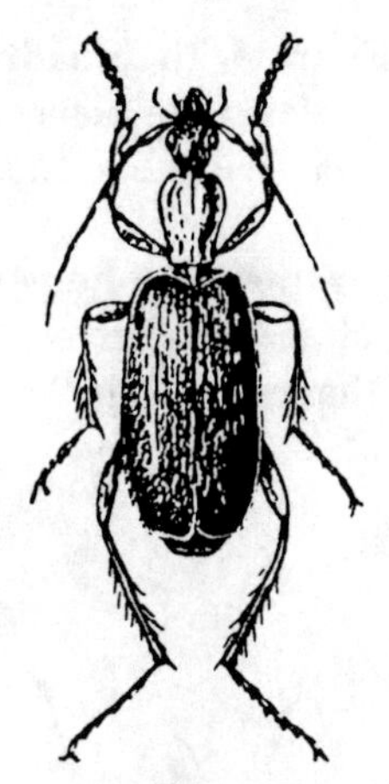

Figure 236

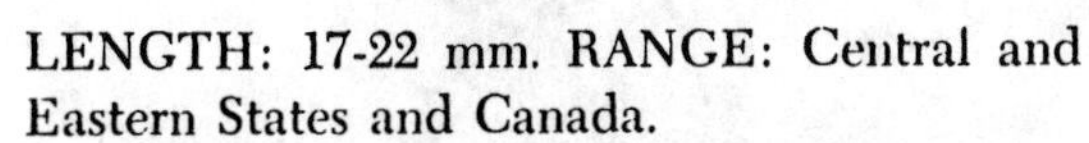

LENGTH: 17-22 mm. RANGE: Central and Eastern States and Canada.

Bluish-black, densely covered with short hairs; legs, palpi, thorax and base of antennae brownish-red. A very common, rather loosely-jointed beetle, often attracted by lights.

G. bicolor Drury is a closely similar species of the same size but apparently more restricted in its distribution. The sides of the head are longer and less rounded behind the eyes and the pubescence near the scutellum stands erect in contrast to *janus*.

One or both species of these interesting beetles are almost certain to be found in spring and summer collections. They are highly valuable in destroying large numbers of caterpillars. 14 No. American species.

39a (38) Head prolonged behind the eyes into a neck which is dilated at posterior into a knob like part. 40

39b Head not prolonged as in 39a. 41

40a (39) Very slender yellowish-brown beetles. Fig. 237. ....
.... *Leptotrachelus dorsalis* (Fab.)

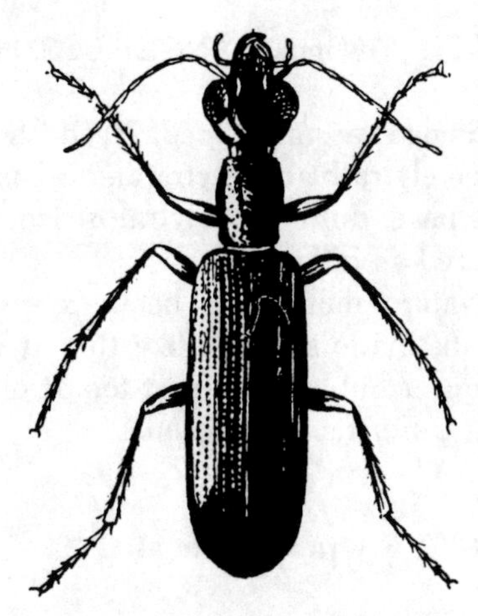

Figure 237

LENGTH: 7-8 mm. RANGE: Central and Eastern States.

Head, thorax and sutural stripe on elytra blackish; antennae, legs and elytra golden brownish-yellow; elytral striae deep and distinctly punctured. This and the following species do not much resemble ground beetles. Both are oddities and attract attention. This is now and then caught in light traps.

40b Slender black beetles with elytra dull red with black spots. Fig. 238. ....
.... *Colliuris pennsylvanica* (L.)

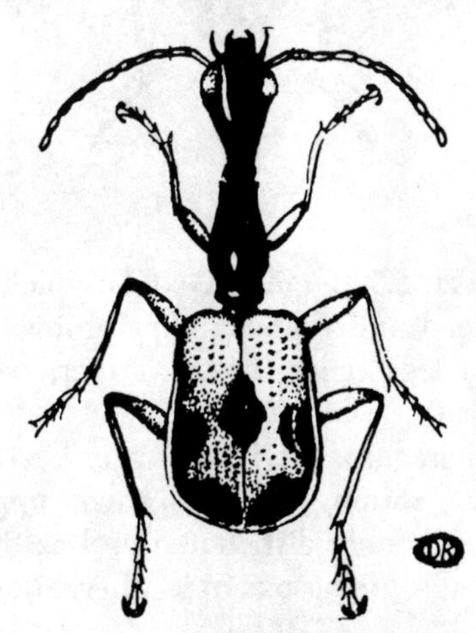

Figure 238

LENGTH: 7-8 mm. RANGE: U. S. and Canada.

Somewhat ant-shaped. Head, thorax and spots on elytra black; elytra and 3 basal joints of antennae dull red; elytral striae coarsely punctured on basal half.

Nature must have been experimenting when she made a beetle like this. It not only looks queer but acts the part too. It often flies to light and is fairly common.

41a (39) Elytra truncate at tip. 48

41b Elytra obliquely sinuate at tip. 42

42a (41) Tarsal claws serrate. Fig. 239. .... *Calathus opaculus* Lec.

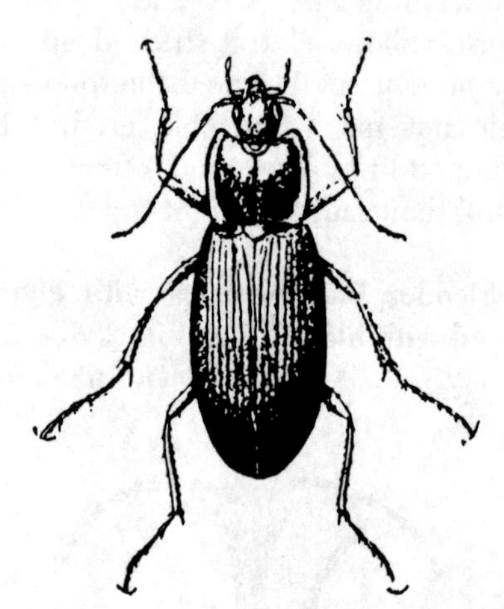

Figure 239

LENGTH: 8.5-10 mm. RANGE: much of U. S.

Head and thorax reddish-brown; antennae and legs paler; elytra darker, with very fine striae.

C. gregarius Dej. is a larger beetle, 10-11 mm with shining reddish-brown upper surface. It is widely distributed and as the name suggested is gregarious in it hibernation.

42b Tarsal claws not serrate. 43

43a (42) Elytra with broadly rounded humeral angles. Side pieces of metathorax longer than wide. 44

43b Elytra oval without humeral angles; side pieces of metathorax scarcely, if at all, longer than wide. Fig. 240. .... *Platynus hypolithos* (Say)

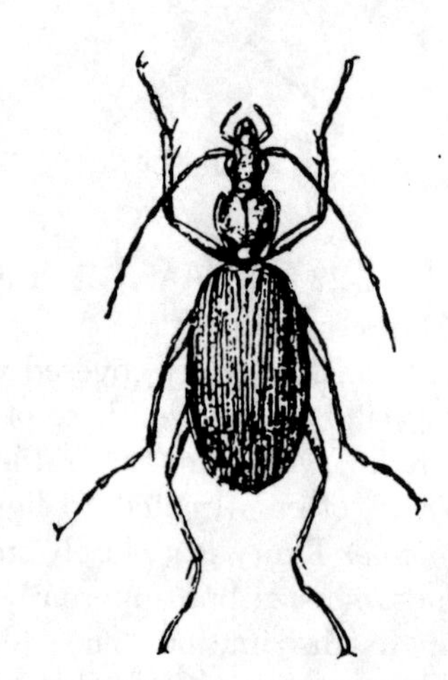

Figure 240

LENGTH: 13-15 mm. RANGE: New York to Indiana.

Black shining; legs and antennae pale reddish-brown; Elytral striae moderately deep with intervals convex. Fairly large punctures on the sides of alternating intervals.

P. caudatus (Lec.) may be readily recognized by the prolonged elytral suture which extends into the diverging tips of the elytra. The third joint of the elytra is nearly twice as long as the fourth. It is dark reddish-brown with antennae and legs paler, and measures 12-13 mm.

44a (43) Hind angle of thorax much rounded. 46

44b Hind angles of thorax not rounded. 45

45a (44) All the tarsi with distinct grooves on the sides; thorax and scutellum reddish-yellow. Fig. 241. ....
.... *Agonum decorum* (Say)

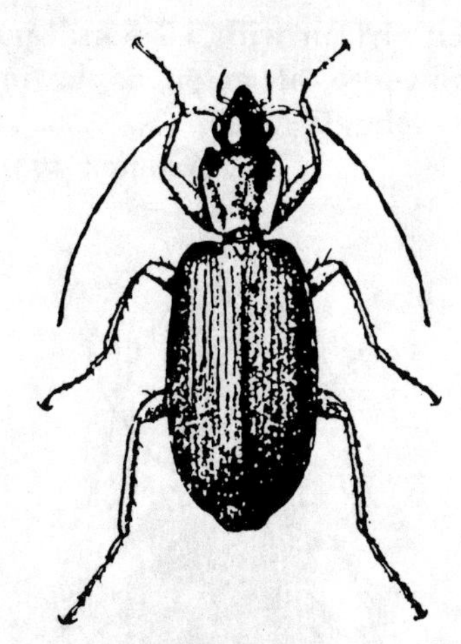

Figure 241

LENGTH: 8-8.5 mm. RANGE: East of Rockies including Canada.

Head green, often bronzed; thorax, scutellum, legs and base of antennae reddish-yellow. Elytra blackish, often tinged with green at margins, the striae shallow and not punctured.

A. extensicollis (Say), is a trifle larger. Its head and thorax are both metallic green and the elytra dull green or purplish.

It has a variety, *viridis* in which the elytra and thorax are bright green.

45b Only the middle and last tarsi with grooves on the side, the front tarsi smooth. Fig. 242. ....
.... *Platynus decentis* (Say)

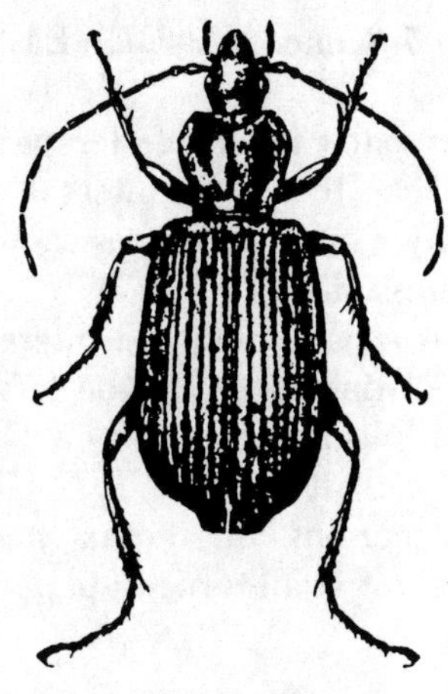

Figure 242

LENGTH: 10-11 mm. RANGE: Northeast United States and Canada.

Black, shining. Elytral striae but moderately deep, with fine punctures.

P. cincticollis (Say) is often abundant and gregarious in its range. It is wholly black. The elytral striae are not punctured.

46a (44) Side margins of thorax wider towards the base and reflexed; green with elytra and disk of thorax bronzed. Fig. 243. ...... *Agonum octopunctatum* (Fab.)

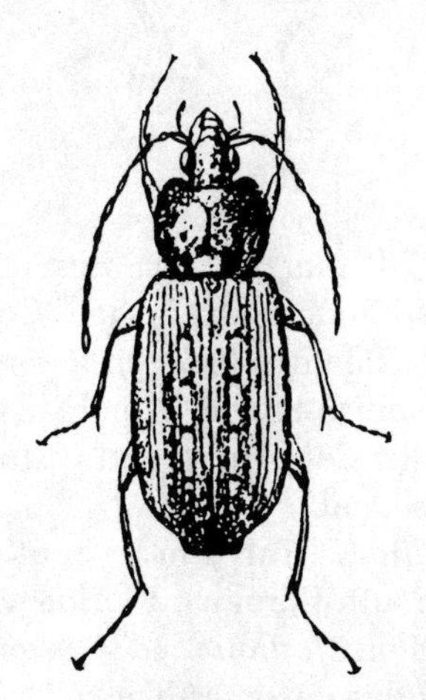

Figure 243

LENGTH: 7-8 mm. RANGE: Eastern U. S. and Canada.

Upper parts green; under parts shining greenish-black; the third interval with four (occasionally 3 or 5) punctures each in a large square impression.

This is a handsome and interesting beetle readily distinguished by the large impressions. The adults hibernate.

46b Side margins of thorax narrow and scarcely if at all turned up. 47

47a (46) Each elytron with 3 dorsal punctures; pubescence of antennae beginning on 4th joint. Fig. 244. .... *Agonum punctiformum* (Say)

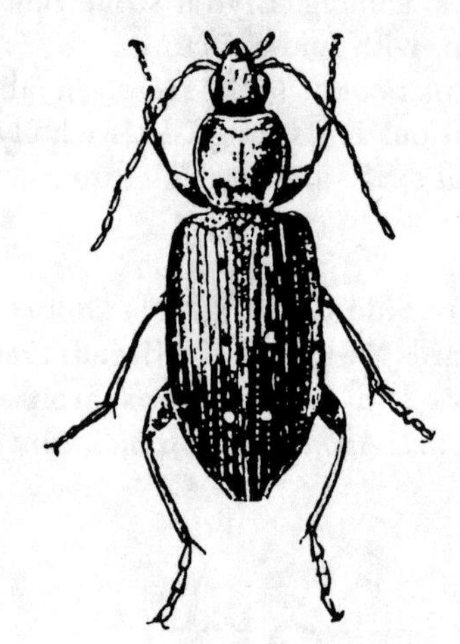

Figure 244

LENGTH: 7-9 mm. RANGE: East of the Rockies in United States and Canada. Common.

Black, shining, base of antennae, tibia and tarsi reddish-brown, femora darker. Elytra with rather deep striae, and with fine punctures at basal end.

A. pallipes (Fab.) has a southern range (Texas to South Carolina, Florida). Its size is similar to *punctiformum*. It is often reddish-brown though sometimes black. The legs, palpi and base of antennae are yellowish brown.

Agonum and *Platynus* are both large genera, the former with more than 100 and the latter more than 50 species in the U. S. and Canada.

47b Each elytron with 4-6 dorsal punctures; pubescence of antennae beginning on third joint. Fig. 245. .... *Agonum thoreyi* Dej.

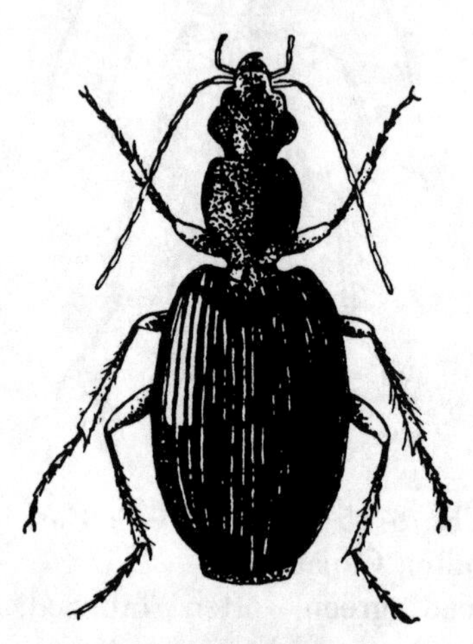

Figure 245

LENGTH: 6-7 mm. RANGE: North Central and Middle States.

Slender; head and thorax black; antennae, legs and elytra brownish-yellow; elytral striae of moderate depth, not punctured; intervals nearly flat. Four to six dorsal punctures on third interval. It is often quite abundant and is gregarious in winter.

48a (41) Front tibiae rather stout, broadening gradually to tip; paraglossa horn-like. Fig. 246. .... *Helluomorphoides bicolor* (Harris)

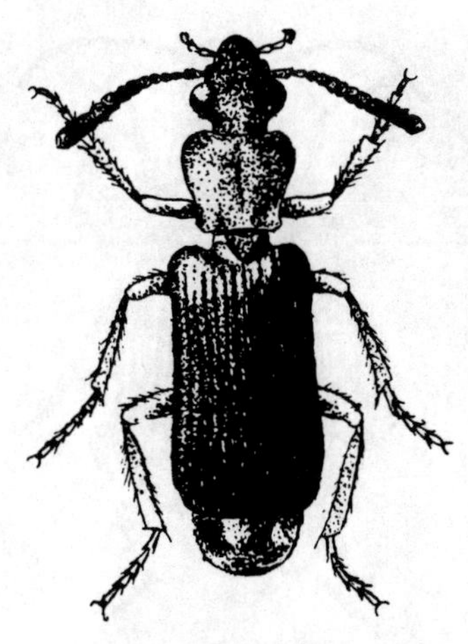

Figure 246

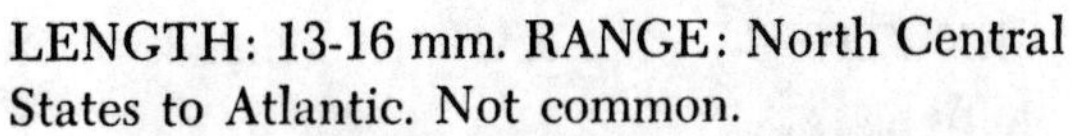

LENGTH: 13-16 mm. RANGE: North Central States to Atlantic. Not common.

Head, thorax and legs reddish-brown, elytra darker. Elytral intervals with punctures in three confused rows.

H. texana Lec. more southern in its range and a bit larger is distinguished by uniform reddish-brown color and but two rows of punctures on the elytral intervals.

48b **Front tibia slender; paraglossae membranous. 49**

49a **(48) Spurs at end of tibiae short or normal. 50**

49b **Spurs at end of tibiae very long, and with fine teeth. Fig. 247. .... *Tetragonoderus fasciatus* Hald.**

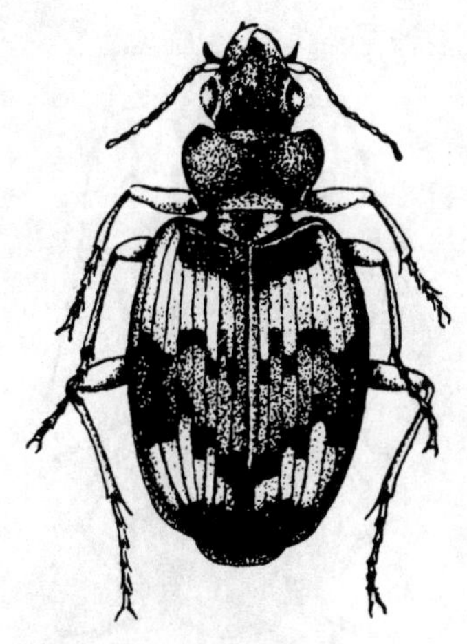

Figure 247

LENGTH: 4-5 mm. RANGE: California to New York.

Head and thorax dark, bronzed, elytra dull yellow or grayish marked with dark bands as pictured. Males with 3 joints of front tarsi dilated. It seems to be found only in sandy areas where its markings gives it protective coloration.

Nemotarsus elegans Lec. is a little larger, brownish-yellow with picious elytra, each marked with a pale spot at base and at apex. It ranges from Illinois, east and is the only member of its genus.

50a **(49) Head constricted back of the eyes. 51**

50b **Head not constricted back of the eyes; tarsi hairy above. Fig. 248. .... *Cymindus americana* Dej.**

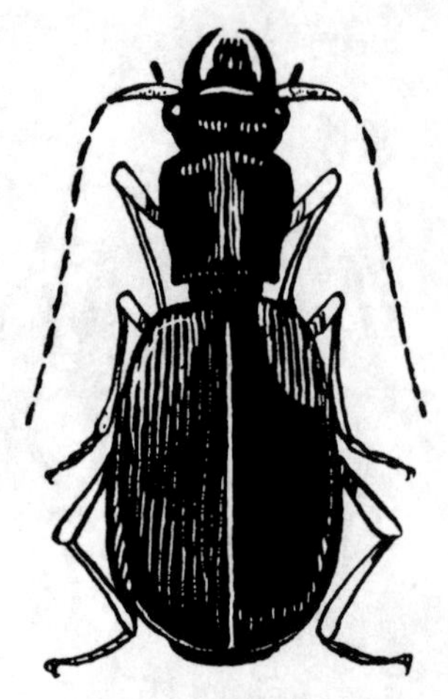

Figure 248

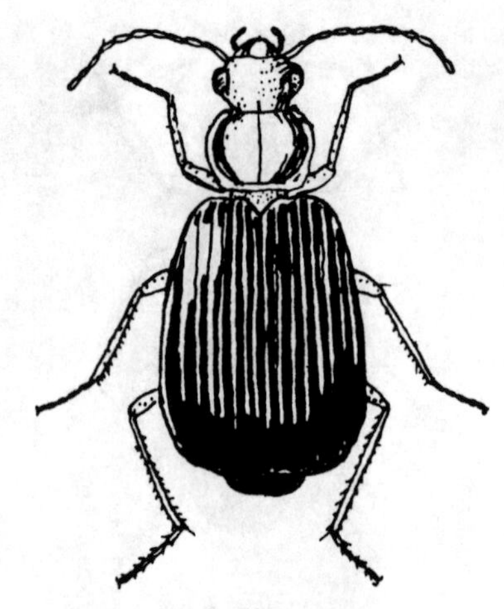

Figure 249

LENGTH: 12-15 mm. RANGE: Iowa, east to Atlantic. Not common.

Piceous, feebly shining, legs, antennae, humeral spot and narrow side margins of elytra pale reddish-brown; sides of thorax curved in front. Striae of elytra deep, finely punctured, with coarser punctures on the rather flat intervals.

Calleida punctata Lec. is a characteristically marked slim-bodied beetle. The head is dark blue, thorax reddish-yellow, elytra bright green, and legs yellowish. It measures 7-8 mm and ranges from Kansas to New York.

51a (50) Front tarsi of male not obliquely dilated; both head and thorax not reddish-yellow. 52

51b Front tarsi of male obliquely dilated; both head and thorax reddish-yellow; elytra dark blue. Fig. 249. .... *Lebia atriventris* Say

LENGTH: 6-7 mm. RANGE: general. Abundant.

Elytra dark blue; head and thorax reddish-yellow. Three basal joints of antennae pale, striae on elytra fine with fine punctures. *L. grandis* Hentz is quite similar but larger 8-10 mm. The antennae are entirely pale and elytral striae deep. It is common and has a wide distribution.

This is an interesting genus with many characteristically marked species. There are about 40 species in our fauna.

52 (51) Elytra with pale stripes; mentum not toothed. 54

52b Elytra without pale stripes; mentum with a distinct tooth. 53

53a (52) Head with lengthwise wrinkles; not over 5 mm long. Fig. 250. .... *Lebia analis* Dej.

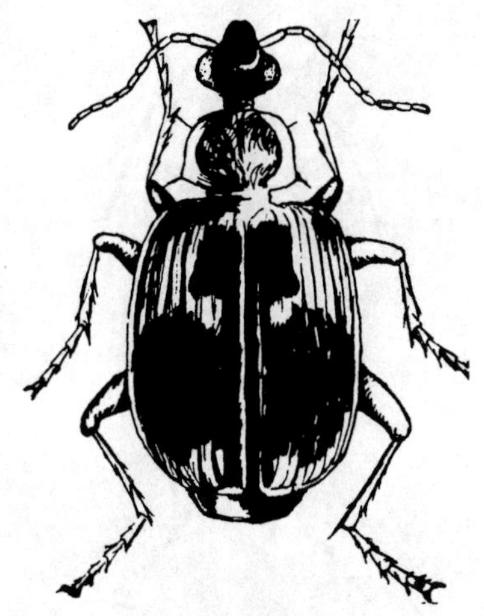

Figure 250

LENGTH: 4.5-5 mm. RANGE: general. Common.

Head black; four basal joints of antennae paler; thorax reddish-yellow with pale margins; elytra black with yellowish markings.

L. viridis Say ranges abundantly throughout the United States and Canada. It is 4.5 to 5.5 mm long and shining green or purplish-blue throughout. Its head is smooth and the elytral striae shallow.

53b Head practically smooth; length 6 mm or more. Fig. 251. .... *Lebia fuscata* Dej.

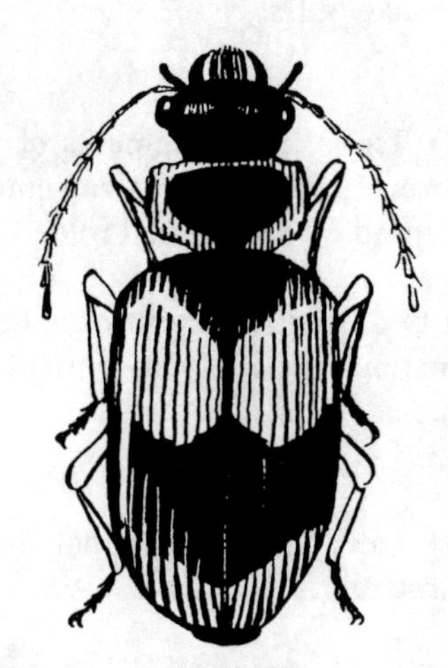

Figure 251

LENGTH: 6-7 mm. RANGE: General.

Elytra and margins of thorax pale brownish-yellow. Head, disk of thorax and markings on elytra black or nearly so, antennae, legs and underparts reddish-yellow.

L. ornata Say ranges quite generally over the United States. It is 4.5 to 5 mm. long. Head, thorax and elytra piceous, with the margins of both thorax and elytra as well as two large basal and two small apical spots on elytra dull yellow.

54a (52) Head and thorax reddish-yellow; thorax with wide margins. Black stripes of elytra broad. Fig. 252. .... *Lebia vittata* Fab.

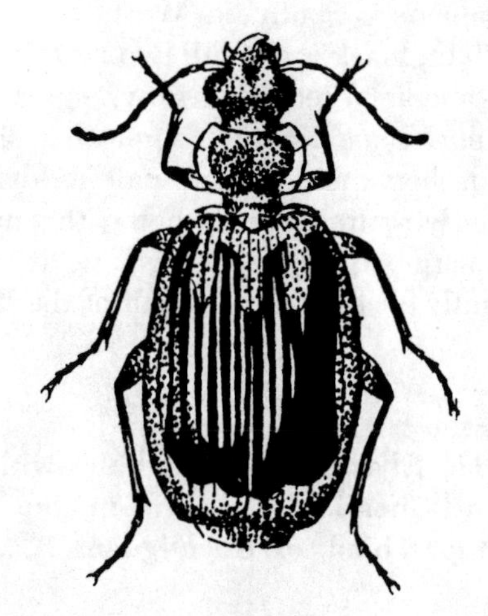

Figure 252

LENGTH: 5.5-6 mm. RANGE: General.

Head and thorax reddish-yellow; legs black; elytra and margin of thorax pale yellow. Elytra marked with black as pictured, or with narrow separated stripes.

54b Head black, thorax reddish-yellow, its margins narrow. Fig. 253. .... *Lebia bivittata* (Fab.)

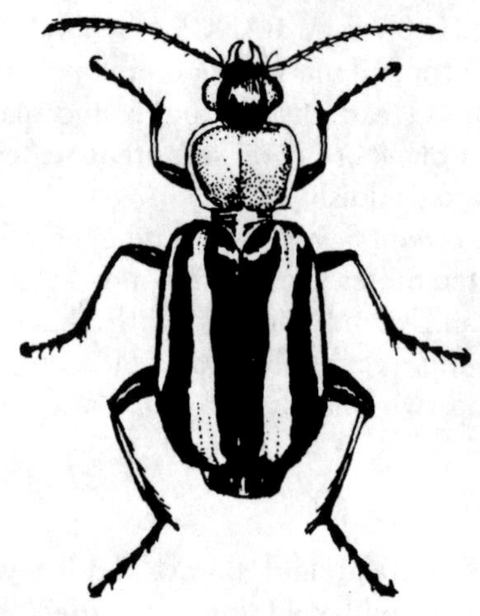

Figure 253

LENGTH: 5.5-6 mm. RANGE: rather general; most common in South and West.

Elytra black each with two narrow white stripes, abdomen reddish-yellow; legs black.

Lebia solea Hentz is somewhat similar. Its head, thorax and legs are pale reddish-yellow; the elytra are piceous, each with a median and a marginal stripe pale yellow. It ranges abundantly over the eastern half of the United States.

55a (11) Elytra entire; mandibles without a bristle-bearing puncture in the outer groove, hind coxae contiguous. 56

55b Elytra truncate at apex; mandibles with a bristle-bearing puncture; hind coxae often separated. Fig. 254. .... *Brachinus americanus* Lec.

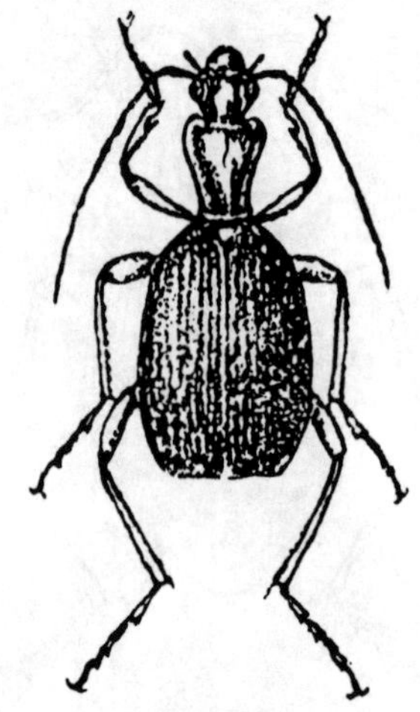

Figure 254

LENGTH: 10-12 mm. RANGE: Central and Southern States. Common.

Elytra bluish-black; four basal joints of abdomen pale, the others dusky.

There are over 50 species of this genus. They vary in size from 4 to 15 mm or more. All have the narrow head and thorax, together with the legs colored reddish-yellow and the elytra dark in bluish or blackish shades. Their defense scheme of "firing" several rounds of a volitile gas at their pursuer have won them the name "bombardier beetles." Even their human enemy is disconcerted by these sounds and smoke-like puffs.

56a (55) Three basal segments of antennae glabrous; margin of elytra somewhat interrupted by an internal fold. 57

56b But two basal segments of antennae glabrous; margin of elytra without interruption. 60

57a (56) Third joint of antennae longer than the fourth. 58

57b **Third joint of antennae not longer than the fourth. Fig. 255. .... *Chlaenius tricolor* Dej.**

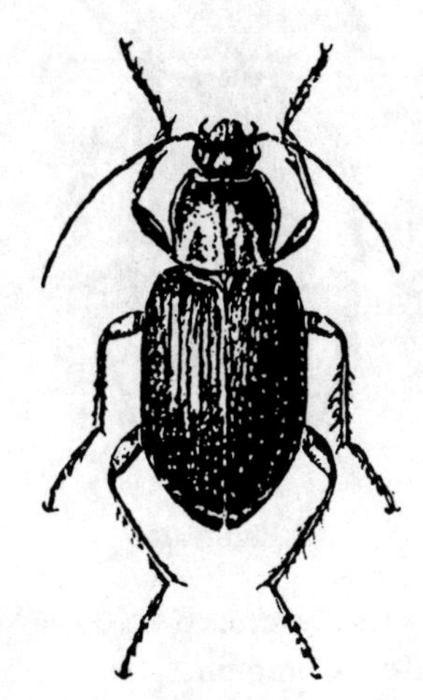

Figure 255

LENGTH: 11-13 mm. RANGE: rather general throughout the United States and Canada.

Elytra blackish-blue; head and thorax metallic green. Elytral striae rather deep and fine with fine punctures.

This genus has some 50 species, all greenish or bluish-black and usually covered with a fine pubescence. They are abundant in damp places under logs, stones, etc. and are attractive.

58a **(57) Middle tibiae of male with a pubescent area near the tip; abdomen with a few punctures at middle and more at the sides. 59**

58b **Middle tibiae of male without a pubescent area near the tip; abdomen without punctures. Fig. 256. .... *Chlaenius tomentosus* (Say)**

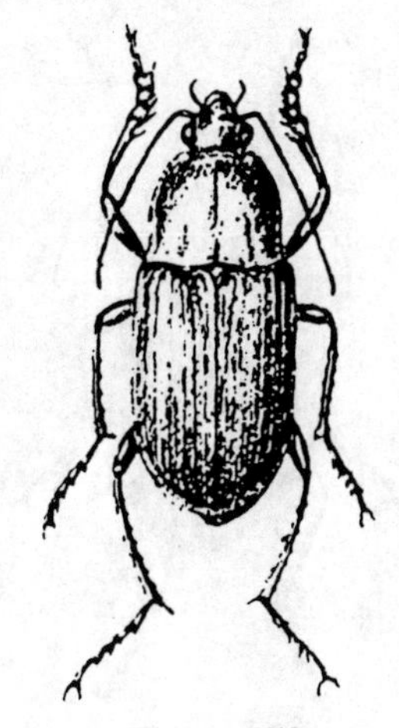

Figure 256

LENGTH: 13-15 mm. RANGE: east of Rockies. Abundant.

Blackish, bluish or greenish, somewhat bronzed. Antennae black, the two basal joints pale. Elytral striae shallow with rather coarse round punctures.

C. purpuricollis Rand. is one of the smaller members of this genus measuring 8.5-9.5 mm. It is a dark violet-blue above and black beneath with black antennae. The elytral striae are very fine with widely spaced fine punctures. It ranges from New England, west through the North Central States.

59a **(58) Episterna of metasternum long, the outer side longer than the front one. Bright green to bluish; legs pale. Fig. 257. .... *Chlaenius sericeus* Forst.**

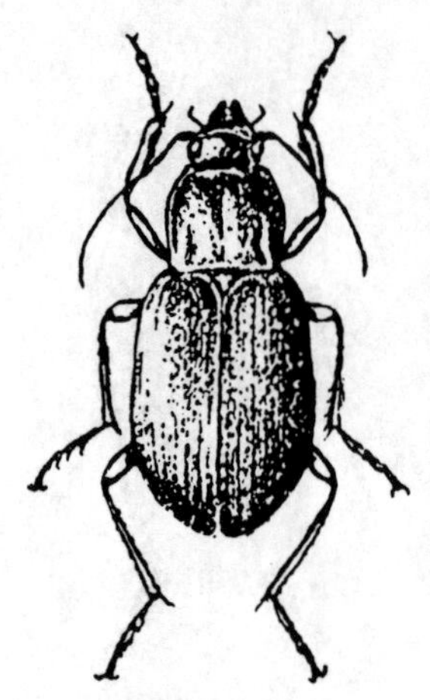

Figure 257

LENGTH: 13-17 mm. RANGE: common throughout the United States and Canada.

Bright green or bluish, under surface black; antennae pale but darker toward tip. Striae of elytra fine with distant fine punctures, the intervals flat with many fine punctures. Under stones and debris near water.

C. fuscicornis Dej. is one of the larger species, with a length of 21 to 23 mm. It ranges through our Central and Southern States. Blackish with a bluish tinge, its antennae are brown, the basal joints paler.

59b Episterna of metasternum short, the outer side shorter than the front one. Fig. 258. .... *Chlaenius platyderus* Chaud.

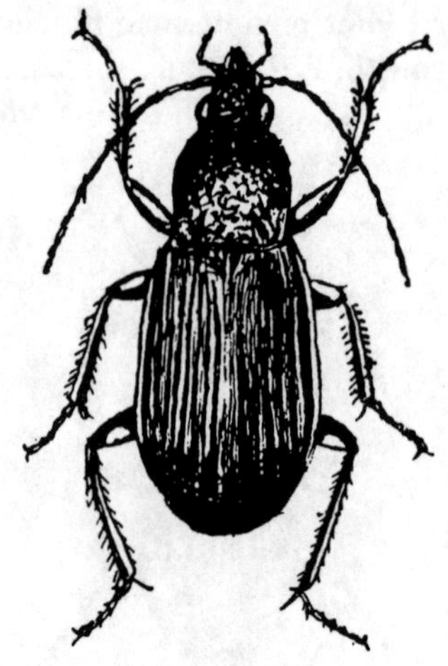

Figure 258

LENGTH: 13-15 mm. RANGE: Mexico and Central States. Common.

Thorax violet or purplish-blue; elytra dark bluish tinged with green; antennae brown with three basal joints paler.

C. solitarius Say though seldom abundant ranges through much of the western two-thirds of the United States and Canada. It is bright green above, with the underparts reddish-brown, the last ventral segment being margined with dull yellow. Its length is 12-14 mm.

60a (56) Next to the last segment of the labial palpi with two bristles; about the same length as the last segment. Fig. 259a. 66

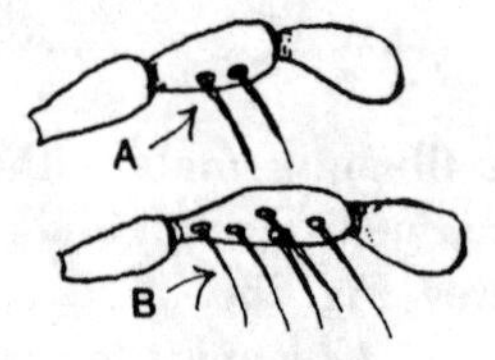

Figure 259

60b Next to the last segment of the labial palpi with more than two bristles, longer than the last segment. Fig. 259b. 61

61a (60) **Front tibia with its outer apical side produced into a plate for digging.** 62

61b **Front tibia normal; not used for digging.** 63

62a (61) **Mandibles large, body somewhat peduncled. Fig. 260.** .... ***Geopinus incrassatus*** **(Dej.)**

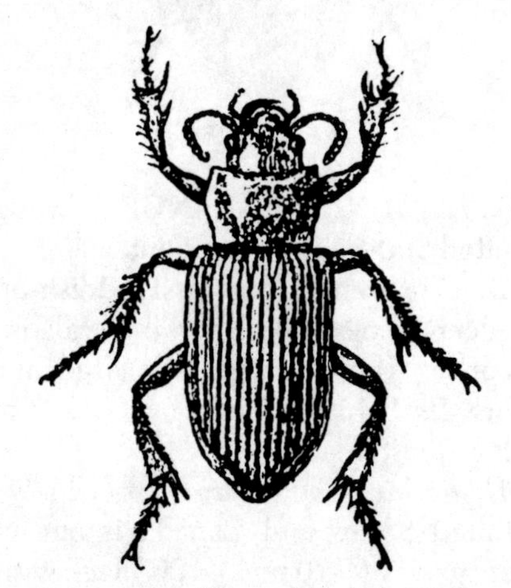

Figure 260

LENGTH: 13-15 mm. RANGE: Middle and North Central States, East. Canada.

Somewhat cylindrical. Brownish-yellow; front of head and disks of thorax and elytra often darkened. Striae of elytra fairly deep and not punctured; intervals but slightly convex.

There is but the one species of this genus in our country. It burrows in damp, sandy soil and sometimes flies to lights. It carries but little resemblance to most of the ground beetles and at a casual glance looks more like a Scarabaeid.

62b **Mandibles not prominent; body not pedunculate. Fig. 261.** .... ***Euryderus grossus*** **Say**

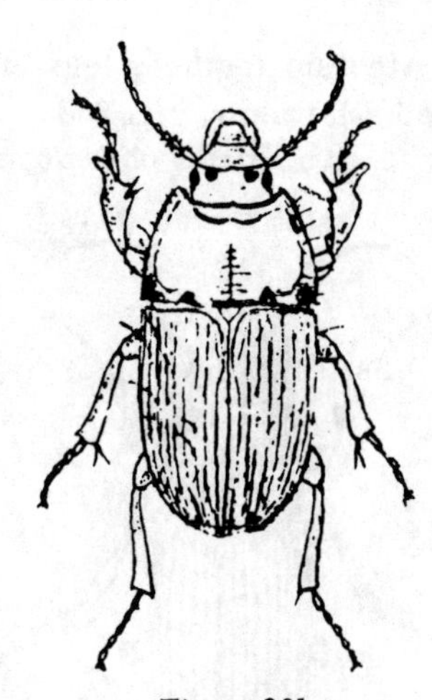

Figure 261

LENGTH: 14-15 mm. RANGE: Central States to the Pacific. Not common.

Heavy, convex. Black, shining; basal margin and hind angles of thorax depressed. Striae of elytra fine, not punctured; 5 to 8 distinct bristle-bearing punctures on each of the 3rd, 5th and 7th intervals.

Only one species of this genus occurs in our region. There are many other small genera in this family that are seldom collected and but poorly known.

63a (61) **Dilated front tarsi of male spongy pubescent or brush-like beneath. Fig. 262.** 65

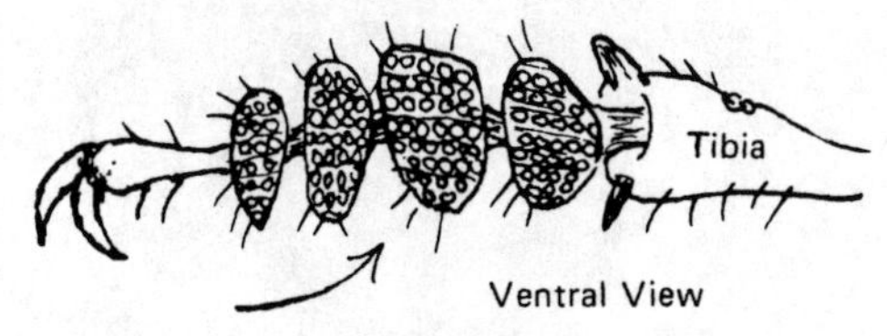

Figure 262

63b **Front tarsi not spongy but with two rows of small scales beneath.** 64

64a (63) Mentum toothed; legs and antennae reddish-yellow. Fig. 263. .... *Harpalus pensylvanica* DeG.

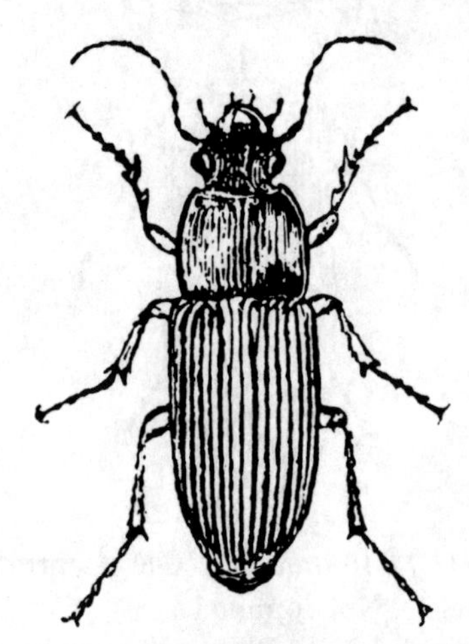

Figure 263

LENGTH: 13-16 mm. RANGE: general in United States and Canada. Common.

Black, moderately shining; under-surface reddish-brown; striae of elytra moderately deep.

This genus is a large one with well over 100 species known for the region. They are for the most part highly beneficial though a few feed on seeds.

64b Mentum without a tooth; legs black. Fig. 264. .... *Harpalus caliginosus* (Fab.)

Figure 264

LENGTH: 21-25 mm. RANGE: throughout the United States, very abundant.

Black; antennae and tarsi reddish-brown; elytra deeply striate with the intervals moderately convex. It often appears in large numbers at lights. Its habit of feeding on seeds makes its value questionable.

H. herbivagus Say ranging widely over the United States and Canada is one of the smaller species (8-10 mm). It is black with reddish-brown antennae and legs.

65a (63) Abdomen punctate over its entire surface. Fig. 265. .... *Amphasia interstitialis* (Say)

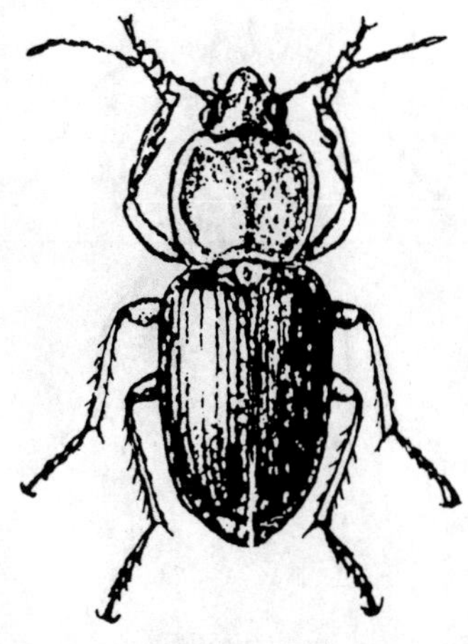

Figure 265

LENGTH: 9-10 mm. RANGE: North-central States to the Atlantic.

Head, thorax, antennae and legs reddish-yellow; elytra piceous; intervals of elytra subconvex, densely and rather coarsely punctate. Flies to lights; common.

Amphasia sericea (Harris) in shape and size is much like the above. The upper side is wholly black; the legs are paler. It too is widely distributed and often very abundant at lights.

65b Abdomen with only basal punctures; terminal spur of front tibiae with three points. Fig. 266. .... *Anisodactylus rusticus* Say

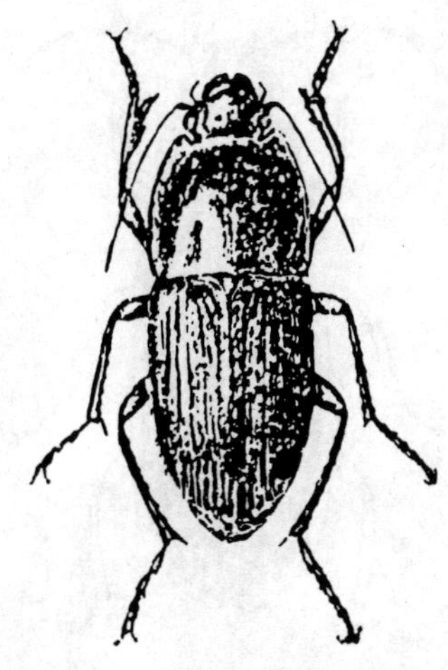

Figure 266

LENGTH: 9-14 mm. RANGE: Central and Eastern States. Common.

Brownish-black; base of antennae and part of hind thoracic angles reddish-brown; legs piceous. Elytral striae deep, intervals convex, the third with one to four punctures behind the middle.

A. dulcicollis Laf. A shining black species 11 mm long and having black legs, ranges from the Gulf north to Iowa.

66a (60) Submentum with a distinct tooth; antennae with two glabrous segments. Fig. 267. .... *Bradycellus rupestris* Say

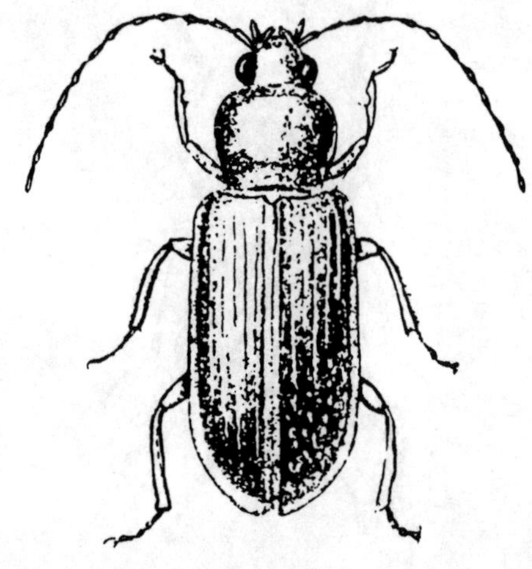
Figure 267

LENGTH: 4.5-5 mm. RANGE: throughout the United States especially in the southern half.

Reddish-brown, shining; head and disk of elytra usually piceous; two basal joints of antennae and the legs pale. Striae of elytra deep, their intervals convex.

B. tantillus (Dej.) is still smaller (3 mm) and is black or piceous, with pale antennae and legs. It ranges in the Gulf States and North. More than 60 species found in our area.

66b Submentum without a tooth. 67

67a (66) First segment of hind tarsi shorter than the 2nd and 3rd segments together. 69

67b First segment of hind tarsi fully as long as segments 2 and 3 taken together. 68

68a (67) Sutural striae long, joining the first dorsal; thorax with rather deep mid impression. Fig. 268. .... *Stenolophus ochropezus* (Say)

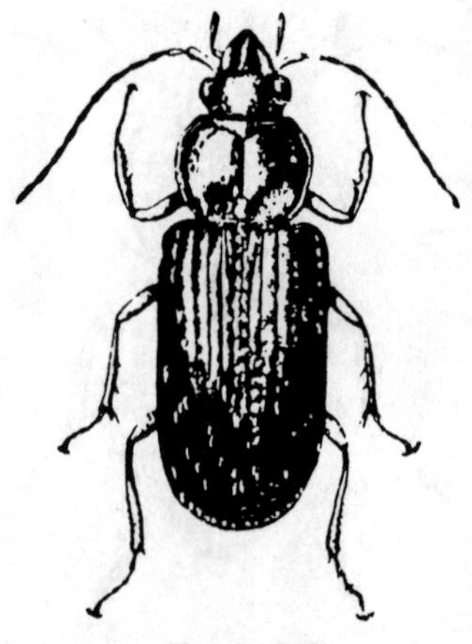
Figure 268

LENGTH: 5.5-6 mm. RANGE: much of the United States and Canada. Common.

Black or piceous, the elytra frequently iridescent; legs and base of antennae pale; striae of elytra deep; intervals flat.

S. carbonarius (Dej.) is larger, 7-7.5 mm. The color is black, with tibia, tarsi and first antennal joint brown. Its range includes the Central and Eastern States.

68b Sutural striae fine, short and not joining the first dorsal. Fig. 269. .... *Stenolophus plebejus* Dej.

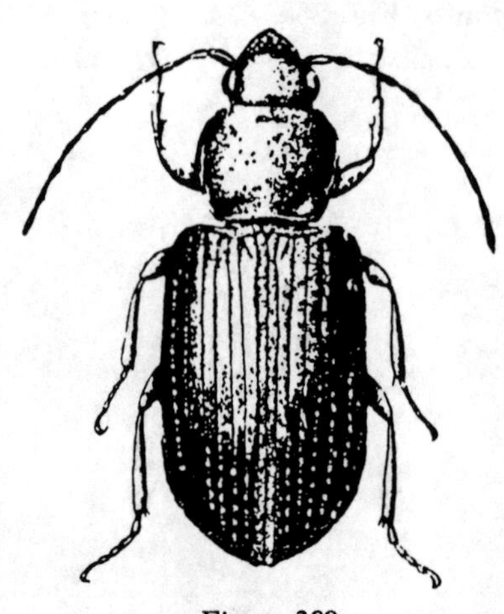
Figure 269

LENGTH: 4.5-5 mm. RANGE: United States and Canada east of the Rockies.

Piceous black, shining; narrow margins of thorax, legs and base of antennae brownish-yellow.

S. conjunctus (Say) is a little fellow (3.5-4.5 mm) which ranges widely throughout the United States and Canada. It is piceous with reddish-brown legs and antennal base. We often see it flying around our study lamp.

69a (67) Thorax pale without a dusky spot; size 2.5-4 mm. 70

69b Thorax usually with a large dusky spot; size 5.5-7 mm. Fig. 270. .... *Stenolophus lecontei* Chaud.

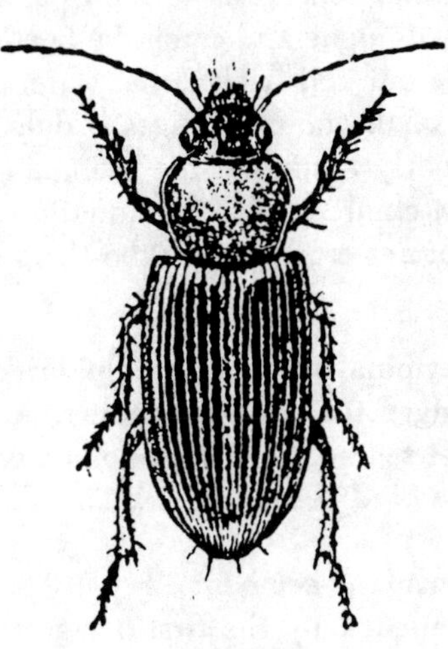

Figure 270

LENGTH: 5.5-7 mm. RANGE: Central and Southern States.

Yellowish-brown with an indefinite wide blackish stripe on each elytron (this sometimes faint or practically wanting); head black.

S. comma (Fab.) closely similar to *pallipes* averages a little larger with wider elytral stripes.

These two species are known as the Corn Seed Beetle. In cold damp seasons they feed on the slowly germinating seed. They fly in great abundance at lights so that an example can be had readily in their season, when they are one of our most common beetles. *Stenolophus* includes about 35 species in our fauna.

70a (69) Basal impressions well marked with a few coarse punctures, thorax dusky beneath. Fig. 271. .... *Acupalpus partiarius* Say

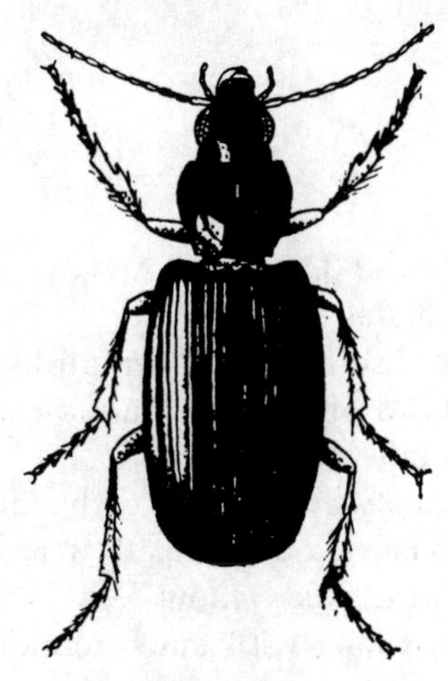

Figure 271

LENGTH: 3.5-4 mm. RANGE: Central and Eastern States to Texas. Common.

Head black, antennae brownish, legs pale. Elytral striae deep; intervals convex.

A. indistinctus (Dej.), has much the same range and size. The elytra are blackish with margins and suture brownish-yellow.

70b Basal impressions more poorly marked, with but few if any punctures; thorax pale beneath. Fig. 272. .... *Acupalpus pauperculus* Dej.

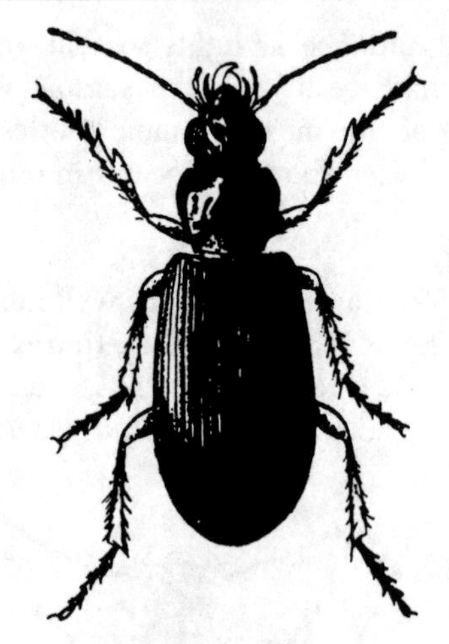

Figure 272

LENGTH: 3-3.5 mm. RANGE: Central and Eastern States. Common.

Reddish-brown; elytra dusky or piceous, the margins and suture pale. Striae of elytra fine, the intervals flat.

A. testaceus (Dej.), with a similar range is still smaller 2.5-3 mm. It is reddish-brown above, with pale yellow legs. The thorax is as wide as long with much rounded hind angles.

FAMILY 5. AMPHIZOIDAE
The Trout Stream Beetles

This is a small family of beetles made up of the single genus *Amphizoa.* These insects are elongate oval and somewhat convex beetles, dull black or brownish in color, and varying in size from 11 to 16 mm. Both the adults and larvae live in the swift, cool water of mountain streams, usually being found clinging to driftwood and other debris found in the eddies of these streams. Amphizoids are incapable of prolonged submersion, the adults having to surface for air and the larvae breathing through the eighth abdominal segment. There are four North American species of this genus found along the Pacific Coast from British Columbia to California and eastward to Colorado and Montana. An adult is shown in Fig. 36.

FAMILY 6. HALIPLIDAE
The Crawling Water Beetles

Though rather poorly equipped for swimming, these beetles prefer to spend much of their time in the water where they crawl about over the aquatic plants or swim feebly in the open. They are usually pale colored with dark spots. Perhaps the best way to get a larger number of species is to stand out in a water course quietly and use a small aquatic net to dip up the choice ones seen active down in the water. Another scheme is to haul out on land a quantity of algae and catch the beetles as they scramble out and back to the water. The family is a small one but ranges widely. The legs seem poorly equipped for swimming but are useful in climbing through aquatic vegetation. Some species crawl out on the shore at night.

1a **Terminal segment of palpi large, conical, longer than the preceding. All but the last segment of the abdomen covered by the hind coxae. 2**

1b **Terminal segment of palpi small, awl shaped; only the first 3 segments of abdomen covered by the hind coxae; pronotum widest at the base. Fig. 273.**

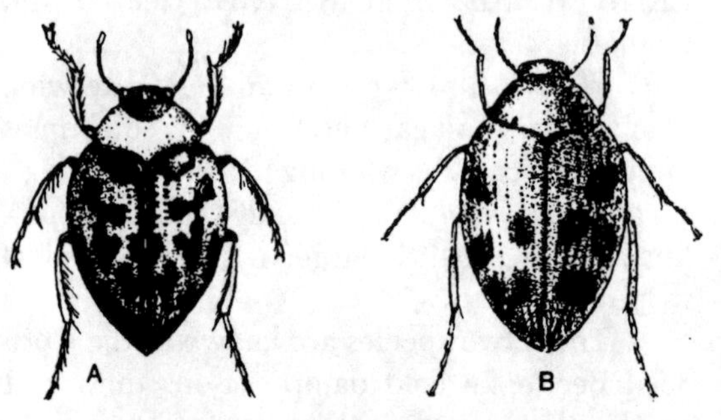

Figure 273

(a) ***Haliplus triopsis* Say**

LENGTH: 3.5-4.25 mm. RANGE: Maine and Ontario west to Wisconsin and Colorado, south to Georgia and New Mexico.

Pale to reddish-yellow with black markings; eyes circular in outline. Legs pale yellow; the markings on elytra are variable.

(b) ***H. immaculicollis* Harr.**

LENGTH: 2.5-3 mm. RANGE: North America, common.

Rufous to reddish-yellow marked with black; may be recognized by paired basal impressions on the pronotum and by the deeply grooved prosternum.

Haliplus with 40 species is the largest genus of this interesting family.

H. borealis Lec. measuring 3 mm and ranging through the northern half of the Central States and on into Canada, resembles *immaculicollis* but lacks its basal impressions. It is dull reddish-yellow with five black spots on the elytra.

2a (1a) Vertex with a prominent black spot. Fig. 274. .... *Peltodytes edentulus* (Lec.)

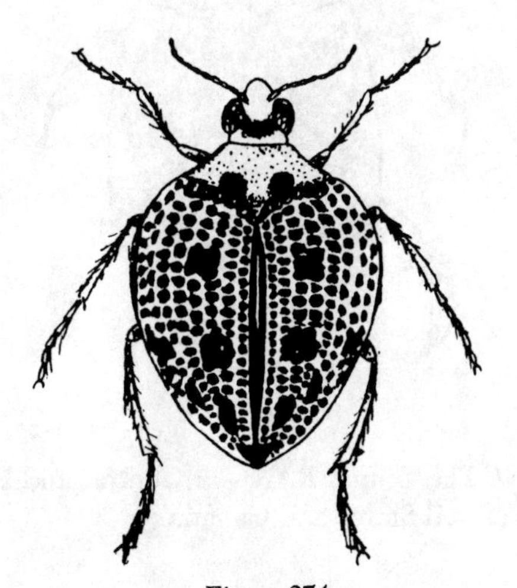

Figure 274

LENGTH: 3.5-4 mm. RANGE: Central and Eastern United States and Canada.

Pale yellow with black markings. Head and antennae dull yellow-brown, with a crescent-shaped black spot on the vertex.

Brychius horni Cr. with palpi and hind coxae like the genus *Haliplus* can be distinguished from it by its quadrate pronotum, widest in front instead of being widest behind. It measures 3.5 to 4 mm and is pale yellow marked with black. It has been found in Montana and California.

P. callosus (Lec.) is distinguished from other members of the genus by a prominent tubercle at the place of the anterior black spot. It has a length of 3.5 mm and is found in California, British Columbia and New Mexico. It is pale yellow and black.

2b Head wholly pale. Fig. 275. .... *Peltodytes duodecimpunctatus* (Say)

Figure 275

LENGTH: 3.5-4 mm. RANGE: East half of United States and Canada.

Dull yellow, with black markings as pictured. Often very common.

P. muticus (Lec.) widely distributed species closely resembles the above but may be distinguished by its wholly darkened femora and broadly rounded hind coxal plates. 15 species are in this genus.

FAMILY 7. DYSTISCIDAE*
The Predaceous Diving Beetles

These beetles vary widely in size and coloring. Blackish or dull green is a common color but markings of white or subdued yellows are frequent. Some species display rather brilliant colors. The mandibles are hollow for sucking the juices of their prey. The 11-segmented antennae are filiform.

The Dytiscids are excellent swimmers. In contrast with other water beetles both members of a pair of legs move together instead of alternately. The adults bury themselves in mud at the bottom or sides of the water-courses to spend the winter. The front tarsi of many of the males bear suction cups to better enable them to cling to the female in mating.

1a Episternum (es) of metathorax reaching the middle coxal cavity (cc). Fig. 276a. 3

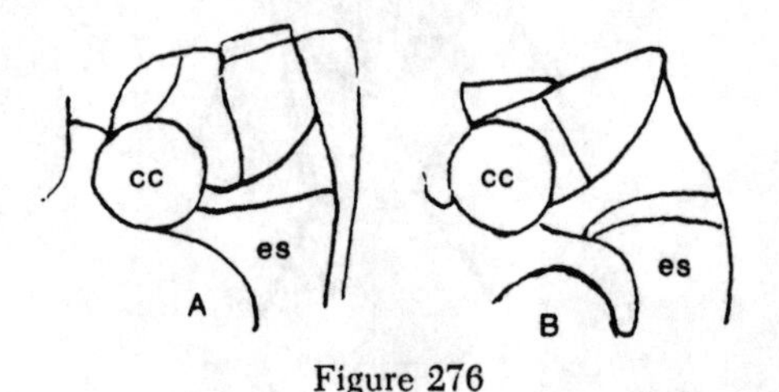

Figure 276

1b Episternum (es) not reaching the middle coxal cavity (cc). Fig. 276b. 2

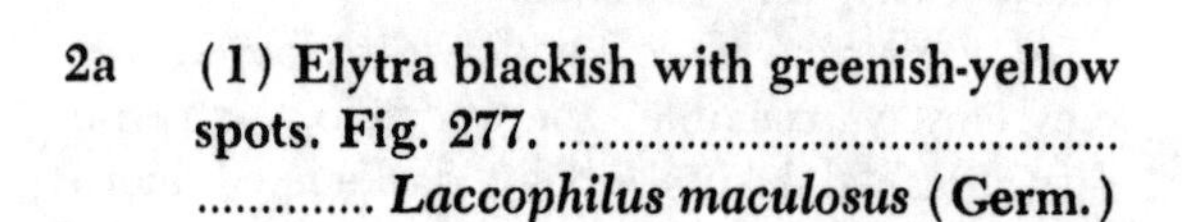
2a (1) Elytra blackish with greenish-yellow spots. Fig. 277. .... *Laccophilus maculosus* (Germ.)

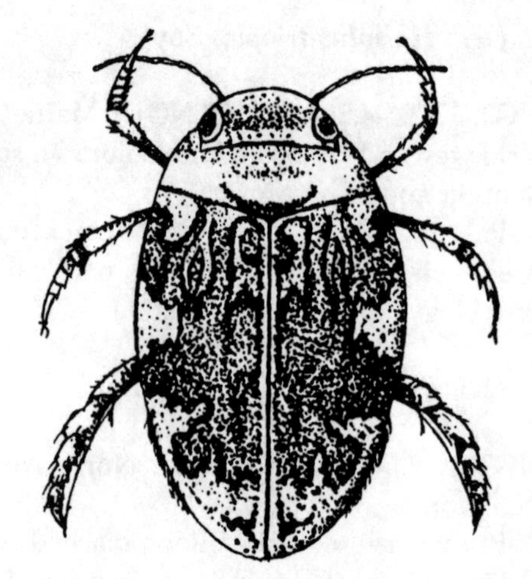
Figure 277

LENGTH: 6 mm. RANGE: Central and Eastern United States. Common.

This is a widespread genus with 14 species in our fauna.

2b Elytra dull yellow with blackish bar behind middle. Fig. 278. .... *Laccophilus fasciatus* Aube.

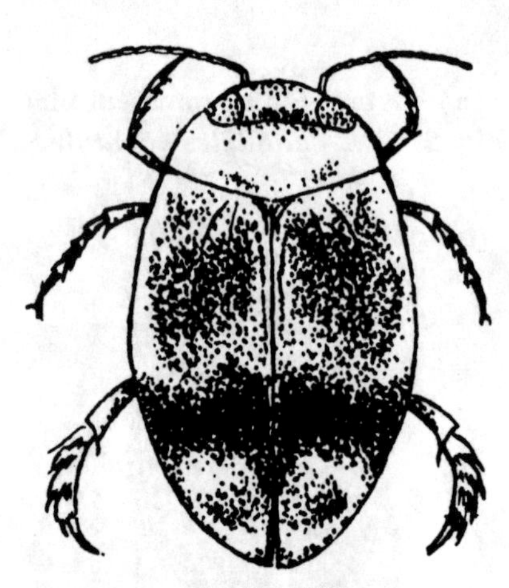
Figure 278

LENGTH: 5 mm. RANGE: Central and Eastern United States and Canada.

*See also Family 8.

Head, thorax and underparts dull brownish-yellow; elytra greenish-yellow marked in black.

L. undatus Aubé. is recognized by its black elytra which are marked with two yellowish cross bars. It measures a scant 5 mm.

3a (1) Spine at back of prosternum bent down into a different plane from the sternum; less than 6 mm. 4

3b Spine at back of prosternum not bent; length at least 7 mm. 6

4a (3) At least 3.5 mm long. 5

4b Small. 2 mm or less in length. Fig. 279. .... *Uvarus lacustris* (Say)

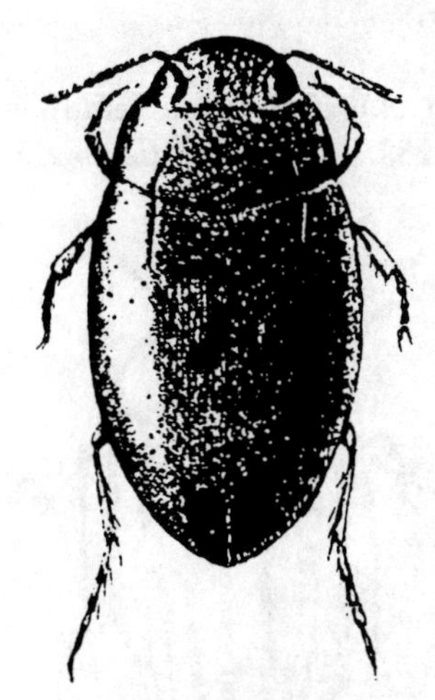

Figure 279

LENGTH: 1.4-1.8 mm. RANGE: likely much of United States east of the Rockies. Common.

Head and thorax dull reddish-yellow; elytra brownish-yellow, with darker clouding. Underparts black. Nine species of this genus found in U. S. and Canada. (*Bidessus* of American authors is not New World.)

Desmopachria convexa (Aubé.) is a tiny (1.7 mm) brownish-red water beetle that has much the same range as the above but is more rounded in form.

This is one of a half dozen species found in our area.

5a (4) (a, b, c) Hind coxal cavities contiguous. Thorax distinctly margined. Fig. 280. .... *Hydroporus undulatus* Say

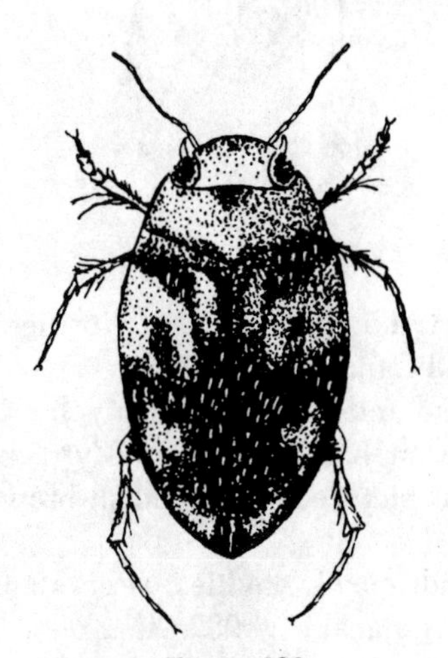

Figure 280

LENGTH: 4-4.5 mm. RANGE: Eastern half of the United States. Common.

Head and thorax reddish-yellow, elytra blackish with reddish-brown markings, rather variable.

H. stagnalis G. & H. measuring 3-4 mm and with northern range, is uniformly colored a dark reddish-brown. Each elytron has a row of four to six coarse punctures on the basal half, with a few distinct punctations scattered rather uniformly.

This is a large genus of more than 100 species which are difficult to separate.

5b As in 5a, except thorax with narrow indistinct margin. Fig. 281. .... *Hydroporus consimilis* Lec.

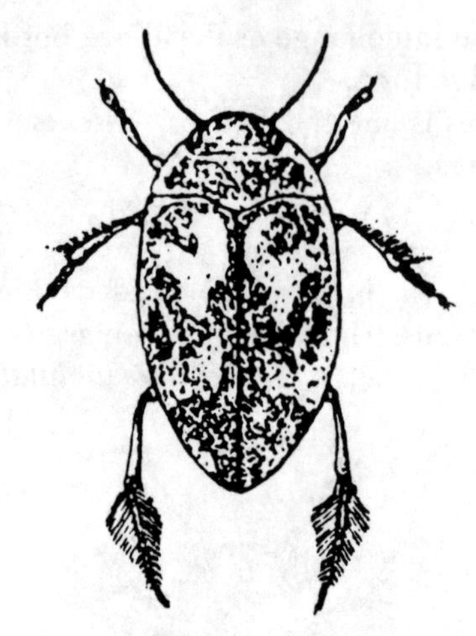

Figure 281

LENGTH: 4.5 mm. RANGE: Northern United States and Canada. Common.

Head and thorax reddish-yellow, the latter darker at base and apex. Elytra blacklish, marked as pictured with reddish-brown.

5c **Hind coxal cavities separated; under parts black. Fig. 282. .... *Hydroporus wickhami* Zaitz.**

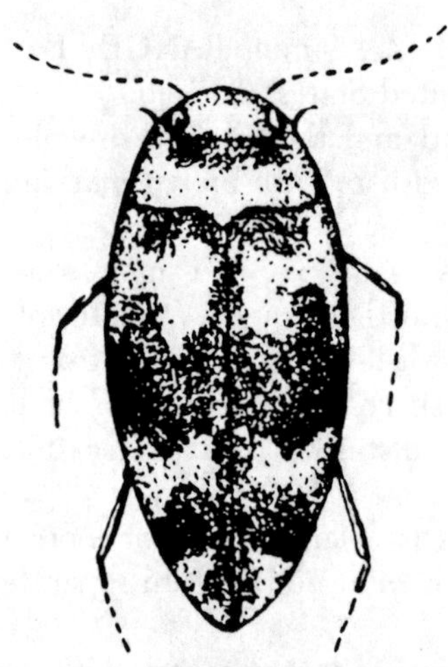

Figure 282

LENGTH: 3.2-3.75 mm. RANGE: Eastern half of the United States. Common.

Head and thorax reddish-brown; elytra black with reddish-brown cross bands, interrupted at suture.

H. niger Say, a common widely distributed species ranging through the eastern half of our region. It measures 4.4 to 5 mm and is black or nearly so on under parts, and above, with some rufous spots on head, thorax and elytra.

6a **(3) Margin of front of head making a notch in outline at front of eye. Length 7 to 16 mm. 7**

6b **Outline of eye not notched. Length 10 to 40 mm. 11**

7a **(6) Hind claws of unequal length; the outer claw more sharply bent at tip than inner. 8**

7b **Hind claws of equal length and shape. Fig. 283. .... *Agabus disintergratus* (Cr.)**

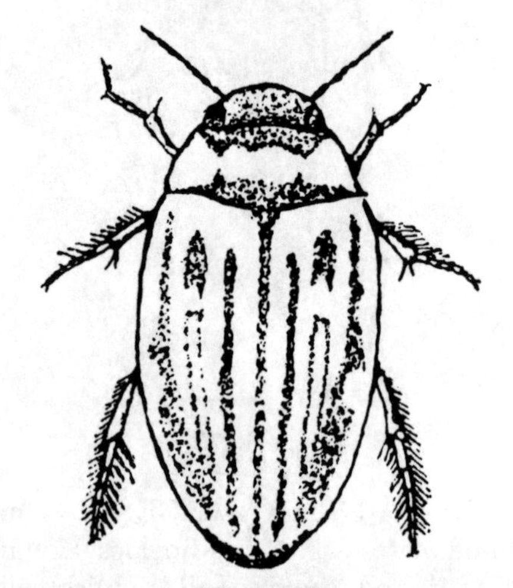

Figure 283

LENGTH: 7-8 mm. RANGE: much of the United States and Canada. Common.

Head and thorax reddish-yellow, markings black; elytra dull yellow with black stripes. Underparts mostly pale.

A. lugens (Lec.), ranging throughout much of the Rocky Mountain and Pacific coast region is very common there. It measures 8.3-9.5 mm and is black with small yellowish spots on the elytra.

Another widely distributed black species is *A. erichsoni* G. & H. ranging from Labrador to British Columbia and down into the Northern United States. It is one of the largest species (9.3-11.1 mm).

The genus Agabus contains about 25 species in our fauna. They are difficult to separate as most are black.

8a (7) Hind tibiae with a line of cilia on the inner half of the apical angles. Fig. 284. .... *Ilybius biguttulus* Germar

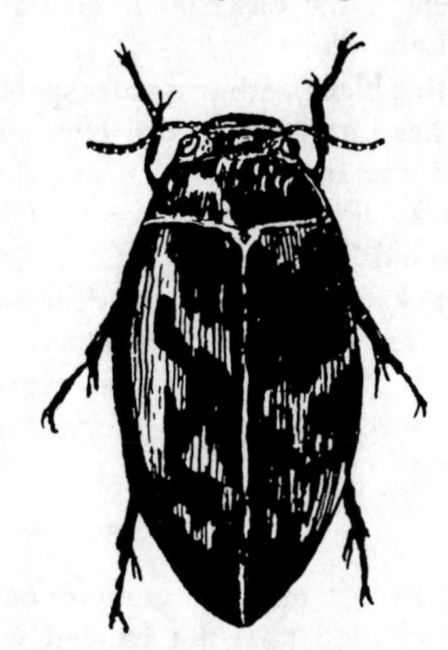

Figure 284

LENGTH: 10-11 mm. RANGE: Northeastern United States and Canada.

Black, slightly bronzed; elytra margined with reddish; antennae and front and middle legs reddish-brown, hind legs and under surface piceous. This genus includes some 15 species.

Matus bicarinatus (Say), 8-9 mm and ranging through much of the eastern half of the United States is a shining reddish-brown. The head is unusually wide.

8b Hind tibia not as in 8a. 9

9a (8) Side lobes of the metasternum broad and wedge-shaped. 10

9b Side lobes of the metasternum narrow; linear. Fig. 285. .... *Coptotomus interrogatus* (Fab.)

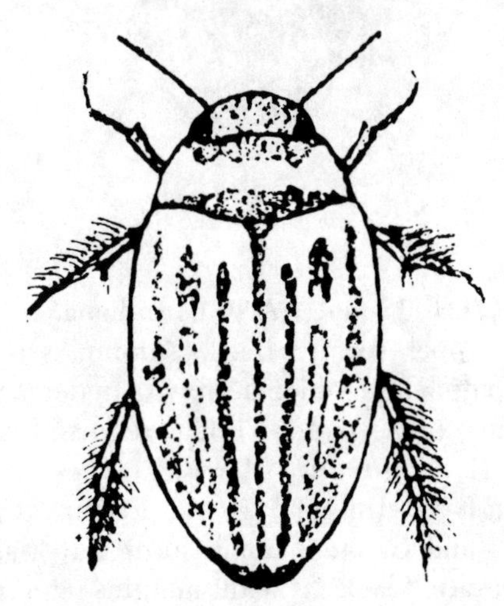

Figure 285

LENGTH: 7 mm. RANGE: much of Canada and the United States east of the Rockies. Common. There is also a Western species, *C. longulus* LeC. and a Southern one, *C. serripalpis* Say.

Head, thorax and underparts reddish-brown, the thorax with a stripe of black at front and back; vertex black. Elytra pitch brown with obscure lighter markings. Often seen at lights.

10a (9) Metasternum with a broad distinct pit-like depression on its front border between the coxae; pronotum usually margined. Fig. 286. .... *Rhantus confusus* Blatch.

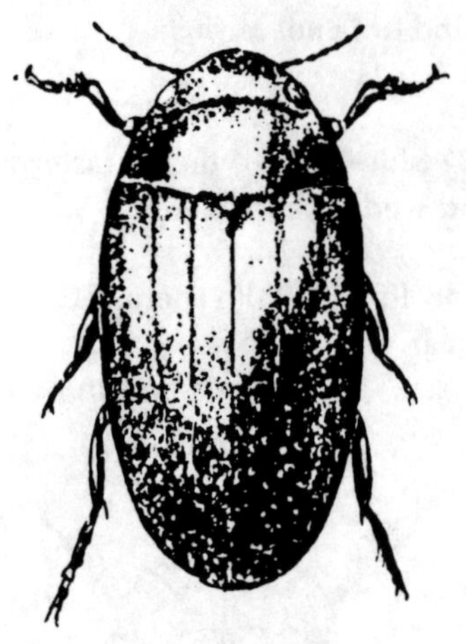

Figure 286

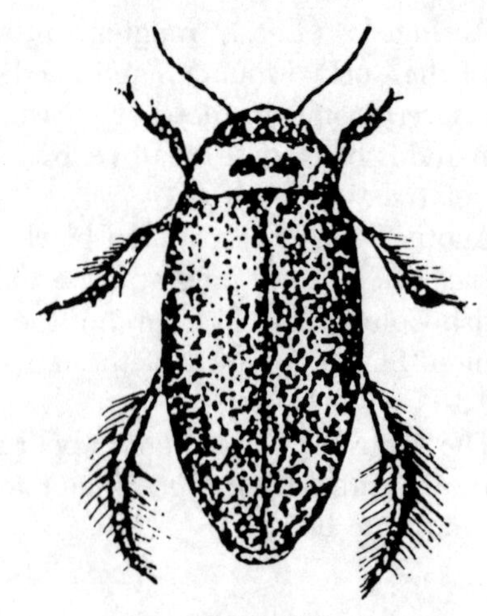

Figure 287

LENGTH: 12 mm. RANGE: Indiana.

Upper surface black, shining; antennae and palpi light reddish-brown; under surface piceous; tibiae and tarsi dark reddish-brown.

R. suturellus Harris, ranges widely through Central and North Eastern United States and Canada. It measures 9-10 mm. Its elytra are blackish with margins and many lines through the disk, dull yellow. The front and vertex are dull yellow. *Rhantus* is a widespread genus with 13 species in our area.

10b Metasternum flat between the coxa or sometimes with a faint longitudinal impression; thorax without margins. Fig. 287. .... *Colymbetes sculptilis* Harr.

LENGTH: 15-16 mm. RANGE: California, North Central and Eastern United States, Canada and Labrador.

Vertex black with two paler spots; thorax, front of head and margins of elytra dull yellow; bar across disk of thorax black. Under surface black; legs reddish-brown. 10 similar species found in N. U. S. and Can.

Our knowledge of the distribution of many species is highly fragmentary. Many of the "ranges" given here are based on actual collecting records but should be understood to be incomplete.

11a (6) Length one inch or more; outer margin of hind tarsi not fringed with flattened hairs. 12

11b Length but little over ½ inch at most; hind tarsi fringed on outer margin with golden-yellow flattened hairs. 13

12a (11) Labrum distinctly notched at middle; abdominal segments reddish-brown with hind margins darker. Fig. 288. .... *Dytiscus fasciventris* Say

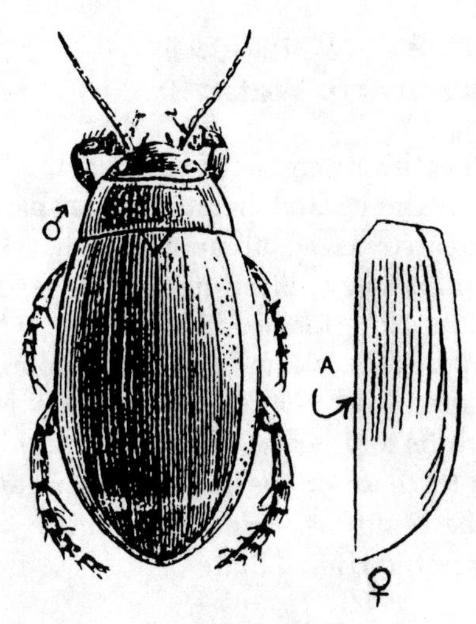

Figure 288

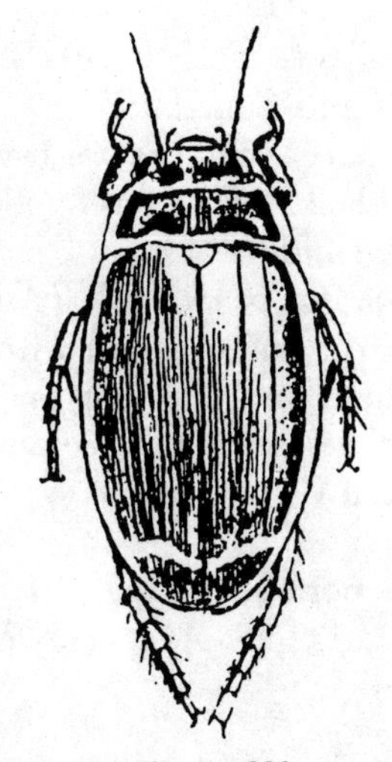

Figure 289

LENGTH: 25-28 mm. RANGE: North Central and Eastern United States.

Greenish-black above; thorax margined with yellow at sides and sometimes more faintly at front and back. Elytra margined with yellow. Elytra of female each with ten grooves reaching beyond the middle (a).

D. marginicollis Lec. similarly colored but 30 mm long is very common in California and the West.

The members of this genus are sometimes known as water tigers. They are highly destructive to young fish and tadpoles. 13 species found in our area.

12b Labrum practically truncate at middle; all margins of thorax broadly yellow. Fig. 289. .... *Dytiscus harrisi* Kby.

LENGTH: 38-40 mm. RANGE: N.E. United States and Canada.

Elytra with yellow marginal line narrowing at apex and with a crossbar as pictured. Segments of abdomen reddish-yellow margined with piceous.

The members of this genus as well as many of the other members of the family are readily taken in light traps.

13a (11) Hind femora reddish-brown. Fig. 290. .... *Acilius semisulcatus* Aubé.

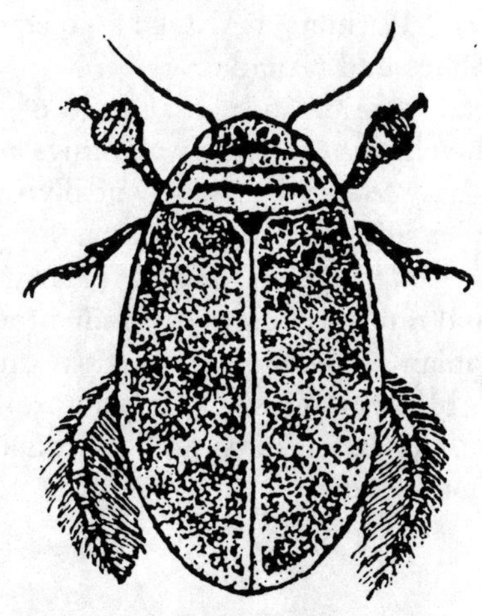

Figure 290

LENGTH: 12-14 mm. RANGE: Northern United States and Canada.

Upper parts dull brownish-yellow with markings of black. Elytra with yellowish cross bar behind the middle.

Thermonetus ornaticollis (Aube.), 11-13 mm long, is dull yellow above with two transverse black lines on the thorax and many fine black dots on the elytra. Very common at lights in Southeastern U. S.

13b Hind femora black. Fig. 291. .... ***Acilius mediatus* (Say)**

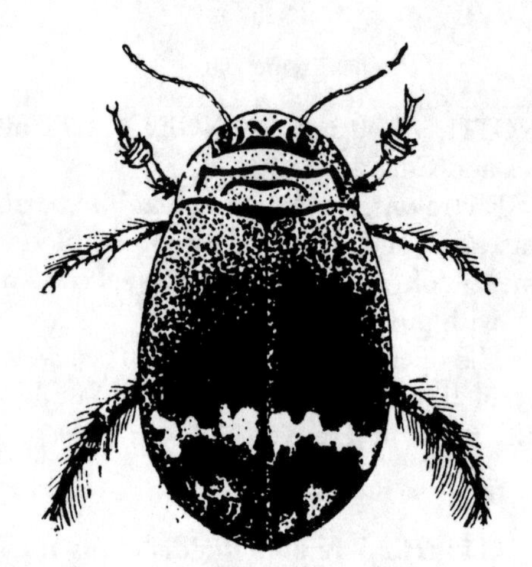

Figure 291

LENGTH: 12 mm. RANGE: Northeastern United States and Canada.

Head, thorax and basal half of elytra brownish-yellow with black markings as pictured. Wavy cross bars at tips of elytra light buff. Under surface reddish-brown.

Graphoderus fasciatocollis (Harris), 13-15 mm and ranges across the continent around our Canadian boundary. Its thorax is dull yellow with black margins. The elytra are blackish with numerous dots and marginal and sutural lines yellow.

FAMILY 8. NOTERIDAE
The Burrowing Water Beetles

These beetles resemble the Dytiscidae in general appearance and habits. In the past they were treated as a subfamily of that family. Structurally they differ by having no visible scutellum. Noterids are small beetles with the size ranging between 1.2 and 5.5 mm. They are shaped and colored like the Dytiscidae. Both adults and larvae are aquatic with habits similar to those of the predaceous water beetles. The family is widely distributed with 13 species in the United States.

1a Size greater than 4 mm; apex of prosternal process very broad, being at least 2½ to 3 times wider than its width between the coxae.

Hydrocanthus iricolor Say, Fig. 292, is fairly common. Its head, thorax and underparts are reddish-yellow; elytra dark reddish-brown. It is 4 to 5 mm long and is found in the Central and Eastern United States. Comes to lights.

1b Size usually less than 3 mm; apex of the prosternal process narrower being only twice as wide as the distance between the procoxae.

Suphisellus bicolor (Say) is a small species (2.5 mm) with yellow head and thorax and dark brown elytra, having a yellowish cross bar near middle. RANGE: Southeastern U. S.

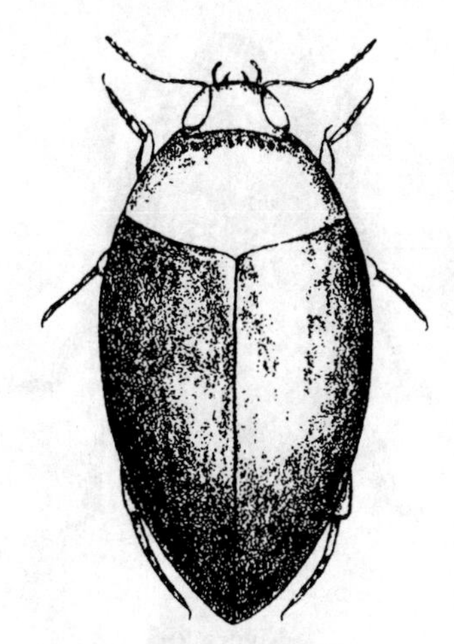

Figure 292

FAMILY 9. GYRINIDAE
The Whirligig Beetles

These small to medium sized beetles have world-wide distribution. Their surface is smooth and hard; black is the usual color. The short, thick, odd-shaped antennae have eleven segments. They get their common name because of their gyrations over the water when disturbed.

These highly gregarious beetles are always giving a skating carnival. Just how they maintain so much activity without some serious collisions is a mystery. Structurally, they are oddities. Flattened bodies, divided eyes and very queer antennae are characters that make them distinct. Catch one (if you can) and it gives off a milky secretion with a rather unpleasant odor. Some of them are called "apple bugs" because of this. Up at McGregor, Iowa the kids call them "penny bugs." They have been told if they put one under their pillow on going to bed, in the morning they'll find a "scent" there.

1a Scutellum distinct; length less than 8 mm; elytra with 11 striae. 3

1b Scutellum hidden; length 9 or more mm; elytra with 9 striae or smooth. 2

2a (1) Under surface brownish yellow. Fig. 293. .... *Dineutes discolor* Aubé.

Figure 293

LENGTH: 11.5-13 mm. RANGE: Central and Eastern United States.

Upper parts black with bronze reflection, shining; under surface brownish-yellow to straw color; elytra with side margins at outer apical angle and tips slightly sinuate.

Dineutes ciliatus Forsberg may be readily told by its larger size (12-15.5 mm). Its upper parts are black with bronze stripes near the margins and the under surface dark brown or pitchy. It ranges through the Atlantic and Southern States.

2b Under surface black or bronzed. Fig. 294. .... *Dineutes americanus* Linnaeus

Figure 294

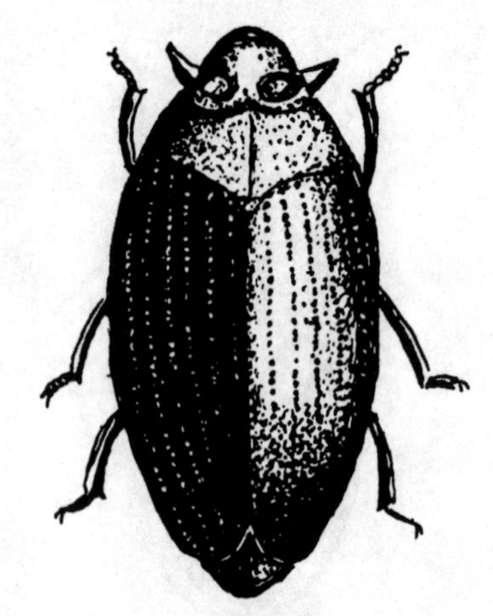

Figure 295

LENGTH: 10-11 mm. RANGE: pretty much throughout the United States and Canada. Common.

Black above and below, much bronzed and very shiny; legs brownish-yellow. Femur without a tooth; punctures shallow and not closely placed.

Dineutes carolinus Lec. is a smaller species 9-10 mm which ranges from the South Atlantic States to Texas. Its surface is black, sometimes bronzed, and rather dull.

This genus includes 13 species generally distributed throughout our area.

3a (1) Scutellum with a distinct but very fine longitudinal ridge at base. Small. Fig. 295. .... *Gyrinus minutus* Fab.

LENGTH: 3.5-4.3 mm. RANGE: United States, Canada, Alaska (also Central and Northern Europe and Asia). Common.

Upper parts black, sides bronzed, dull; under parts brownish-yellow; abdomen black; thorax with a fine ridge at middle.

Gyrinus analis Say is a common and very widely dispersed species ranging from Nova Scotia to Kansas and south to Florida and Louisiana. It measures 4.5-5.5 mm and is bronzed above and beneath; the legs are reddish brown.

3b Scutellum flat, under surface reddish-brown. Fig. 296. .. *Gyrinus ventralis* Kby.

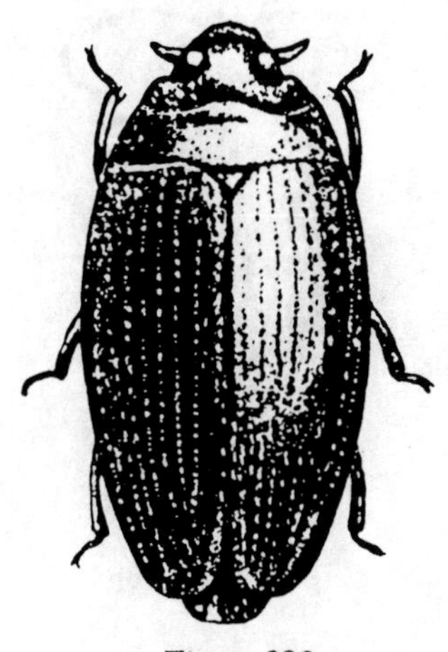

Figure 296

LENGTH: 4.8-6 mm. RANGE: Northern United States and Canada.

Black, highly polished, often with bluish reflections, sides bronzed.

Gyrinus affinis Aubé. is very common in its range, apparently over much of the United States and Canada. It is shining black with faint bluish reflections. In length it is 5.6-7 mm.

In our area the genus *Gyrinus* includes some 35 small, black beetles. Identification to species has to be based on the structure of the male genitalia.

SUBORDER MYXOPHAGA

FAMILY 10. SPHAERIIDAE
The Minute Bog Beetles

This is a small family of tiny beetles. They are subhemisphaerical in shape, .5 to .75 in length, and shiny black, sometimes with pale markings. These beetles are found under stones near water, among the roots of plants, or in moss in boggy places. *Sphaerius politus* Horn found on the West Coast is one of the several species of this group. Fig. 296a.

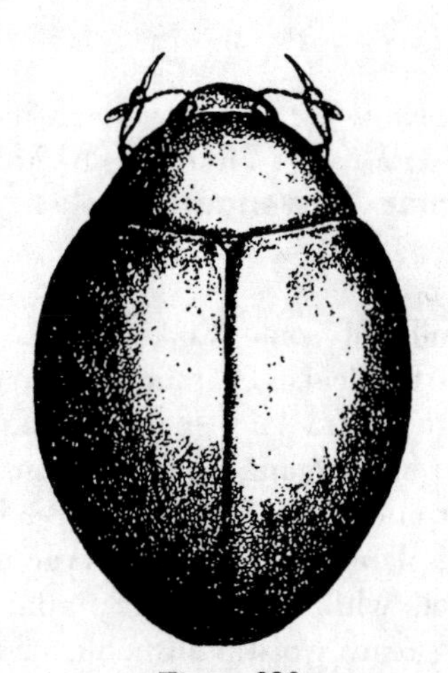

Figure 296a

FAMILY 11. HYDROSCAPHIDAE
The Skiff Beetles

These small beetles are somewhat depressed, fusiform, about 1.5 mm in length, tan or brown with darker margins, fine pubescence, and with the elytra leaving several segments of the abdomen exposed. The 9-segmented antennae are short with the apical segment as long as the preceding four segments.

Both adults and larvae are abundant in streams on the filamentous algae growing on rocks. One species, *Hydroscapha natans* LeC., is found in the West between southern Idaho and Arizona and the Coast. Fig. 296b.

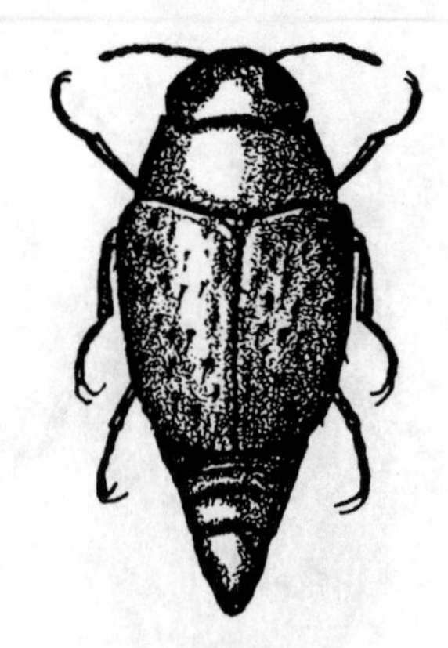

Figure 296b

SUBORDER POLYPHAGA

This suborder contains most of the families of beetles. The many thousands of species are very diverse, showing adaptations for all terrestrial habitats, as well as aquatic.

Sweeping submerged plants with a water net is a good means of collecting. Many species fly to lights.

FAMILY 12. HYDROPHYLIDAE*
The Water Scavenger Beetles

This family of some 1,000 species, many of which are tropical, is characterized in many species by the extra long maxillary palpi which serve as sense organs, and which may be mistaken for antennae. The adults live largely on decaying plants; some of the larvae utilize this same food while others are predacious. The adults are readily distinguished from the Dytiscidae by being more convex on the dorsal side and their antennae ending in a club. In swimming, the legs move alternately instead of as pairs as in the Dytiscidae. A film of air spread over the ventral side from a bubble "picked up" at the surface by the antennae supplements that carried under the elytra in supporting respiration while the adults remain under water.

*See also Family 13.

1a **Thorax narrower than elytra; elongated with surface roughened; less than 6 mm; elytra with 10 rows of punctures. 2**

1b **Thorax at base as wide as elytra. 3**

2a **(1) Antennae with 9 segments. Fig. 297. .... *Helophorus lineatus* Say**

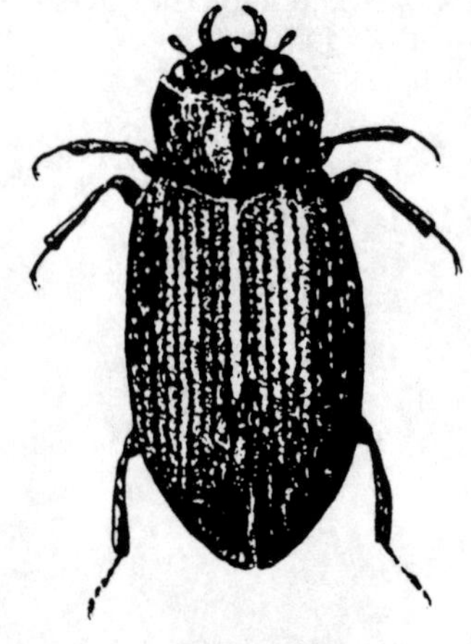

Figure 297

LENGTH: 2.8-3.5 mm. RANGE: United States east of the Rockies, and Canada.

Above light brown with greenish tinge, antennae and legs pale; elytral striae with deep transverse punctures.

H. lacustris Lec. can be distinguished from the above by its obtuse hind thoracic angles. It is larger (4-4.5 mm), blackish-brown and ranges in our Central States.

2b Antennae with 7 segments. Fig. 298. .... *Hydrochus squamifer* Lec.

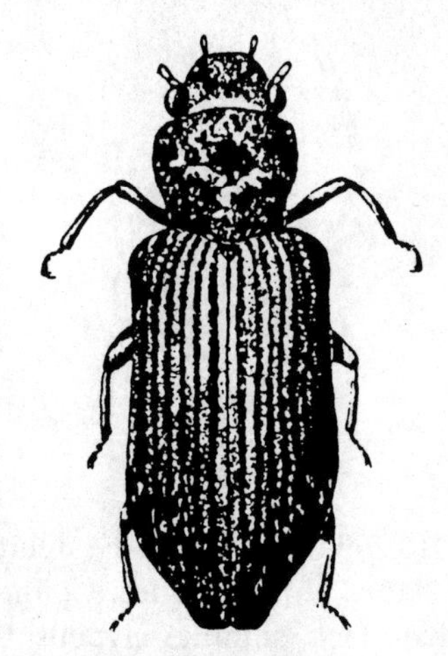

Figure 298

LENGTH: 3.5-4 mm. RANGE: North Central United States.

Grayish-bronzed or coppery; head and thorax darker than elytra and tinged with greenish. Both of these genera contain about 20 species each in our fauna. They are similar in appearance and hence difficult to separate.

3a (1) First segment of middle and hind tarsi short. Fig. 299a. 4

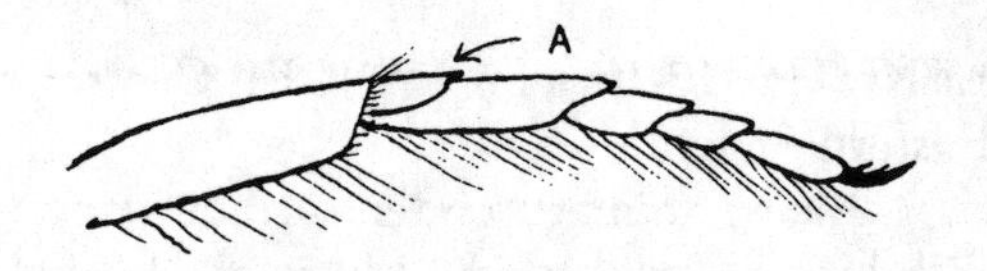

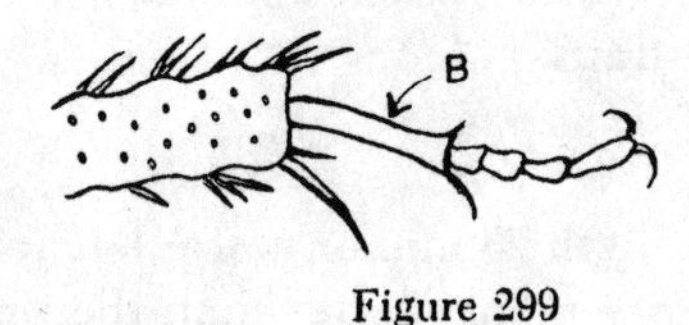

Figure 299

4a (3) Metasternum elongated into a distinct spine; tarsi compressed. 5

4b Metasternum not elongated; tarsi not compressed. 7

5a (4) Posternum marked with grooves; metasternal spine long. 6

5b Prosternum with a ridge, metasternal spine short; length 13-16 mm. Fig. 300. .... *Hydrochara obtusata* (Say)

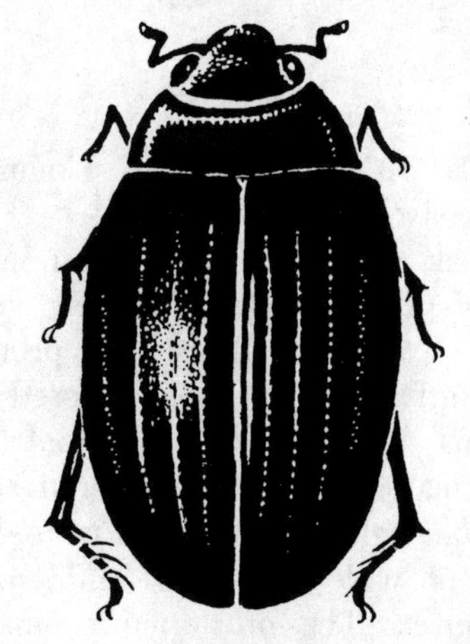

Figure 300

LENGTH: 13-16 mm. RANGE: Central and Eastern States.

Black, shining; under surface dark reddish-brown, pubescent. Elytra each with 4 rows of distinct punctures, the outer row double. Spine of metasternum not reaching beyond middle coxae. Often abundant at lights and in light traps.

6a (5) Length 25 mm or more; last joint of maxillary palpi shorter than the preceding. Fig. 301. .... *Hydrophilus triangularis* (Say)

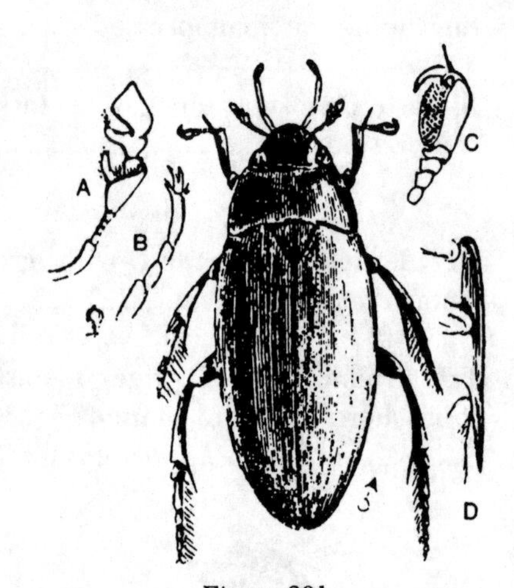

Figure 301

LENGTH: 34-37 mm. RANGE: Common from coast to coast.

Black with olive tinge, shining; under surface dark often triangular yellowish spots at sides of abdomen. Prosternal prominences (receiving front end of spine) closed in front; a, antenna; b, front tarsus of female; c, front tarsus of male; d, side view of sternal spine.

Dibolocelus ovatus G & H is similar to the above, but wider, shorter (31-33 mm) and more convex. The prosternal prominence is open in front. It ranges through the eastern third of the United States and like the above is often attracted to lights. More common in the South.

6b Length less than 12 mm; last joint of maxillary palpi equal to the preceding or longer. Fig. 302. .... *Tropisternus lateralis* (Fab.)

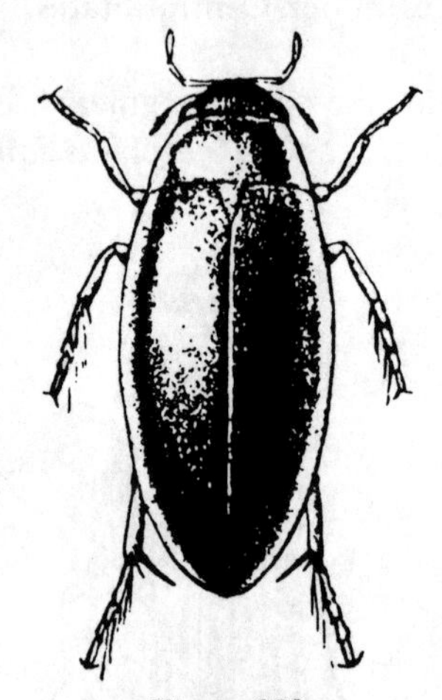

Figure 302

LENGTH: 8-9 mm. RANGE: Quite general over the entire United States. Common.

Olive-black, shining; clypeus, thorax and margins of elytra pale yellow; under surface black or piceous. Legs yellow, femora black at base.

T. collaris striolatus (Lec.) is a southern species, a bit larger in size (9-10.5 mm) than the above and readily distinguished by the yellow stripes on the disk of the elytra.

Tropisternus contains about 15 species in our fauna.

7a (4) Last joint of maxillary palpi longer than the third; elytra with striae or punctures in rows. Fig. 303. .... *Hydrobius melaenum* Germ.

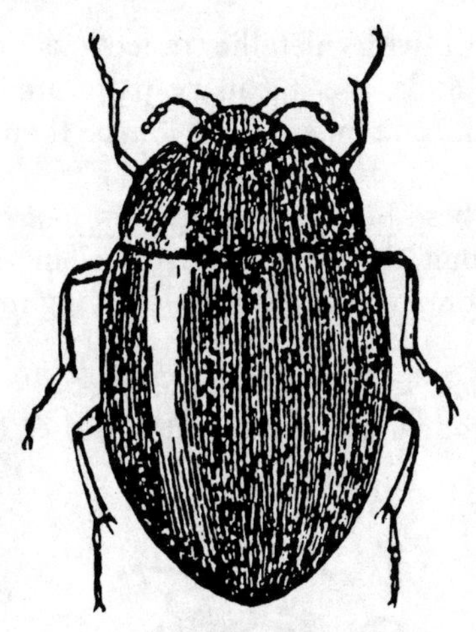

Figure 303

LENGTH: 7.5 mm. RANGE: Northeast United States.

Piceous-black, shining, with bronze reflections. Scutellar striae distinct.

Paracymus subcupreus (Say) colored much like the above; is small (1.5-2 mm) but common throughout much of the United States and in parts of Canada.

7b Last joint of maxillary palpi shorter than the third. Fig. 304. .... *Enochrus orchraceus* (Melsh.)

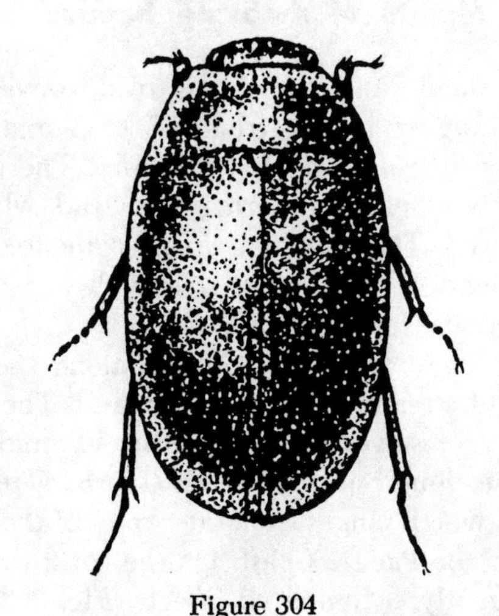

Figure 304

LENGTH: 3.5-4 mm. RANGE: Central and Eastern United States. Often common.

Dull smoky-brown, shining, head darker with a pale space in front of each eye; under parts piceous, the tibia and tarsi paler.

Berosus striatus (Say) is a dull greenish-yellow water beetle 4-5 mm long which differs from the above in having its last ventral abdominal segment notched. It is common over much of the eastern half of the United States.

8a (3) Length 5 mm or over. Scutellum elongate; antennae with 8 segments. Fig. 305. .... *Sphaeridium scarabaeoides* (L.)

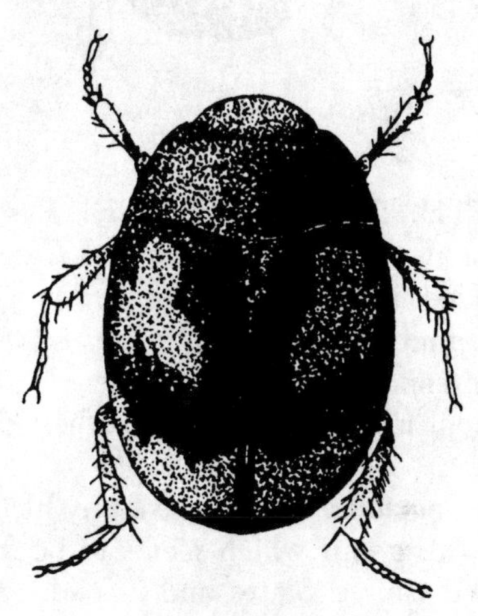

Figure 305

LENGTH: 5.5-6.5 mm. RANGE: much of North America, also Europe.

Upper parts black, shining; elytra with reddish sub-basal spot and the apical fourth yellowish; underparts piceous. Elytra punctured but without striae.

This European import, seems to have taken seriously the admonition "don't go near the water." It is found abundantly in fresh dung of cattle and may thus be mistaken for a Hister beetle or a Scarabaeid. It and the other members of its tribe do not have swimming legs. This is one of three similar species.

8b Length not over 3 mm; scutellum equilateral; antennae with nine segments. Fig. 306. .... *Cercyon praetextatus* (Say)

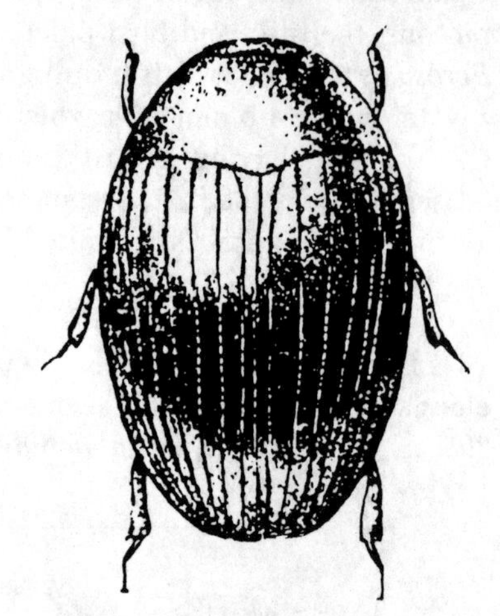

Figure 306

LENGTH: 2.5-3 mm. RANGE: East of the Rockies in the United States and Canada.

Piceous black, shining; front angles of thorax, and margins and tip of elytra yellowish; antennae dull yellow, the club darker. Common in dead fish and other decaying debris.

C. navicularis Zimm. is a tiny black member of this genus which seems to be found all over the United States and Canada. It measures 1.5-1.7 mm and is found in damp leaves near water pools.

38 species of *Cercyon* are known from the U. S. and Canada. They are often common in decaying fungi, manure, and the like.

FAMILY 13. LIMNEBIIDAE
The Minute Moss Beetles

This small family of beetles was formerly included in the *Hydrophilidae*. Members of this family are elongate or broadly oval ranging in size from 1.2 to 2 mm. They are black or brownish with metallic reflections. As with hydrophilids, the maxillary palpi are well-developed, sometimes being longer than the antennae.

These beetles are found in moist vegetation along the edge of streams, in decaying sphagnum moss in swamps, or living in brakish pools.

Ochthebius puncticollis LeC. is a small dark beetle living in the southern U. S. Fig. 307.

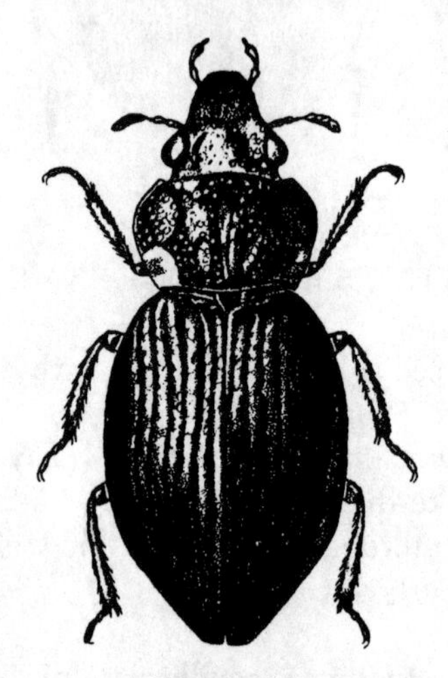

Figure 307

FAMILY 14. GEORYSSIDAE
The Minute Mud-Loving Beetles

This small family consists of oval, convex beetles ranging in size from 1.5 to 3 mm, dark brown in color with no vestiture. The pronotum is elongate covering the head which is deflexed. The antennae are 9-segmented with segments 1 and 2 stout and with 7, 8, and 9 forming a stout oval club.

These beetles are found along the margins of streams in the mud or sand. They conceal themselves with a coating of mud. One species found in the U. S. is *Georissus pusillus* LeC. which ranges from the center of the country to the Pacific Coast. On the coast they are frequently encountered insects. Fig. 308.

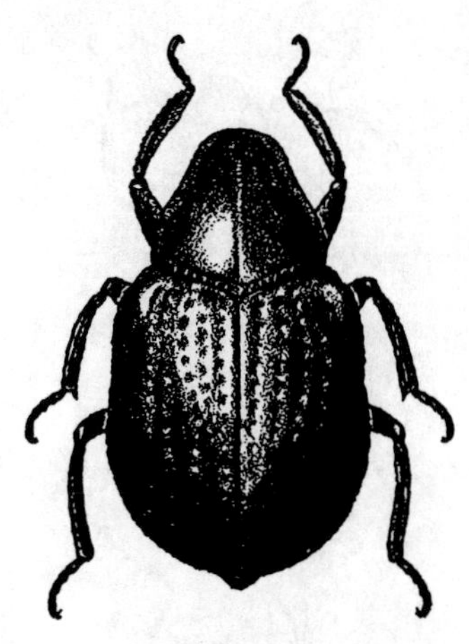

Figure 308

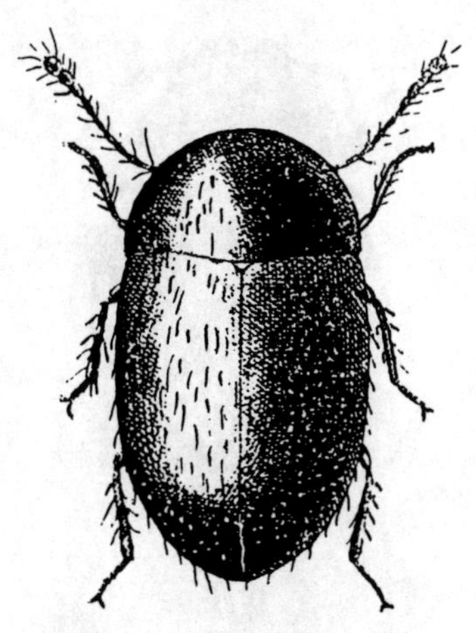

Figure 309

FAMILY 15. PTILIDAE
The Feather-Winged Beetles

This family consists of minute, .25 to 1 mm, round or oval beetles, dark in color, and covered with dense pubescence. These characters along with the long slender whorls of hair on the apex of each antennal segment and the long fringes of hair on the edges of the wings easily distinguish this family. The 11-segmented antennae have the last two or three segments forming a loose club. These, the smallest of all beetles, live in dung, fungi, rotting wood, in ant's nests, and under logs. In the U. S. and Canada there are about 100 species of the family.

Fig. 309. *Nossidium americanum* Mots. LENGTH: 1 mm. RANGE: Much of the United States.

Head and thorax shining, piceous; elytra reddish-brown, clothed with rather long yellowish hairs; legs and antennae yellow.

Some of the smallest known beetles belong to this family. The true wings when present are usually fringed with long hairs. They live in decaying wood. Another species is *Acratrichis moerens* Matth., .6-.7 mm. and ranging widely, shining black.

FAMILY 16. LIMULODIDAE
The Horse-Shoe Crab Beetles

These small beetles, .7 to 3.2 mm, are wedge-shaped or horse-shoe crab-shaped, compact, pale yellow to dark in color, and covered with fine hairs. They are closely allied to the *Ptilidae* and share many common structural characters. However, these beetles are all found in the nests of ants where they feed on the exudations of the body of the ant and of that also of the larvae and pupae.

One species, Fig. 310, *Limulodes paradoxus* Matth. shining reddish-yellow with grayish pubescence, 1 mm, ranging widely.

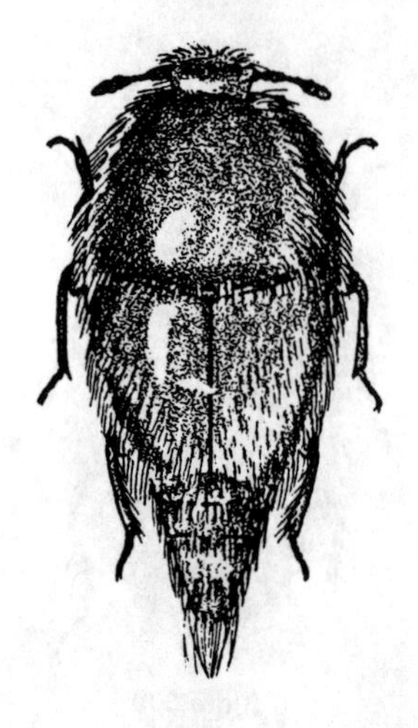

Figure 310

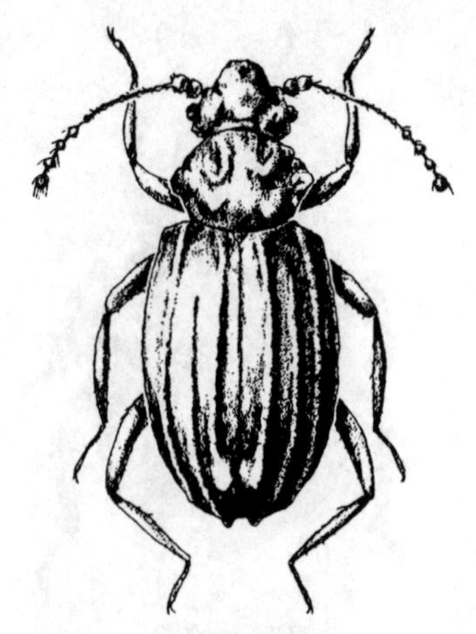

Figure 310a

FAMILY 17. DASYCERIDAE
Minute Scavenger Beetles

This is another small family (two species in the U.S.) of tiny beetles, about 2 mm, containing the single genus *Dasycerus.* Members of this group are elongate oval, rather convex, brownish in color with three-segmented tarsi. The procoxae are open behind, the antennae have the two basal segments enlarged and the outer ones making up a three segmented club. The elytra are strongly costate.

Dasycerus carolinensis Horn is found in North Carolina and Georgia and *D. angulicollis* Horn in California. Fig. 310.

FAMILY 18. LEPTINIDAE
The Mammal Nest Beetles

Members of this family are small flattened louse-shaped beetles. In size they range between 2 and 5 mm and in color are testaceous or reddish-brown. They possess six visible abdominal segments, globular anterior coxae, and reduced wings and eyes. They are the only parasitic beetles.

Leptinids are taken in the fur and nests of various mammals. The following species are taken in the U. S. and Canada: *Leptinus testaceus* Mueller with the prosternum not extending posterior to the anterior coxae and not appearing to separate them, 2 mm, in the nests of mice, moles, and shrews and occasionally in those of social *Hymenoptera* (Fig. 311); *Leptinillus validus* (Horn) with the prosternum produced posteriorly, fringed with long setae, 4-5 mm; associated with the American beaver in the U. S. and Canada; *Leptinillus aplodontiae* Ferris with the prosternum extending posteriorly but with no setae, 3-3.5 mm, on the fur of the mountain beaver on the Pacific slope from Washington to California; and *Platypsillus castoris* Rits. with the prosternum forming a broad flat plate ending posteriorly in a median lobe fringed with long setae, blind, abdomen with five tergites exposed, living as ectoparasites on beavers in Europe, Asia, the U. S. and Canada.

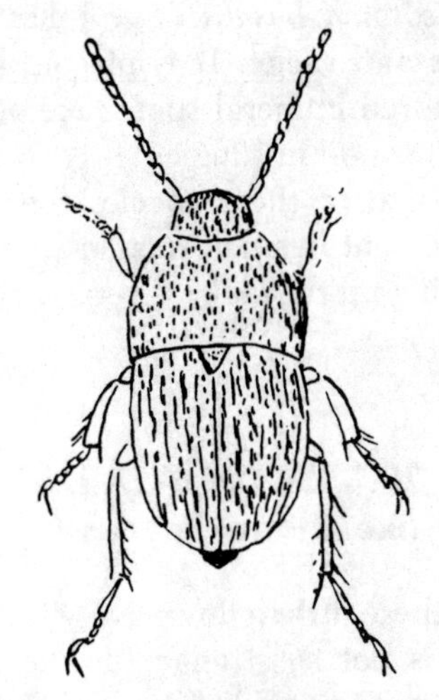

Figure 311

FAMILY 19. LEIODIDAE
The Round Fungus Beetles

Members of this family are convex, oval, shiny beetles ranging in color from shiny black to testaceous. Many species are contractile and may roll themselves into a ball. The antennae are composed of 10 or 11 segments and usually end in a loose club, often with segment 8 reduced when the club is 5-segmented. The pronotum is wider than the head and with the anterior margin broadly and deeply emarginate. The elytra are smooth or striato-punctate, infrequently transversely strigose.

Adults and larvae are found in fungi, slime molds, or under bark. The family is of moderate size with 17 genera and about 125 species found in North America. The following are typical:

1a **Head with antennal grooves beneath. Fig. 312. .... *Agathidium oniscoides* Beauv.**

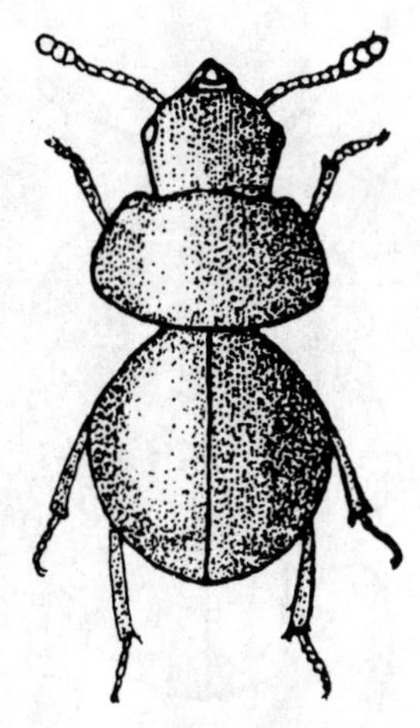

Figure 312

LENGTH: 3.5-4 mm. RANGE: Central and Eastern United States and Canada.

Black or piceous, smooth and shining. Lives beneath the bark of fungus bearing logs. For protection it rolls itself into a tiny ball with legs hidden and does not at all resemble a beetle. The hind legs have but 4 tarsi.

A. politum Lec. is a more convex and smaller species (2-2.5 mm). It is a shining reddish-brown and is often abundant within much the same range and habitat as the preceding.

1b **Head without antennal grooves. 2**

2a **(1) Antennal club with 3 segments. Fig. 313. .... *Colenis impunctata* Lec.**

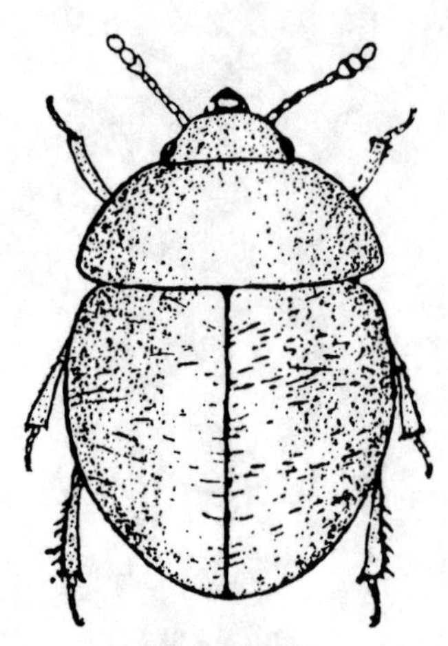

Figure 313

LENGTH: 1.5-2 mm. RANGE: Central and Eastern United States.

Pale reddish-brown; elytra with fine transverse lines. Its tarsi number 5-4-4. It is often found in fleshy fungi and is rather common. It does not contract itself into a ball when disturbed.

2b Antennal club with 5 segments. Fig. 314. .... *Leiodes valida* (Horn)

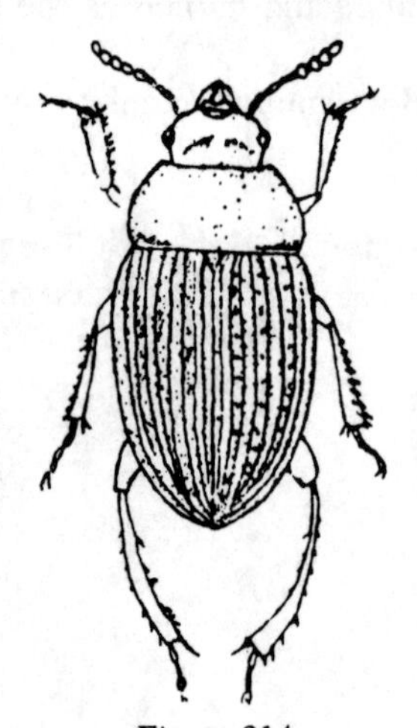

Figure 314

LENGTH: 3.5-6 mm. RANGE: Northern United States and Canada.

Piceous to paler, moderately shining. Antennae 11 segments. Hind tarsi with but 4 segments.

Anisotoma basalis (Lec.) has about the same size and range. It is piceous-black with an orange-red humeral spot. Like others of its genus it is found in fungi.

In both of these species the eighth antennal segment is much smaller than the other four making up the club.

FAMILY 20. SCYDMAENIDAE
The Ant-like Stone Beetles

Many of these little fellows live with the ants,—but that is not so strange for the ants often maintain quite variable "ranches" with all sorts of insect guests, pets, pests, etc., hanging around. It is a large family with considerably over a thousand known species. They are small, usually brownish or blackish beetles, shining and oval in shape. Some are only a fraction of a millimeter in length.

1a Fourth joint of maxillary palpi awl-shaped; neck narrow, eyes in front of middle of head. Fig. 315. .... *Pycnophus rasus* (Lec.)

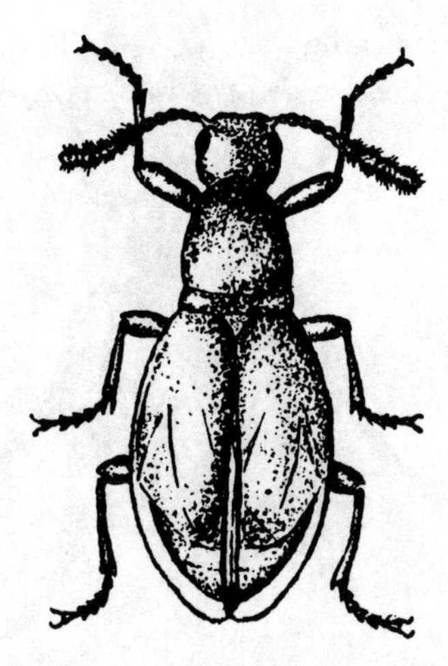

Figure 315

LENGTH: 1.6-1.8 mm. RANGE: East half of United States and Canada.

Shining. Pale reddish-brown. Thorax with well defined small pit at each basal angle. Found in ants' nests and under logs.

Napochus fossiger (Lec.) 1.6 mm, ranging widely, is black with elytra reddish except at tips. It is heavily clothed with long pale hairs.

1b Fourth joint of maxillary palpi obtuse; hind coxae oval; femora abruptly clubbed. Fig. 316. .... *Scydmaenus motschulskyanus* Csiki

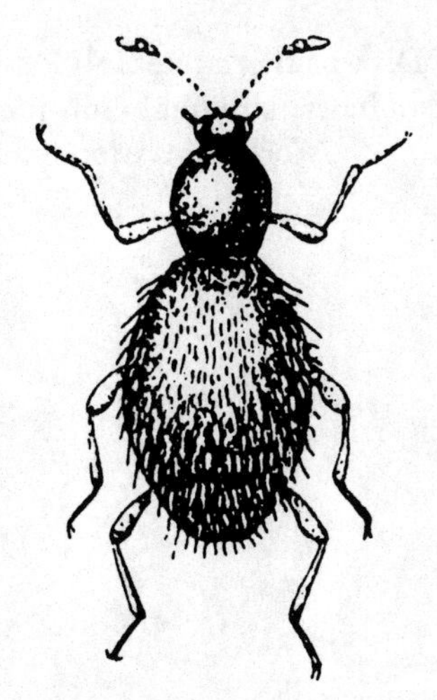

Figure 316

LENGTH: 1.7 mm. RANGE: East half of United States.

Dark chestnut brown with recurved yellowish pubescence. Elytra very convex.

Acholerops zimmermanni (Schaum) is only 1.5 mm long. The first joint of the hind tarsi is longer than in the above. It is dark reddish-brown and is found in the central parts of the United States.

Extremes in sizes are always interesting. *Napochus fulvus* (Lec.) measures only .8-.9 mm. It is shining, pale chestnut-brown with legs and antennae lighter. The antennae are as long as the head and thorax combined and have a 3-segmented club. *Euthiodes latus* Bndl. is still smaller, measuring a scant .7 mm. Its color is brownish-yellow.

FAMILY 21. SILPHIDAE
The Carrion Beetles

A number of fairly large, attractively marked and quite abundant beetles belong to this family making it somewhat a favorite with collector in spite of the disgusting ways of its members.

While some feed on vegetable matter and a few are even predaceous, they, for the most part, live on decaying animals.

The family is widely scattered and numbers over 1500 species.

1a Antennae with 11 segments; antennae slender or gradually clubbed; form rather flattened. Genus *Silpha*. 5

1b Antennae with 10 segments; the last 4 forming an abrupt club; form robust. The Burying Beetles, *Nicrophorus*. 2

2a (1) Tibia straight. 3

2b Hind tibiae (and frequently the middle one) curved. Fig. 317. .... *Nicrophorus americanus* (Oliv.)

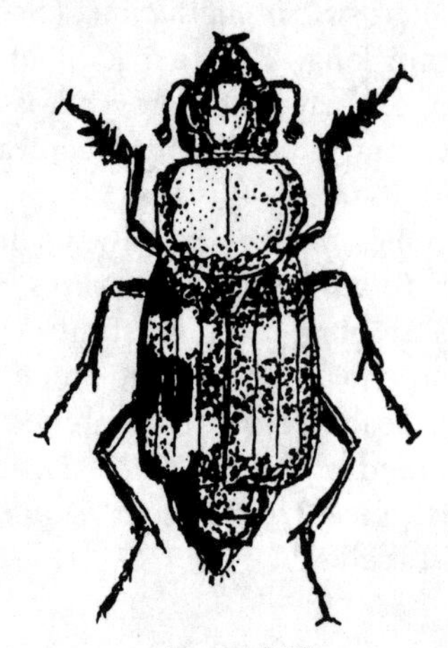

Figure 317

LENGTH: 27-35 mm. RANGE: Central and Eastern United States.

Black shining; vertex, disk of thorax, epipleural fold and 2 spots on each elytron orange-red; antennae black with orange-red club.

N. marginatus Fab., is colored much the same but is smaller (20-27 mm) and only the first segment of the antennal club is red. The thorax is wholly black. Under dead snakes is a good place to look for it. It ranges over the entire United States and much of Canada.

3a (2) Thorax oval, wider than long with wide side and basal margins. 4

3b Thorax a circular disk, about as long as wide. Fig. 318. .... *Nicrophorus orbicollis* Say

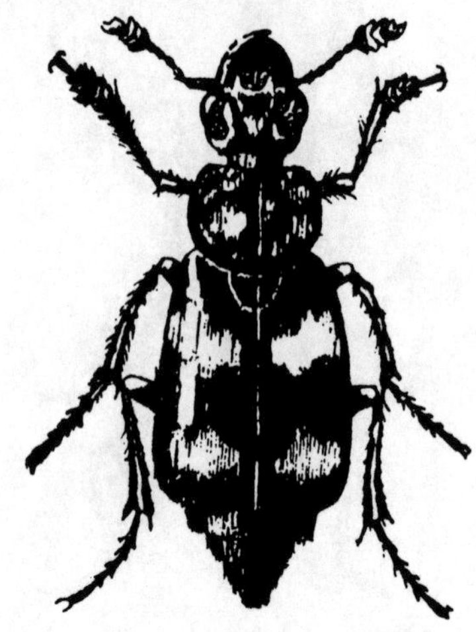

Figure 318

LENGTH: 20-25 mm. RANGE: Central and Eastern North America from Hudson Bay to Florida.

Black shining marked with orange-red on elytra and last three segments of antennal club; epipleural fold black.

The members of this genus are known as Burying or Sexton Beetles because of their habits of digging under small dead animals, reptiles or birds until the cadaver is buried. Before the process is completed eggs are laid on the animal which then furnishes the food for the rapidly developing larvae.

4a (3) Disk of thorax densely covered with golden-hairs; antennal club piceous. Fig. 319. .... *Nicrophorus tomentosus* Web.

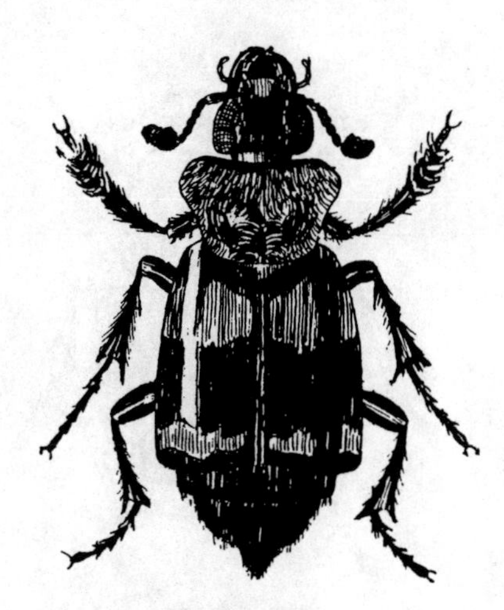

Figure 319

LENGTH: 15-20 mm. RANGE: Central and Eastern United States.

Black, shining, markings as pictured and the epipleural fold orange-red; antennal club piceous.

Nicrophorus guttulus Mots. ranging from Alaska to our southwestern border is known as the Yellow-bellied Burying Beetle because of the mass of golden hairs on the ventral parts of the thorax. It measures 12-20 mm and is shining black.

4b Disk of thorax without hairs; all but first segment of antennal club orange. Fig. 320. ...... *Nicrophorus pustulatus* Hersch.

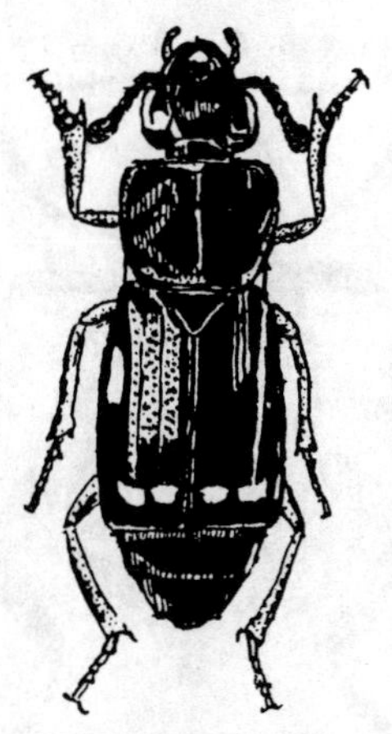

Figure 320

LENGTH: 17-18 mm. RANGE: Much of North America. Black, shining; a small basal spot and double spots at apex of each elytron and last 3 segments of antennae orange-red.

N. investigator (Zett.) which is found throughout the West from Alaska to the Mexican border is recognized by a red epipleural fold.

N. nigritus (Mann.) belongs to the west from Texas to California, north to British Columbia. It is wholly black except the 3 segments of the club.

5a (1) Form oval; eyes not prominent. 6

5b Form elongate; eyes large, prominent. Fig. 321. .... *Necrodes surinamensis* Fab.

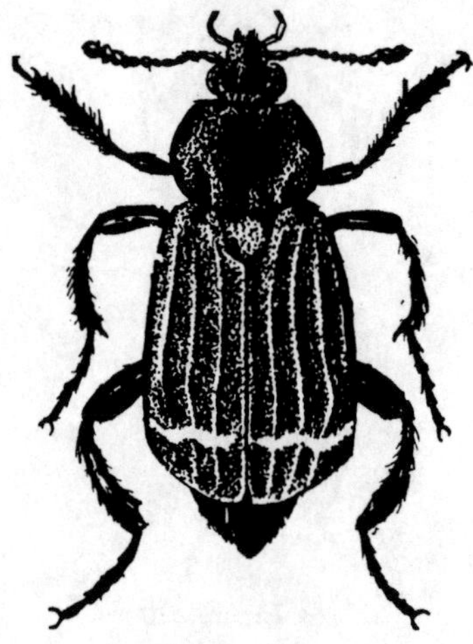

Figure 321

LENGTH: 15-24 mm. RANGE: United States.

Elongate, depressed. Black or piceous; apical cross bar of elytra red, often broken into spots or occasionally wholly wanting.

If the collector wishes to use "wholesale methods" in gathering members of these two genera he has only to put dead fish or other animals in a depression and cover them with a large board so that little if any light gets in. They are nocturnal but apparently pay no attention to the clock as long as it is dark. A day or two later hold your nose, turn the board back and harvest your crop!

6a (5) Thorax wholly black or blackish. .. 8

6b Thorax partly yellow or red. 7

7a (6) Thorax bright yellow with brown or black spot at center; third joint of antennae shorter than second. Fig. 322. .... *Silpha americana* L.

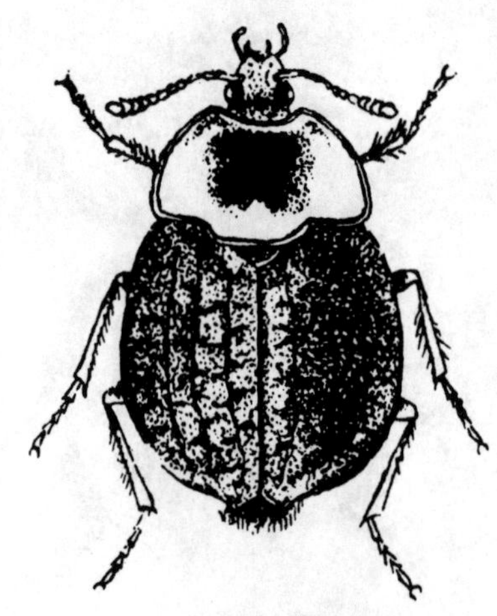

Figure 322

LENGTH: 16-20 mm. RANGE: Central and Eastern United States and Canada.

Elytra brownish, the ridges darker. Often common at carrion and sometimes in garbage cans. When clean this is a very attractive beetle. The larvae of Silphids are flattened and wedge shaped and are often found with the adults.

7b Thorax marked with red or reddish-yellow; third joint of antennae as long or longer than the second. Fig. 323. .... *Silpha noveboracensus* Forst.

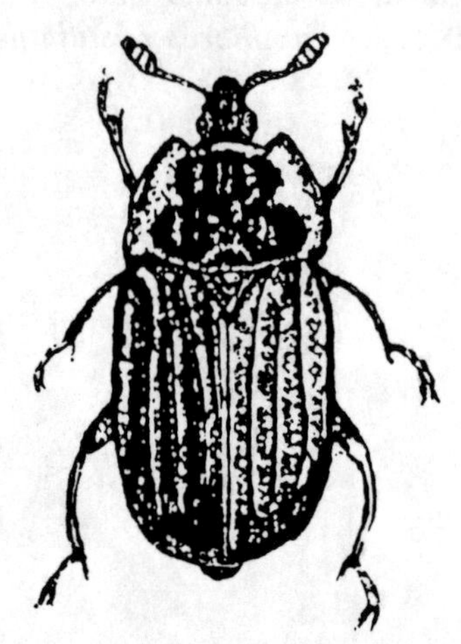

Figure 323

LENGTH: 12-15 mm. RANGE: Much of the United States.

Elytra and thorax brownish or piceous; three ridges on disk of each elytron.

S. lapponica Hbst. measuring 12-13 mm, ranges through our West to the Missouri River or beyond. It is grayish black with head and thorax thickly clothed with yellowish hairs.

8a (6) Each elytron with three distinct ridges; the outer the most prominent. Fig. 324. .... *Silpha inaequalis* Fab.

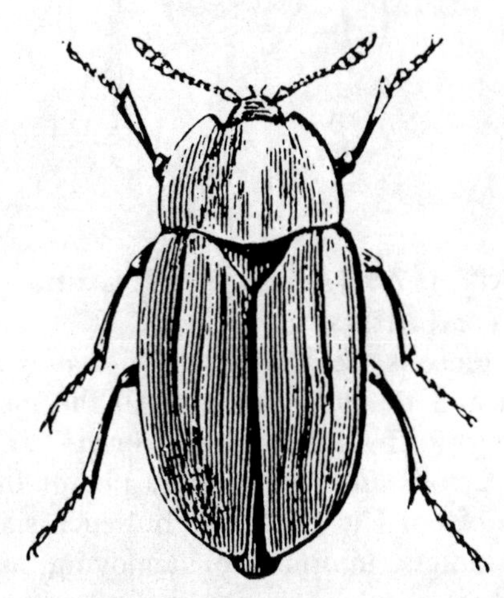

Figure 324

LENGTH: 10-14 mm. RANGE: United States.

Dull black. Thorax with broad depressed truncate lobe at middle. Elytra rounded at apex, longer than wide, the disk with three prominent ridges, as pictured. Both adults and larvae are abundant in carrion and hibernate together. It is often found associated with *S. noveboracensis.*

8b With but one ridge on each elytron. Fig. 325. .... *Silpha bituberosa* Lec.

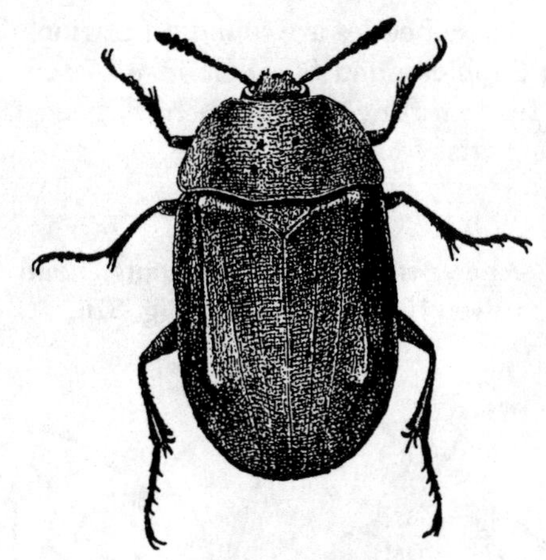

Figure 325

LENGTH: 10-13 mm. RANGE: From Kansas west.

Depressed. Black. This spinach carrion beetle damages spinach, squash, beets, pumpkin, alfalfa and other cultivated plants. (U. S. D. A.)

Two other members of the genus ranging from Alaska through western Canada and the United States are sometimes destructive to garden crops. The Beet Carrion Beetle *S. opaca* L. (12 mm) is shining black; the Garden Carrion Beetle *S. ramosa* Say (12-18 mm), is velvety black.

In our area there are about 20 species of *Nicrophorus* and 9 of *Silpha.*

FAMILY 22. LEPTODIRIDAE
The Small Carrion Beetles

Members of this family are elongate, oval, rather depressed beetles, ranging in size from 2 to 5 mm, varying from testaceous to black in color, covered with moderately dense pile, and possessing 11-segmented antennae. Segment 8 is narrower than both 7 and 9. The pronotum is much broader than the head and often wider than the elytra.

These beetles are found on carrion along with Silphidae and Staphylinidae. A few species feed on fungi and some have been taken in the nests of ants.

1a Abdomen with six segments; head narrowed behind the eyes. Fig. 326.

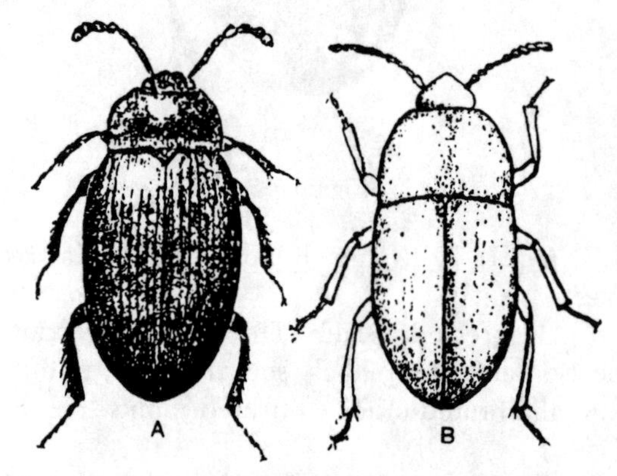

Figure 326

(a) ***Catops basillaris*** **Say**

LENGTH: 3-4 mm. RANGE: Central United States.

Elytra dark brown; head and thorax piceous.

(b) ***Catopocerus cryptophagoides*** **(Mann.)**

LENGTH: 2-3 mm. RANGE: General. Shining. Chestnut-brown. One of three members of its genus, all of which are blind. They live in rotten wood, etc.

Nemadus parasitus (Lec.) measuring 2 mm, oval, and shining dark reddish brown is interesting since it is fairly common in the nests of the large Carpenter Ants, *Camponotus* spp.

1b Abdomen with but 5 segments, (sometimes 4); head not narrowed behind. Fig. 327. .... *Colon magnicolle* Mannh.

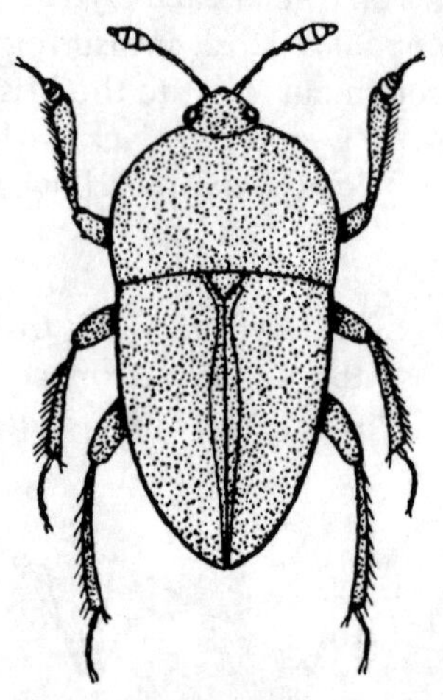

Figure 327

LENGTH: 2.5-3 mm. RANGE: Central United States and Alaska.

Piceous, somewhat dull, heavily punctured and finely pubescent. All the members of the genus have 11 segments in the antennae.

C. pusillum Horn likely ranges through much of our Central States but such small insects unless injurious or annoying are not readily found. It measures about 2 mm.

FAMILY 23. SCAPHIDIIDAE
The Shining Fungus Beetles

These polished oval, boat-shaped beetles may be found in fungi, decayed wood, under bark and in debris. There are about 50 species in our area.

1a Antennae not definitely clubbed, slender scutellum very small or invisible; punctures on elytra scattered or absent. 2

1b Antennae with a broad, somewhat flattened, five-jointed club; scutellum distinct; elytra with punctures in rows. Fig. 328. .. *Scaphidium quadriguttatum* (Say)

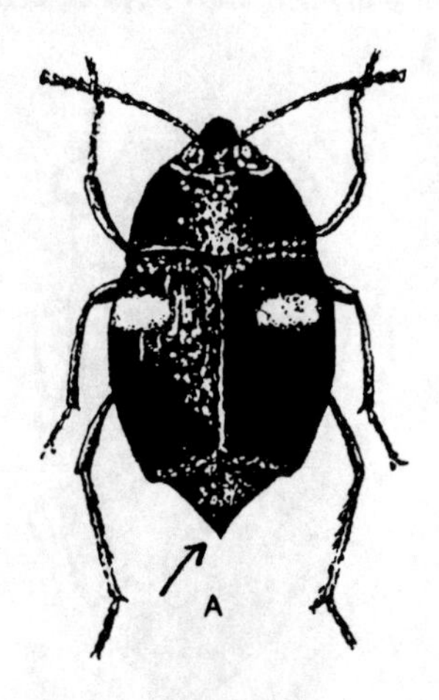

Figure 328

LENGTH: 3.8-4.5 mm. RANGE: Central and Eastern states. Common.

Black, highly shining; elytra with transverse reddish spots; elytra with a twice curving row of coarse punctures along the basal margin.

If the red spots are lacking, this wholly black beetle is variety *piceum* Melsh.

Another variety is *obliteratum* Lec. in which the elytral spots are smaller, and yellow instead of red. The disk of the elytra also lack the coarse punctures.

2a (1) Body oval; sutural striae reaching the base. Fig. 329. .... *Cyparium falsata* Achard

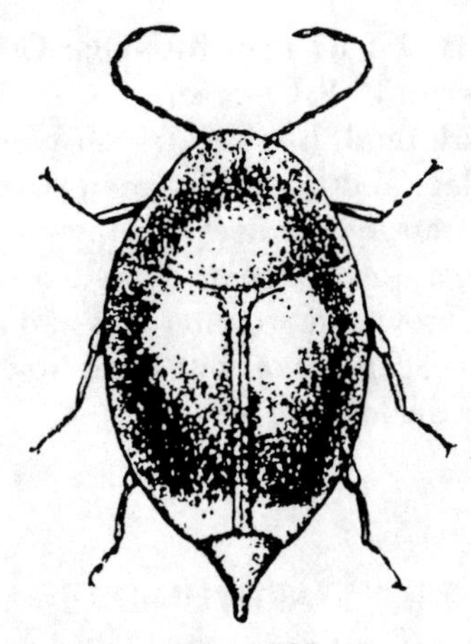

Figure 329

LENGTH: 2.7 mm. RANGE: Central and Eastern States.

Black, highly polished; legs, antennae and tips of elytra and abdomen reddish-piceous. No scutellum; thorax with scattered coarse punctures.

Scaphisoma convexum Say about 2.5 mm long, shining black or chestnut-brown is rather common in its range of the eastern half of the United States. It is found on fungi on logs. The genus is distinguished by a very short third antennal segment.

2b Body narrowed; sutural striae not reaching the base. Fig. 330. .... *Toxidium compressum* Zimm.

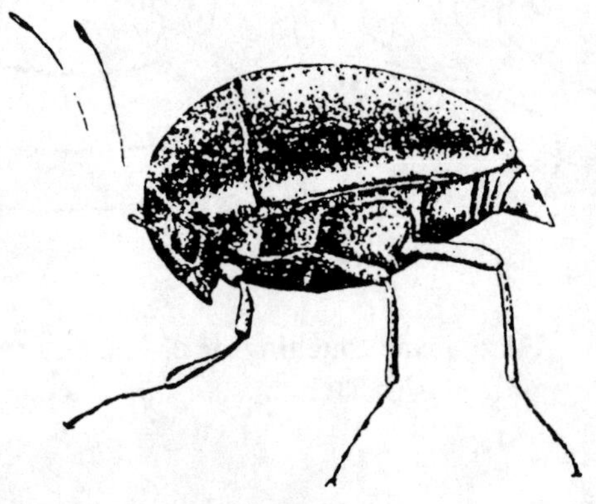

Figure 330

LENGTH: 1.4-1.7 mm. RANGE: Central and Southeastern U. S. Common.

Dark reddish-brown, much polished; antennae, legs and tip of abdomen paler. Metasternum coarsely and deeply punctured.

Cyparium concolor Fab. 3.5 mm long, chestnut brown with lighter legs and antennae, resembles *Scaphidium* but has a row of small spines on the hind tibia.

Family 24. STAPHYLINIDAE
The Rove Beetles

The short wing covers of these usually slender beetles together with their habit of turning up the abdomen as though about to sting, make them readily recognized. More than 20,000 species, small to large, have been named. Some members of this family rival all other beetles in strong and swift flight.

1a Hind coxae widely separated, rounded, small. Fig. 331a. 2

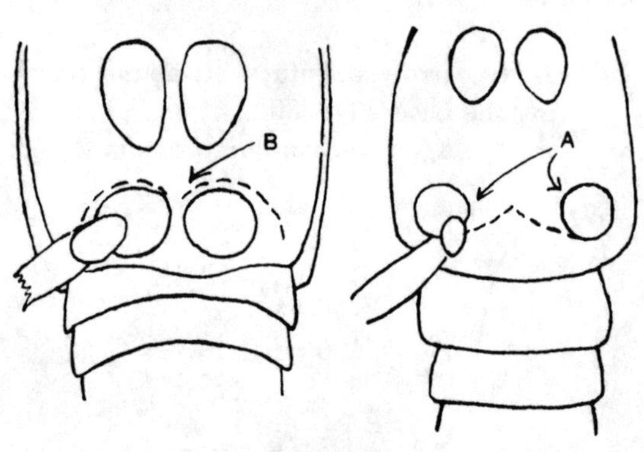

Figure 331

1b Hind coxae touching or at least close together. Fig. 331b. 3

2a (1) Antennae 9-segmented, capitate and inserted at the side of the head. Next to last segment of abdomen with an elevated ridge. Fig. 332. .... *Micropeplus cribratus* Lec.

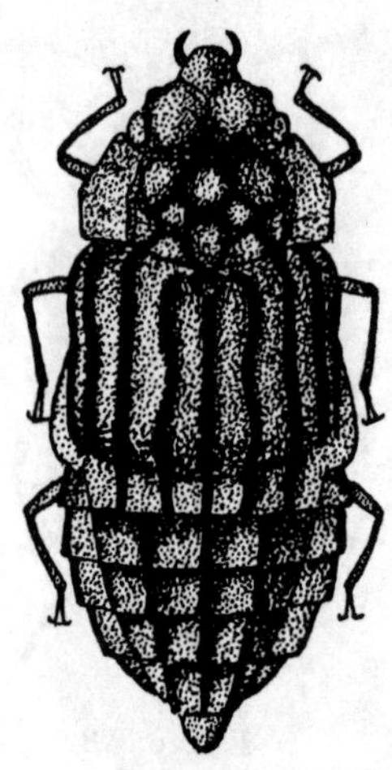

Figure 332

LENGTH: 2 mm. RANGE: Southeast United States.

Blackish; sides of thorax orange-brown; elytra with wavy ridges, and the intervals coarsely punctate. Antennae pubescent. (There are but 3 tarsal segments.)

Kalissus nitidus Lec. Smooth and shining and has no ridges on its abdomen. It is known from Western B. C. and Wash. It, too, has but three segments to the tarsi.

2b Antennae inserted between the eyes, 11 segmented, cheeks very small. Fig. 333. .... *Stenus flavicornis* Er.

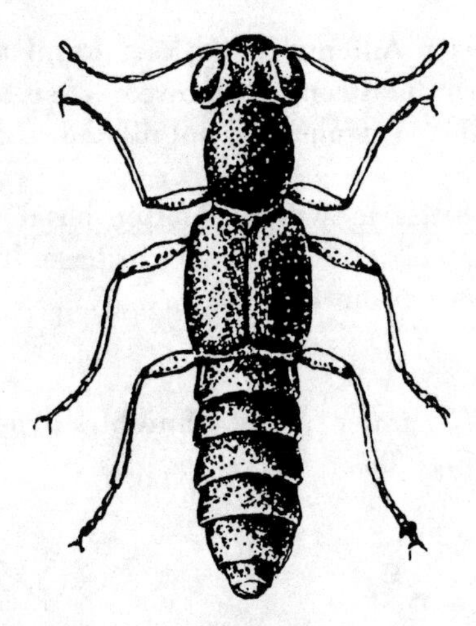

Figure 333

LENGTH: 4.6-4.8 mm. RANGE: Eastern U. S.

Shining black, thinly clothed with gray hairs; elytra wider than abdomen; thorax widest in front of middle.

This large genus numbers more than 150 American species. Several species have a reddish spot on the elytra.

Dianous nitidulus Lec. represents a small genus differing from the above in having large cheeks and a large red-yellow spot on each elytron. It measures 4.5 mm and ranges widely.

3a (1) Hind coaxae conical. 4

3b Hind coxae transverse. 23

3c Hind coxae triangular. 13

4a (3) Front coxae short; tarsi with but 4 segments; abdomen with margin; body punctured. Fig. 334. .... *Euaesthetus americanus* Er.

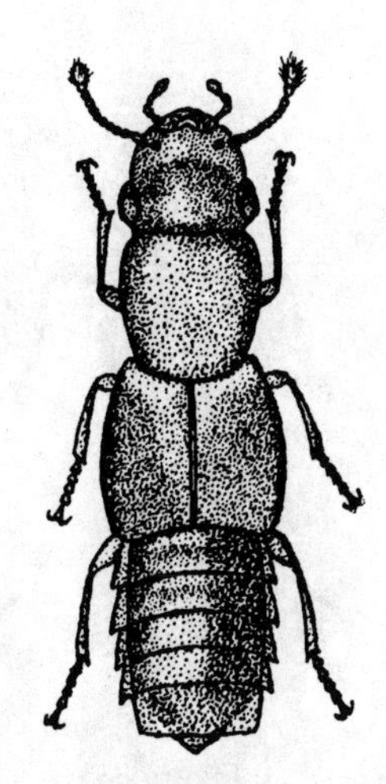

Figure 334

LENGTH: 1.2-1.5 mm. RANGE: Eastern third U. S.

Piceous or dark reddish-brown, antennae and legs paler. Abdomen as wide at base as elytra, punctate.

Stictocranius puncticeps Lec. has no margin on the abdomen. It resembles the ants with which it lives. Eastern U. S.

4b Front coxae long. Tarsi 5 segmented. 5

5a (4) Last segment of maxillary palpi obliquely hatchet-shaped and as long as the preceding one. Abdomen not margined. Fig. 335. .... *Palaminus testaceus* Er.

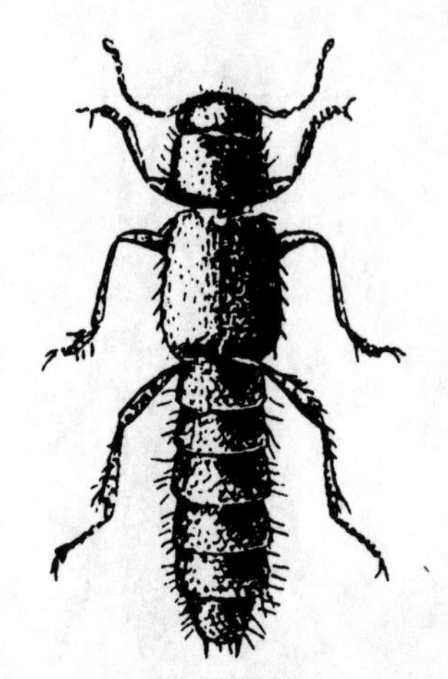

Figure 335

LENGTH: 3.5-4 mm. RANGE: Eastern half U. S.

Shining, pale reddish-yellow; abdomen reddish-brown; antennae and legs pale yellow.

Pinophilus latipes Grav. ranging throughout eastern U. S., has its abdomen definitely margined. It is black with reddish-yellow antennae and legs and measures 13-15 mm.

5b **Last segment of maxillary palpi very small and awl-shaped. 6**

6a **(5) Prosternum shortened between and under the coxae to form an acute point which does not reach the mesosternum. 7**

6b **Prosternum lengthened behind to reach the mesosternum, but more or less acute and not dilated under the coxae; neck very slender. 11**

6c **Prosternum lengthened as in 6b but also greatly dilated laterally under the coxae. 12**

7a **(6) Antenna with long basal segment, rather strongly elbowed; (See e of Fig. 336a); front tarsi not dilated. 8**

7b **Antenna with a shorter basal segment and most flexible near its base; front tarsi of various forms. 9**

8a **(7) Inner face of mandibles two-toothed. Fig. 336.**

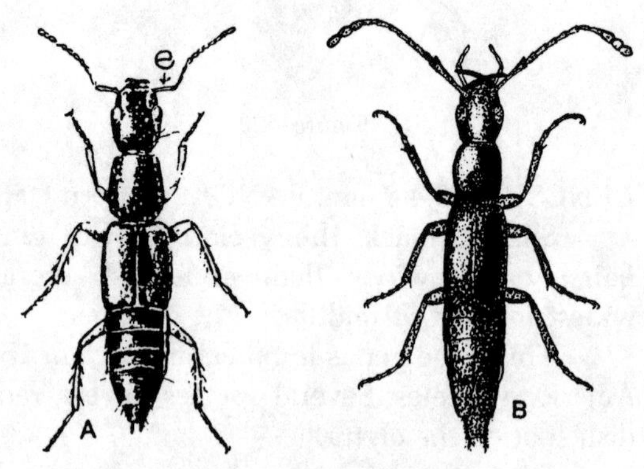

Figure 336

(a) *Homaeotarsus sellatus* (LeC.)

LENGTH: 8-9 mm. RANGE: Central States.

Shining, black; elytra reddish-yellow marked with black as pictured; sides of head behind eyes obliquely sloping without angles.

(b) *H. pallipes* (Grav.)

LENGTH: 8-11 mm. RANGE: Central and Eastern United States.

Shining, black or blackish, including elytra; legs dull yellow; head with distinct hind angles.

8b **Inner face of mandible and 3 teeth, the 2 lower ones on a common base equal in size and larger on the left mandible. Fig. 337. .... *Homaeotarsus bicolor* (Grav.)**

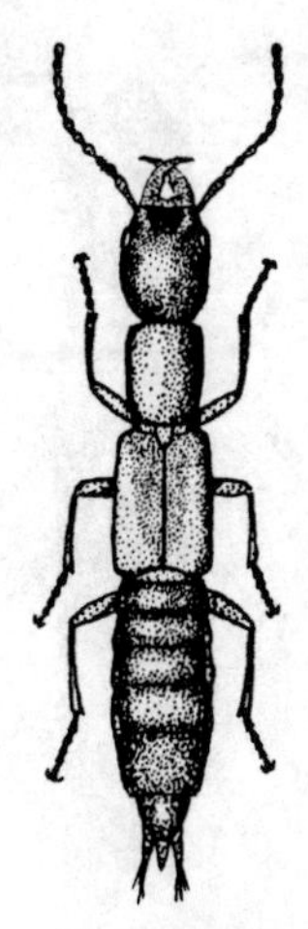

Figure 337

LENGTH: 7-10 mm. RANGE: Central and Eastern States.

Head black; labrum, antennae, thorax, elytra and last two segments of abdomen pale reddish-brown; legs pale yellow.

H. pimerianus (Lec.) reddish-brown to blackish, with head darker than thorax and elytra and abdomen lighter, measures 8-11 mm and ranges from California to Indiana.

H. badius (Grav.) measuring 10-13 mm is uniformly reddish-brown with the head slightly darker. The thorax is much narrower than the head and one-fourth longer than wide. The elytra are about one-half wider than the thorax. It is common throughout much of the eastern one-third of the United States. There are about 40 species of this genus in the U. S. and Can.

9a (7) Labrum dentate at middle. Tip of ligula densely fringed with hairs; surface dull and closely sculptured and haired. Fig. 338. .... *Lithocharis ochracea* (Grav.)

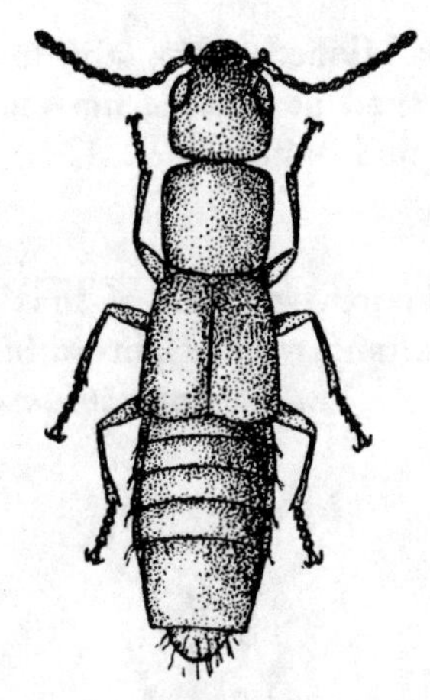

Figure 338

LENGTH: 3-4 mm. RANGE: Europe and North America.

Reddish-brown, antennae and elytra lighter; legs yellowish; antennae monilliform; front tarsi not dilated.

Achenomorphus corticinus (Grav.) dark reddish brown, with antennae and legs paler and head darker, differs from the above in having its front tarsi dilated and padded beneath, and the first segment of the hind tarsi much longer than the second. It measures 6-7 mm and ranges through the eastern half of our country.

The genus *Lobrathium* differs from the above in having a longitudinal fold or raised line on the elytra, and having the first and second segments of the hind tarsi each longer than the third or fourth. *L. longiusculum* (Grav.) has dull red elytra with black head and thorax. It is fairly slender (6.5-7.5 mm), with broad neck and ranges over much of the eastern half of the United States. *L. dimidiatum* with similar range, is very slender, has a slim neck and measures 3.8-4.5 mm. The head and abdomen are black, while the thorax and elytra are dull red. Both of these species are fairly common. Our fauna contains about 70 species of this genus.

9b **Ligula bilobed at tip; labrum not strongly toothed at middle; not closely sculptured and haired. 10**

10a **(9) Fourth segment of maxillary palpi conical and naked; labrum bilobed. Fig. 339. .... *Lathrobium simile* Lec.**

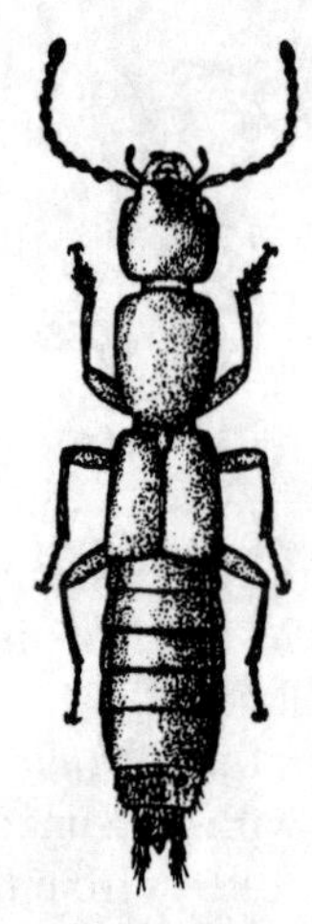

Figure 339

LENGTH: 7.5-9 mm. RANGE: Central and Eastern United States.

Shining, black; elytra and abdomen piceous; antennae and tip of abdomen reddish-brown; legs paler.

Lobrathium collare (Er.) with range similar to above, has head black, antennae and thorax reddish-brown; legs paler and abdomen piceous, with its tip pale. It measures 4.5-5.5 mm.

This is a large group of about 60 very similar species.

10b **Fourth segment of maxillary palpi compressed, linear at apex and thickly clothed with fine pubescence; thorax ovate; elytra subquadrate. Fig. 340. .... *Paederus littorarius* Grav.**

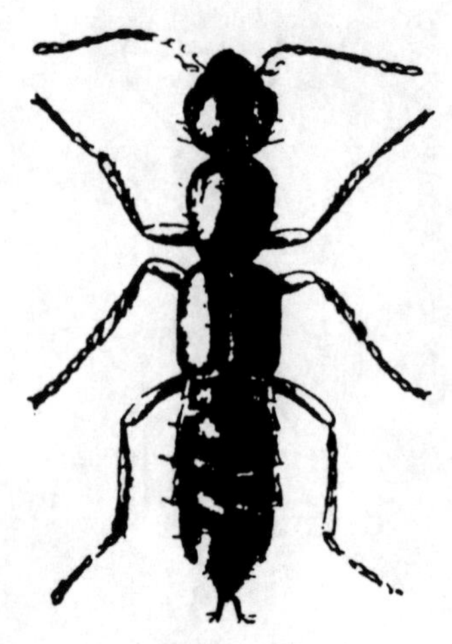

Figure 340

LENGTH: 4.5-5 mm. RANGE: North and Eastern United States.

Shining, reddish-yellow; elytra dark blue; antennae dark at middle, paler at each end.

P. palustris Aust. differs in having elytra that are shorter than wide. In coloring, range and size it is much like *littorarius.* There are about 15 other members of this genus in our area.

11a **(6) Hind tarsi rather short and thick with the basal joint at most but slightly longer than the second. Fig. 341. .... *Scopaeus exiguus* Er.**

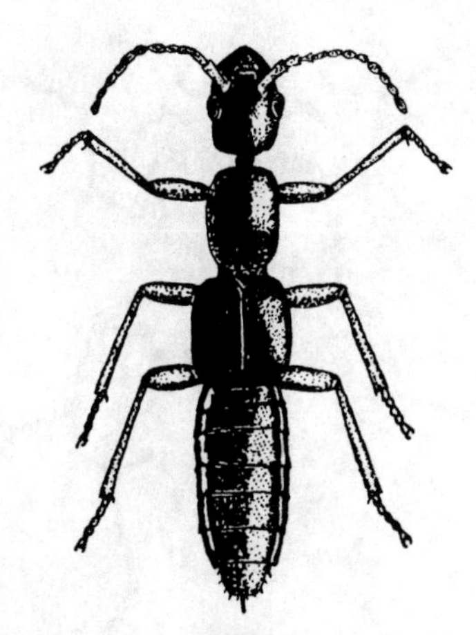

Figure 341

LENGTH: 2-3 mm. RANGE: Eastern third United States.

Head and elytra blackish-piceous; thorax dusky yellow; abdomen blackish, paler at tip, antennae and legs pale yellow.

11b Hind tarsi long slender the basal segment distinctly longer than the second; gular sutures practically united most of their distance. Fig. 342. .... *Scopaeus opacus* (Lec.)

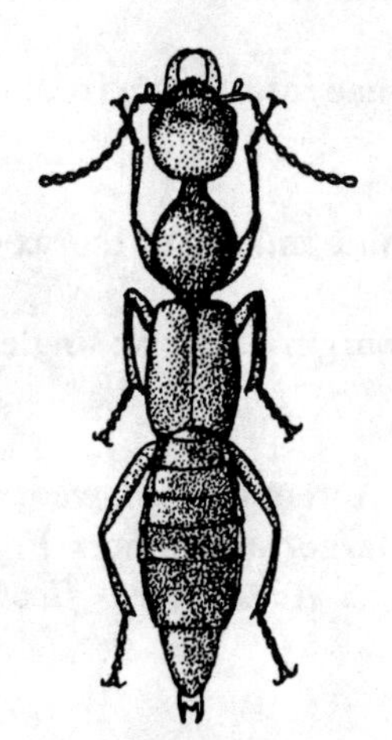

Figure 342

LENGTH: 4-5 mm. RANGE: Eastern United States.

Very slender. Dark reddish-brown, legs and antennae yellowish.

S. *duryi* Csy. piceous black with head, thorax and antennae dusky yellow has been taken in Ohio. It measures 4 mm. There are about 35 other members of this genus in U. S. and Can.

12a (6) Fourth segment of hind tarsi bilobed; last joint of maxillary palpus small and slender. Fig. 343. .... *Astenus longiusculus* (Mannh.)

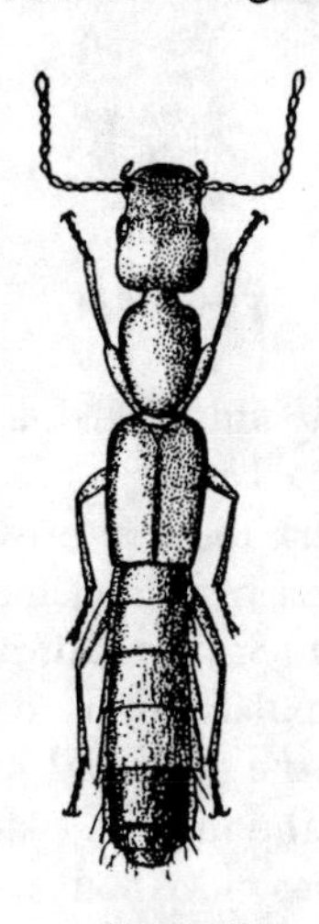

Figure 343

LENGTH: 3.5-4.5 mm. RANGE: Western United States.

Light reddish-brown. Very slender with head, elytra and abdomen all wider than the oval thorax.

A. *binotatus* (Say) measuring 3.5 to 4 mm. and ranging through the eastern half of the United States is reddish-yellow with the head, a large spot on elytra and on the abdomen black; antennae legs and tip of abdomen pale yellow. Our fauna contains 21 other similar members of this genus.

12b **Fourth segment of hind tarsus not lobed beneath; antennae distinctly longer than head; neck very slender. Fig. 344. .... *Rugilus angularis* (Er.)**

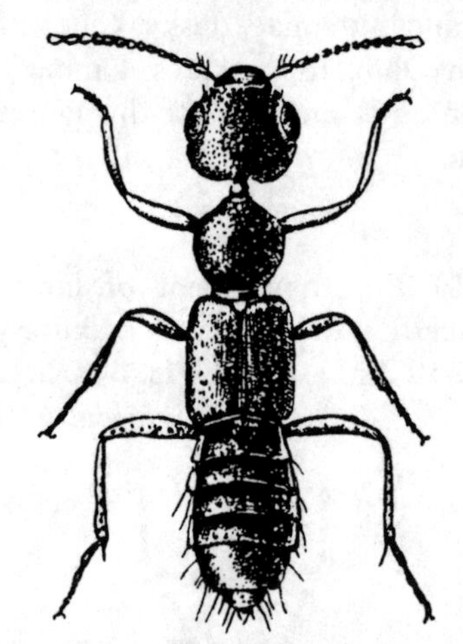

Figure 344

LENGTH: 3.8-4 mm. RANGE: Eastern and Southern States to Texas.

Shining, dark reddish-brown; tips of elytra pale; abdomen pisceous; legs pale yellow. *R. dentatus* (Say) substantially the same size and with similar range, differs from *angularis* in having the head and thorax strigosely punctured. The abdomen is black and the antennae reddish-brown. About a dozen *Rugilus* are known in the U. S. and Can.

13a **(3) Antennae inserted on sides of front. Fig. 345. .... *Habrocerus schwarzi* Horn**

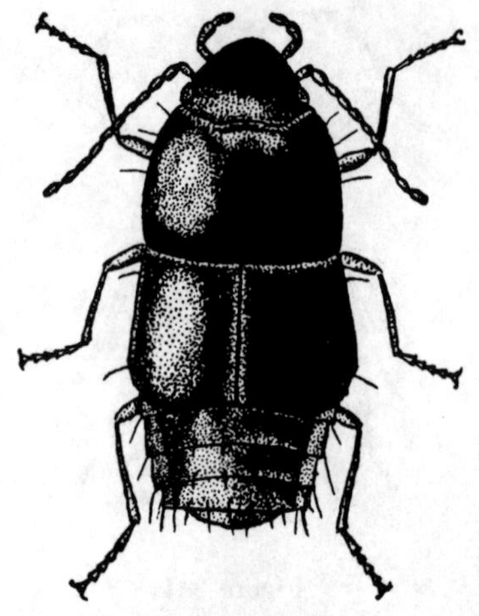

Figure 345

LENGTH: 1.5-2 mm. RANGE: Michigan and Connecticut.

Dark reddish-brown; elytra and tip of abdomen more reddish; legs yellowish-brown. Antennae thickly set with long hairs; a few long heavy bristles arising from anterior end and more from abdomen.

Only two other species of this genus *H. magnus* (Lec.) known along Lake Superior and *H. capillaricornis* Grav. from the Pac. Northwest and Europe.

13b **Antennae inserted on front of front. 14**

14a **(13) Side margin of thorax double. 16**

14b **Side margin of thorax single. 15**

15a **(14) Antennae elbowed; front tarsal claws larger than others. Fig. 346. .... *Acylophorus flavicollis* Sachse**

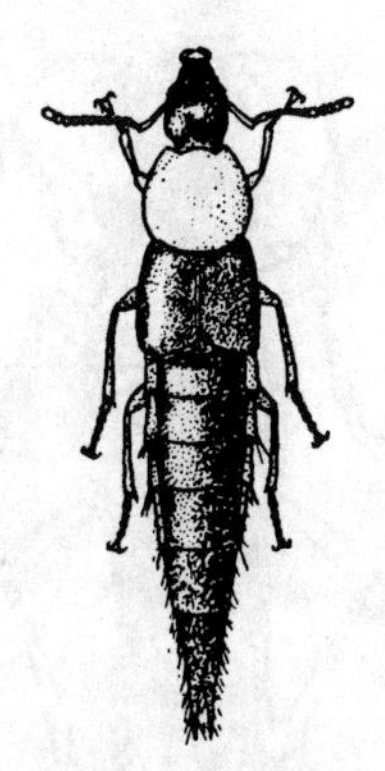

Figure 346

LENGTH: 5-6 mm. RANGE: Pennsylvania to Kansas and Texas.

Shining, black, scant pubescence on elytra and abdomen; thorax and legs reddish-yellow; antennae dusky, paler at the base.

A. pronus Erichs. much the same size and ranging rather commonly throughout most of the United States has antennae piceous and legs dull yellow to piceous, but otherwise closely resembles *flavicollis.* A small genus of 7 N. Amer. species.

15b Antennae not elbowed; palpi filiform; all tarsal claws similar. Fig. 347.

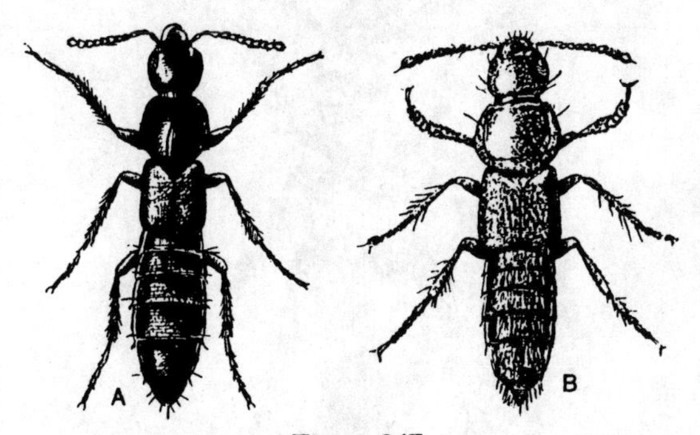

Figure 347

(a) ***Quedius capucinus*** (Grav.)

LENGTH: 6-9 mm. RANGE: Much of North America.

Shining, black or piceous; antennae legs and sometimes the elytra dark reddish-brown; thorax broader than long; abdomen iridescent.

(b) ***Q. spelaeus*** **Horn**

LENGTH: 10-14 mm. RANGE: Colorado to Florida.

Pale reddish or chestnut brown, the elytra sometimes a bit darker; antennae reaching base of thorax; thorax wider than elytra. Found in caves. In the most recent study of Quedius 88 species were found in the U. S. and Can.

16a (14) Antennal bases rather widely separated. 18

16b Antennae with their bases close together; elytra often overlapping. 17

17a (16) Elytral suture beveled and abnormal; fourth segment of maxillary palpi longer than third. Fig. 348. .... *Hyponygrus fracticornis* Mueller

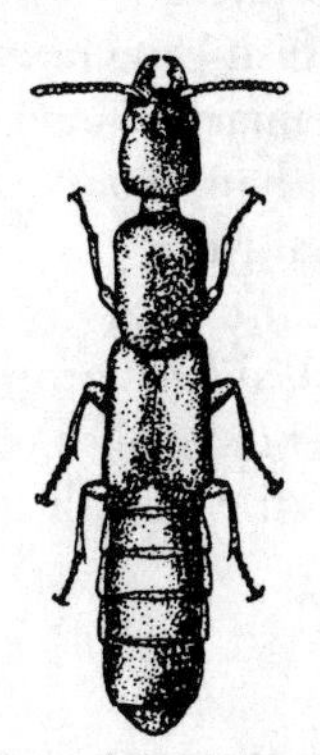

Figure 348

LENGTH: 6.5 mm. RANGE: Eastern half United States.

Black; legs dark reddish-brown. This genus contains 26 species in our area.

Nudobius cephalus (Say) a widely distributed species differs from the above genus in the margins of the pronotum being rapidly deflexed. It is shining black with reddish-yellow elytra and legs. It measures 6.5-7.5 mm.

17b **Elytral suture normal and with a definite stria; last segment of maxillary palpi much longer than the preceding one. Fig. 349. .. *Atrecus macrocephalus* (Nordm.)**

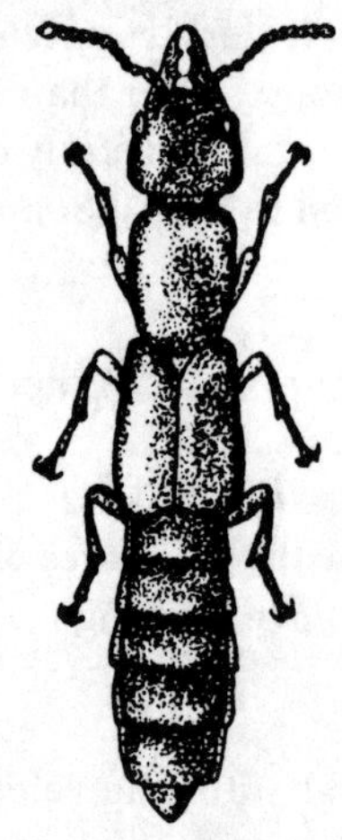

Figure 349

LENGTH: 6-7 mm. RANGE: Alaska, Canada and N. E. United States.

Shining, reddish-brown; elytra more reddish; legs and antennae yellowish-brown. Head and elytra wider than thorax; abdomen with prominent marginal ridges.

Parothius californicus (Mann.), taken from B. C. to Cal. differs from the above in the sutural striae being obsolete and the last two segments of the maxillary palpi being equal.

18a **(16) Fourth segment of maxillary palpi shorter than third. 19**

18b **Fourth segment of maxillary palpi equal to or longer than the third. 20**

19a **(18) Thorax punctured and pubescent. Fig. 350. .... *Ontholestes cingulatus* (Grav.)**

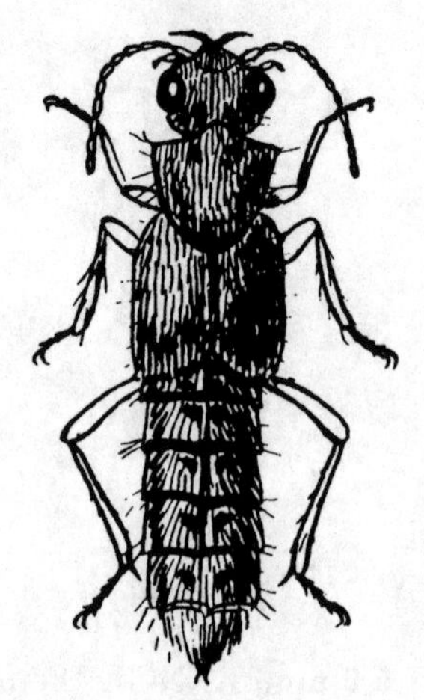

Figure 350

LENGTH: 13-18 mm. RANGE: Widely distributed and common.

Heavy. Dark brown or piceous. A heavy pubescence forms irregular black spots on head thorax and abdomen, and makes gold the metasternum and tip of abdomen.

19b **Thorax smooth, middle coxae remote; suture straight. Fig. 351. .... *Creophilus maxillosus* (L.)**

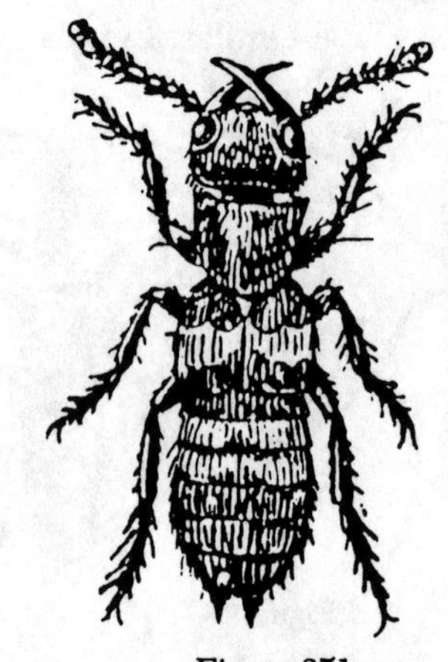

Figure 351

LENGTH: 12-21 mm. RANGE: United States and Canada.

Robust shining, black; second, third and sometimes fourth abdominal segments widely banded with yellowish-gray hairs; elytra marked as pictured with similar hairs.

This very common species may be readily trapped in large numbers with carrion of almost any kind. Davis '15, found both larvae and adults feeding upon the maggots of the flesh flies ever present in carrion.

20a (18) Ligula emarginate (notched); form usually robust; 11 mm or over. Fig. 352a. 21

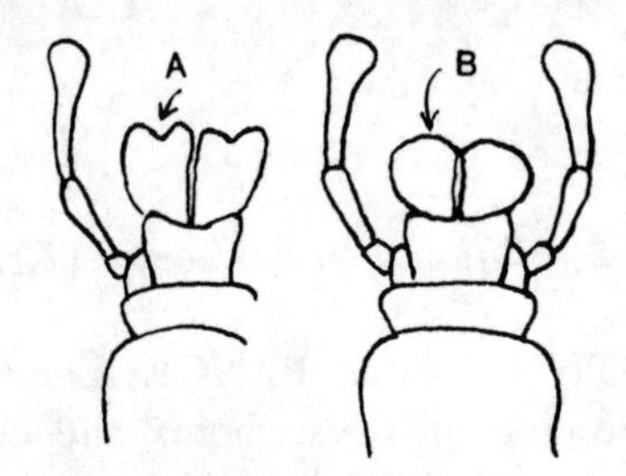

Figure 352

20b Ligula entire; form slender; usually smaller than in 20a. Fig. 352b. 22

21a (20) Abdomen narrowed at tip; middle coxae slightly separated. Fig. 353.

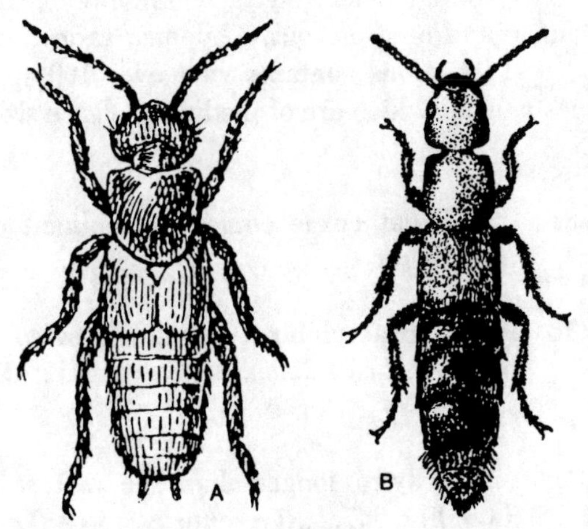

Figure 353

(a) ***Staphylinus maculosus* Grav.**

LENGTH: 18-25 mm. RANGE: Widely distributed and very common.

Dark brown; elytra and upper part of abdomen marked with fuscous spots; base of antennae, pale; femora piceous, the edges paler.

(b) ***S. cinnamopterus* Grav.**

LENGTH: 12-14 mm. RANGE: Much of United States.

Head, thorax, elytra, tibiae, tarsi, apical margins of abdomen and entire last segment brownish-red. Antennae, femora and much of abdomen piceous.

S. violaceus Grav. is a very attractive species. In shape and size is much like *cinnamopterus* but is shining black with metallic violet blue head thorax and elytra.

The genus *Staphylinus* contains 32 large and often colored species in our fauna.

Thinopinus pictus Lec. is flightless, possessing no wings. It is 15-18 mm long, brownish-yellow, ornamented with black on the upper surface. The larvae strongly resemble the adults but differ in having no teeth on the mandibles. This "Pictured Rove Beetle" is found on the sand beaches of the Pacific Coast from British Columbia to southern California.

21b Sides of abdomen parallel; middle coxae touching. Fig. 354. .... *Ocypus ater* (Grav.)

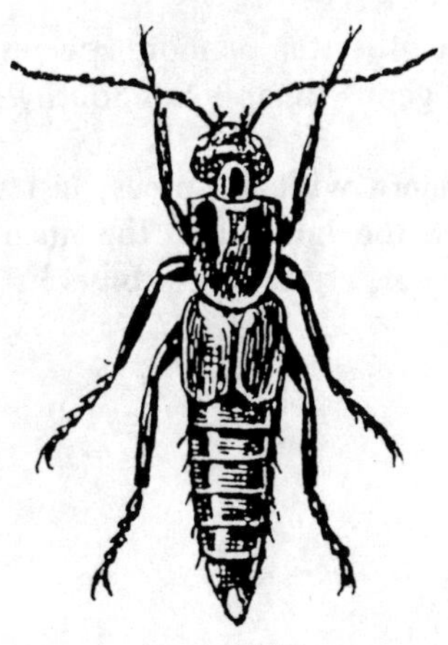

Figure 354

LENGTH: 15-18 mm. RANGE: Europe and North America.

Shining, black; tibia, tarsi and basal half of antennae piceous. Several other species have been introduced on the East and West Coasts.

22a (20) Femora with a row of fine spines beneath. Fig. 355. .... *Belonuchus rufipennis* Fab.

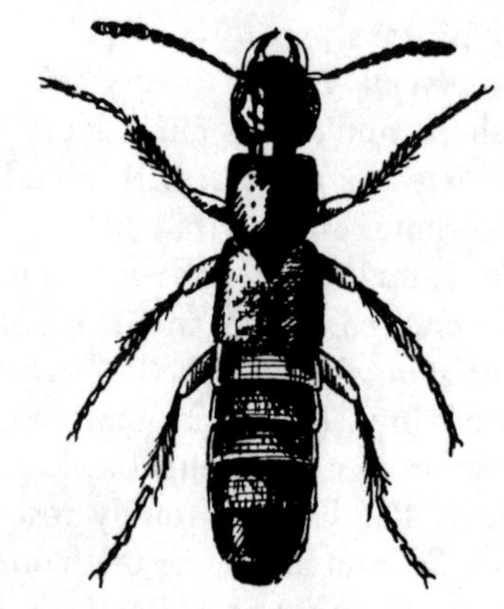

Figure 355

LENGTH: 6.5-7.5 mm. RANGE Atlantic States to Arizona.

Reddish-yellow; head, thorax and last two segments of abdomen shining black. Antennae fuscous; piceous at base and pale at apex.

The other ten or more species of this interesting genus inhabit the Southwest.

22b Femora without spines; last segment of both the labial and the maxillary palpi slender; elytra red or blue. Fig. 356.

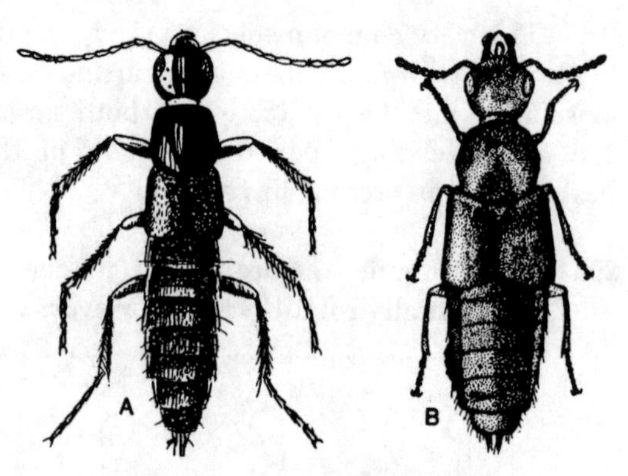

Figure 356

(a) *Philonthus thoracicus* (Grav.)

LENGTH: 7.5-8 mm. RANGE: Central States.

Shining, piceous; thorax and elytra dull red; legs and three basal joints of antennae brownish-yellow. All joints of antennae longer than wide.

(b) *P. cyanipennis* (Fab.)

LENGTH: 12-15 mm. RANGE: Europe and North America.

Shining, black; elytra metallic-blue; antennae and tarsi piceous. Abdomen iridescent.

This genus contains well over 100 species most of which are of medium to large size.

23a (3) Front coxae conical, prominent or transverse. 26

23b Front coxae globose, winged; eyes complete. 24

24a (23) Elytra longer than the metasternum. Fig. 357. .... *Trigonurus crotchi* Lec.

Figure 357

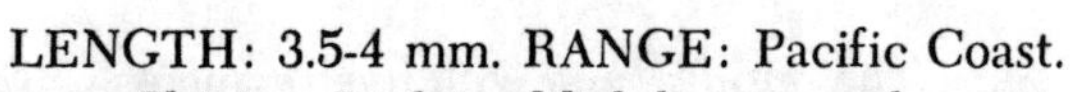

LENGTH: 3.5-4 mm. RANGE: Pacific Coast.

Shining. Light reddish brown, rather uniformly colored throughout. Antennae thickened at apex and pubescent; elytra with longitudinal rows of heavy punctures.

Six other known species of this genus are also western in their range.

Siagonium americanum Melsh. reddish-yellow and 5-7 mm long, ranges widely. Range Eastern half of the U. S. Two other Western species exist.

24b Not as in 24a. 25

25a (24) Abdomen with a fine margin; front tibia with spines or teeth. Fig. 358. .... *Eleusis pallida* Lec.

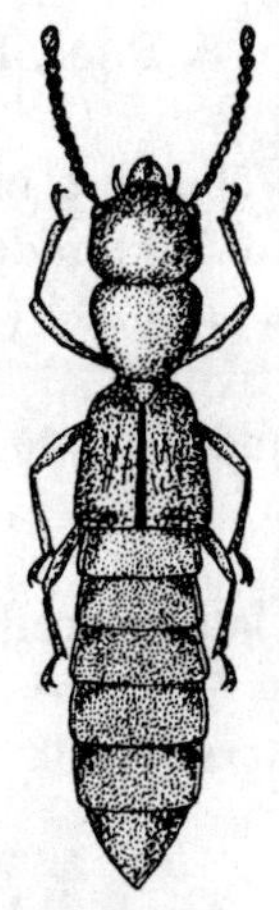

Figure 358

LENGTH: 3.5 mm. RANGE: New York to Colorado.

Shining, reddish-brown. It lives in the nests of ants.

Charhyphus picipennis (Lec.) ranging from Quebec to Kansas, differs in having the abdomen widely margined. It is blackish with paler antennae and legs. It measures 3 mm. This flat beetle is found under tight-fitting bark of fallen trees.

25b Abdomen without a margin; tarsi with but 3 segments. Fig. 359. .... *Thoracophorus costalis* Er.

Figure 359

LENGTH: 2-2.5 mm. RANGE: Much of North America.

Dull, dark chestnut brown; head margined. Each elytron with four ridges.

Clavilispinus exiguus (Er.), a cosmpolitan species, has five tarsal segments. It measures 2-2.5 mm. and is shining chestnut brown.

26a (23) **With lateral ocelli. 47**

26b Without lateral ocelli. 27

27a (26) **With seven visible ventral abdominal segments. 43**

27b With but six ventral segments on abdomen. 28

28a (27) **Antennae arising between the eyes. 29**

28b Antennae arising at sides of head. 38

29a (28) **All tarsi of four segments; antennae 10-segmented. Fig. 360. .... *Oligota exiguua* Say**

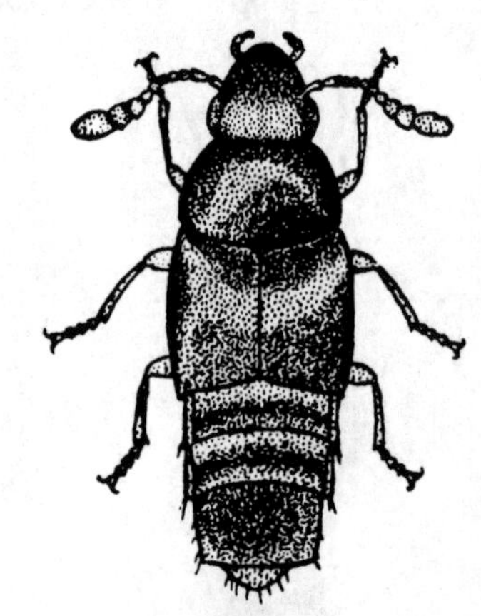

Figure 360

LENGTH: .8-1 mm. RANGE: Found in Indiana.

Shining. Piceous; tip of abdomen lighter. Antennae clubbed; thorax much widened behind; elytra widening still more from front to back.

This genus contains about 10 species most of which are western in their range. A closely similar genus, *Holobus* has six known species.

29b At least the hind tarsi with five segments. 30

30a (29) **Only the hind tarsi with five segments, the others with four. 31**

30b The middle tarsi with five segments. .. 32

31a (30) **Front of head prolonged forming a beak; labial palpi of but 3 segments. Fig. 361. .... *Myllaena vulpina* Bnhr.**

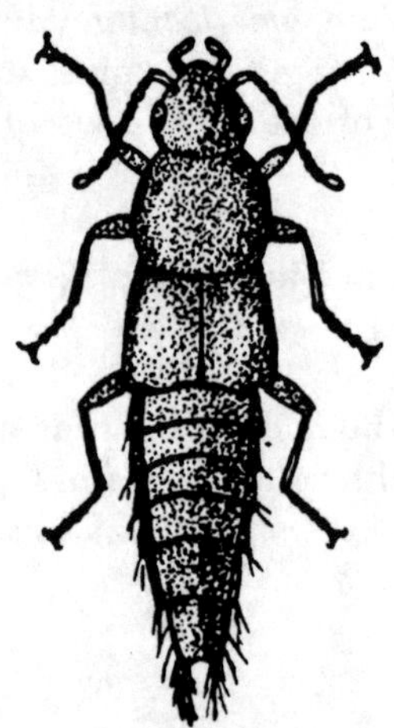

Figure 361

RANGE: Indiana to West Virginia. It is piceous and measures 2.2 mm.

31b Head not forming a beak: labial palpi of but 2 segments; middle coxa widely separated. Fig. 362. .... *Phanerota fasciata* (Say)

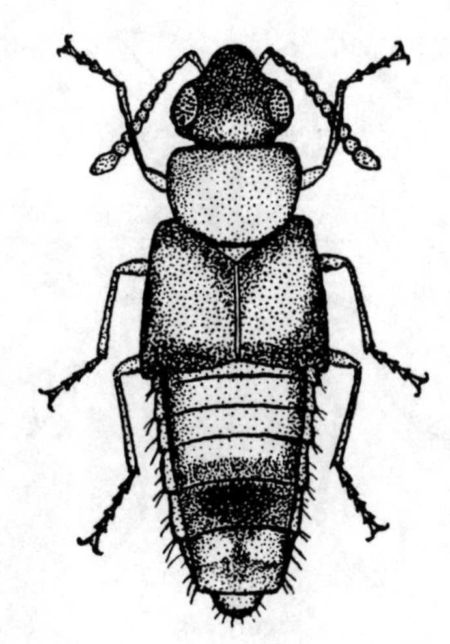

Figure 362

LENGTH: 1.5-2.5 mm. RANGE: Eastern United States.

Pale reddish-yellow; head, apical third of elytra and fourth and fifth abdominal segments black or piceous.

P. dissimilis (Er.) 1.5-2.3 mm in length is pale dull yellow with head, apical angles of elytra and abdomen blackish.

It is found in the Central and Eastern States.

Eumicrota corruscula (Er.) is a rather common little species (1-1.5 mm) found on leathery fungi from Texas to New York. It differs from the above in the antennae having a long cylindrical club of 7 loosely-jointed segments. It is shining black with pale yellow antennae and legs. The elytra are a bit wider and longer than the thorax and are sometimes brownish.

32a (30) Front tarsi with but four segments; labial palpi with 4 segments; maxillary palpi with 5 segments. 33

32b Front tarsi with 5 segments; antennae with eleven segments. 36

33a (32) Sides of anterior, dorsal, abdominal segments with dense tufts of yellow hair. (a) Fig. 363. ...... *Xenodusa cava* (Lec.)

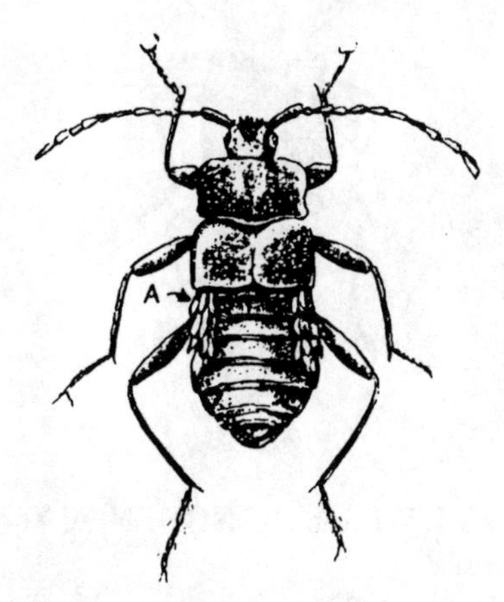

Figure 363

LENGTH: 5-6 mm. RANGE: Central and Eastern States.

Shining. Reddish-brown with scattered pubescence; antennae reaching middle of abdomen, the third segment twice as long as the second.

This and the few other species of the genus, all of which are western are found in ants nests sharing the food brought in by their hosts and in turn apparently furnishing some secretion the ants crave.

33b Not as in 33a. 34

34a (33) Maxillary cavity separated from the eyes by large cheeks and reaching backward beyond the eye. Fig. 364. .... *Myrmoecia lauta* (Csy.)

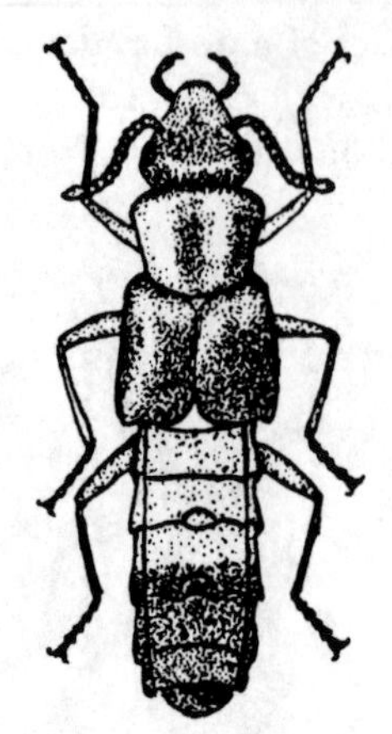
Figure 364

LENGTH: 2.5-3 mm. RANGE: New York and Massachusetts.

Shining. Yellowish-orange; head, short; elytra and four apical segments of upper abdomen purplish-black. Legs and antennae dusky; lateral margins of abdomen dark.

Zyras caliginosa (Csy.) found from Iowa to New York and measuring 3 mm is black with fuscous elytra, and reddish-yellow antennae and legs. The eyes are large.

Z. planifer (Csy.), measuring 3-3.5 mm and reported from North Carolina and Indiana, is dark reddish-brown, shining, with black abdomen, and legs and base of antennae pale reddish-brown. The thorax is one-half wider than long, but not as wide as the elytra.

34b Not as in 34a. 35

35a (34) Head constricted into a narrow neck (a) about a fourth as wide as head. Fig. 365. ...... *Aleodorus bilobatus* (Say)

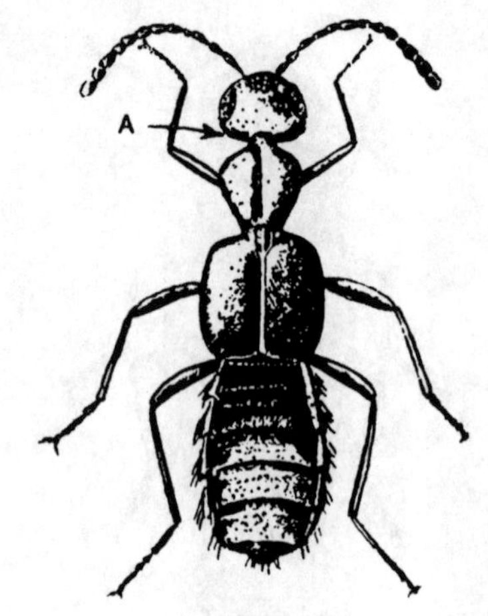

Figure 365

LENGTH: 3-3.3 mm. RANGE: Eastern half United States.

Shining. Piceous or dark brown; legs and basal part of antennae brownish-yellow. Scutellum not carinate. Thorax heart-shaped, a bit wider than the head.

Falagria dissecta (Er.) shining, black or piceous and measuring 2 to 2.5 mm. Has sharp ridges (carina) on its scutellum. It is scattered widely through the U. S. and Canada.

One could spend years in a systematic study of the Staphylinidae alone without becoming too well acquainted with them. It's quite out of the range of a book of limited scope to attempt to treat even all the important genera of so large a family. One other in this tribe will be mentioned.

Meronera venustula Er. measuring 1.6-1.8 mm is a common species. The head and last three segments of the abdomen are piceous, while the base of the abdomen and the antennae are dull yellow; the other parts are brown. It apparently ranges throughout much of the eastern half of our country.

35b Head but slightly if at all constricted behind; maxillary palpi with four-labial palpi with but three segments. Fig. 366.

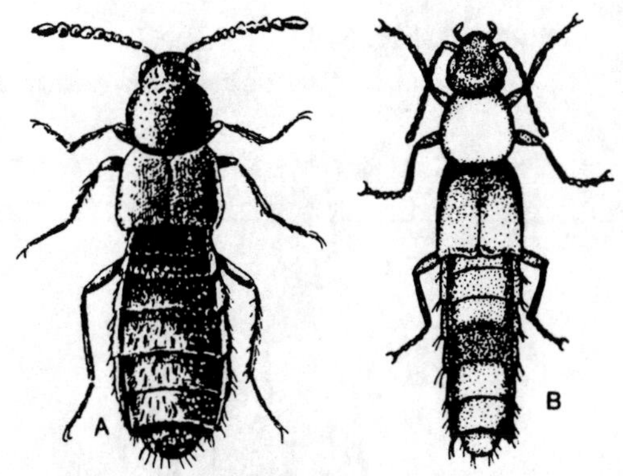

Figure 366

(a) ***Atheta species***

LENGTH: 3-3.2 mm.

This and related genera contain a large number of species that are very difficult to separate.

(b) ***Pancota redundans*** (Csy.)

LENGTH: 1.6-1.8 mm. RANGE: southeastern N. Y.

Head and abdomen black, the latter pale at base and tip; pronotum rufopiceous; elytra testaceous; legs and antennae pale.

36a (32) Maxillary palpi with 5 segments. Fig. 367.

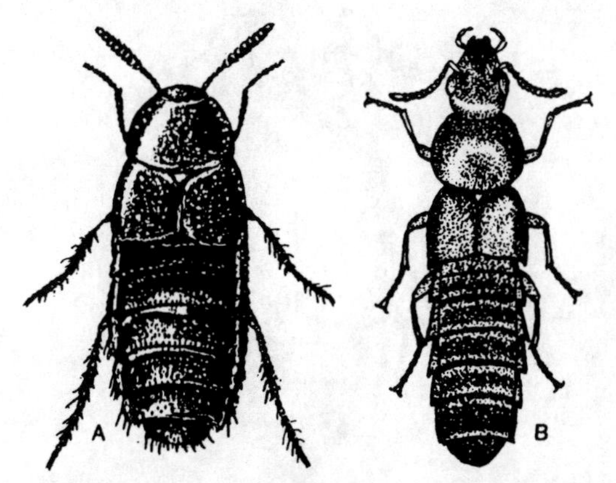

Figure 367

(a) ***Aleochara lata*** **Grav.**

LENGTH: 5-7.5 mm. RANGE: Cosmopolitan.

Shining. Black; clothed with scattered gray pubescence, tarsi reddish-brown. Common in carrion.

(b) ***Baryodma bimaculata*** **(Grav.)**

LENGTH: 4-6.5 mm. RANGE: Much of United States and Canada.

Shining. Black; elytra paler near apex; tibiae, tarsi and tips of posterior abdominal segments dull brownish-red.

36b Maxillary palpi with 4 segments. 37

37a (36) Head prominent, narrowed at base; 1st segment of hind tarsi shorter than following two taken together. Fig. 368. .... *Phloeopora sublaevis* Csy.

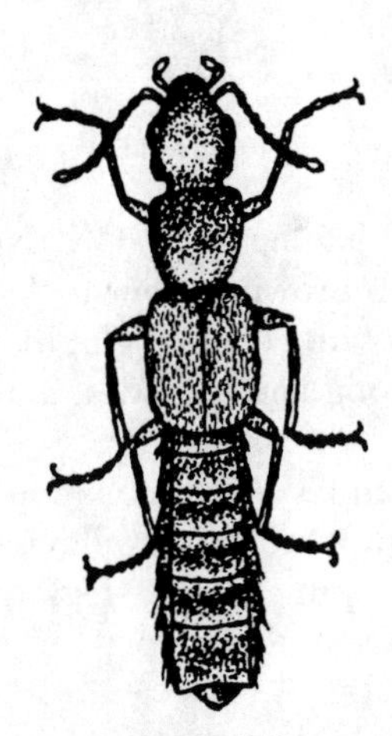

Figure 368

LENGTH: 1.8-2 mm. RANGE: Eastern half United States.

Dark brown; head and abdomen piceous; antenna fuscous, the basal part and the legs yellow. Thorax as long as wide; elytra longer and wider than the thorax.

Amarochara fenyesi Blatch. a brilliantly shining chestnut brown little species (1.8-2 mm). The legs are dull yellow. Its first segment of hind tarsi is longer than the following two taken together.

37b Head retracted, not narrowed at base; ligula divided; labial palpi gradually narrowing from base to apex. Fig. 369. .... *Oxypoda convergens* Csy.

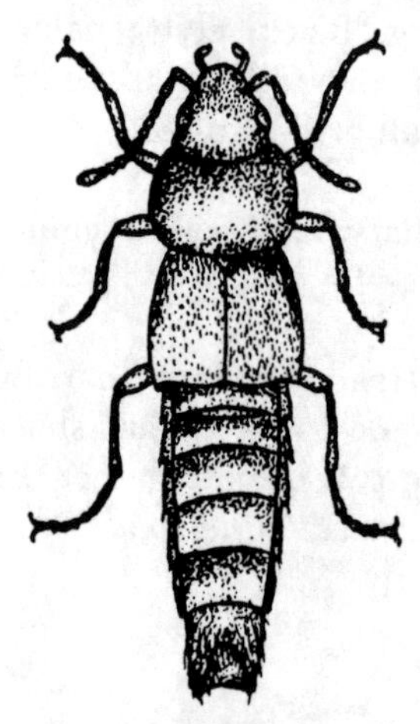

Figure 369

LENGTH: 3-3.5 mm. RANGE: New York.

Reddish-brown; antennae, legs and elytra yellowish-brown. Thorax globular; abdomen with rather long hairs at sides; antennae bead-like.

O. amica Csy. This common little fellow (2-2.2 mm) dull brownish-yellow, clothed with long shaggy pubescence; legs pale, ranges through the Central States.

38a (28) Body slender; head with jaws directed forward. 39

38b Body broader; head inclined. 40

39a (38) Front coxae conical, prominent; maxillary palpi awl-shaped; mandibles toothed. Fig. 370. .... *Pseudopsis obliterata* Lec.

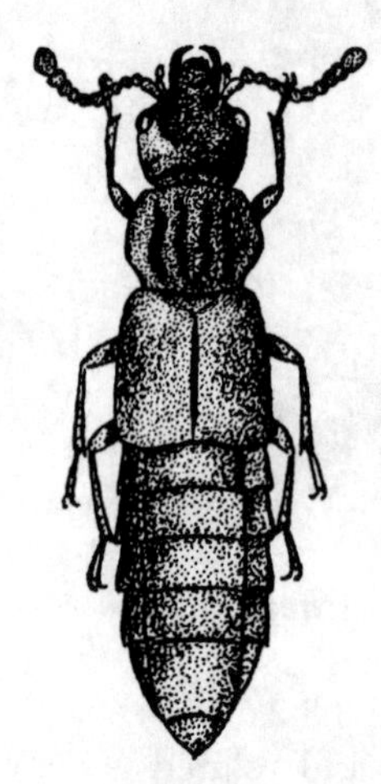

Figure 370

LENGTH: 2 mm. RANGE: Oregon, Idaho, Colorado to California.

Dull, brown; head and elytra darker; legs and antennae lighter; elytra with broad rounded costa; abdomen margined.

P. sulcata Newn. reddish-brown and measuring 2.5-3.5 mm., ranges widely in both Europe and America.

39b Front coxae transverse; antennae inserted on sides of front. Fig. 371. .... *Megarthrus americanus* Sachse

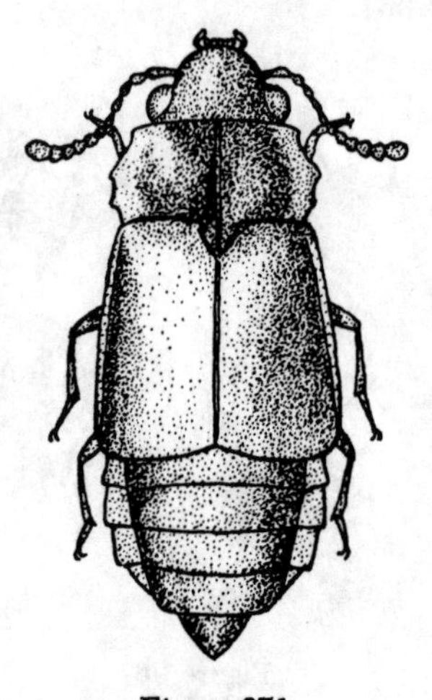

Figure 371

LENGTH: 2.3-2.7 mm. RANGE: General, throughout North America.

Dusky yellow; head black; legs dull yellow. Abdomen sparsely and rather coarsely punctate.

M. excisus Lec., black with sides slightly angulated, and measuring 2.5 mm is known from our central area.

Proteinus automarius Er. scattered throughout Europe and our country, is blackish with dull yellow antennae and measures a strong 1 mm. Structurally, it has the last three antennal segments enlarged whereas *Megarthrus* has only the last segment larger.

At least three other European species belonging to this group have infiltrated into our country; *M. sinuatocollis* Boisd. *P. limbatus* Makl. and *P. brachypterus* (Fab.). All three species are found in the Pacific Northwest.

40a (38) Head margined; antennae with 11 segments. Fig. 372. .... *Lordithon cinctus* (Grav.)

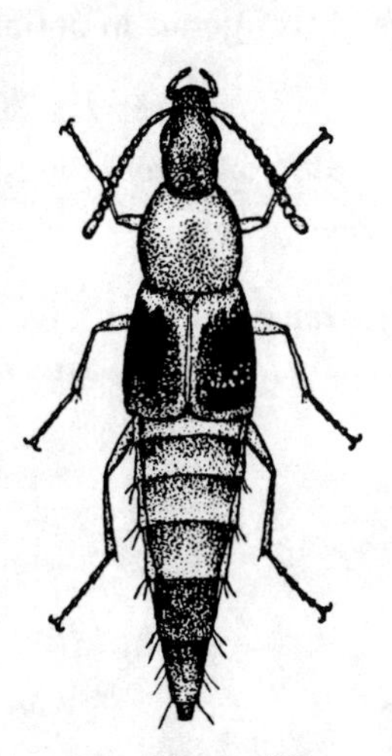

Figure 372

LENGTH: 4.5-7 mm. RANGE: Europe and North America.

Shining. Reddish-yellow; head, under surface, a large spot near outer apical angle of elytra and last two segments of abdomen shining black; antennae piceous, with apical and basal parts dull yellow. About 30 species in this genus in our area.

Mycetoporus splendidus (Grav.) measuring 3-3.5 mm and ranging through both Europe and America has thorax and elytra reddish-yellow, abdomen reddish-brown and head piceous. The two genera are separated by the maxillary palpi which is awl-shaped in this genus and filiform in *Lordithon.*

Still another common member of the genus is *M. americanus* Er. It is about the same size as *splendidus* but differs in having two discal punctures on the thorax just behind the middle. It is reddish-brown to piceous. The elytra have equal length and breadth. It is known to occur from Iowa eastward.

40b Head not margined. 41

41a (40) Mesosternum ridged (carinate) maxillary palpi filiform. 42

41b Mesosternum not carinate. Fig. 373. .... *Coproporus ventriculus* (Say)

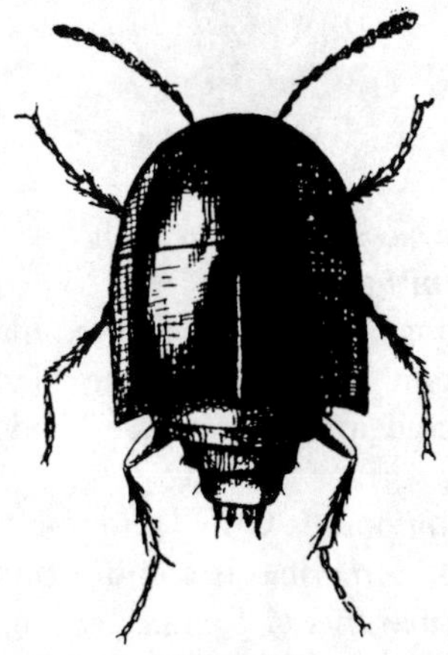

Figure 373

LENGTH: 2-2.5 mm. RANGE: Europe and United States.

Highly shining. Black, elytra and abdomen with a reddish-piceous tinge; antennae and legs dark reddish-brown. Upper surface finely punctate. (*C. laevis* (Lec). is substantially the same except the upper surface is smooth.)

Sepedophilus is an important genus. The members differ from *Coproporus* in not having the mesosternum carinate. *S. crassus* (Grav.), piceous with pale brown pubescence and a narrow reddish part at base of thorax and elytra, 3-5 mm is widely distributed.

42a (41) Maxillary palpi filiform. Fig. 374. .... *Tachinus fimbriatus* Grav.

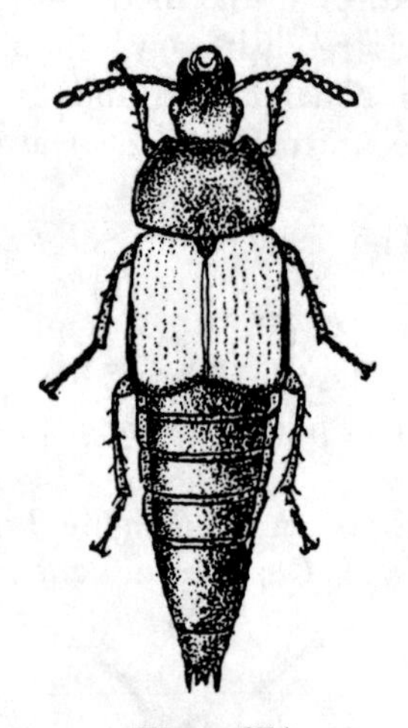

Figure 374

LENGTH: 7-9 mm. RANGE: Canada and Eastern United States.

Shining. Head and thorax black; elytra pale chestnut brown; antennae black, 4 basal segments and apical one pale; abdomen, legs and under surface piceous.

T. fumipennis Grav. common throughout Europe and America, is shining piceous black with legs, margins of thorax, base and sides of elytra and 4 basal joints of antennae, reddish yellow. It measures 5-6 mm.

42b Maxillary palpi awl-shaped. Fig. 375. .... *Tachyporus chrysomelinus* (L.)

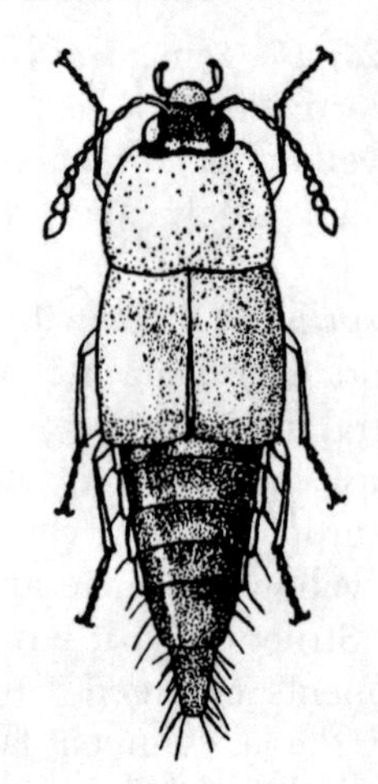

Figure 375

LENGTH: 3-4 mm. RANGE: Europe, Siberia and much of United States.

Shining. Piceous; head black; thorax, elytra, and legs reddish-yellow; antennae dull yellow.

T. nitidulus Fab. 2.5-3 mm, reddish-yellow with head piceous and antennae and legs dull yellow, is a cosmopolitan species. Fourteen members of this genus in our area.

43a (26) Middle coxae touching each other. 44

43b Middle coxae separated. Fig. 376. .... *Oxyporus major* Grav.

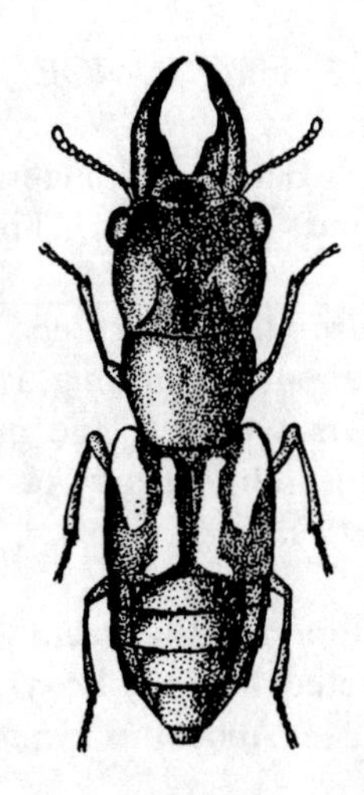

Figure 376

LENGTH: 10-14 mm. RANGE: Central and Eastern United States.

Shining. Black; elytra marked as pictured with golden yellow; tarsi pale yellowish; tibia with golden brown hairs; legs otherwise black.

O. vittatus Grav. has reddish-yellow legs instead of black. It is shining black or piceous with clay-yellow elytra with a black stripe on sides and at the suture. It measures 6 mm. Eleven other members of these attractive beetles found in the U. S. and Can.

44a (43) Antennae with but 10 segments; eyes very large. Fig. 377. .... *Megalopinus caelatus* (Grav.)

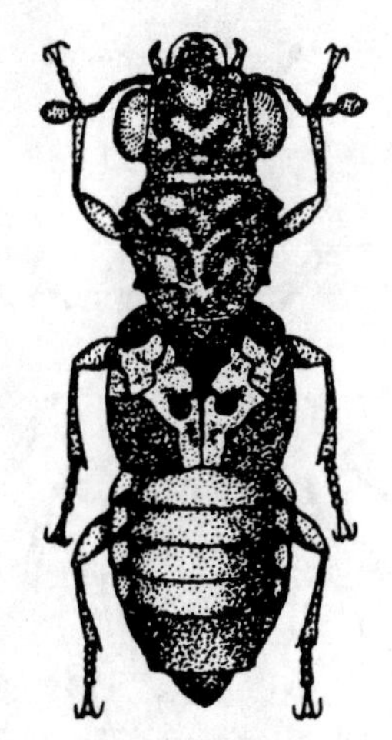

Figure 377

LENGTH: 4-4.5 mm. RANGE: Indiana to Florida.

Shining. Black; elytra with an oblique reddish stripe from humerus to suture at apex; legs and antenna reddish-brown; antennae shorter than head, which is wider than the thorax.

The genus *Osorius* differs in that the abdomen is not margined and the front tibia is spined. The shining blackish species *latipes* (Grav.), 4-5.5 mm long, is rather common.

44b Antennae with 11 segments. 45

45a (44) Front tibiae with two rows of spines; antennae elbowed. Fig. 378. .... *Bledius borealis* Blatch.

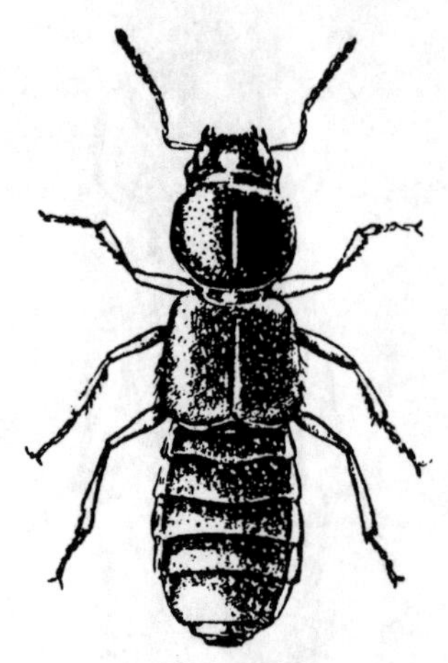

Figure 378

LENGTH: 6-6.5 mm. RANGE: Indiana.

Shining. Black; legs and elytra chestnut brown, antennae dusky towards apex.

B. emarginatus Say, common along streams in the eastern half of the U. S., is black, very slender and measures 2 mm or a little more. Around 100 members of this genus are found in our area. They are difficult to separate.

45b Tibia pubescent. 46

46a (45) Scutellum invisible. Fig. 379. .... *Carpelimus quadripunctatus* (Say)

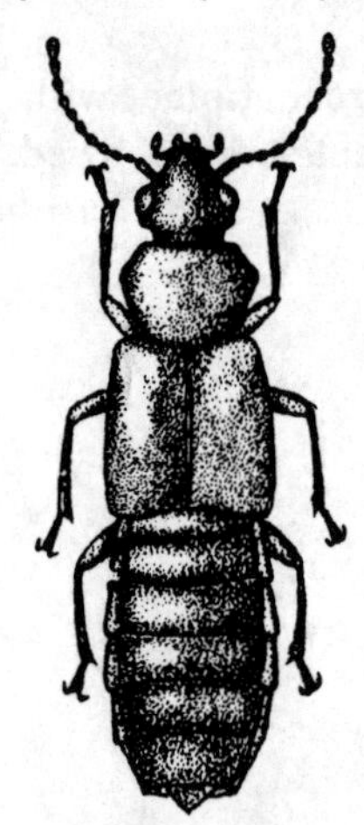

Figure 379

LENGTH: 3 mm. RANGE: Eastern half United States.

Shining. Black; antennae and legs piceous; knees and tarsi paler. Thorax 2/3 wider than long.

Thinodromus arcifer Lec., ranging rather widely and about 3 mm long, is uniform dark reddish-brown with legs and antennae paler. This is another large genus of small, hard to separate beetles.

46b Scutellum readily seen; head strongly constricted behind; body glabrous. Fig. 380. .... *Apocellus sphaericollis* (Say)

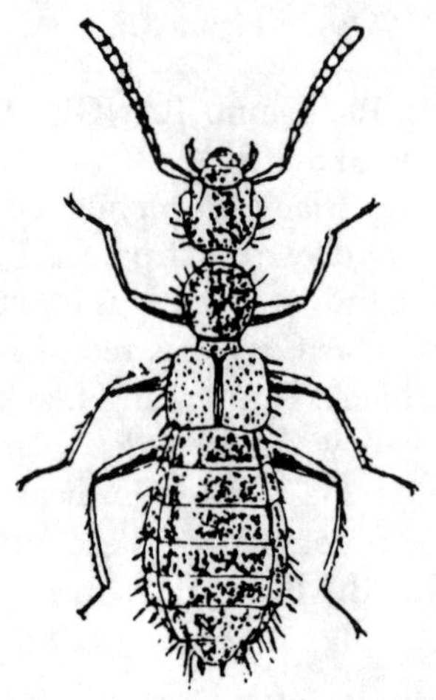

Figure 380

LENGTH: 2.7-3 mm. RANGE: North America.

Shining. Dark reddish-brown; head and abdomen usually darker; legs and first three joints of antennae pale. Eyes very small. Antennae longer than the head; thorax variable both in color and size.

A related species, *Thinobius fimbriatus* Lec. is a tiny (.7 mm), slender beetle, dark reddish-brown with paler legs. It is pubescent and the head is not constricted behind.

47a (26) Segments 1 to 4 of hind tarsi unequal, segments 1 and 2 being elongated

and equal; antennae slender; tibiae pubescent. Fig. 381. .. L
.... *Olophrum obtectum* Er.

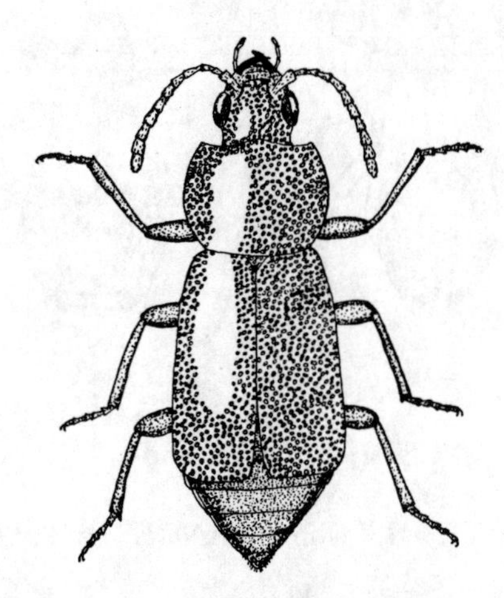

Figure 381

LENGTH: 5-6 mm. RANGE: Central and Eastern States.

Shining. Piceous; antennae, legs and very narrow margin of thorax, reddish-brown. Head closely and coarsely punctuate.

Anthobium sordidum Er. dull yellow with piceous abdomen and 2.7 mm long ranges from Iowa to Virginia. Its antennae are thickened apically.

47b Segments 1 to 4 of hind tarsi very short and equal. 48

48a (47) Tibiae with fine spines; mesosternum carinate (ridged). Fig. 382. ....

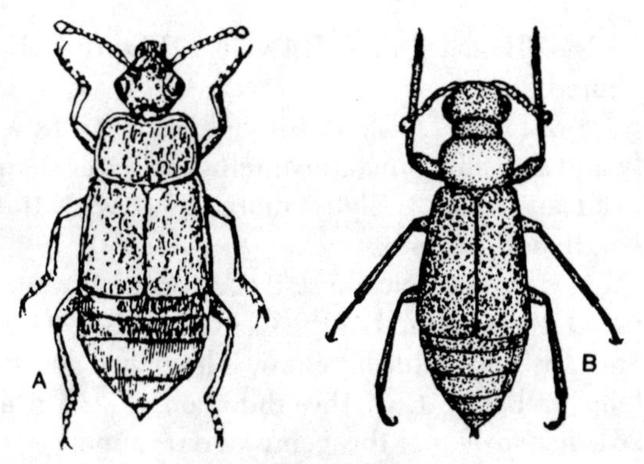

Figure 382

(a) *Elonium diffusum* Fauv.

LENGTH: 3.5-4 mm. RANGE: Shores of Lake Superior.

Somewhat flattened, with elytra reaching beyond the middle.

(b) *Omalium hamatum* Fauv.

LENGTH: 2.3 mm. RANGE: Central States.

Dull reddish or dusky yellow, sparsely pubescent; head and tips of elytra fuscous. Hind tibiae deeply notched on the outer side.

48b Tibiae pubescent. Fig. 383. ....
.... *Eusphalerum horni* (Fauv.)

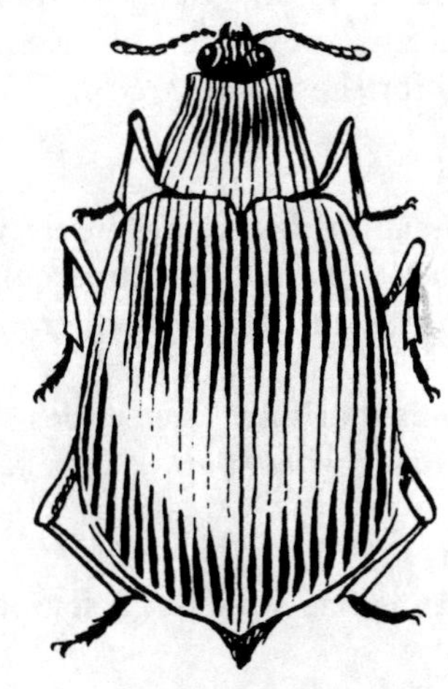

Figure 383

LENGTH: 2-2.4 mm. RANGE: Eastern half-United States.

Shining. Dull reddish-yellow. Head finely and sparsely punctate; antennae shorter than head and thorax. Elytra more than twice the length of the thorax.

E. convexum Fauv. 2 mm or a bit more in length and found in the Central and Eastern States is dull reddish yellow. The elytra are as long or longer than the abdomen. There are 26 other species of this genus in our fauna.

The related genus, *Arpedium,* differs in that the hind tarsal segments 1-4 are unequal instead of very short and equal as in *Anthobium. Arpedium cribatum* Fauv., 4.5-5 mm, is known from Iowa eastward. It is elongate-oval in form, with thorax and elytra reddish-brown, head black and legs and margins of the elytra and thorax paler.

FAMILY 25. PSELAPHIDAE
The Ant-loving Beetles

These tiny beetles, all less than 5.5 mm, are found beneath bark, stones, and in the nests of ants. Yellow, red, brown and fuscous are the usual colors. The elytra are very short; the eyes are coarse or sometimes wanting. Some 500 species are known for our country and about 3,000 for the world.

1a **Antennae arising from under two closely placed horizontal tubercules, their basal segments close together. 2**

1b **Antennae arising from sides of head, their bases wide apart. 5**

2a **(1) Antennae with but 2 segments. Fig. 384.**

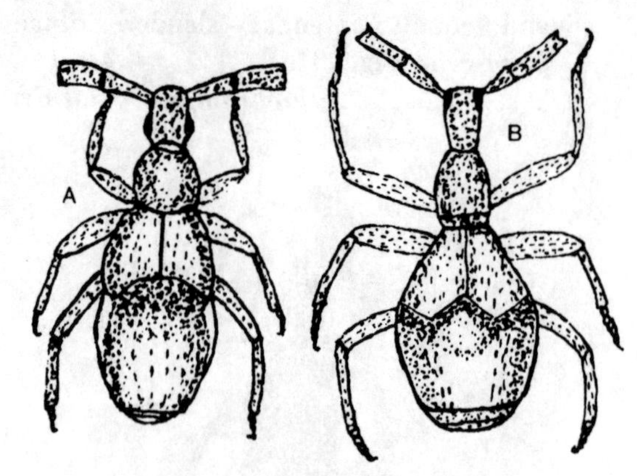

Figure 384

(a) ***Fustiger fuchsi*** **Bndl.**

LENGTH: 1.7 mm. RANGE: Ind., Tenn., Ariz.

Dark brownish yellow; head and thorax rather deeply impressed with a netted design. Basal segment of abdomen large and wide.

(b) ***Adranes lecontei*** **Bndl.**

LENGTH: 2.5 mm. RANGE: Ind. and Miss.

Like the few other species of the genus it differs from those of the genus *Fustiger* in having no eyes. The head and thorax are elongated and slim. It is colored brownish-yellow and covered with a fine pubescence.

2b **Antennae of more than three segments. 3**

3a **(2) Tarsi with two equal claws. 4**

3b **With but one claw on each tarsus; last segment of maxillary palpi long and hatchet shaped (a). Fig. 385. .... *Cylindrarctus longipalpis* (Lec.)**

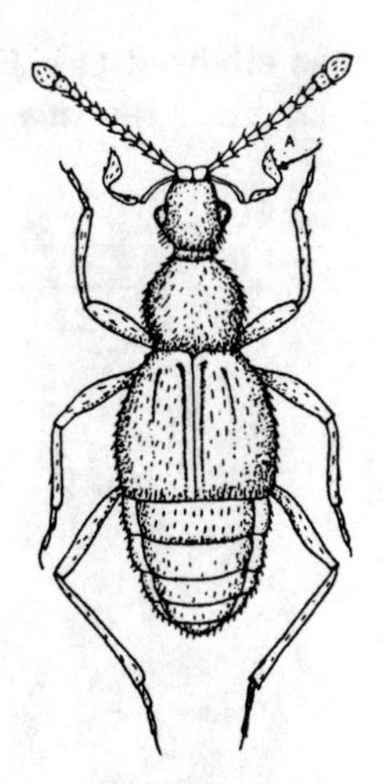

Figure 385

LENGTH: 2 mm. RANGE: widely scattered from Florida to B. C.

Somewhat depressed, pale reddish-brown with scattered long and somewhat erect hairs; abdomen with narrow margins.

Lucifotychus minor Lec. a widely distributed species measuring about 1.5 mm. differs from the above in being darker colored and in having five fovea at the base of the thorax instead of four.

4a (3) Antennae with bead-like segments, not club-like; last two joints of maxillary palpi transverse (a). Fig. 386. .... *Ceophyllus monilis* Lec.

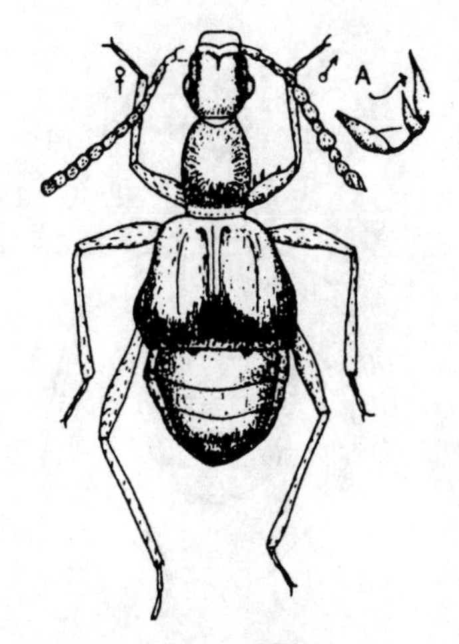

Figure 386

LENGTH: 3.3 mm. RANGE: Central and Eastern U. S.

Reddish brown, without punctures; antennae half as long as the body; all segments of female antennae about equal; male as pictured.

Atinus monilicornis (Bndl.) measuring 2.8 mm and reddish brown; differs in the maxillary palpi which is small and has the last two segments forming a globular club. It has been taken in Va., Ohio, and Tenn.

Closely related but differing from the above two species in having clubbed antennae, *Cedius ziegleri* Lec. a bit under 3 mm. and known from Iowa eastward is fuscous-brown with appressed pubescence. *C. spinosus* Lec. is much like *ziegleri* but is smaller (2 mm or less) and has less pubescence. It, too, is found in Iowa and eastward.

4b Last joints of antenna gradually larger forming a distinct club (a); long bristle-like appendages on maxillary palpi (b). Fig. 387. .... *Pilopius lacustris* Csy.

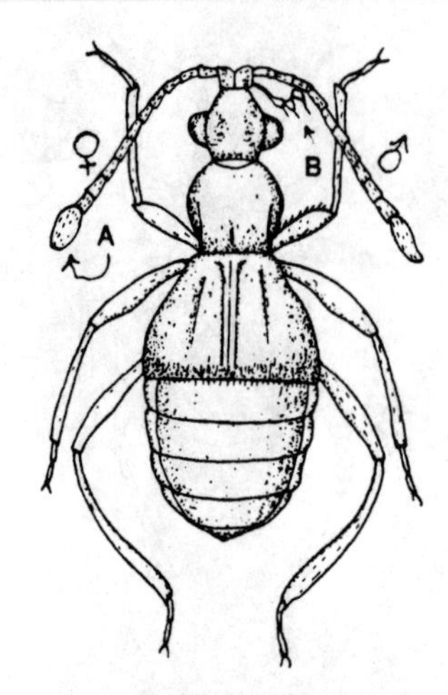

Figure 387

LENGTH: 1.8 mm. RANGE: Central U. S. and Canada.

Dark reddish-brown, elytra, legs, and antennae paler; antennae of male ¾ as long as body; of female shorter; top of head with two distinct pits between the eyes.

Tmesiphorus carinatus Say differs in having only short appendages on the palpi. It ranges through our Central and Eastern States and measures 2.5 mm. It is pale reddish brown.

Another member of this latter genus, *T. costalis* Lec., is common over a rather wide range. It is shining, piceous, and covered with short, appressed, yellowish hairs. The eyes are prominent and the antennae of the male over half as long as the body; the antennae of the female are shorter. It is 3.3 mm long.

5a (1) Hind coxae conical, prominent and close together. 6

5b Hind coxae transverse, not prominent and widely separated. 8

6a (5) Antennae not elbowed; first joint not elongated; tarsi with but one claw. 7

6b Antennae elbowed (a). Fig. 388. .... *Rhexius insculptus* Lec.

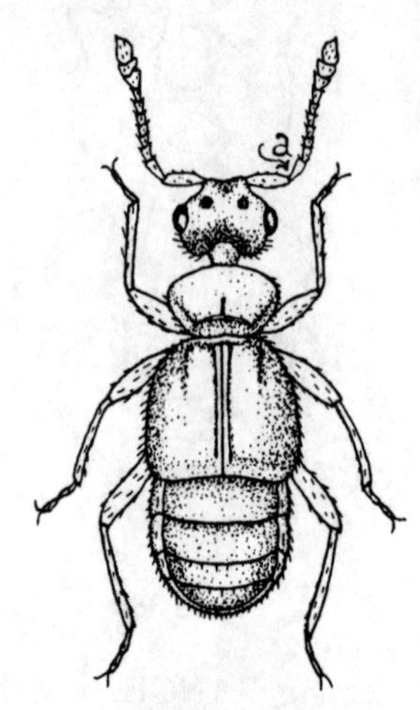

Figure 388

LENGTH: 1.2 mm. RANGE: eastern half of U. S.

Pale reddish-brown covered with short erect hairs; front with 3 pits; top of head ridged, antennal joints 3-8 transverse and equal. Tarsi with two unequal claws.

Sonoma tolulae (Lec.) from Georgia, Virginia and Pennsylvania has two equal tarsal claws. Its length is 2.1 mm and is dark brown and covered with long dense pubescence. The antennae and legs are yellowish.

7a (6) Last joint of antennae much larger than the others. Fig. 389. .... *Melba parvula* (Lec.)

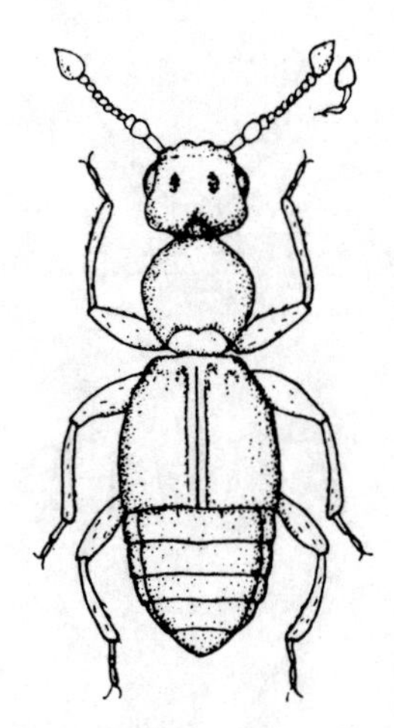

Figure 389

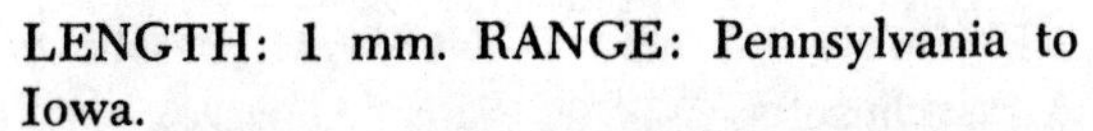

LENGTH: 1 mm. RANGE: Pennsylvania to Iowa.

Brownish yellow, shining, pubescent; thorax without punctures; top of head with foveae.

A very tiny (.7 mm) related species is *Trimiomelba convexula* (Lec.) The first dorsal segment of the abdomen is much elongated. It is deep yellow and has been found in Florida, Pennsylvania, and Indiana.

7b **Antennal club formed gradually, the last joint not radically larger than the adjoining ones. Fig. 390. .... *Euplectus confluens* Lec.**

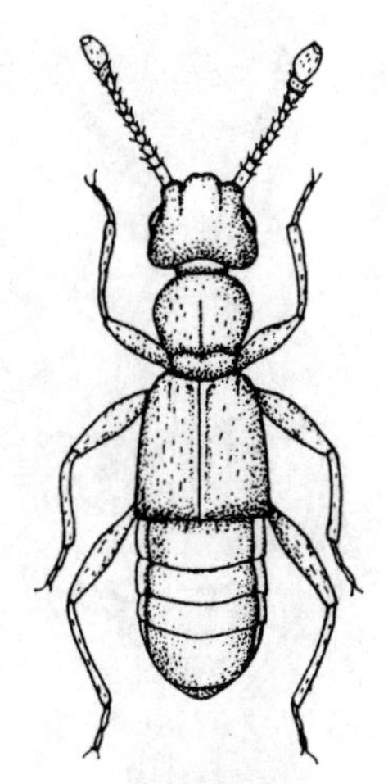

Figure 390

LENGTH: 1.2-1.5 mm. RANGE: Central and Eastern U. S.

Reddish-brown, finely pubescent. Abdomen longer than elytra; antennae reaching middle of thorax.

Thesiastes fossulatus (Bndl.), known to range from Iowa to Indiana is distinguished from the above by a head narrower than the thorax. It is a shining dark brown and measures 1.2 mm.

8a **(5) Antennae with 11 segments. 9**

8b **Antenna with but 10 segments three of which form the club. Fig. 391. .... *Decarthron brendeli* Csy.**

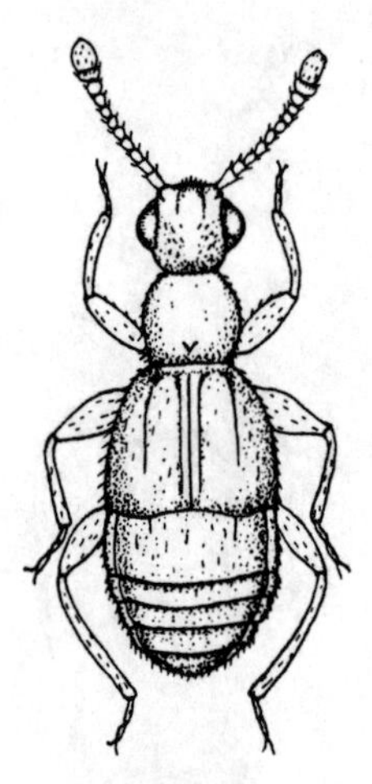

Figure 391

LENGTH: 1.4 mm. RANGE: Central States and Southwest.

Piceous brown; elytra dull red; antennae and legs paler brown; antennae scarcely as long as head and thorax; eighth and ninth segments transversely oval, tenth one-half longer than wide and three times as thick as eighth, truncate at base.

D. abnorme (Lec.) is black. Its two large pits on the top of head are connected by grooves. It measures 1.4 mm and is found from Canada to Florida and west to Iowa.

9a (8) Abdomen margined; tarsi with but one claw. Elytra with three small basal pits and thorax with two large and one small pit. Fig. 392. .... *Reichenbachia propinqua* Lec.

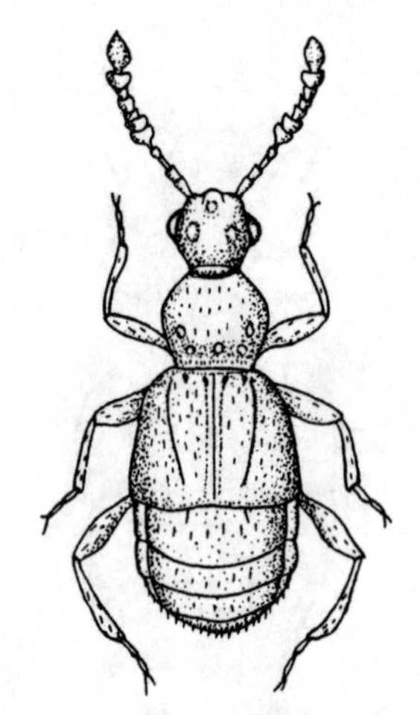

Figure 392

LENGTH: 1.4 mm. RANGE: Minnesota to Massachusetts.

Piceous black; elytra dark blood-red; antennae of male with fifth segment strongly dilated, the sixth a bit longer than the seventh; pubescence short and recurved.

R. rubicunda Aube. is common in much of its range throughout the eastern half of United States.

It is piceous black, not punctured; legs, antennae, and elytra dull red and measures around 1.5 mm. The eighth antennal segment is globular and the tenth subglobular and one-half thicker than the ninth with the eleventh one-half wider than the tenth.

9b Abdomen without margin; tarsi with 2 unequal claws; body long, cylindrical and narrow; hind tibiae with a long terminal spur. Fig. 393. .... *Batrisodes spretus* Lec.

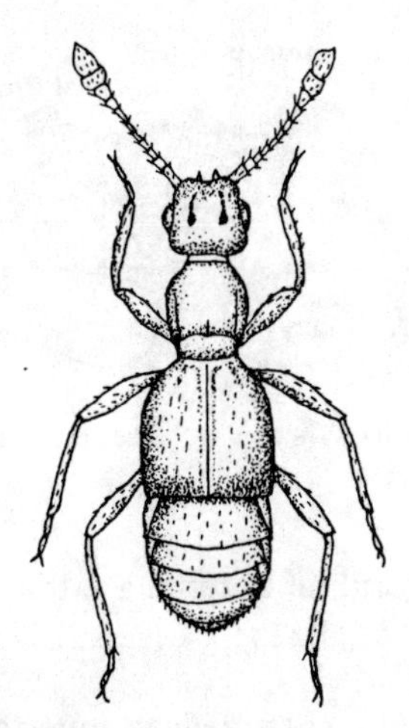

Figure 393

LENGTH: 1.7 mm. RANGE: Eastern half of U. S. Common.

Piceous black; elytra dark red, sparsely and finely pubescent. Antennae reaching base of thorax, segments 2-8 oblong; 9th longer and wider, transverse; 10th large squarish; 11th narrower than the 10th but twice as long.

Batriasymmodes monstrosus (Lec.) is 2.4 mm long, reddish-brown throughout and distributed through the Central and Eastern States. It is often common.

FAMILY 26. HISTERIDAE
The Clown Beetles

These polished, hard, usually ovoid beetles, (some live under bark and are much depressed), with truncate elytra are easily recognized. They frequent carrion where they are apparently predacious on the carrion feeders. More than 3,000 species are known, with over 500 from our area.

1a **Head bent downward and usually retracted; form convex; mandibles not especially prominent. 2**

1b **Head extended forward; form much flattened; mandibles conspicuous. Fig. 394. .... *Hololepta aequalis* Say**

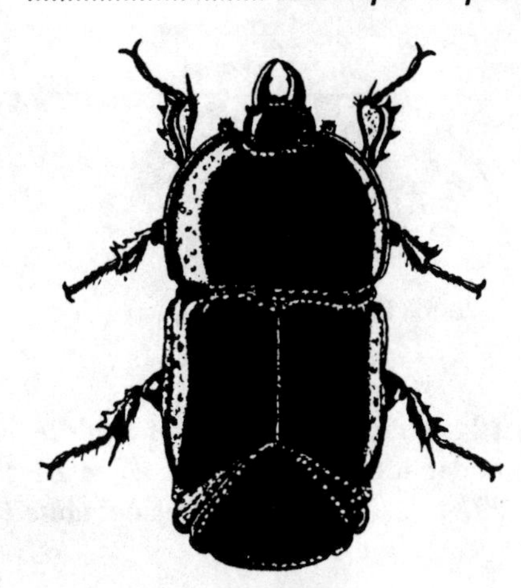

Figure 394

LENGTH: 7-10 mm. RANGE: eastern half of the U. S. and parts of Canada.

Fairly common under the tight bark of logs, especially poplar, elm and tulip trees. Very flat and thin. Black, shining. Thorax punctured at the sides.

Hololepta yucateca Mars. ranging from Texas to Southern California is found in decaying cacti. It is black and measures 8-10 mm.

Hololepta vicina Lec. belongs in the west from Washington to Southern California; inland to Arizona.

2a **(1) Prosternum distinctly lobed in front. Fig. 395. .... 3**

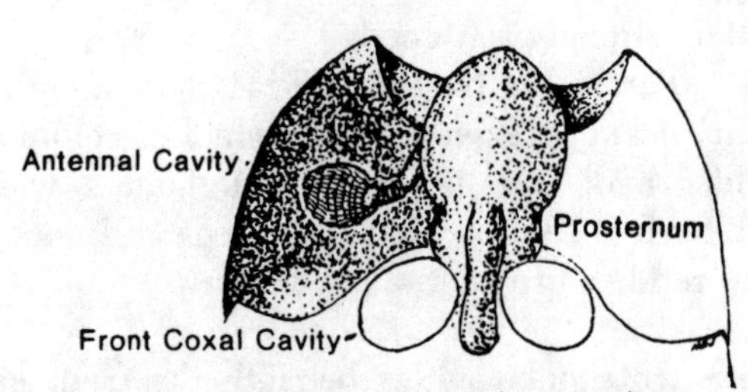

Figure 395

2b **Prosternum not lobed in front. Fig. 396.** **8**

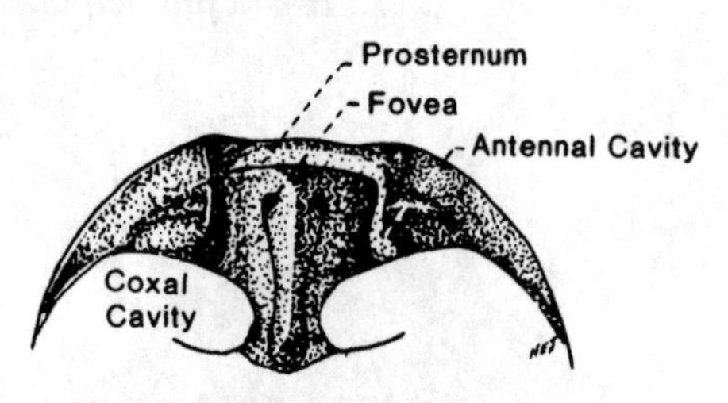

Figure 396

3a **(2) Antennal cavities at middle of the underside of the thorax. (See Figs. 395, 397.) .... *Platylomalus aequalis* (Say)**

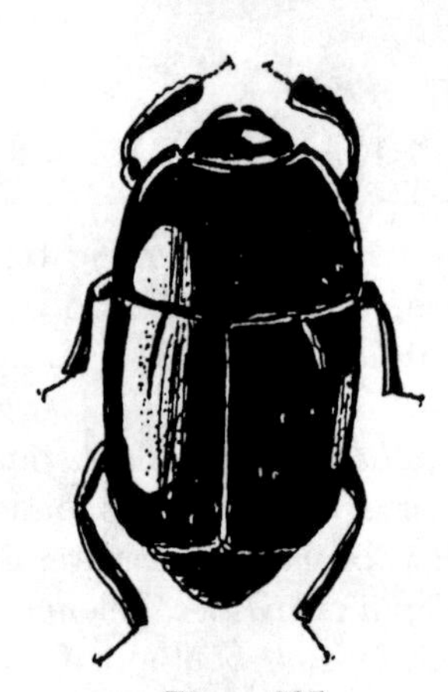

Figure 397

LENGTH: 2.5-3 mm. RANGE: eastern half of the United States and Canada.

Common under bark on poplar and other logs. Black, shining elytra with but few or no striae; strongly flattened.

Paromalus bistriatus Er. is a smaller (2 mm), less depressed histerid which is common under bark and in fungi throughout a wide range. It is shining black with legs and antennae reddish brown.

3b **Antennal cavities beneath the front angles of the thorax, open in front. Fig. 398.** **4**

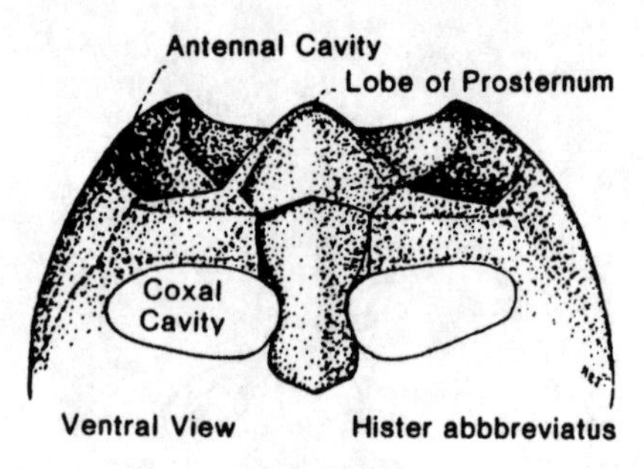

Figure 398

4a **(3) Club of antennae rather oval, pubescent.** **5**

4b **Club of antennae an inverted cone, glabrous; thorax with a lobe on each side. Fig. 399.** .... ***Hetaerius brunneipennis* Rand.**

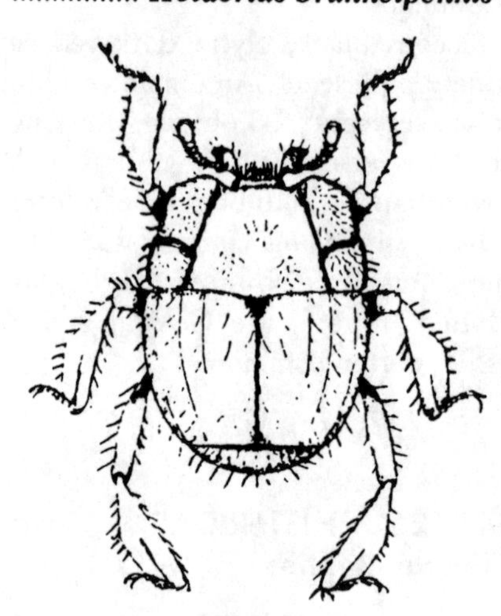

Figure 399

LENGTH: 1.5 mm. RANGE: Connecticut to Iowa.

Pale chestnut-brown shining; thorax with a lobe on each side which is separated from the disk by a groove; elytra with scattered erect fine yellowish hairs along striae.

This interesting tiny beetle like several other species of its genus lives in intimate association with ants in their nests. *H. exiguus* Mann. is thus found in several of the western states. Several species of these ant-loving hister beetles are known only from California.

5a (4) **Tarsal groove of front tibiae deep, S-shaped, and plainly margined. 7**

5b **Tarsal groove of front tibiae straight and with margin only on inner edge. 6**

6a (5) **Margin of thorax with a fringe of short hairs; elytra with a prominent red spot. Fig. 400. ....**
.... *Spilodiscus biplagiatus* (Lec.)

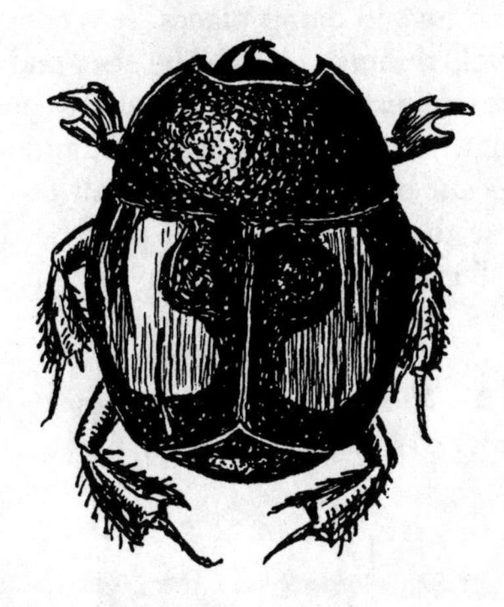

Figure 400

LENGTH: 5-6 mm. RANGE: North Central and Eastern United States.

Black with red marks as pictured; disk of thorax smooth; pygidium with but a few fine punctures.

6b **Margin of thorax not fringed with hairs; prosternum convex; mesosternum notched in front. Fig. 401. ....**
.... *Hister abbreviatus* Fab.

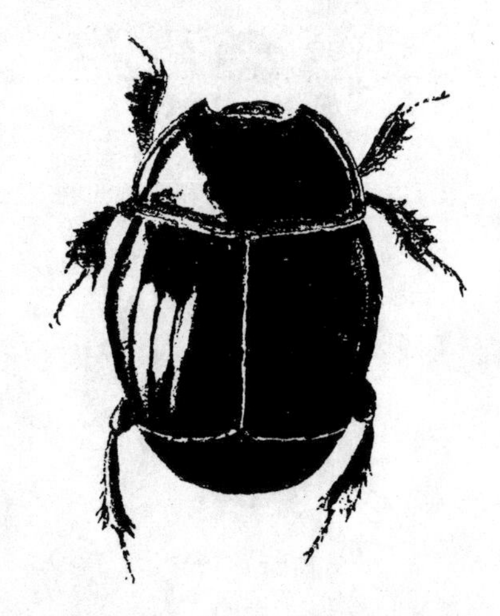

Figure 401

LENGTH: 3.5-5.5 mm. RANGE: eastern half of U. S. Common.

Black, shining; elytra with four entire dorsal striae, fifth short and two overlapping subhumerals; dorsal striae punctured.

H. bimaculatus L. is a member of the genus marked with red, having a large diagonal orange-red spot covering much of the outer half of each elytron. It is 4.5 mm long and is known in Europe as well as much of North America.

Margarinotus foedatus (Lec.) ranging, often abundantly, in Canada and entirely across the United States measures 4-6.5 mm. Its subhumeral and 3 dorsal striae are entire. The front tibia has at least six teeth.

Atholus americanus (Payk.) is a comparatively small (3-4 mm) member of the genus, but one of the most common hister beetles and one that has a wide range. It is black; the inner striae of the thorax are entire and the elytra possess five entire striae. The front tibiae have three teeth, the apical one prominent and divided into two fine points.

7a (5) **Body at least twice as long as wide, form cylindrical. Fig. 402. ....**
.... *Platysoma basale* Lec.

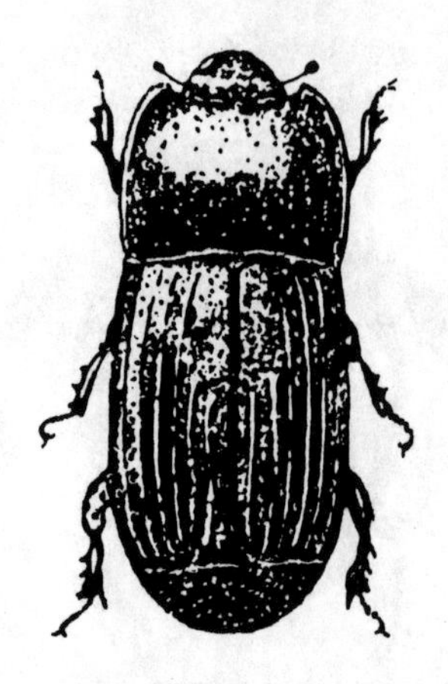

Figure 402

LENGTH: 4.5-5.5 mm. RANGE: North Central States under bark, not common.

Black, shining; antennae reddish brown; legs piceous; disk of thorax nearly smooth; elytra with four dorsal striae entire.

P. coarctatum Lec. appears to be a small edition of *basale.* It measures but 2.5 to 3 mm and has the thorax evenly punctured throughout. It seems to have a wider range and to be more common. On pine.

7b Body broad and plainly depressed. Fig. 403. .... *Platysoma lecontei* Mars.

Figure 403

LENGTH: 2.5-3 mm. RANGE: Atlantic States to So. Cal. Common under bark of logs and stumps.

Black, shining; thorax with scattered coarse punctures at sides, smooth at middle; elytra with but three entire dorsal striae; front tibiae with 4 teeth; middle tibia with 3.

Phelister subrotundus Say, a bit smaller and quite convex, is also common under bark. It has a reddish-brown cast with legs and antennae definitely that color. It too has a wide range.

Phelister vernus (Say) a widely distributed and very common species, lives under bark and logs in damp places. It is convex, oblong-oval, shining, black. The legs and antennae are reddish-brown. The elytra possess 4 full-length striae; the sutural stria and still another occur only on the apical half of the elytron; the thorax has no marginal stria. It measures 2.5 mm.

8a (2) Front of head without a margin. Fig. 404. 9

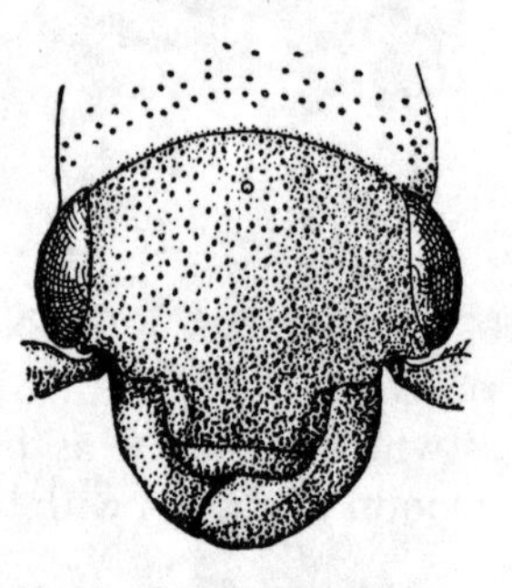

Head—Saprinus assimilis

Figure 404

8b Front of head margined (Fig. 405). Prosternum with a ridge. Fig. 406. .... *Hypocaccus patruelis* (Lec.)

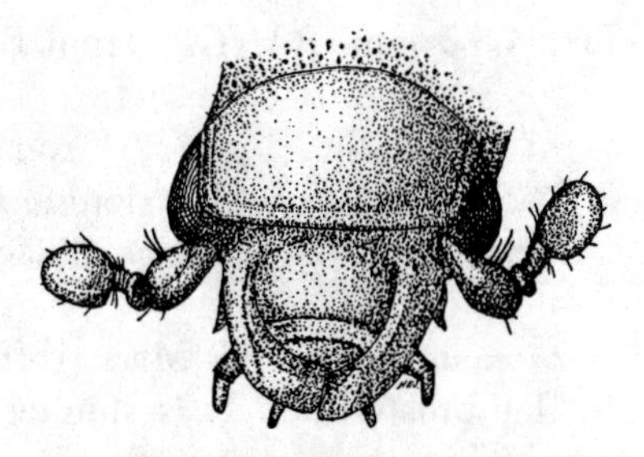

Head—Hypocaccus patruelis

Figure 405

Figure 406

LENGTH: 3-4.5 mm. RANGE: North Central States.

Shining with bluish-green or bronze lustre; hind tibiae with three rows of long spines; elytra with scattered coarse punctures. Front tibiae with six teeth, the apical three coarser.

H. fraternus (Say) has "mirror spots" on the elytra similar to *lugens*. It is much smaller however (3-4 mm) and has a distinct margin on the front of head (see Fig. 304). It ranges from Newfoundland to Iowa.

Acritus exiguus (Er.) has its antennae arising from the front instead of from under the frontal margin. It is only 1 mm long, dark reddish-brown and is often common under bark throughout much of the United States.

9a (8) Prosternum without a pit or fovea on each side close to front margin; short but distinct striae on front above the eye. 10

9b Prosternum with a small but distinct fovea on each side terminating the striae (see Fig. 295); no striae above eye. Fig. 407. .... *Euspilotus assimilis* (Payk.)

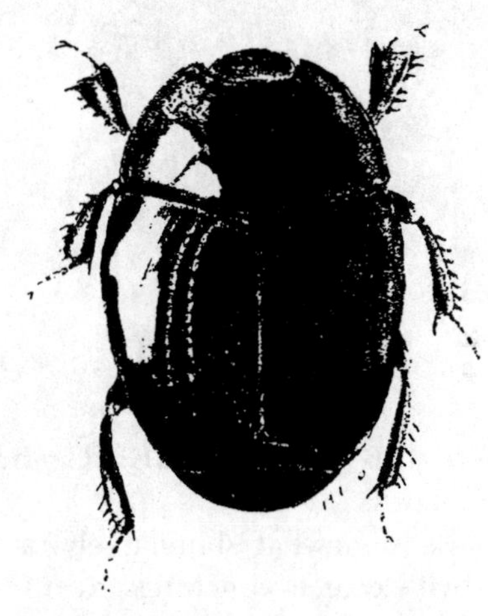

Figure 407

LENGTH: 4-5.5 mm. RANGE: eastern half of U. S. Common.

Black, shining; sides of thorax with coarse punctures, disk smooth; coarse punctures on apical third of elytra; teeth on front tibia fine.

E. conformis (Lec.) is told from *assimilis* by its lack of a deep marginal groove, at the apex of the pygidium. It is shining black, ranges through the eastern half of the United States and is 3.5-4.5 mm long.

Geomysaprinus obsidianus (Csy.) 3-4 mm in length apparently ranges widely through the eastern half of our region. It is broadly oval, black and highly polished. The thorax carries many punctures. The elytral striae are all abbreviated, with the sutural and the first being the longest; the fourth is joined to the sutural at the base by an arch. The front tibiae are armed with several fine teeth.

10a **(9) Fourth dorsal striae but weakly arched and not joining the short sutural stria. Fig. 408. ...... *Saprinus lugens* (Er.)**

Figure 408

LENGTH: 4.6 mm. RANGE: Central and Western States.

Black, somewhat shining; elytra wholly covered with coarse punctures except a large somewhat oval scutellar spot and a small subhumeral area which are smooth.

10b **Fourth dorsal stria strongly arched at base, joining the sutural which is entire. Fig. 409. .... *Saprinus pennsylvanicus* Payk.**

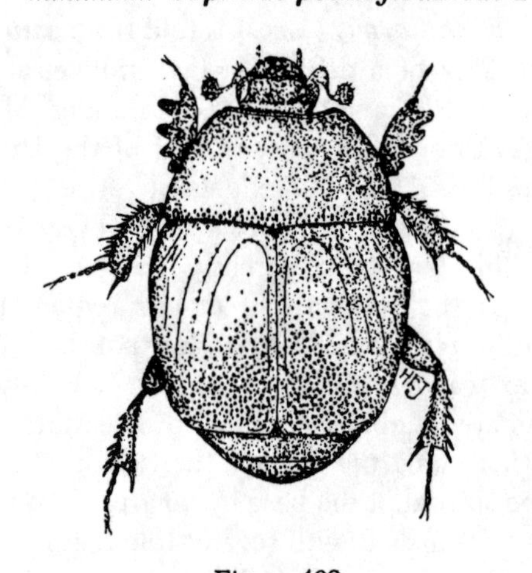

Figure 409

LENGTH: 4.5-5 mm. RANGE: from Rockies east.

Bright metallic green, sometimes bronzed. Thorax smooth with a double row of coarse punctures at base. Front tibiae with coarse teeth.

Gnathoncus communis Mars. differs in having a flat prosternum. It is shining black with brown legs and antennae. It ranges throughout Europe and North America. Taken in bird nests and roosts and measures 2.5-3.5 mm.

FAMILY 27. SPHAERITIDAE
The False Clown Beetles

Only three species are known of this family. Members of the family resemble hister beetles in general characteristics, but the body is not so compact. They are oval, convex, about 6 mm in length, and of a dark greenish-metallic color. The truncate elytra cover all but the last dorsal tergite. The antennae terminate in a three-segmented club. The head is much narrower than the pronotum and the latter is deeply emarginate anteriorly and evenly arcuate laterally. The elytra striae are represented by rows of punctures. There is a single American species, *Sphaerites politus* Mann., that is found in the Pacific Northwest from B. C. to Alberta, south to Idaho and Oregon. They have been recorded from fungi, bear dung, and other decomposing vegetable matter.
Figure 410 *Sphaerites politus* Mann.

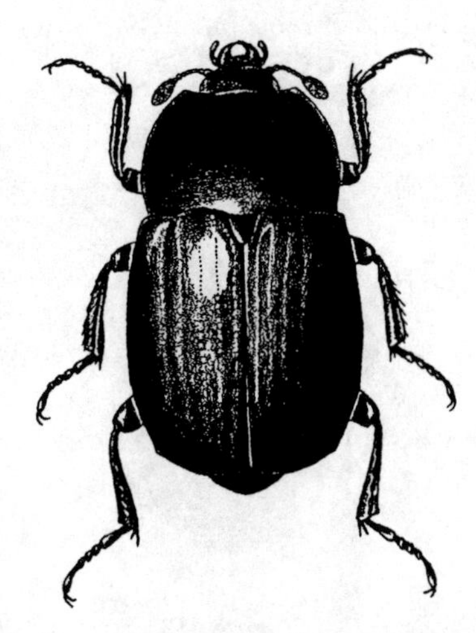

Figure 410

FAMILY 28. LUCANIDAE
The Stag Beetles

This family of usually large mahogany or black beetles is often characterized by great mandibles. The larvae live in decaying wood. They are widely distributed and number nearly a thousand species, with but a few in our region.

1a Elytra smooth; front tibia with large teeth on outer side. 2

1b Elytra with striae and punctures. 3

2a (1) Mandibles with one tooth on inner side; femora light brown. Fig. 411. .... *Pseudolucanus capreolus* (L.)

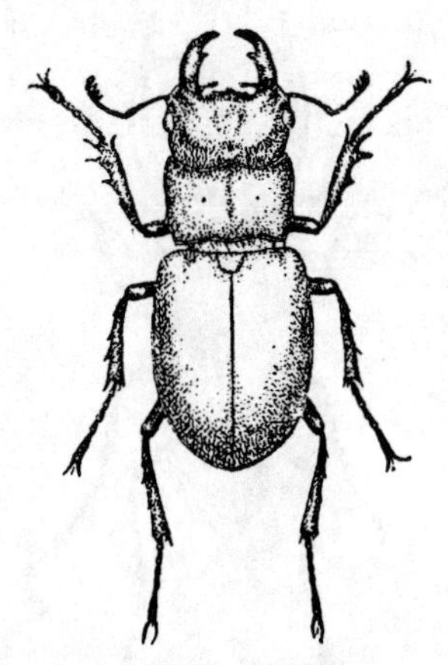

Figure 411

LENGTH: 22-35 mm. RANGE: Eastern half of United States.

Reddish-brown. Mandibles of male twice as long as those of female. The larvae live in wood. The adults are known as "Pinching bugs."

P. placidus (Say) 19-32 mm, is nearly black. Both sexes have more than one tooth on inside of mandible. East from Ill. to Pa.

P. mazama (Lec.) breeds in cottonwood from Utah to Arizona and New Mexico. It is brown with black femora and measures 24-32 mm.

2b Female with black legs; male with mandibles as long as abdomen. Fig. 412. .... *Lucanus elaphus* Fab.

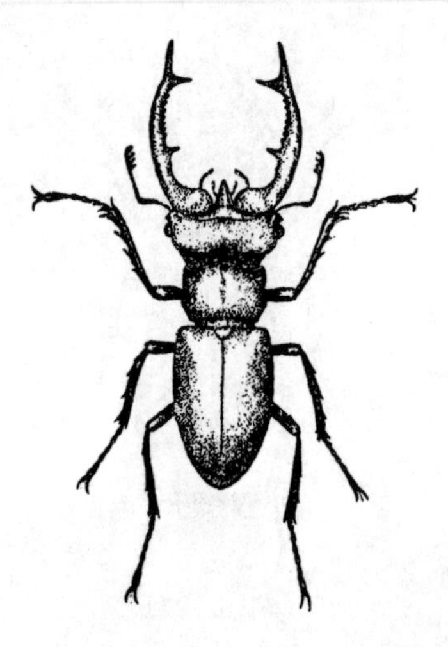

Figure 412

Figure 413

LENGTH: male including mandibles 45-60 mm, female 30-35 mm. Ill., Ind., Okla. to Va. and N. C.

Shining. Dark chestnut brown. Legs and antennae nearly black.

Closely related are some tropical species which attain extraordinary size. Likely largest of all is an East Indies species *Odontolabis alces Fab.* This bulky beetle may be fully 4 inches or more in length. India and Java present the Giraffe Stag Beetle, *Cladognathus giraffa* (Fab.); it is reddish-brown and black. Chile's "giant" *Chaisognathus granti* Steph. is colored metallic dark green with an iridescence of red. Both of these last two species have great branched mandibles as long as the rest of the body, giving them an over-all length of about 4 inches.

These "pinch beetles" never fail to attract attention even from chickens, cats and dogs.

3a (1) Eyes notched. Fig. 413. ....
.... ***Dorcus parallelus*** **Say**

LENGTH: 15-26 mm. RANGE: Eastern half United States.

Nearly black. Males with head about as broad as thorax. Elytra with deep striae and punctures. It is known as the "antelope beetle."

The Rugose Stag Beetle, *Sinodendron rugosum* Mann. breeds in rotting oak, alder, laurel and willow from British Columbia to California. It measures up to 18 mm, is shining black and has a short horn (more prominent on the male) on the head.

3b Eyes not notched. Fig. 414. ....
.... ***Platycerus virescens*** **(Fab.)**

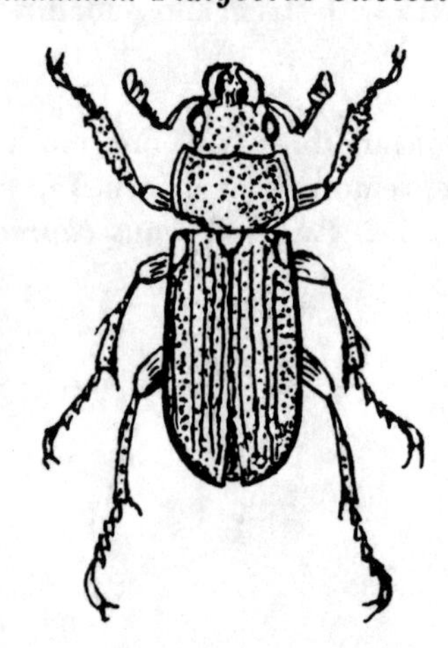

Figure 414

LENGTH: 10-12 mm. RANGE: Northeastern half United States.

Black to dark reddish-brown, sometimes with metallic sheen. Male with much larger mandibles than female. Striae with scattered punctures but prominent.

The Oak Stag Beetle, *Platyceroides agassii* Lec. black and 10 mm in length, breeds in oak from coastal Oregon south while *P. latus* (Fall) a heavy-set, nearly-black species, 9-11 mm is found within dead incense cedar in the same state.

FAMILY 29. PASSALIDAE
The Peg or Bess Beetles

Members of this family are large, elongate, somewhat flattened, black, shiny beetles. Ventrally they are clothed with dense yellow hairs. The head is narrower than the pronotum and sometimes bears a median dorsal horn that is recurved anteriorly. The antennae are three-segmented with the last three segments forming a loose lamellate club. The quadrate pronotum is much wider than the head and the elytra are elongate, parallel-sided with deep well-developed striae.

The adults and larvae live in colonies in rotting logs and stumps. The two forms communicate with each other by the use of stridulating organs. The larvae are fed food that has been chewed and mixed with saliva.

Almost 500 species of this family have been described. Most of these are tropical. Three species are found in the U. S., two of these being in Texas. One species is widely distributed and common over the eastern half of the U. S.

Figure 415 *Odontotaenius disjunctus* (Ill.)

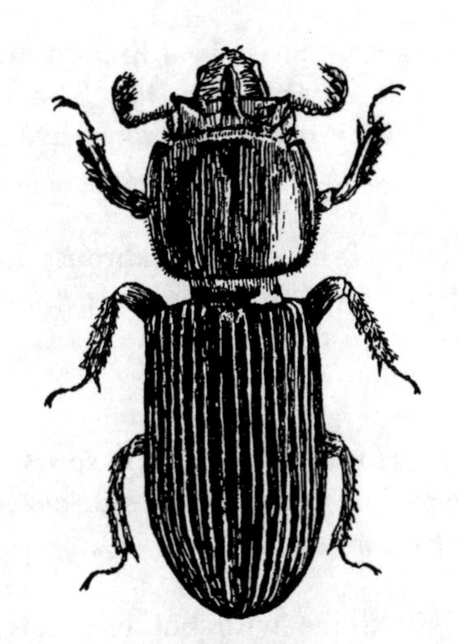

Figure 415

LENGTH: 30-36 mm. RANGE: Eastern half United States.

Somewhat flattened, shining. Black, head with short bent horn. Elytral striae deep and finely punctured.

This is known as the "Horned Passalus" or the "Bess-beetle." Both adults and larvae are often very abundant in decaying logs. The light brown adults sometimes found, have of course but recently emerged and have not had time to turn black.

FAMILY 30. SCARABAEIDAE
The Lamellicorn Beetles

Here is a great group of some 30,000 species of small to very large beetles. Some systematists would prefer to make it into several families. *Megasoma actaeon* Linn. from South America, is likely the largest and heaviest beetle known. The Atlas Beetle *Chalcosoma atlas* (L.), found in Asia, is an attractive, bulky beetle measuring 4 inches or more. This family has produced a good number of extra large, heavy species, the largest of which are found in tropical areas.

1a Spiracles placed in a line on membrane at side of abdomen; last 3 segments of club gray tomentose; most live in dung. 2

1b Club of antennae glabrous; not more than 1 or 2 spiracles on membrane, not living on dung. 12

2a (1) Hind tibiae with 2 spurs at apex; pygidium usually covered; coxae touching. 7

2b Hind tibiae with but one apical spur; pygidium exposed; hind coxae usually separated. 3

3a (2) Middle and hind tibiae much widened at tip; males often with horn on head or thorax. 4

3b Middle and hind tibiae slender, curved; no horn on head or thorax. Fig. 416. .... ***Canthon pilularius*** **(L.)**

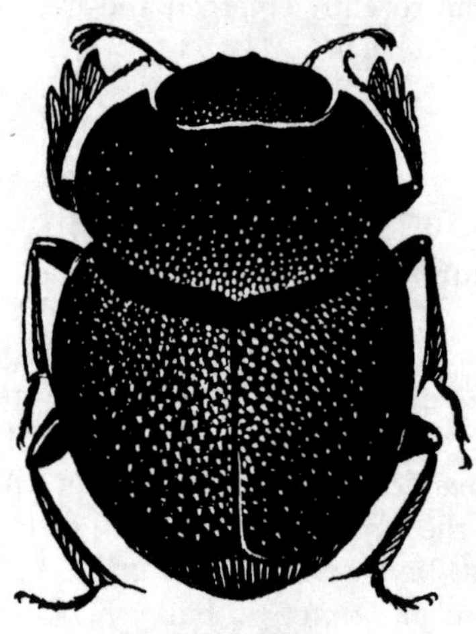

Figure 416

LENGTH: 11-19 mm. RANGE: United States.

Usually, dull black with coppery tinge; (in south and west becoming deep blue or bright green). The ball which it rolls receives an egg and contains enough dung to mature a grub. It is presently buried for protection while the grub hatches and matures. Some 18 species of this genera are found in our fauna.

4a (3) Labial palpi 3 jointed. 5

4b Labial palpi 2 jointed; not over 9 mm, scutellum not visible. Fig. 417. .... ***Onthophagus hecate*** **Panz.**

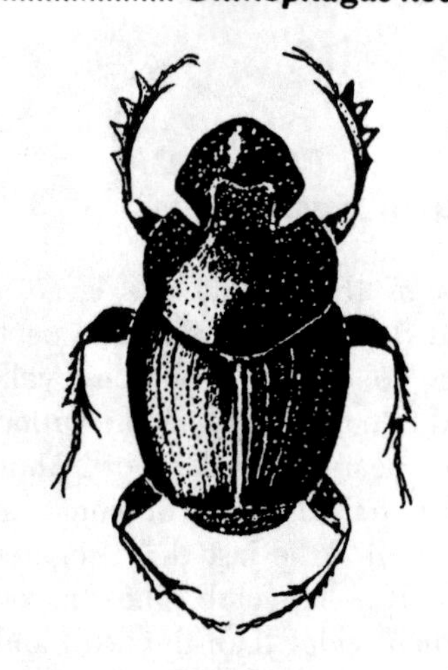

Figure 417

LENGTH: 6-9 mm. RANGE: United States and Canada.

Dull. Black with scattered grayish hairs; thorax of male with scoop like horn, variable; much smaller on female. Common in carrion and dung.

O. subaeneus P. de B., Atlantic Coast to Kan. and Tex. is a deep metallic purple and much smaller (4.8 mm). Both sexes display the same convex thorax, without a horn or other protuberance. This genus consists of more than 30 species in the U. S. and Can.

5a (4) Large beetles; 10 mm or more; front coxae short. 6

5b **Front coxae transverse; not over 8 mm. Fig. 418. .... *Ateuchus histeroides* (Web.)**

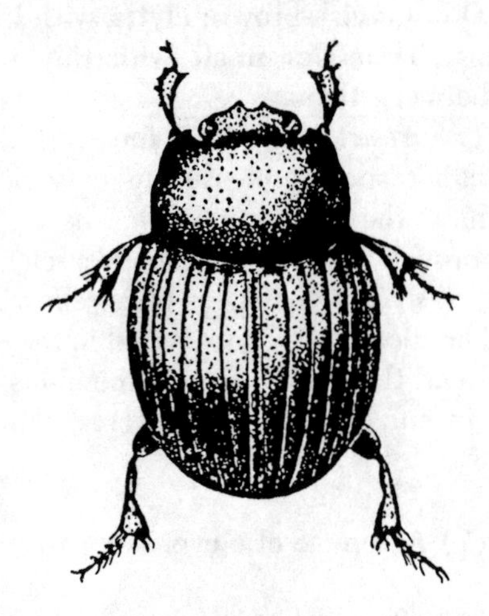

Figure 418

LENGTH: 6-7 mm. RANGE: Eastern half of United States.

Convex. Bronzed-black above; under parts chestnut-brown. Clypeus notched, usually with 2 teeth. Elytral striae fine. Often common in dung. Two other species are found in our fauna.

6a **(5) Metallic green; no tarsi on front legs of male. Fig. 419. .... *Phanaeus vindex* MacL.**

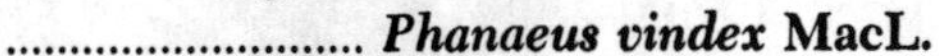

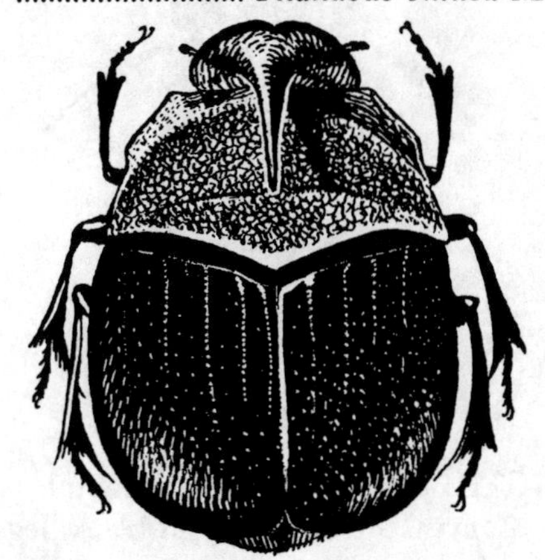

Figure 419

LENGTH: 14-23 mm. RANGE: from Rockies eastward.

Robust, flattened above. Head bronzed; thorax coppery; elytra bluish-green or green. Clypeus of male with long curved horn; with a tubercle in female. Showy flattened disk on thorax of male. This is a much decorated beetle that seldom fails to attract attention. Nine species are found in the Eastern half of the U. S.

6b **Black; front legs with tarsi. Fig. 420. .... *Dichotomius carolinus* (L.)**

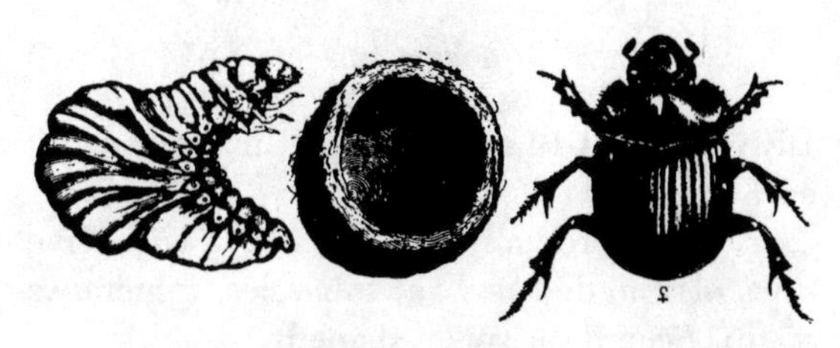

Figure 420

LENGTH: 20-30 mm. RANGE: widely distributed.

Robust. Shining. Black. Elytra each with seven shallow striae. Street light attracts many insects. This species is often among them.

This is our most bulky member of the family and is found east of the Rockies and south of the Great Lakes. Some tropical Scarabaeids rival small mammals in size.

7a **(2) Second antennal segment arising from the apex of the first segment. 8**

7b **Second antennal segment arising from a point inward from the apex of the first segment; elytra rough with irregular ridges and elevations.**

(a) Elytra with very distinct rows of tubercles, covered with fine hairs. Fig. 421. ...... *Trox monachus* Hbst.

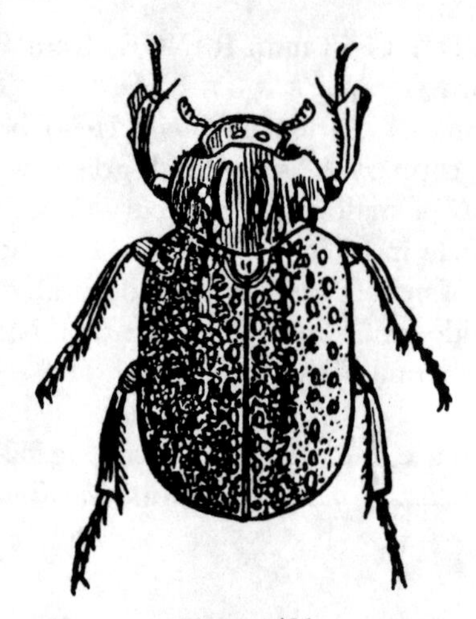

Figure 421

LENGTH: 13-16 mm. RANGE: much of United States.

Dark brownish. Elytra each with five rows of rounded or oval tubercles, tomentose at tip. Scutellum spear shaped.

The Trox beetles are so different in appearance from any other beetles as to be readily recognized. They are found on carrion in advanced stages of decomposition.

(b) Tubercles on elytra less distinct and without hairs. Fig. 422. .... *T. suberosus* Fab.

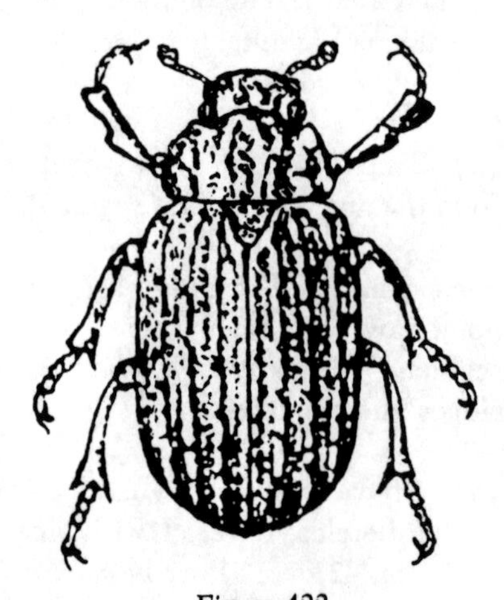

Figure 422

LENGTH: 12-17 mm. RANGE: United States and Canada.

Dull grayish-brown; elytra with blackish mottling. Tubercles small with tiny tufts of hairs between them.

T. terrestris Say and *T. foveicollis* Harold are smaller species measuring only 5-6 mm. Both have the scutellum oval. The latter has erect brownish hairs on the tubercles while with *terrestris* the hairs are pale yellowish and scale-like. Both species are found in the eastern U. S. In all there are about 40 members of this genus in our fauna.

8a (7) Antennae of eleven segments. 11

8b Only 9 segments in antennae. 9

9a (8) Outer terminal angle of hind tibiae obtuse. 10

9b Outer terminal angle of hind tibiae elongated into a spine; color black. Fig. 423. .... *Ataenius spretulus* (Hald.)

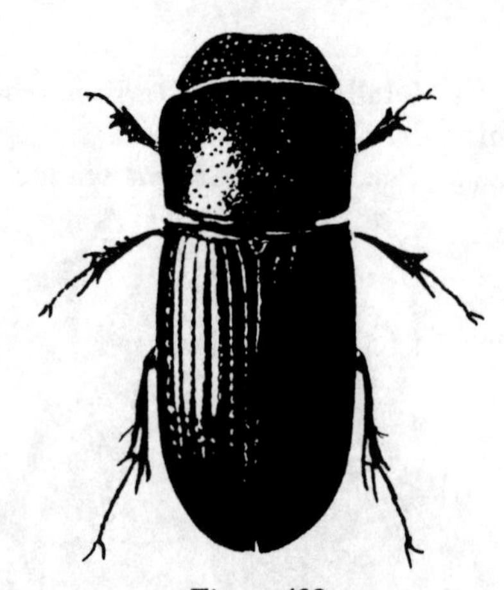

Figure 423

LENGTH: 4-5 mm. RANGE: general.

Convex. Shining. Piceous-black; legs and narrow front margin of thorax reddish-brown;

clypeus with several fine wrinkles. Common in dung and fungi and at lights. Some 40 rather similar appearing species are in the U. S. and Can.

Psammodius interruptus Say is a central and western species (3-4 mm). Thorax black with head and elytra brown and under parts reddish-brown. This genus is distinguished from the preceding by its triangular tarsal segments instead of cylindrical ones. There are 15 members of this genus in our fauna.

10a (9) Head with 3 small tubercles on vertex, hind tibia fringed at apex, with short equal spinules. Fig. 424.

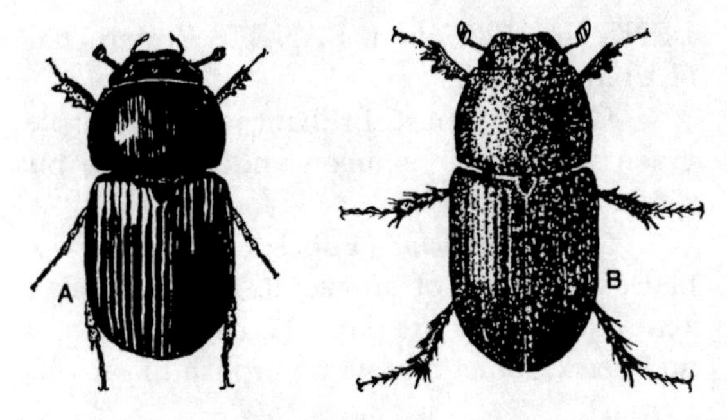

Figure 424

(a) ***Aphodius fimetarius* (L.)**

LENGTH: 6-8.5 mm. RANGE: Europe and North America.

Convex. Shining. Black; elytra brick red; thorax reddish-yellow near front angles. Very common in half-dry cow dung.

(b) ***A. granarius* (L.)**

LENGTH: 4-6 mm. RANGE: cosmopolitan.

Convex. Shining. Piceous; legs reddish-brown, antennae paler. Common in dung and fungi.

10b Spinules on apical edge of hind tibiae unequal, head ordinarily without tubercles. Fig. 425. .... *Aphodius distinctus* (Mull.)

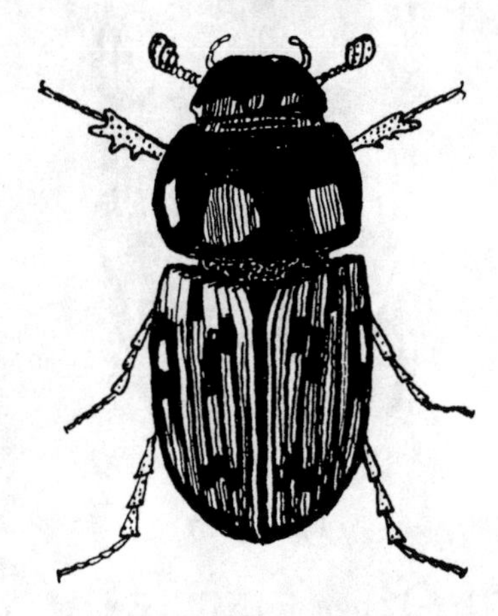

Figure 425

LENGTH: 4.5-5.5 mm. RANGE: Europe and North America.

Convex. Head and thorax piceous; elytra dull yellow marked with black as pictured. Very common.

More than a hundred species of this genus have been named. *A. femoralis* Say is a widely distributed and very common native species (4.5-6.5 mm). Head and thorax are black, the latter with yellowish margins; the elytra are brownish with paler margins; legs and antennae reddish-brown.

11a (8) Antennal club large, round, and convex on both sides; eyes entirely divided. Fig. 426. .... *Bolbocerosoma farctum* (Fab.)

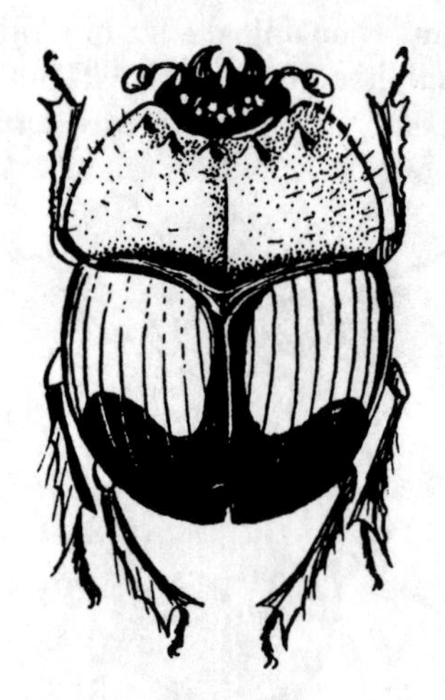

Figure 426

LENGTH: 8-12 mm. RANGE: East of Rockies.

Robust. Convex. Yellowish-red with variable black markings on whole upper surface. An interesting and attractive beetle, not very common.

With eyes but partly divided, *Eucanthus lazarus* (Fab.) a shining, chestnut-brown species, 6-11 mm, is rather widely scattered but is more common. It is occasionally fairly abundant at lights. In this genus and closely related ones some 40 species are found.

11b Antennae with smaller club, composed of leaf-like plates. Fig. 427. ....
.... *Geotrupes splendidus* (Fab.)

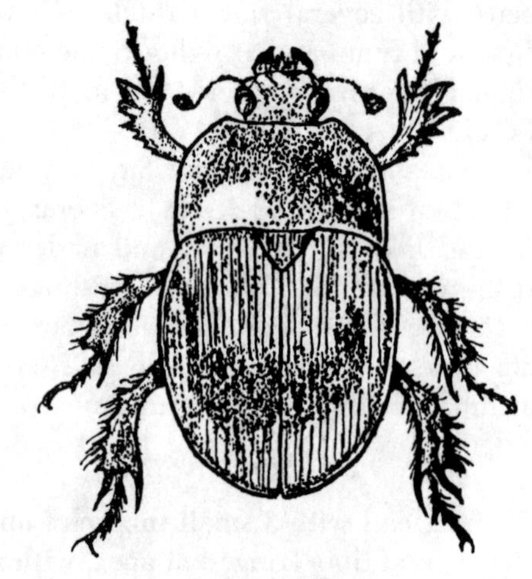

Figure 427

LENGTH: 12-17 mm. RANGE: Eastern half of United States.

Convex, robust. Brilliant metallic purple, green or bronze. Common under manure but sometimes on carrion.

G. blackburni (Fab.) (13-16 mm) is black with sheen of bronze; its first hind tarsal joint equals the next three. *G. opacus* Hald. is dull black, sometimes with purplish tinge, (12-15 mm).

Seven other members of this genus are found in our territory.

12a Mandibles plainly visible from above, bent, leaf-like; front coxae transverse. 22

12b Mandibles scarcely visible from above, not leaf-like. 13

13a (12) Claws of equal length or no spur on hind tibiae. 15

13b Claws, especially of hind leg almost always unequal in length; 2 apical spurs on hind tibiae. 14

14a (13) Elytra with membranous border; hind tibiae longer than the femur. Fig. 428. ....

Figure 428

(a) ***Popillia japonica* Newm. Japanese Beetle**

LENGTH: 7-13 mm. RANGE: Asia and steadily spreading in United States.

Robust. Head and thorax shining greenish-bronze; elytra tan or brownish with green margins; abdomen metallic green with white spots along sides and two white spots on pygidium. This pest was introduced into our country in 1916. The adults eat the leaves and fruit of many plants while the grubs feed on the roots.

(b) ***Anomala orientalis* Waterh. Oriental Beetle**

LENGTH: 6-9 mm. RANGE: Eastern United States.

Yellowish brown with black markings as pictured; highly variable, the upper surface being sometimes almost wholly black.

14b Elytra without a membranous border. Fig. 429.

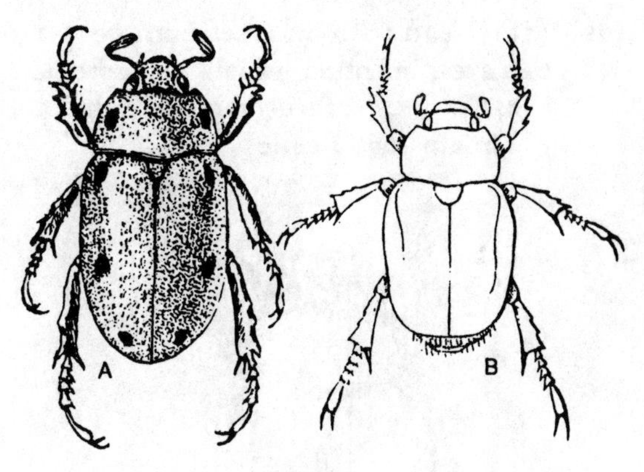

Figure 429

(a) ***Pelidnota punctata* (L.)**

LENGTH: 20-25 mm. RANGE: North America.

Convex, rather dull brownish-yellow with black markings. Under parts blackish or greenish-bronze. The adults feed on grape leaves while the grubs feed on the roots of various trees.

(b) ***Cotalpa lanigera* (L.)**

LENGTH: 20-26 mm. RANGE: Eastern half of United States and Canada.

Robust. Head, thorax and scutellum greenish or yellow with a strong metallic sheen; elytra yellowish, less metallic; under parts dark, densely clothed with long wooly hairs. Feeds on willow. Popularly known as the "Goldsmith Beetle." Occasional sport forms show areas of dark green or chestnut.

15a (13) Clypeus notched at sides in front of eyes, exposing base of antennae. 25

15b Clypeus not as in 14a. 16

16a (15) **Head with an erect horn between the eyes; mentum nearly semicircular. Fig. 430. .... *Pleocoma fimbriata* Lec. Fimbriate June Beetle**

Figure 430

LENGTH: 18-28 mm. RANGE: California.

Females; robust, shining, brown; heavily clothed on under side with long fine hairs. Males; considerably smaller than females and black.

Twenty-eight other species of this genus are found in California and adjoining states.

16b **Mentum large, nearly square. 17**

17a (16) **Only one spur at most on middle and hind tibia; hind tarsae with but one claw; body scaly. Fig. 431. .... *Hoplia trivialis* Harold**

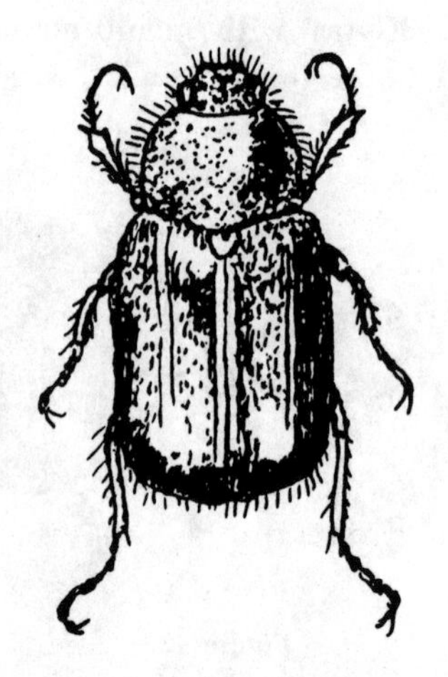

Figure 431

LENGTH: 6-7 mm. RANGE: Central and Eastern United States.

Dull black; upper surface covered with hair-like scales; pygidium and abdomen covered with silvery scales.

The Grape-vine *Hoplia* of the west coast, *H. callipyge* Lec. is destructive to many other plants as well as grape. It measures around 7 mm and is mottled brown. This genus includes 14 other widely distributed species.

17b **Two spurs on both middle and hind tbiae; tarsi with 2 claws. 18**

18a (17) **Brownish, (sometimes iridescent) heavy, robust beetles. 19**

18b **Dull yellow or metallic green or bronzed elongated beetles. 21**

19a (18) **With five ventral segments; with punctured striae or elytra rather evenly punctured. Fig. 432. .... *Diplotaxis harperi* Blanch.**

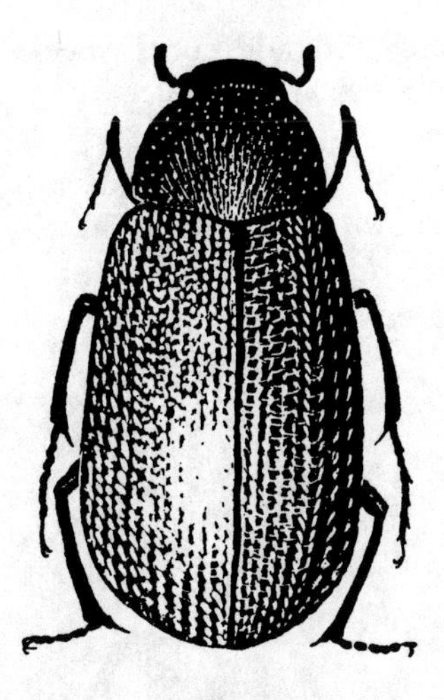

Figure 432

LENGTH: 8-10 mm. RANGE: Eastern half of United States.

Convex, shining. Reddish-brown to blackish; elytra each with three indistinct ridges each of which bears scattered punctures with a row of punctures on each side.

D. tenebrosa Fall (7.5-10 mm) ranges throughout the West. It is black, and damages young fruit trees. Some 125 species in this genus, widely distributed, and very similar in appearance.

19b With six ventral segments. 20

20a (19) Not over 10 mm with small regular grooves on disk. Fig. 433. .... *Serica evidens* Blatch.

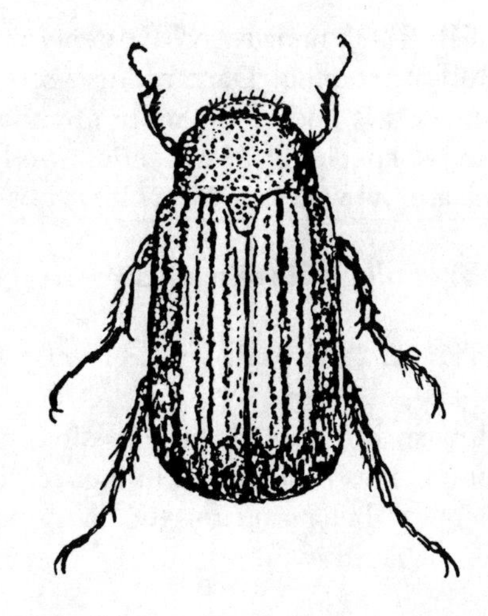

Figure 433

LENGTH: 8-10 mm. RANGE: Central States.

Brown, feebly shining; under parts reddish-brown.

S. anthracina Lec. 7.5 mm, dark brown to black defoliates fruit trees from British Columbia to California.

Some 73 species in all are known for this genus, of which S. *sericea* (Ill.) is one or the most common and widely distributed. It is piceous or purplish-brown with a marked iridescence. Several species have a notch at each side of the clypeus; this and a few others do not. It is 8-10 mm long.

20b 11.5 mm or larger; without striae or grooves. Fig. 434.

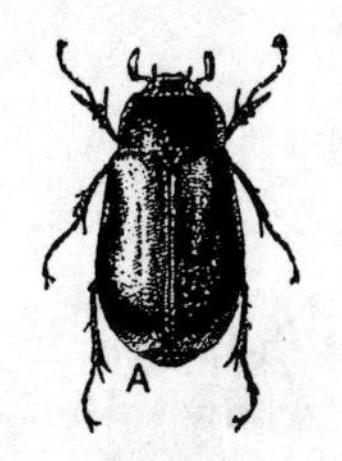

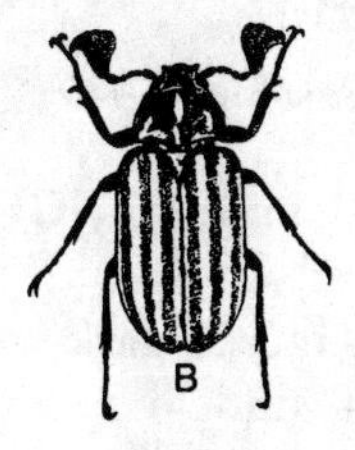

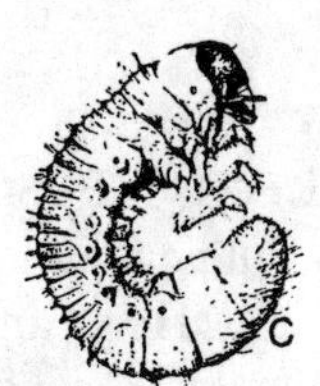

Figure 434

(a) *Phyllophaga fervida* (Fab.)

LENGTH: 17-23 mm. RANGE: general.

Robust, shining. Dark reddish-brown to blackish. This is one of the more abundant of our over 140 species of May beetles, the larvae of which are the well known "white grubs" (c).

(b) *Polyphylla crinata* Lec.

LENGTH: 22-28 mm. RANGE: Pacific Coast States.

Robust, brown, striped with silver scales as pictured. This is one of a number of "Lined June Beetles" belonging for the most part to the West or South-west.

21a (18) Elytra covered with hairs, not scaly. Fig. 435. .... *Dichelonyx linearis* (Gyll.)

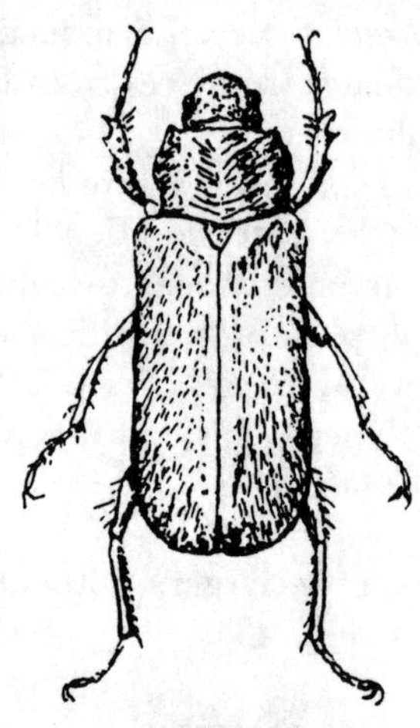

Figure 435

LENGTH: 8-10 mm. RANGE: Eastern half United States and Canada.

Head and thorax dark reddish-brown or blackish; elytra dull brownish-yellow often with greenish, bronze or purplish sheen.

Crotch's Green Pine Chafer *D. fulgida* Lec. a slender reddish-black beetle measuring about 9 mm has bright metallic green elytra. It feeds on pine leaves in the mountains from Montana and British Columbia to California. In our area, 26 additional species of these greenish beetles are known.

21b Elytra densely covered with elongate yellowish scales. Fig. 436. .... *Macrodactylus subspinosus* (Fab.) Rose Chafer

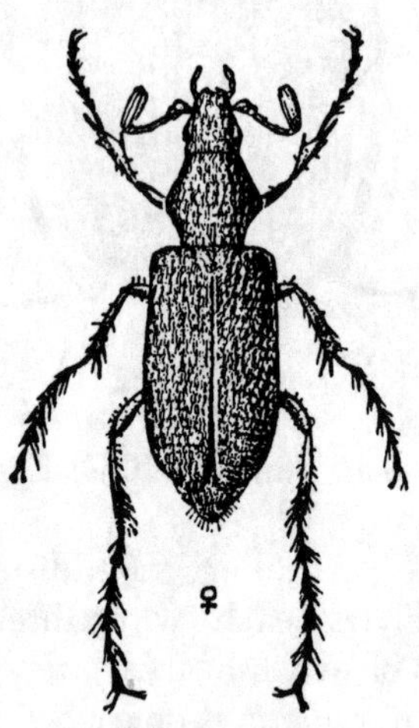

Figure 436

LENGTH: 8-10 mm. RANGE: United States.

Dull brownish-yellow or reddish-brown thickly clothed with yellow scales. Under parts darker. (U.S.D.A.)

The Western Rose Chafer, *M. uniformis* Horn about 10 mm long and yellowish brown, resembles *subspinosus* in structure and habits.

22a (12) Head or thorax, or both, with either a horn or tubercle. 23

22b Without horn or tubercles as above. Fig. 437. .... *Cyclocephala borealis* Arrow

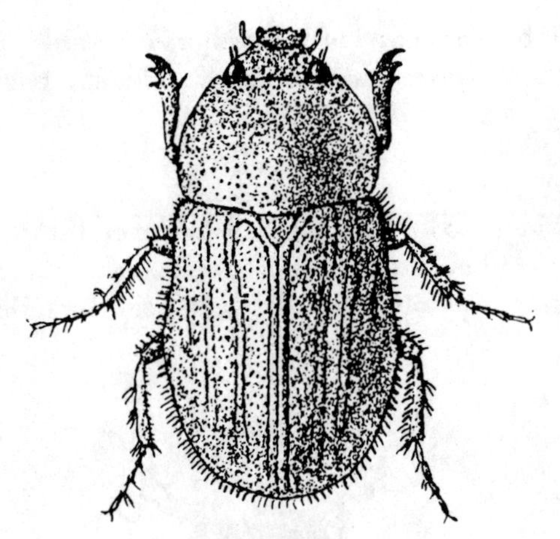

Figure 437

LENGTH: 11-14 mm. RANGE: United States. Convex.

Pale yellow to dull brown clothed with scattered fine hairs. This "Desert June Beetle" ranges throughout the United States. Its grub is foe to the roots of grasses and cereals. The adults feed on pigweed. Being active at night it is often abundant at lights. 10 similarly colored species are recognized in the U. S.

23a (22) Greenish-gray with black spots; front tarsi of males elongated. Fig. 438. .... *Dynastes tityus* (L.)

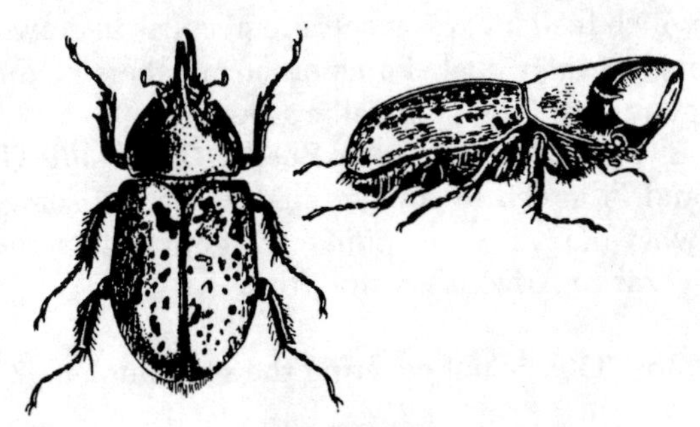

Figure 438

LENGTH: male 45-60 mm, female 40-48 mm. RANGE: South, Central and Southeast United States.

Heavy. Usually gray with variable black or brown markings; rarely wholly chestnut brown. Male with horns as pictured; female with small tubercules only on head.

This genus has the distinction of including one of the world's largest known beetle *D. hercules* (L.) It lives in the West Indies and northern South America. The males may measure 6 inches or more. It is shaped much like *tityus*. The female is dull brownish or blackish and roughened with coarse punctures and a small tubercle on the head. The male is shining black; its elytra are spotted with black and dull green. A fringe of rust-red hairs ornament the long upper horn on its ventral side.

23b Dark reddish-brown or blackish; males with front tarsi not elongated. 24

24a (23) Vertex with long horn (male) or tubercle (female); hind tibiae with blunt rounded teeth at tip. Fig. 439. .... *Xyloryctes jamaicensis* (Drury)

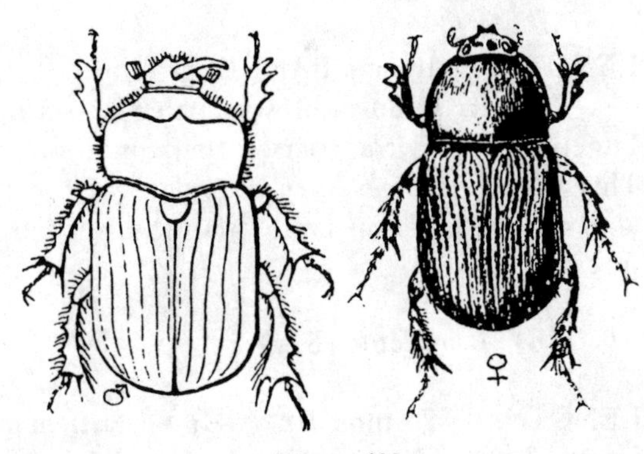

Figure 439

LENGTH: 25-30 mm. RANGE: general.

Robust. Reddish or blackish-brown; under parts paler and covered thickly with red-brown hairs. Commonly known as "Rhinoceros beetle" but several other species also bear that name.

A very interesting mahogany-brown species. *Strategus antaeus* (Drury), inhabits our southern states. The male in addition to a prominent horn on its head has two other large horns, one at each side of the thorax. It is called the "Ox-beetle."

24b Hind tibiae with hairs at apex but no teeth; head with a transverse ridge. Fig. 440.

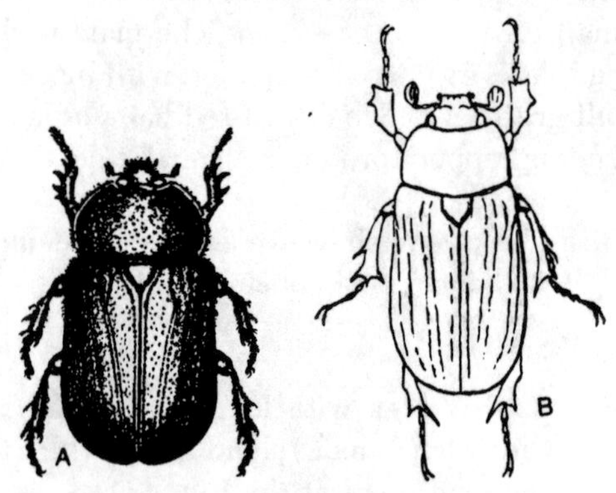

Figure 440

(a) ***Bothynus gibbosus* (DeG.) Carrot Beetle**

LENGTH: 11-16 mm. RANGE: general.

Robust. Reddish-brown, paler beneath. Punctures on elytra coarser than on thorax. This beetle becomes very abundant at times and may be taken in large quantities in light traps.

(b) ***B. relictus* (Say)**

LENGTH: 18-23 mm. RANGE: Eastern half United States.

Robust. Shining. Brownish-black; transverse carina of head broken at middle.

25a (15) Epimera of mesothorax not visable from above; sides of elytra not sinuate behind the shoulders. 28

25b Epimera of mesothorax visable from above; sides of elytra sinuate back of shoulder. 26

26a (25) Thorax lobed to cover the scutellum. Fig. 441. ....
.... *Cotinis nitida* (L.) Green June Beetle

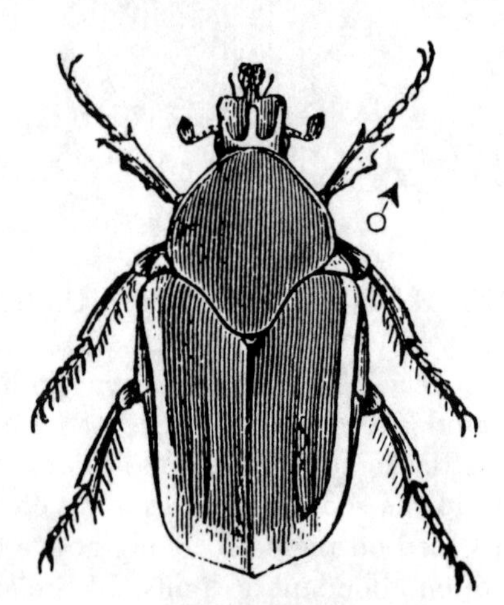

Figure 441

LENGTH: 20-23 mm. RANGE: Eastern and Southern States.

Robust. Dull velvety green marked with yellowish-brown as shown; under parts with metallic green and reddish-yellow. The larvae which feed on grass roots are peculiar in crawling on their back. Some other members of the genus are shining metallic green.

The Green Peach Beetle, *C. palliata* G. and P., green to coppery and violet and somewhat larger than *nitida,* is abundant from Texas to Arizona on ripe fruit.

26b Thorax not covering the scutellum. 27

27a (26) Color wholly black; mentum cup-shaped; clypeus broad. Fig. 442. ....
.... *Cremastocheilus castaneae* Knoch.

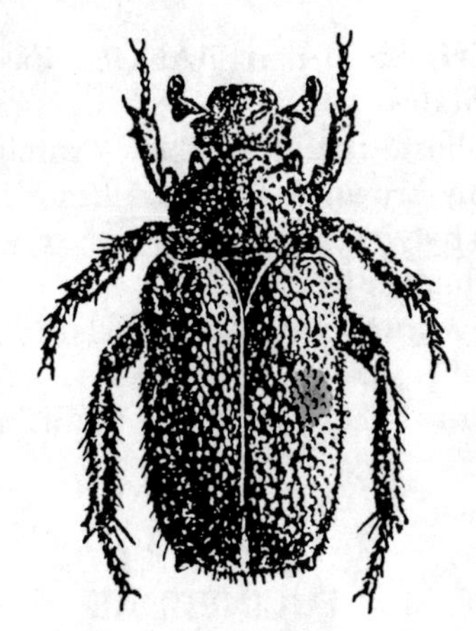

Figure 442

LENGTH: 9-10 mm. RANGE: from Colorado eastward.

Black, slightly shining. Elytra somewhat flattened with shallow impressions. Pygidium very coarsely punctate. The members of this genus of nearly 50 species are associated with ants (perhaps even enslaved by them).

27b Color never wholly black; clypeus narrow; mentum normal. Fig. 443. .... *Euphoria inda* (L.) Bumble Flower Beetle

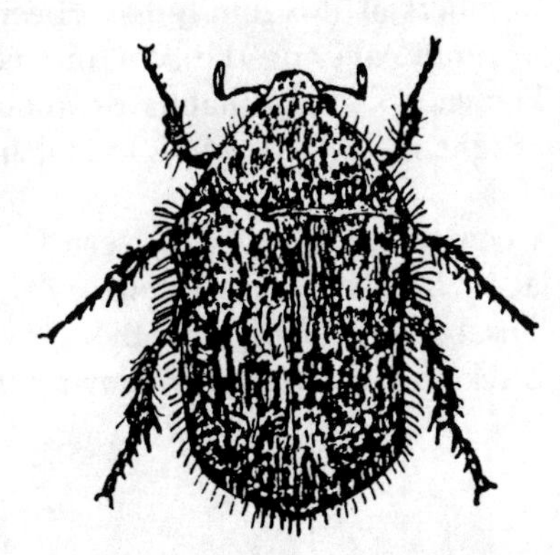

Figure 443

LENGTH: 13-16 mm. RANGE: general.

Heavy. Head and thorax dark, somewhat bronzed; elytra brownish-yellow with black spots. The whole insect densely hairy. It gets its name because of the buzzing noise it makes as it flies near the ground. The adults feed on fruit. Some 30 species are found in this colorful and related genera.

28a (25) Hind coxae touching; thorax not grooved. 29

28b Hind coxae separated; thorax with deep groove at middle. Fig. 444. .... *Valgus seticollis* (Beauv.)

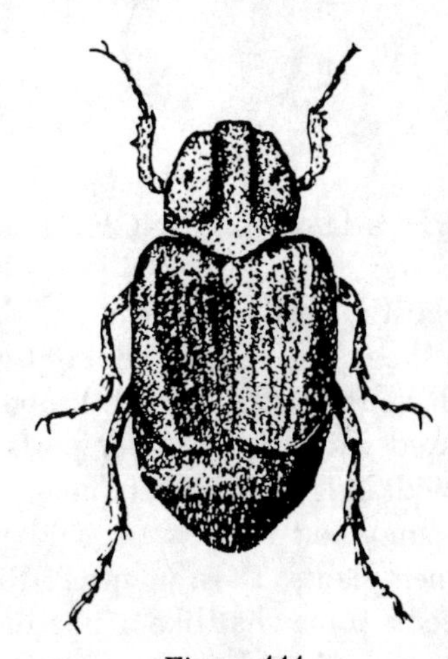

Figure 444

LENGTH: 6-7.5 mm. RANGE: Eastern half United States.

Blackish-brown. Thorax coarsely punctate with large teeth on margins; elytra and pygidium granulate. Found on flowers during the Spring. In the same range and more common is *V. canaliculatus* (Fab.); smaller, less than 6 mm; front tibiae with 3 or more widely separated, slender teeth; 5-6 closely placed, stout rounded ones in *V. seticollis;* common on flowers of shrubs such as *Spiraea.*

29a (28) Body very hairy; not over 13 mm. Fig. 445. .... *Trichiotinus piger* (Fab.)

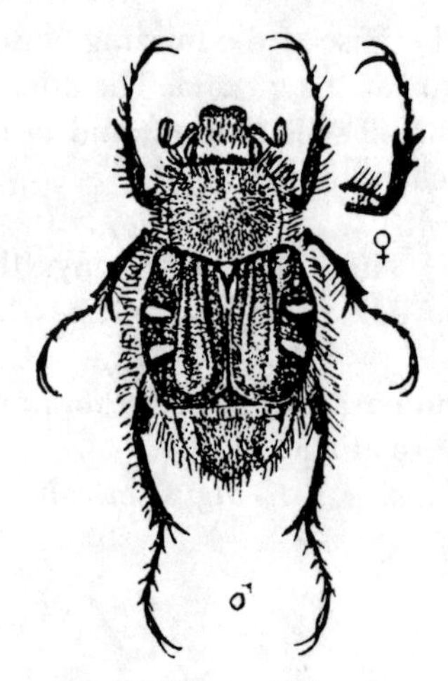

Figure 445

LENGTH: 9-11 mm. RANGE: Eastern half United States.

Robust. Head and thorax dark with greenish sheen and usually covered with erect yellowish hairs; elytra reddish-brown to piceous, marked with white; under parts bronzed, clothed with long white silken hairs.

Common on flowers of wild roses and many other plants. It is an interesting beetle which seems somewhat like a bee in both appearance and behavior. Seven other similar species are found in our fauna.

29b Body almost wholly without hairs. Fig. 446. .... *Osmoderma eremicola* Knoch.

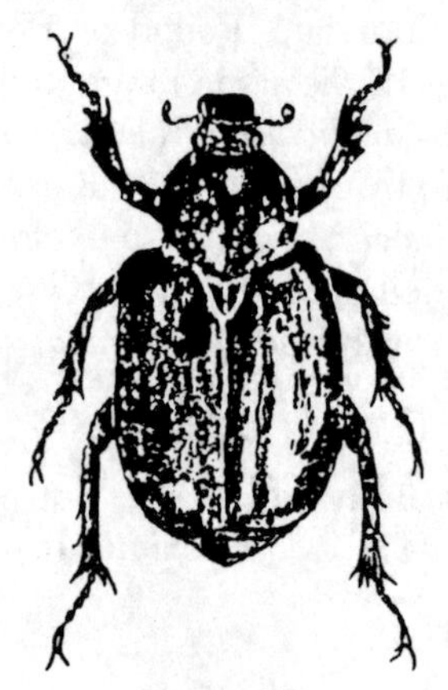

Figure 446

LENGTH: 25-30 mm. RANGE: Eastern half United States.

Robust, flattened above, shining. Dark mahogany-brown to blackish; head deeply excavated between the eyes. Thorax with deep median line on basal half.

O. scabra (Beauv.) is similar in color and range but is smaller. 18-25 mm.

Both of these are found in rotting logs.

FAMILY 31. EUCINETIDAE
The Plate-Thigh Beetles

Members of this family are elongate, oval, convex beetles ranging in size from 2.5 to 3 mm. They vary in color from brown to black and are pubescent. A distintinguishing characteristic of the family is the greatly enlarged hind coxal plate similar to that found in the *Haliplidae*. The elongate head is deflexed and rests on the front coxae. The antennae are 11-segmented. The pronotum is broader than the head, wide, short, and narrowed in front. The entire elytra are rounded apically and are minutely transversely strigose.

Members of this family have been taken in or on fungi covering the bark of trees. Little is known about the habits or about their larvae. Eight species are found in North America.

A common species of Eastern U. S. and Canada is *Eucinetus terminalis* LeC. (Fig. 447). The body is black with the apex of the elytra brick-red. The legs are yellowish. Length 3 mm.

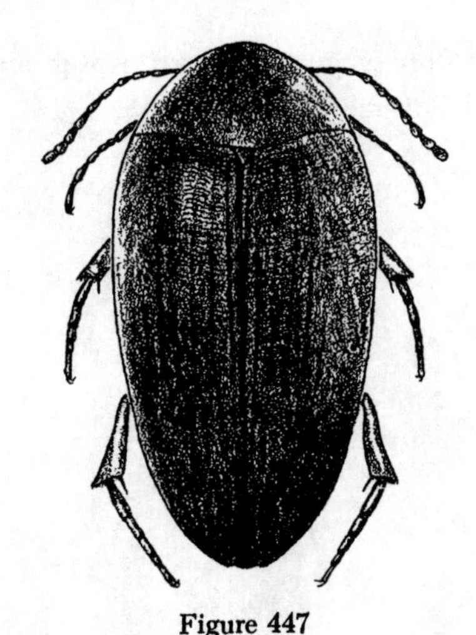

Figure 447

FAMILY 32. HELODIDAE
The Marsh Beetles

This is a small family of convex, oval, or elongate beetles. They vary in size from 2 to 4 mm. The color runs from testaceous to piceous and they may be spotted or striped. Their soft bodies are covered with rather dense and long pubescence. Two genera have the hind femora greatly enlarged for jumping. The above characters plus the conical front coxae separate this family from related forms. The oval head is deflexed with 11-segmented subserrate antennae. The pronotum is broad, short, explanate and partially covering the head. The entire elytra are apically rounded and rather finely punctate.

Adults are taken by sweeping herbage along the banks of lakes and streams and in marshy places. Some species are attracted to lights in large numbers. Thirty-four species are found in the U. S. and Canada. The following are typical:

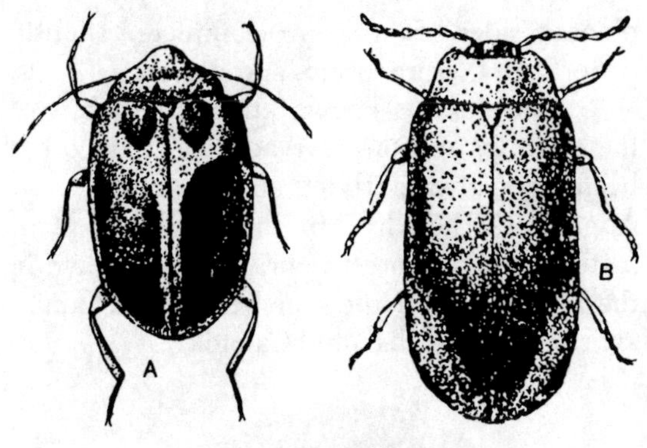

Figure 448

(a) *Elodes pulchella* Guer.

LENGTH: 3.5-5 mm. RANGE: Eastern half United States.

Yellow marked with black. Thorax finely punctate, elytra densely punctured.

(b) *Cyphon collaris* (Guer.)

LENGTH: 3.5-4 mm. RANGE: Eastern half United States.

Black shining, with fine pubescence. Marked with reddish-yellow as shown including base of antennae and tibia and tarsae.

FAMILY 33. CLAMBIDAE
The Minute Beetles

These tiny beetles, about 1 mm in length, are oval, compact, and convex. In color they vary from testaceous to piceous and the body is densely hairy. Very distinguishing features are the expanded hind coxal plates and the fringe of hair on the inner wings. The nature of the body is such that the beetles can roll themselves into a ball. The strongly deflexed head is flat and broadly oval. The antennae have from 8 to 10 segments with segment 1 enlarged and the last two expanded into a distinct club. The pronotum is short, very broad, with explanate sides. Elytra are complete, slightly wider than the pronotum, and convex.

Not much is known about the habits of these beetles and their larvae. They have often been taken at dusk flying above a forest floor. Apparently they live in the decaying plant materials of the forest floor. Some may live in the nest of ants. Nine members of the family are found in the U. S. and Canada.

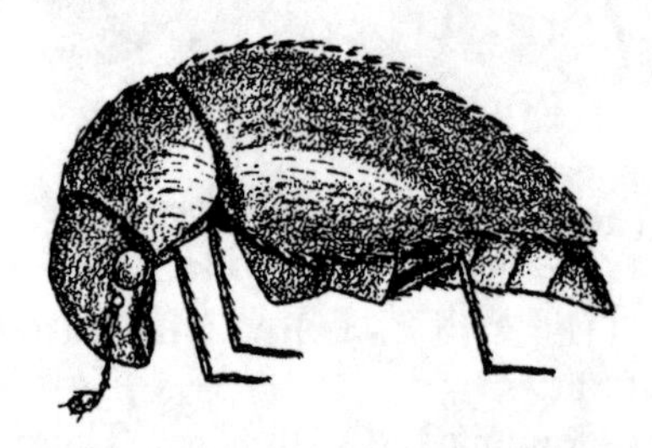

Figure 449

Figure 449 *Clambus pubescens* Redt. Length 1 mm. Range: Eastern half of U. S. and Canada. Piceous, shining, sometimes the elytra reddish-brown, pubescent. Appendages also piceous. In mounted specimens the body is much contracted.

FAMILY 34. DASCILLIDAE
The Soft-Bodied Plant Beetles

This is a small family of beetles that range in size from 3 to 14 mm. They are black, brown, or mottled in color. They may be either elongate and convex or short and broad. The body is covered with hairs that vary from moderate to dense and that are rather long. The broad, triangular head may or may not be deflexed. The 11-segmented antennae vary from subserrate to pectinate or flabellate. The pronotum is much wider than the head, convex, quadrate, with evenly arcuate sides that are posteriorly sinuate.

Not much is known about this family. Some of the larvae live in the soil. Adults are swept from vegetation near water and from trees that are not near water.

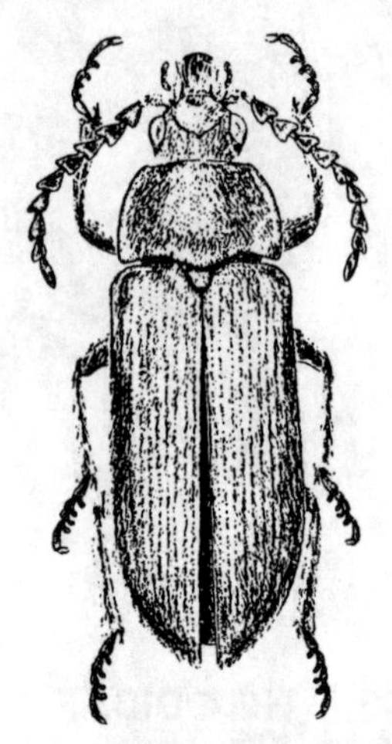

Figure 450

Figure 450 *Dascillus davidsoni* LeC. Length 12-13 mm. Range: California. Elongate, convex; body and legs piceous, elytra brownish; densely covered with pubescence.

FAMILY 35. BYRRHIDAE
The Pill Beetles

The distinctive characteristics of this family are a very convex shape, deflexed head, a carinate labrum, and transverse front coxae. The body form is oval and very convex. Body size ranges from 5 to 10 mm. The integument may be shiny, covered with clavate bristles or erect hairs, or with a dense covering of recumbent hairs that give the surface a velvety appearance. The surface color is very variable and some species are brilliantly metallic. The convex, deflexed head bears 11-segmented antennae that have a club consisting of the gradual enlargement of the last 3 to 7 segments or that may be filamentous.

Both adults and larvae are herbivorus. They live in sand under logs, in moss, etc. Often they are taken under boards and stones where they are in a position with all the appandages retracted. Then they resemble a seed or a small pebble. Hence the common name "pill bug."

Some 40 species are found in the U. S. and Canada.

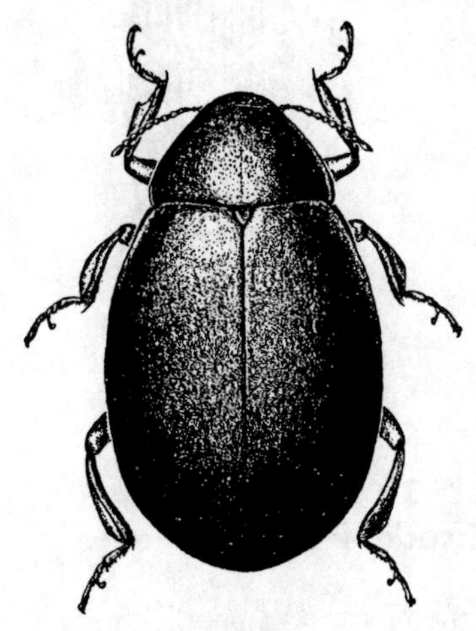

Figure 451

Figure 451 *Aphicyrta chrysomelina* Er. Length 8-10 mm. Range along the coast from western Washington to California. Castaneous with the head and pronotum darker; glabrous, finely closely punctate; elytra widest behind the middle. Larvae damage vegetable seedlings, carrots, beets, onions, and lily bulbs.

FAMILY 36. PSEPHENIDAE
Water-Penny Beetles

Members of this family are elongate-oval beetles ranging in size from 4 to 6 mm. Luteous to piceous black forms occur covered with short, fine, dense, suberect hairs. The head is slightly deflexed and slightly retracted into the pronotum. The serrate antennae possess 11 segments. The pronotum is wider than the head, but anteriorly it is narrowed to the width of the head. The elytra are entire with rounded apices, the surface costate or vaguely sulcate.

The adults and larvae are aquatic. The flattened larvae are disc-shaped (hence water-penny), almost as broad as long with protruding projections.

Adults may be taken by sweeping vegetation along either running or still water or by picking them off the surface of submerged stones. Six or more species are found in the U. S. and Canada.

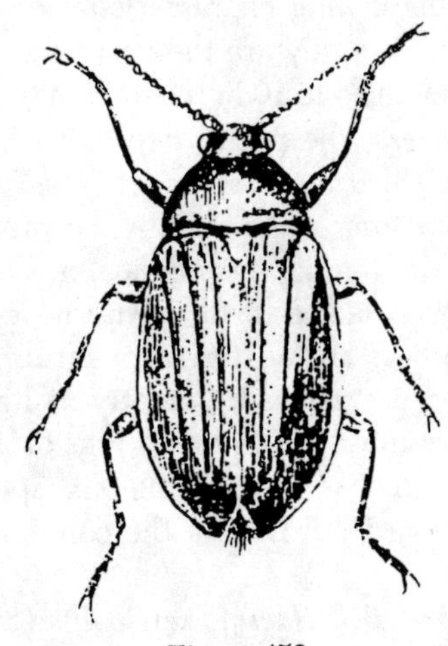

Figure 452

Figure 452 *Psephenus herricki* DeKay. Length 4-6 mm. Range: Eastern U. S. and Canada. Piceous to black, the head and pronotum darker; densely covered with fine hairs; elytra vaguely sulcate.

FAMILY 37. BRACHYPSECTRIDAE
The Texas Beetle

This family contains a single species, *Brachypsectra fulva* LeC. (Fig. 82). These beetles are shaped somewhat like click beetles, rather broad, somewhat depressed, yellow testaceous in color, and covered with fine sparse pubescence. The head is short, broad, coarsely punctate with a large central depression. The 11-segmented antenna has segments five to eleven serrate. The pronotum is short, broader than long, narrowed in front with the basal angles acute. These uncommon beetles are found under bark.

FAMILY 38. PTILODACTYLIDAE
The Toed-Wing Beetles

The distinguishing characteristics of members of this small family are the very long antennae with segments 4 to 10 in the male having a long basal process, the partly depressed head, and the lobed third tarsal segment. The female antennae are long and simple. The pronotum is broader than long, triangular in shape, and partly covering the head when viewed dorsally. The adults are beaten from shrubbery near water and are taken frequently at lights. The larvae live in decaying damp logs or are aquatic. Two genera comprising six species are found in the U. S. One of the commonest species is:

Figure 453 *Ptilodactyla serricollis* (Say)

LENGTH: 4-6 mm. RANGE: Eastern half of United States.

Chestnut brown to blackish with thin pubescence; legs and antennae pale.

P. angustata Horn. A closely related species is known from Florida.

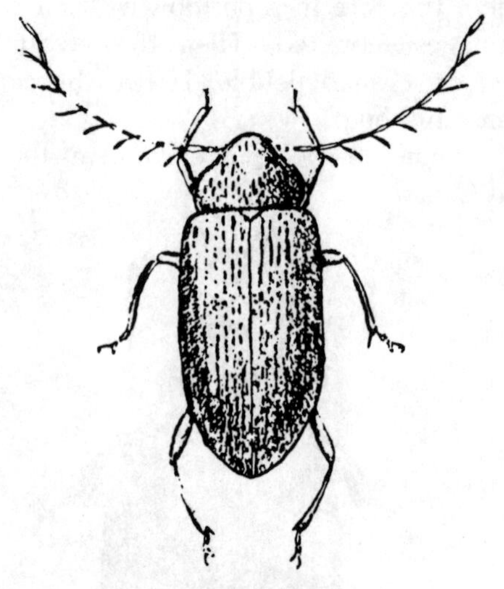

Figure 453

FAMILY 39. RHIPICERIDAE
The Cicada Parasite Beetles

These members of a small family are characterized by the fan-shaped antennae of the males (usually serrate in the females). They are found under bark and logs but are rather rare at best.

Tarsi with distinct lobes; antennáe flabellate in male, serrate in female short. Fig. 454. .... *Sandalus niger* Knoch

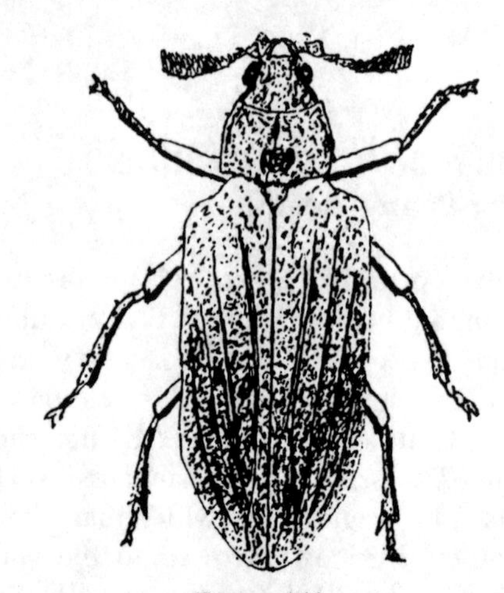

Figure 454

LENGTH: 21-24 mm. RANGE: North and South Central States.

Black, somewhat shining; elytra about twice the width of thorax, with faint ridges.

S. petrophya Knoch with a length of 12-17 mm and ranging through the eastern one-third of the United States is dark reddish-brown to black. The male is noticeably smaller than the female.

FAMILY 40. CALLIRHIPIDAE
The Cedar Beetles

Members of this family were once included in the *Rhipercidae*. On the basis of larval characteristics the two families were separated. Like the previous family, this one includes a single genus.

Tarsi not lobed; antennae serrate, moderately long. Fig. 455. .... *Zenoa picea* (Beauv.)

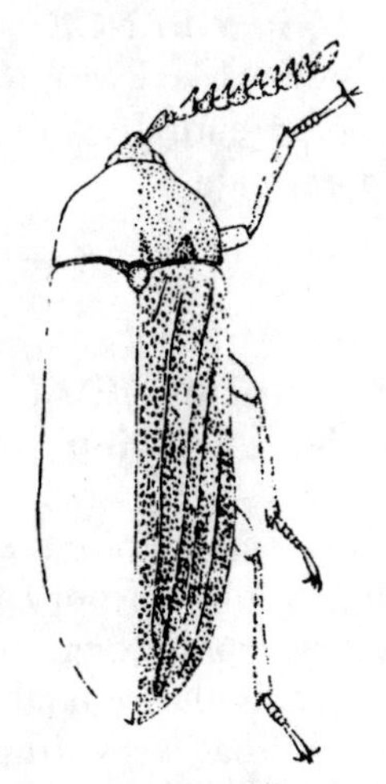

Figure 455

LENGTH: 11-15 mm. RANGE: Indiana to Florida.

Dark reddish or blackish-brown. Elytra each with four raised lines and a short oblique raised line on each side of the scutellum. Found under bark and in logs in dry woods. Comes to light.

FAMILY 41. CHELONARIIDAE
The Turtle Beetles

In this family there is one genus and one North American species, *Chelonarium lecontei* Thom., Fig. 455a, which is found along the Atlantic Coast from North Carolina to Florida. Members of this family are very compact with the head drawn into the pronotum and invisible from above. These beetles are broadly oval, convex, strongly contractile, and between 4 and 5 mm in length. In color they vary from pale brown to black. The elytra have patches of whitish scale-like hairs scattered over the surface.

The 11-segmented antennae are modified for reception into grooves on the pronotum. The pronotum is much broader than the head, possesses a prominent frontal ridge, is evenly arcuate to the front, and has strongly margined borders. The elytra are very compact, punctate, and rounded apically.

Adults are taken from vegetation, probably near water, as the larvae are aquatic.

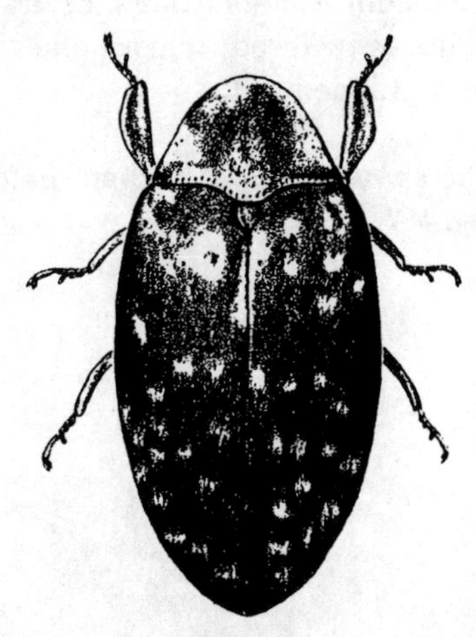

Figure 455a

FAMILY 42. HETEROCERIDAE
The Variegated Mud-Loving Beetles

Here is a small family of somewhat flattened grayish beetles which burrow in sand and mud and are often so mud-covered as to require cleaning before accurate determinations are possible.

1a Thorax with a pale median stripe. Fig. 456. .... *Centuriatus auromicans* Kies.

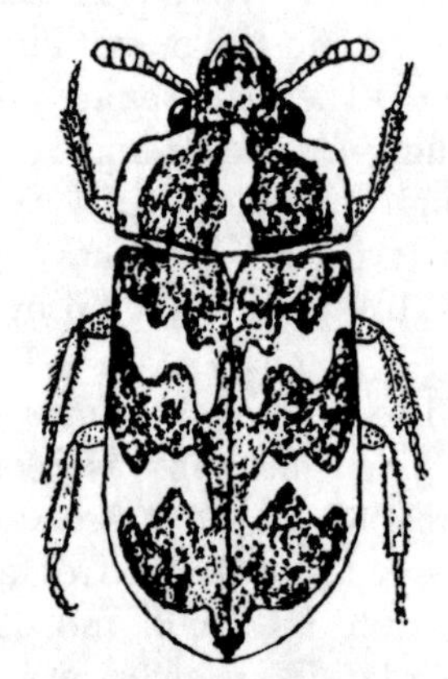

Figure 456

LENGTH: 3.5-4.5 mm. RANGE: general.

Blackish, sparingly clothed with golden short recumbent scale-like hairs. Elytra marked with yellow as pictured; femora and tarsi reddish-yellow, tibiae black.

1b Thorax without a median pale stripe. Fig. 457.

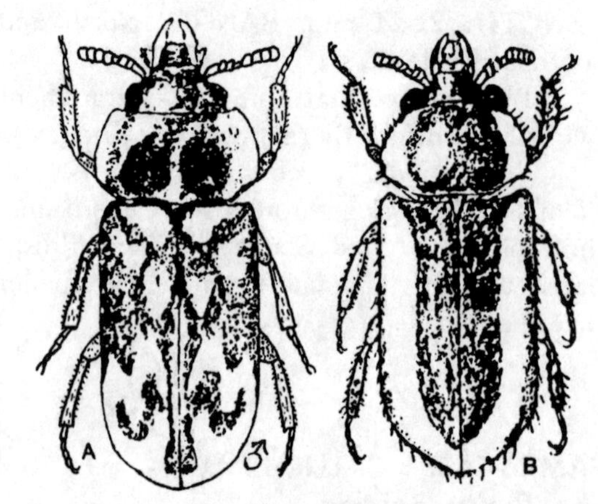

Figure 457

(a) *Dampfius collaris* Kies.

LENGTH: 2.5-4 mm RANGE: general.

Sooty-brown to black, covered with brownish hairs, sides of thorax and elytra marked with pale yellow as pictured; legs reddish-yellow.

(b) *Tropicus pusillus* Say

LENGTH: 2-2.5 mm. RANGE: general.

Dull yellow marked with sooty-brown or sometimes colored entirely with either one of these colors. Legs pale.

FAMILY 43. LIMNICHIDAE
The Minute Marsh-Loving Beetles

Members of this small family are oval, convex beetles varying in length from 1 to 3.6 mm. In color they appear from brown to black, sometimes with a metallic lustre, and covered with fine dense hairs that vary from grayish to golden. The head is small, inserted into the pronotum, with 10-segmented antennae that are short and clavate with most of the segments broad. The pronotum which is wider than long is subquadrate with margined sides.

The elytra are slightly wider than the pronotum and both are punctate.

Adults are taken in or along streams. Some come to lights. The larvae are aquatic. Thirty-two species are found in the U. S.: The largest is shown in Fig. 458. *Lustrochus laticeps* Csy. Length 3-3.6 mm. Range from Arkansas through Ind., Mich., Ohio, to D. C. Black with a slight greenish lustre, densely pubescent; the legs blackish to reddish-brown.

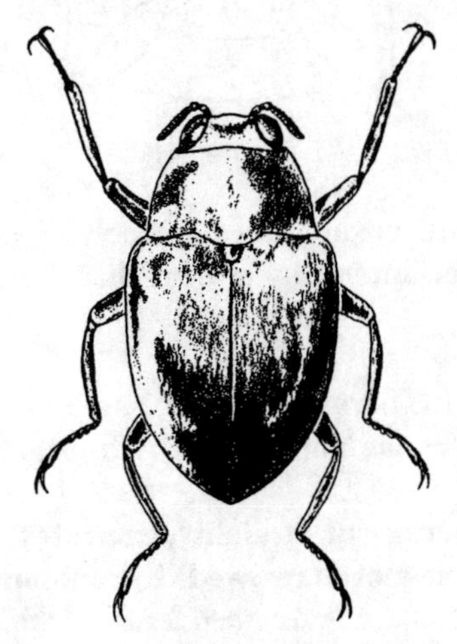

Figure 458

FAMILY 44. DRYOPIDAE
The Long-Toed Water Beetles

Members of this family are elongate, oval, convex beetles varying in size from 1 to 8 mm. In color they are brown or piceous. They may possess conspicuous vestiture or none at all. The head is deflexed and inserted into the pronotum. The 11-segmented antennae that are inserted under a prominent frontal ridge have segments 4 to 11 produced forming a loose lamellate club. The pronotum is wider than the head, is rather ovate with laterally arcuate margined borders, and is roughly punctate. The entire elytra have punctured striae.

The larvae are aquatic and the adults may or may not be. Both are herbivorus. The adults of the aquatic species crawl along the bottoms of streams. When present, they may be taken by lifting up submerged stones or pieces of wood and picking the beetles from them. The family is small with less than 20 known North American Species.

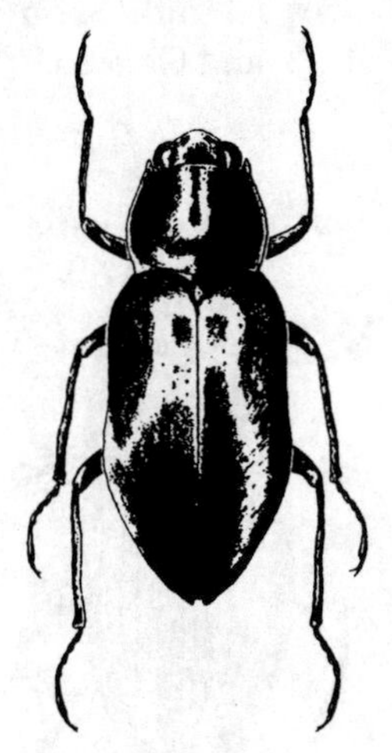

Figure 459

Figure 459 *Helichus lithophilus* (Germ.) Length 5-5.5 mm. Range Eastern Canada and U. S. Brown everywhere covered with a dense, evenly distributed, silky pubescence with a bronze lustre. Last abdominal sternite reddish with no pubescence.

FAMILY 45. ELMIDAE
The Drive or Riffle Beetles

Members of this family are elongate, somewhat depressed beetles ranging in size from 1 to 8 mm. In color they are piceous often with yellow spots or stripes. Both the legs and the 11-segmented antennae are long and slender. The pronotum is broader than the head, irregularly quadrate, produced in front, with the borders laterally crenulate, and having the surface carinate and rugosely punctate. The elytra are rounded apically, carinate, and also rugosely punctate.

The adults are both aquatic and terrestial, while the larvae are aquatic. Most species live in rapidly running, cold streams; however several species have been taken along the edges of lakes and ponds in northern climes. This is a moderately large family with some eighty species in the U. S. and Canada.

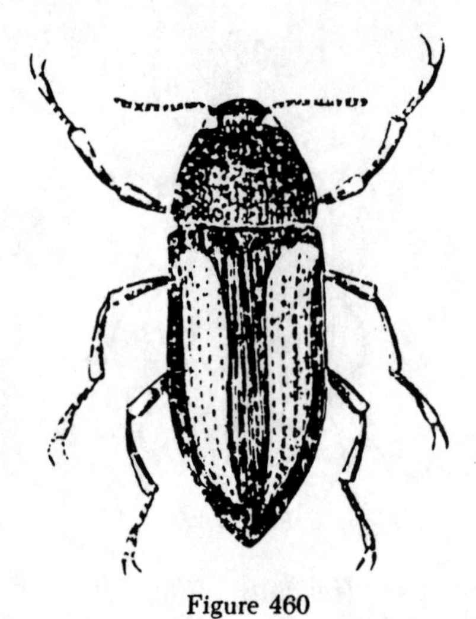

Figure 460

Figure 460 *Dubiraphia bivittata* (LeC.)

LENGTH: 3 mm. RANGE: Eastern and north U. S.

Shiny black, marked with yellowish as pictured.

FAMILY 46. BUPRESTIDAE
The Metallic Woodborers

Engineered much like the click beetles, the some 8000 species belonging to this family appear as though covered with metal; a sort of "robot insect." They are often bronzed or with other metallic lusters. They are destructive borers in woody plants. They may be found sunning themselves on tree trunks, fence posts, etc. and are frequently so much alert as to be difficult to capture.

1a Hind coxal plates distinctly dilated near the base (a), the outer end cut by prolongation of the abdomen; fourth tarsal joint not lobed. Fig. 461. 5

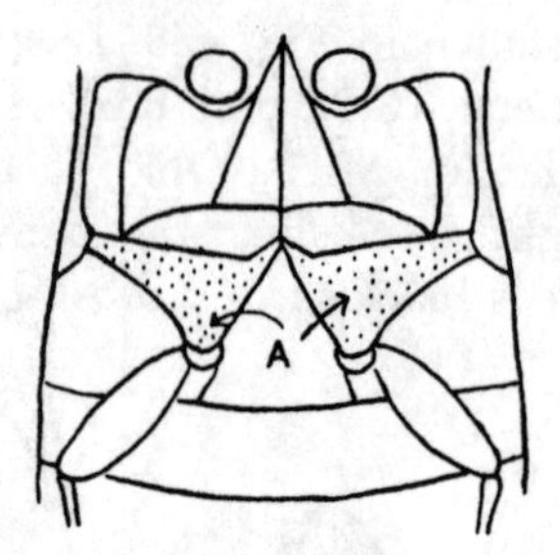

Figure 461

1b Hind coxal plates scarcely if at all dilated internally or near their base. 2

2a (1) Thorax lobed at base, insertion of antennae narrowing the front. 3

2b Thorax cut straight (truncate) at base; front not narrowed by antennae. Fig. 462. .... *Acmaeodera pulchella* Hbst.

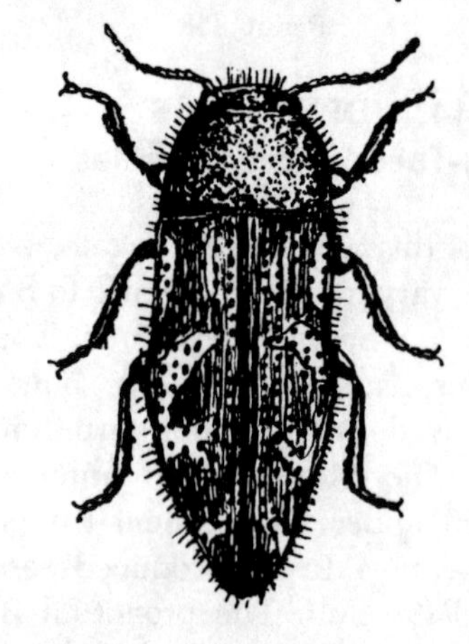

Figure 462

LENGTH: 5-10 mm. RANGE: Eastern and Southern U. S.

Cylindrical, heavy, covered with coarse punctures and brownish hairs. Blackish bronze marked with yellow, markings variable. Rather common on flowers especially New Jersey Tea.

A. tubulus (Fab.) a smaller species 5-8 mm differs in having no yellow on the sides of thorax and with scattering whitish hairs. Its distribution is rather general east of the Rockies. It too is found on flowers, *Crataegus* seeming to be a favorite.

Several species of this genus do severe damage to trees in our western forests. In our fauna there are more than 135 species in this genus.

3a (2) **Antennae at rest in grooves. Body broad. Fig. 463. .... *Brachys ovatus* (Web.)**

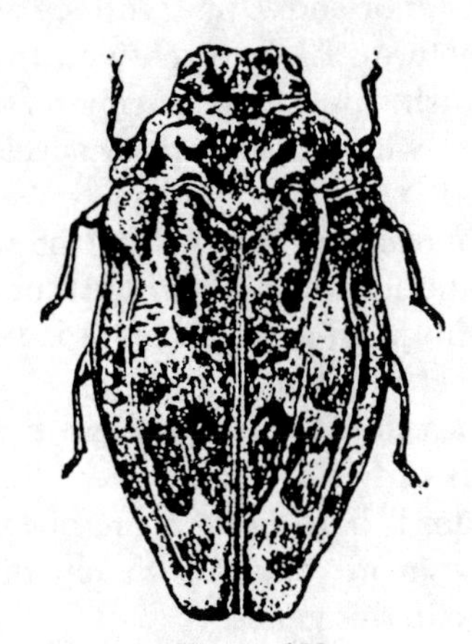

Figure 463

LENGTH: 5-6.5 mm. RANGE: Eastern and Southern U. S.

Flattened. Blackish or purplish bronzed marked with bands of whitish prostrate hairs. The larvae feed within oak leaves.

Taphrocerus gracilis (Say) differs as a genus from the preceding in having the body elongated. This species is a shining blackish-bronze. Its length is 3.5-5 mm and ranges widely throughout Eastern half of U. S. Taken by sweeping.

3b **Antennae without a protecting groove. 4**

4a (3) **Inner prong of tarsal claws (c) turning in and almost touching. Fig. 464.**

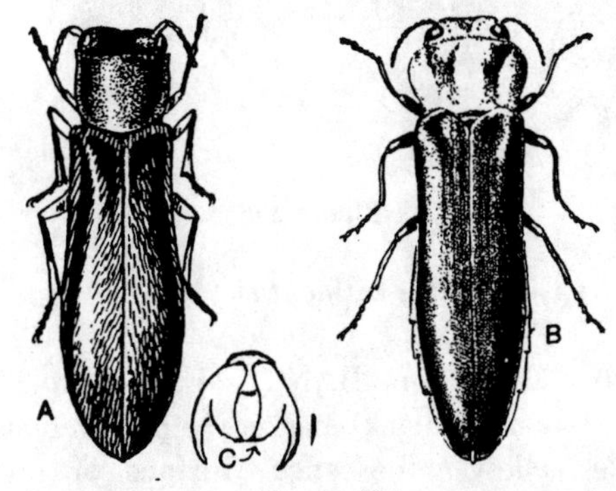

Figure 464

(a) ***Agrilus ruficollis* (Fab.)**

LENGTH: 5.5-8 mm. RANGE: United States and Canada east of Rockies.

Bluish-black; head and thorax coppery-red. Surface of elytra granulate. This is a serious pest of blackberries and raspberries and is known as the Red-necked Cane Borer.

(b) ***A. vittaticollis* Rand.**

LENGTH: 6-7 mm. RANGE: Central and eastern United States.

Head and thorax coppery-red; elytra black. Tarsal claws as pictured.

The Oak Twig-girdler, *A. angelicus* Horn a shining brassy species of California cuts many twigs from oak trees by girding them after its egg is laid in the branch. Its length is 6-7 mm. Over 160 species of *Agrilus* in our fauna.

4b **Inner prong of bifid claw not turning in (c). Fig. 465.**

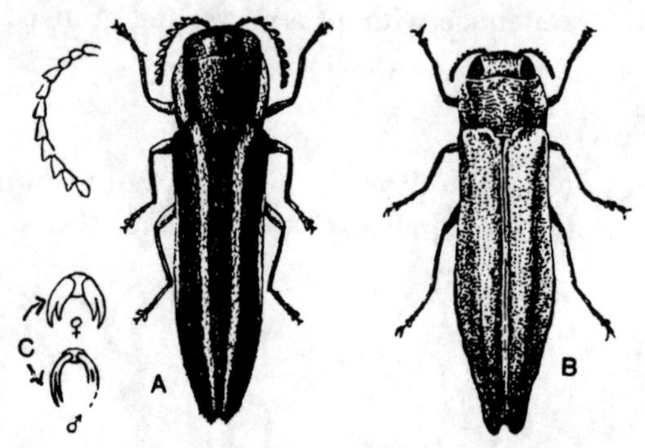

Figure 465

(a) ***Agrilus bilineatus* (Web.)**

LENGTH: 6-9 mm. RANGE: East of Rockies.

Greenish-black, marked as pictured with dense yellow pubescence. Surface of elytra granulate. Feeds on various species of oak.

(b) ***Agrilus anxius* Gory.**

LENGTH: 7-12 mm. RANGE: general.

Olive-green bronzed. Elytra clothed with granulate scales. This Bronze Birch Borer is highly destructive to shade and forest trees.

5a (1) Prosternal spine coming to an acute angle with acute tip behind the front coxae; epimera of metathorax partly covered by the abdomen. 6

5b Prosternal spine obtuse behind the front coxae; epimera of metathorax uncovered, triangular. 7

6a (5) Clypeus small and flattened but not narrowed by insertion of antennae. Fig. 466. .... *Melanophila fulvoguttata* (Har.) The Hemlock Borer

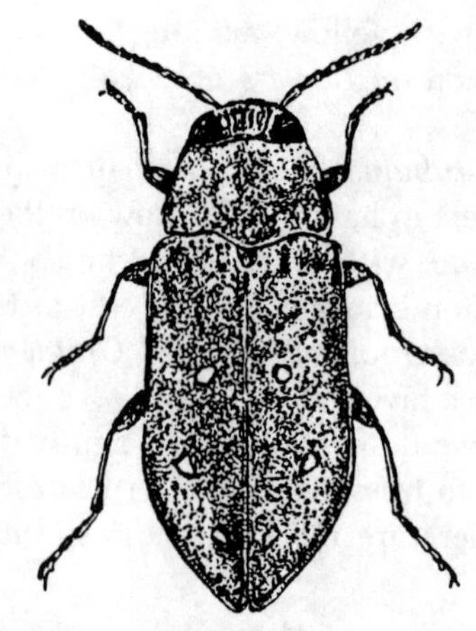

Figure 466

LENGTH: 9-11 mm. RANGE: middle and northern United States.

Blackish bronze with greenish reflections at anterior end. Upper surface with transverse punctures. Elytra each with three white or yellowish spots. An enemy of hemlock, spruce and white pine. Twenty species widely distributed.

Anthaxia quercata (Fab.) of wide general distribution, is shining bluish or purplish, with bright green areas and a broad brownish stripe on each elytron.

It measures 4-6 mm and is a pest of oak. The genus differs from the preceding in that the mentum is horn like instead of leathery in front. Also more than 20 widely distributed species are in this genus.

6b Clypeus reduced in length by insertion of antennae; scutellum large. Fig. 467. .... *Chrysobothris femorata* (Oliv.) Flatheaded Apple Tree Borer

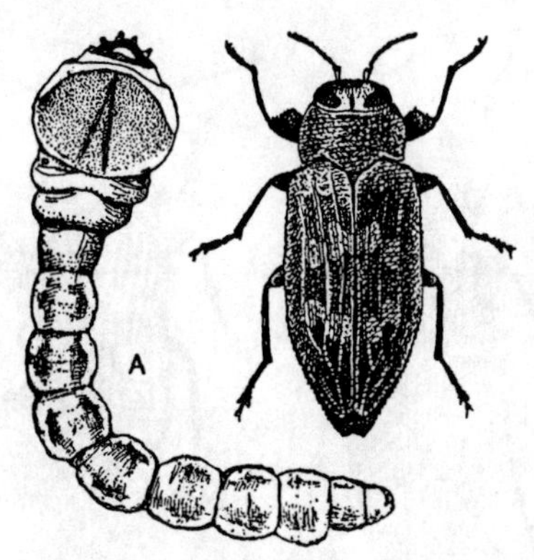

Figure 467

LENGTH: 7-16 mm. RANGE: United States and Canada.

Somewhat flattened. Dark bronze with sheens of green, copper or brass. Third joint of antennae slightly longer than the next two. The name comes from the larva (a).

The Pacific Flatheaded Borer *C. mali* Horn (6-11 mm) may be distinguished from the above by a short lobe on the front margin of the prosternum.

A readily recognized but somewhat smaller species (6-11 mm) is *C. sexsignata* (Say.) It is black above with a faint bronze sheen and green on its under parts. Each elytron is marked with three round brassy spots. The elytra are wider than the thorax. It occurs in central and eastern United States. This is a large genus of more than 125 species in our fauna.

7a (5) Mesosternum and metasternum closely united; 19 mm or more in length. Fig. 468.

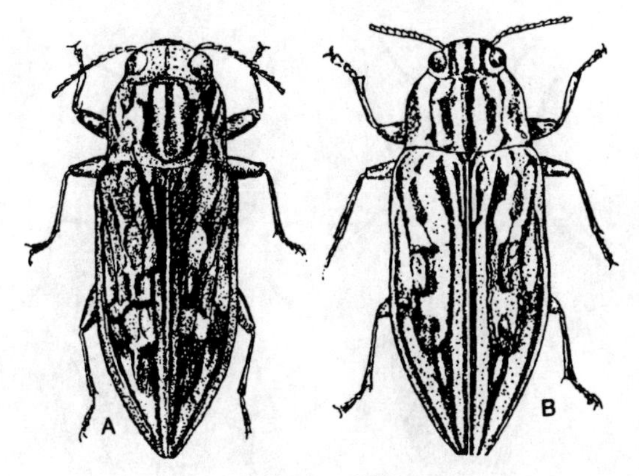

Figure 468

(a) *Chalcophora virginiensis* (Drury)

LENGTH: 23-30 mm. RANGE: Eastern half of United States and Canada.

Dull black, slightly bronzed. Impressions on thorax and elytra as pictured, often brassy. The larvae 1½ inches in length live under the bark of pine trees.

(b) *C. liberta* (Germ.)

LENGTH: 19-25 mm. RANGE: general Eastern U. S. and Can.

Bright coppery or brassy; antennae, legs and markings on head, thorax and elytra blackish-brown. The adults feed on the buds and leaves while the larvae bore within the stems of hard pine.

C. angulicollis (LeC.), shining black with a bronzed luster, smooth, bisulcate pronotum, is common on pine in the West. Three other species found in our area.

7b Mesosternum separated from metasternum by a distinct suture. Usually not over 10 mm in length. 8

8a (7) Mentum wholly horny, tarsi broad, shorter than tibiae. Fig. 469.

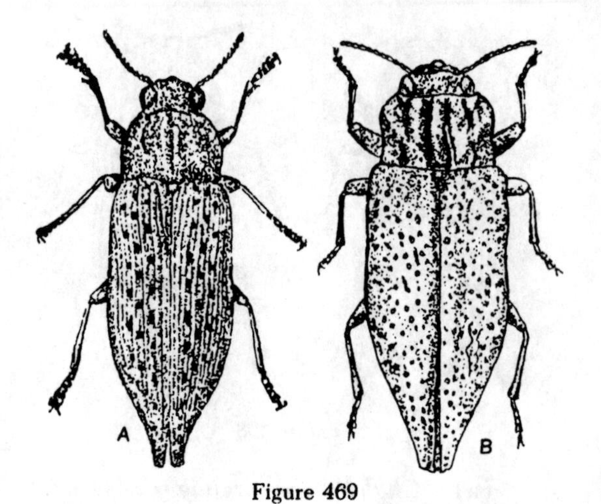

Figure 469

(a) ***Dicerca divaricata*** **(Say)**

LENGTH: 15-21 mm. RANGE: Eastern U. S. and Can.

This rather common flat brassy beetle is recognized by its elongated and diverging elytral tips. Thorax with median impressed line and elytra with numerous irregular impressed lines and short elevated black spots as pictured.

(b) ***D. punctulata*** **(Schon.)**

LENGTH: 10-14 mm. RANGE: Central and eastern United States.

Blackish bronze with coppery sheen. The transverse ridges extending from eye to eye on front are characteristic. About 25 *Dicerca* in our area.

In this same group belongs *Trachykele blondeli* Mars., the Western Cedar Borer, a most beautiful iridescent and gold beetle (14-17 mm). It ranges throughout the mountain states.

9a (8) Elytra marked with yellow. Fig. 470.

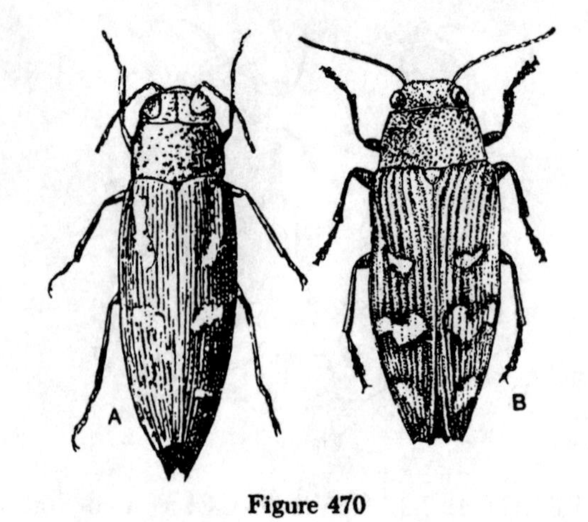

Figure 470

(a) ***Buprestis rufipes*** **(Oliv.)**

LENGTH: 18-25 mm. RANGE: Eastern half of United States.

Above green with brassy sheen; marked with yellow as pictured. Underparts green. Feeds on maple and beech.

(b) ***B. fasciata*** **Fab.**

LENGTH: 23 mm. RANGE: Eastern U. S.

Flattened. Brilliant green, marked with golden yellow on elytra. It seems to feed on maples and poplars.

B. aurulenta L. is another striking species bluish-green with suture and margins banded with gold. It is a native of our western mountain regions where it lives in many of the conifers. Adult beetles are occasionally found in the East where they have emerged from western lumber.

9b Without yellow markings on elytra. Fig. 471. .... *Buprestis maculativentris* Say

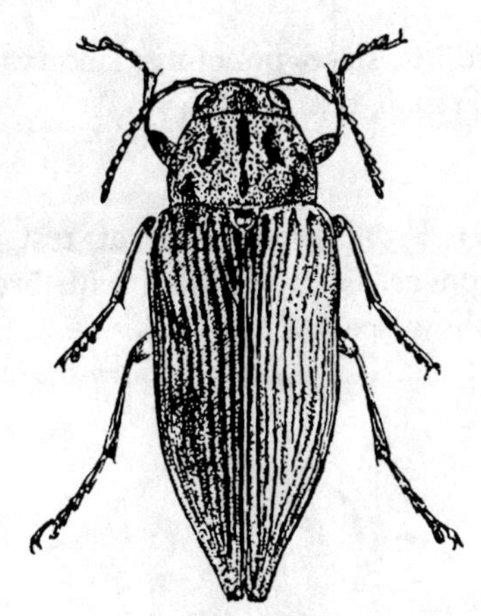

Figure 471

LENGTH: 15-23 mm. RANGE: Eastern U. S. and Can.

Black with brassy sheen. Front angles of thorax and a row of spots on sides of abdomen, reddish yellow.

B. adjecta (Lec.) a short robust species measuring 13-18 mm and wholly metallic green is a native of the West. The tips of the elytra are double-toothed.

B. laeviventris (Lec.) 14-20 mm, is dark brown or black with spots or broken lines of yellow or orange on the elytra; a marginal line at the anterior of the thorax as well as the front of the head are also yellow-orange. Its range is in the mountains of the West.

B. connexa Horn 15 mm long, a brilliant green beetle with bronze sheen on upper parts, has somewhat the same distribution. Both the above species are taken in B. C., Wash., Or., and Idaho. There are over 20 members of this colorful genus in our fauna.

FAMILY 47. CEBRIONIDAE
The Robust Click Beetles

This is a small family of rather large beetles ranging in size from 12 to 18 mm. They are elongate, convex, and fusiform. In color they range from light brown to black and have the body covered with long suberect hairs. The quadrate protruding head has a pair of prominent hook-like mandibles and filiform 11-segmented antennae. The trapezoidal pronotum is broader than the head and narrower than the elytra at the base. The arcuate lateral borders are crenulate.

The males fly at night and sometimes come to lights by the hundreds after a rain. The females are wingless living in the ground along with the larvae. In North America two genera and fourteen species make up the family, most of which are found in the southern U. S. from Texas and Oklahoma to California.

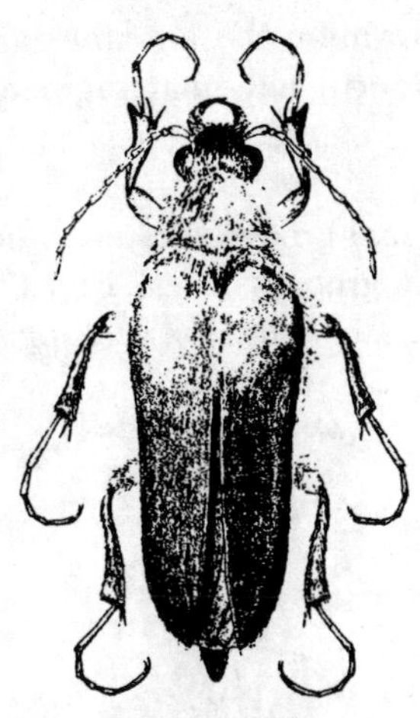

Figure 472

Figure 472 *Scaptolemus lecontei* (Sallé.) Length 12-16 mm. Range: Texas. Light brown with darker head and pronotum; tibiae and tarsi darker; the dense hairs brownish.

FAMILY 48. ELATERIDAE
The Click Beetles

Some 8000 species are included in this large family. The larvae are slender, hard-shelled and are known as wireworms. The adults are of many colors, some beautifully marked, and vary much in size. Their ability to snap themselves into the air and presently land on their feet has given them the names of Snapping beetles, Skipjacks, etc.

1a Prothorax with deep antennal grooves on under side, made by the separation of prosternal sutures. 2

1b Without antennal grooves on under side of thorax as in 1a. 4

2a (1) All or nearly all of the prosternal suture forming the antennal groove; third and fourth antennal segments of equal length. 3

2b Antennal groove shortened behind; front tarsi in grooves at rest. Fig. 473. .... *Colaulon rectangularis* (Say)

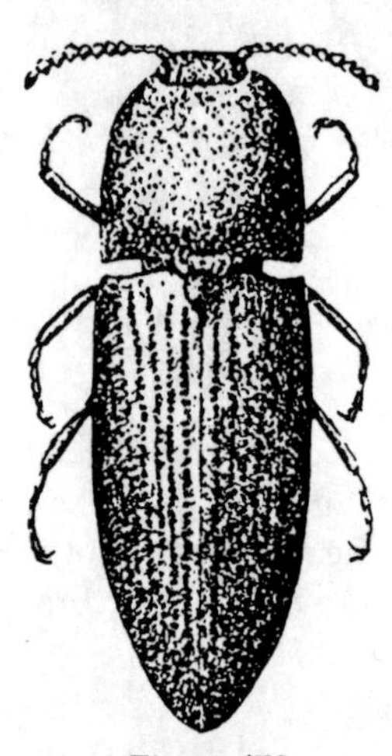

Figure 473

LENGTH: 8-10 mm. RANGE: Eastern and Southern United States.

Dull sooty brown, sparsely clothed with short stiff whitish hairs. Antennae paler; legs reddish-brown. Elytra with rows of rather distant medium sized punctures, intervals wider than the striae, flat.

3a (2) Front tarsi when at rest in deep grooves. Elytra spotted with brown and yellow scales. Fig. 474. .... *Lacon marmorata* (Fab.)

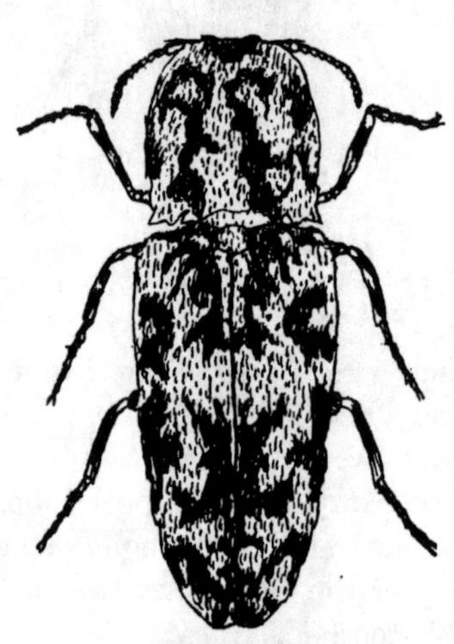

Figure 474

LENGTH: 15-18 mm. RANGE: Central, eastern and southern states.

Dark reddish-brown, dull marked as pictured with dull yellow scales. Gregarious under the bark of dead trees.

Similar in color but smaller (9-12 mm) and having its elytral punctures in rows instead of scattered, *L. impressicollis* (Say) is widely scattered throughout the United States.

3b Tarsal grooves shallow, head and sides of thorax densely covered with golden scales. Fig. 475. .... *Lacon discoidea* (Web.)

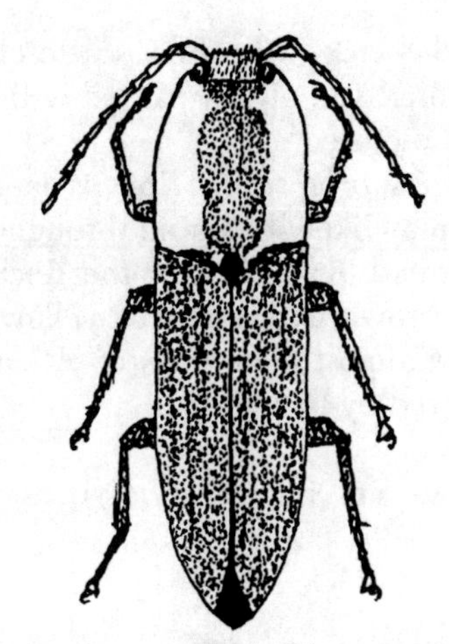

Figure 475

LENGTH: 8-11 mm. RANGE: Eastern half of United States.

Black, feebly shining, head and margins of thorax covered with golden scales. Elytra and body coarsely and densely punctured.

There are about a dozen members of this genus in the U. S. and Can.

4a (1) Claws with one or more bristles at the base; two large velvety oval "eye-spots" on pronotum. Fig. 476. .... *Alaus oculatus* (L.) The Eyed Click Beetle

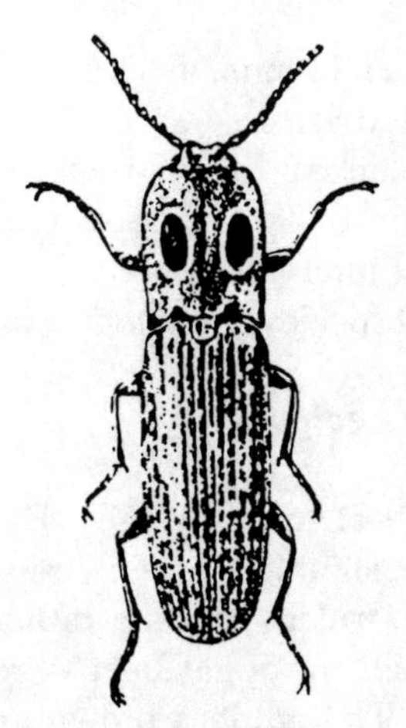

Figure 476

LENGTH: 25-45 mm. RANGE: Eastern United States west to Rockies.

Black shining, sprinkled with silvery-white scales.

This is one of the beetles your friends keep bringing to you. It is fairly common and never fails to arouse interest. The larvae, big husky wireworms, feed in decaying logs and the adults may be found there. Others fly to lights.

There are 8 species of the genus. In the West *Alaus melanops* Lec. (28-35 mm) a dull black species with less white markings, is prevalent. *Alaus lusciosus* Hope about 45 mm long with round-eyespots is native of Arizona while *Alaus myops* (Fab.) 24-40 mm and duller than *oculatus* is native of our Southern States.

4b No bristles at base of claws. 5

5a (4) Anterior of front with a margin. 6

5b Anterior of front without a margin. 11

6a (5) (a, b, c) The claws pectinate (with comb-like teeth). Fig. 477. .... *Melanotus similis* (Kby.)

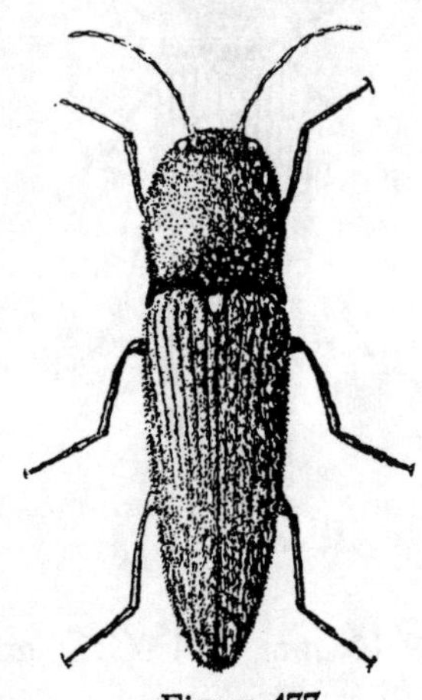

Figure 477

LENGTH: 13-17 mm. RANGE: Southern Canada and U. S. east of Montana, Utah, and Ariz.

Dark smoky brown with scattered pubescence. Third joint of antennae more than twice as long as the second but shorter than the third. The members of this genus are very difficult to separate. Their larvae are some of our most destructive wire worms.

Melanotus castanipes (Payk.) is larger (18-21 mm) and seems to be pretty much distributed as *similis*. It is dark reddish-brown. Its second and third antennal joints are about equal. This is a large genus of 46 similarly colored species that have to be separated using the male genitalia.

6b Claws not pectinate; the coxal plates gradually expanded inward. 7

6c Claws not pectinate; the coxal plates suddenly and strongly expanded inward. .. 8

7a (6) Front margined, prosternal groove single and closed in front. Fig. 478. .... *Athous scapularis* (Say)

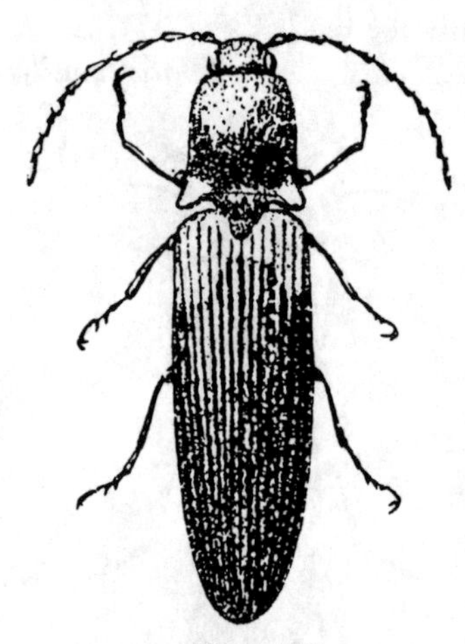

Figure 478

LENGTH: 9-11 mm. RANGE: middle and eastern United States.

Dull black with fine scattered pubescence; thorax and elytra marked with reddish-yellow as pictured.

Athous brightwelli Kby. is larger 11-18 mm and may likely be found throughout much of the United States east of the Rockies. It is pale dull brown with scattered yellowish hairs. There are almost 50 species of *Athous* in the area covered by this book.

7b Front not margined; clypeus truncate. Fig. 479.

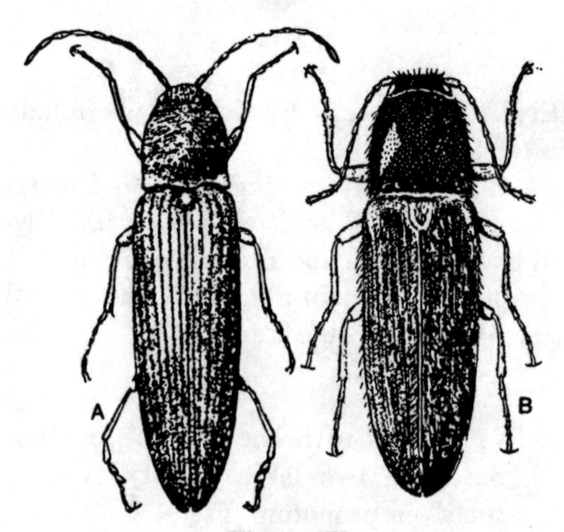

Figure 479

(a) *Limonius dubitans* Lec.

LENGTH: 11-13 mm. RANGE: Central and eastern United States.

Dull blackish-brown, often with a bronze sheen; rather thickly clothed with yellowish hairs; elytral intervals densely punctate. There are about 60 species of *Limonius* in our area.

(b) *L. ectypus* Say

LENGTH: 9-11 mm. RANGE: Eastern U. S.

Thorax shining black; elytra sooty brown.

L. californicus Mann a rather closely related species is the Sugar Beet Wireworm in its larval state. The adult is brown and 9-11 mm in length. Its range is from California to B. C. and Idaho.

8a (6) **Prosternal suture if present bent outward; second and third tarsae lobed. Fig. 480. .... *Negastrius choris* (Say)**

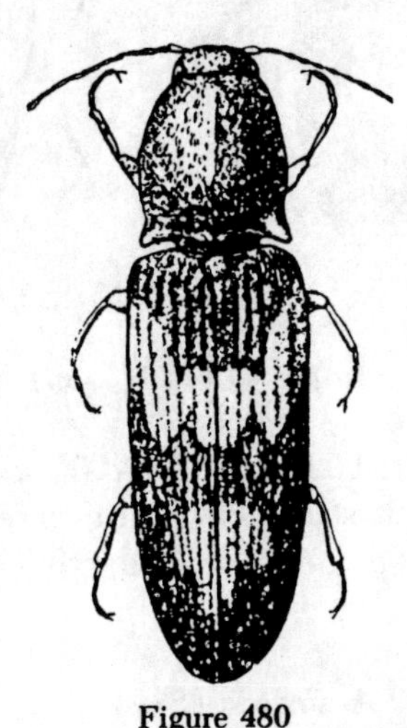

Figure 480

LENGTH: 3-5 mm. RANGE: Central and eastern United States.

Black with scattered yellow hairs; elytra marked with yellow as pictured. It seems to prefer sandy localities.

N. obliquatulus (Melsh.) 2.3-2.7 mm, is rather dull piceous with an oblique yellowish band at middle of each elytron and an oval yellow spot at the tip. It is known from Iowa and points on eastward. This genus includes some 20 species.

Oedostethus femoralis Lec. 4-6 mm long and ranging widely in the United States and Canada is wholly black with scattered yellow hairs. Its claws are toothed whereas those of the preceding are simple. It is the only member of its genus.

8b **Prosternal suture straight or bent slightly inward. 9**

9a (8) **Fourth tarsal segment lobed, first antennal segment longer than usual. Fig. 481.**

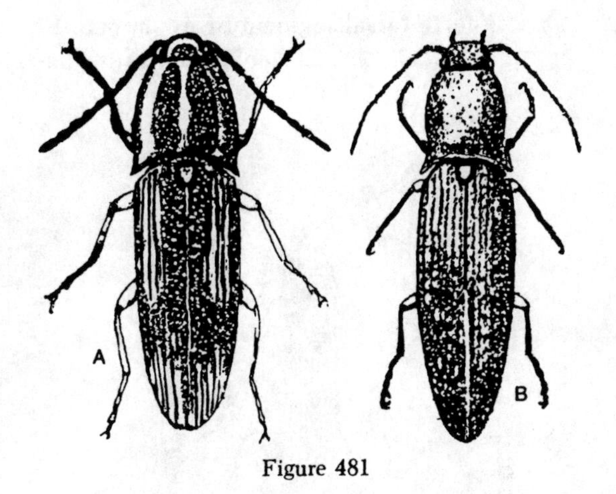

Figure 481

(a) ***Conoderus vespertinus* (Fab.) Tobacco Wireworm**

LENGTH: 7-10 mm. RANGE: Eastern and Southern States.

Usually yellowish marked with brown or occasionally marked with nearly black. Scutellum always light. Elytra with punctate striae. The larvae feed on beans and cotton.

(b) ***Conoderus lividus* (DeG.)**

LENGTH: 14-17 mm. RANGE: Central and Southern States.

Dull brown with a dense covering of short prostrate pubescence; legs yellow. Striae with numerous oblong punctures.

C. auritus (Hbst.) 5-7 mm long, varies in color from red to piceous with varying black marks on elytra and thorax. The legs and two basal joints of the antennae are yellow. The thorax is about as wide as long. It and the following are both common and belong to the Central and Eastern States.

C. bellus (Say) 3.5-4 mm, is black, with hind angles and median on thorax reddish; elytra dull-red, irregularly marked with two or three black crossbars; antennae and legs yellow.

9b Fourth tarsal segment heart shaped. Fig. 482. .... *Aeolus mellillus* (Say)

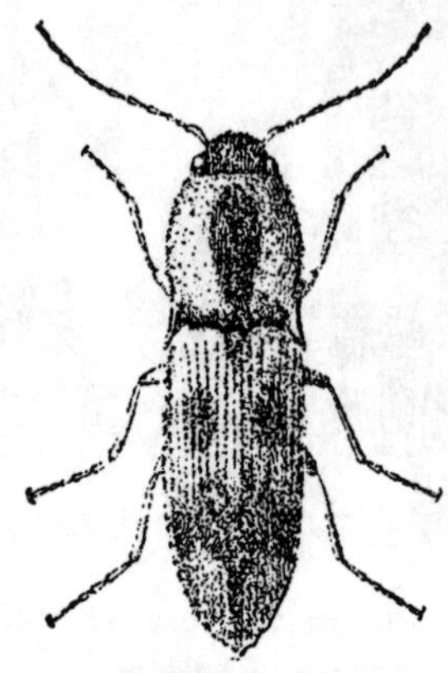
Figure 482

LENGTH: 6-7 mm. RANGE: Much of U. S.

Dull reddish-brown, with blackish markings as pictured. The spot on the thorax is sometimes reduced to a small point.

A. amabilis (Lec.) marked and colored much like the above but only about one-half as long; ranges through the Eastern States.

9c None of the tarsal segments lobed or greatly widened. 10

10a (9) Third joint of antennae triangular; elytra marked with pale yellow and black as pictured. Fig. 483.

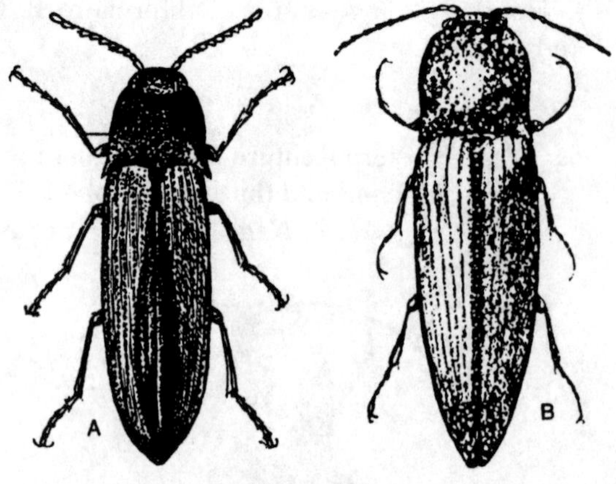

Figure 483

(a) *Ampedus sayi* (Lec.)

LENGTH: 11-13 mm. RANGE: Central States.

Black stripe on suture of elytra broad; third joint of antennae distinctly narrower than fourth.

(b) *A. linteus* (Say)

LENGTH: 7-9 mm. RANGE: Eastern half of United States and Canada.

Head and thorax roughly punctate; black sutural margin narrow.

In these two species the legs are black. *A. nigricollis* (Hbst.) with a length of 8-10 mm and range much the same as *linteus* is distinguished by its yellow legs.

10b Third joint of antennae not triangular. Dark reddish-brown or blackish, marked with dull yellow as pictured. Fig. 484. .... *Ampedus areolatus* (Say)

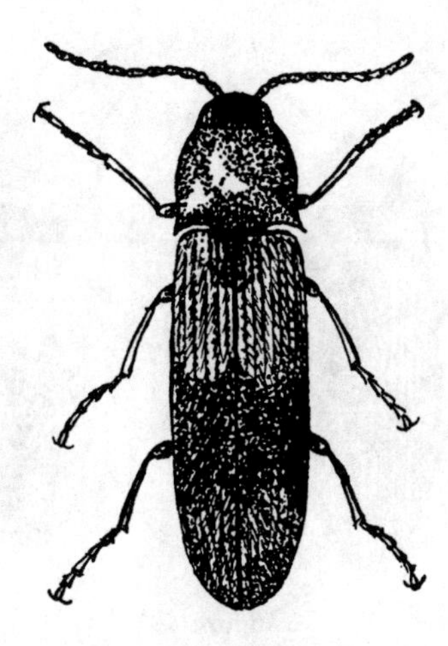

Figure 484

LENGTH: 4.5-5 mm. RANGE: Eastern United States and Canada.

Covered with rather long yellowish hairs. Thorax usually yellowish. Elytral striae shallow with large punctures.

A striking species is *Ampedus sanguinipennis* (Say) with black thorax and elytra scarlet to dull red. It measures 7-9 mm and ranges pretty much over the eastern half of our country. There are about 70 species of this genus in the U. S. and Can.

11a (5) Antennal grooves large, front convex and turned down in front. 13

11b Antennal grooves small; front concave or flattened. 12

12a (11) Tarsi lobed or widened. Fig. 485.

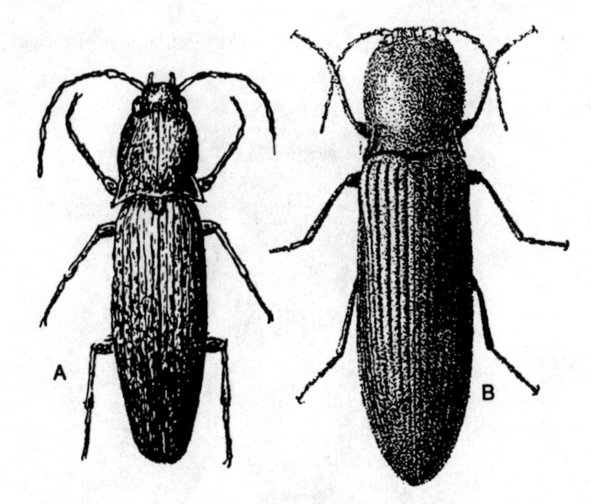

Figure 485

(a) ***Hemicrepidius memnonius* (Hbst.)**

LENGTH: 15-22 mm. RANGE: Eastern half United States.

Pale brown to blackish, legs paler. Elytral striae with coarse shallow punctures, intervals convex.

(b) ***H. decoloratus* (Say)**

LENGTH: 9-15 mm. RANGE: Eastern U. S. and Can.

Black, often somewhat bronzed; legs paler.

H. bilobatus (Say) is a rather slender click beetle, 13-16 mm long. It is shining, chestnut-brown with the antennae and legs paler. The elytra have coarsely punctured striae, the intervals having two rows of rather fine punctures. It occurs in the Eastern half of the U. S.

12b Tarsi normal. Fig. 486. ....
.... *Melanactes piceus* (DeG.)

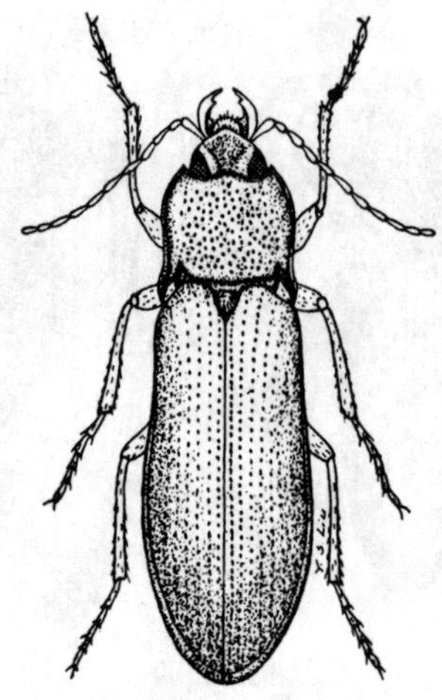

Figure 486

LENGTH: 23-32 mm. RANGE: Eastern and Central States.

Black, shining; elytral striae very shallow with fine punctures. What is supposed to be the larvae of this species is a large wire worm which displays marked phosphorescence at night.

M. consors Lec. ranging through the Central States and having a length of 21-24 mm is distinguished by its deeper striae and coarser punctured thorax.

M. densus Lec. with a length of 20-22 mm is abundant throughout much of California. It is highly polished and wholly black. The elytra have rather deep striations.

13a (11) Coxal plates with their hind margin toothed at their widest place. 14

13b Coxal plates with their hind margin rounded at the widest place. Fig. 487.

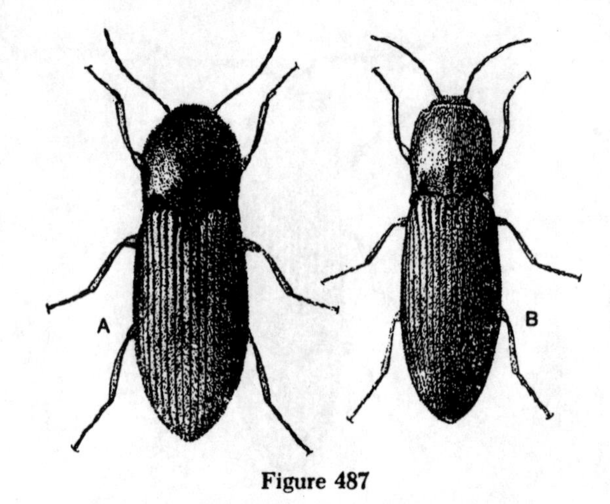

Figure 487

(a) *Agriotes mancus* (Say) The Wheat Wireworm

LENGTH: 7-9 mm. RANGE: Central and eastern United States and Canada.

Dull brownish yellow with scattered short yellow pubescence. Its wireworm larva is a serious pest of corn and small grains.

(b) *Agriotes pubescens* Melsh.

LENGTH: 8-10 mm. RANGE: Central and eastern United States and Canada.

More slender than the preceding. Sooty brown densely covered with grayish-yellow pubescence. Striae with fine punctures; intervals flat, with fine wrinkles.

A. insanus Cand. is shaped like *pubescens* but differs in that the side margins of the thorax are distinct for their whole length, and segments two and three of the antennae are each shorter than the fourth, instead of equal. It is smaller (7-8 mm). Both this and the following are taken in the Eastern half of the U. S. and Can.

A. oblongicollis (Melsh.) which is also brown, differs from all three above in having its hind coxal plates suddenly dilated to more than twice as wide at the inner third as at the outer end. It is 6-9 mm long.

14a (13) Color wholly black. Fig. 488. .... *Ctenicera pruinina* (Horn)

Figure 488

LENGTH: 12-14 mm. RANGE: Northwest United States.

Striae moderately deep, punctured; intervals flat. Its larva is known as the Great Basin Wireworm.

C. inflata (Say), a heavy set click beetle 8-11 mm long, ranges Eastern U. S. and Canada, too, is black and somewhat flattened and is known as the Dryland Wireworm. The larvae feed on potatoes, small grain and corn.

14b Elytra mostly yellow. Fig. 489.

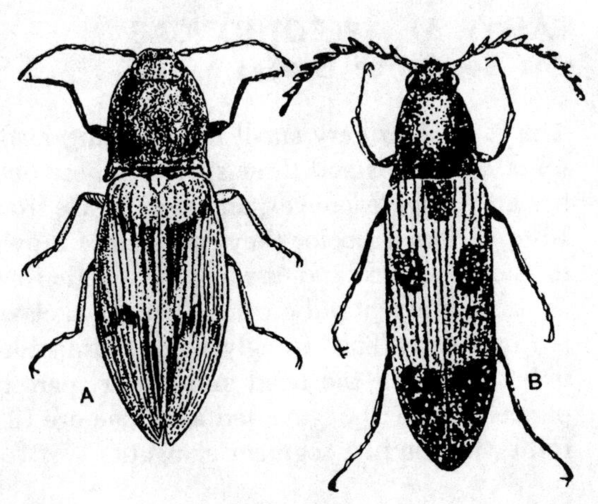

Figure 489

(a) *Ctenicera hieroglyphica* (Say)

LENGTH: 11-13 mm. RANGE: Central and eastern United States.

Bronzed, and densely clothed with fine grayish-yellow pubescence; elytra dull yellow marked with blackish bands as pictured.

(b) *C. vernalis* (Hentz)

LENGTH: 8-11 mm. RANGE: widely scattered through the United States and Canada.

Thorax and markings on elytra black. Antennae serrate. Striae punctate, intervals flat. This genus is a large one, including many rather common and destructive species. About 150 species are found in this colorful genus in our fauna.

FAMILY 49. THROSCIDAE
The False Metallic Woodboring Beetles

This is a small family of beetles that range in size from 2 to 5 mm. In color they are brown to black, with several species having red markings. The body is covered with fine, dense, suberect vestiture. These beetles are oblong-oval in shape, compact, and fusiform. The 11-segmented antennae are serrate or with a loose three-segmented club.

Adults are beaten from foliage or taken from flowers. The larvae are found in decaying, worm-infested wood. In North America the family is made up of four genera comprising about 30 species.

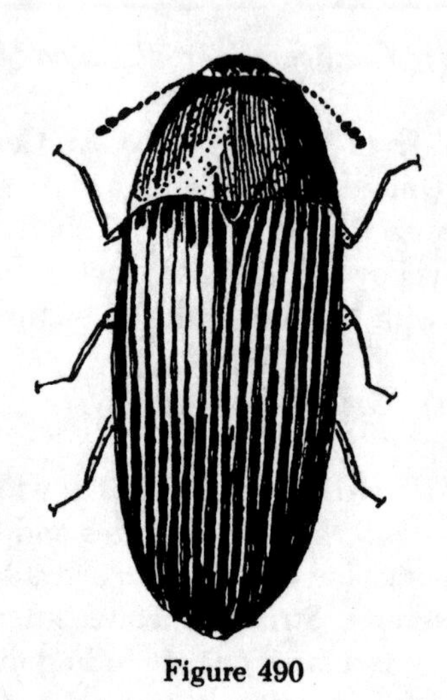

Figure 490

Figure 490 *Trixagus chevrolati* Bonv.

LENGTH: 2.5-3 mm. RANGE: Eastern and central U. S.

Reddish brown, rather densely covered with coarse yellowish pubescence. Clypeus with two parallel ridges.

Drapetes geminatus Say a little larger than the former (4 mm) and widely distributed is black with a crossbar or dot of red on basal half of each elytron. The antennae are serrate in this genus instead of clubbed as in *Throscus*. Ranges from Eastern Can. to Fl. and Tx.

FAMILY 50. CEROPHYTIDAE
The Rare Click Beetles

This is one of the smallest families of beetles, comprising on a world-wide basis one genus and five species. These beetles are elongate, sub-depressed, varying from brown to blackish piceous in color, and with rather sparse vestiture. The 11-segmented antennae, pectinate in the male, serrate in the female, have the first segment long and stout, the second small and triangular, and the eleventh longer than all the others. The pronotum is wider than the head and much broader than long. The elytra have punctate striae and smooth intervals.

Adults are taken by sweeping vegetation in woods, where the larvae live in rotten wood. This insect is very rare.

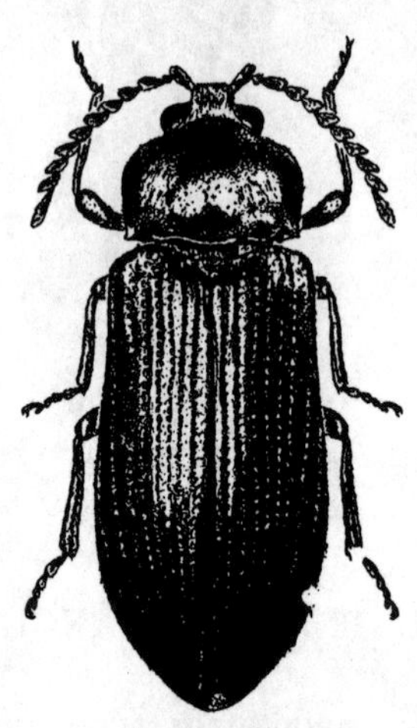

Figure 491

Figure 491 *Cerophytum pulsator* (Hald.)

LENGTH: 5-7.5 mm. RANGE: Ind. to Pa. Piceous to black with the palpi and legs reddish-brown; pubescence golden, especially dense at the base of the pronotum.

FAMILY 51. PEROTHOPIDAE
The Beech-tree Beetles

This is another very small family being made up of one genus and three species. These beetles are elongate, convex, and vary in size from 10 to 18 mm. In color they range from brown to piceous-black and are covered with fine, dense, recumbent pubescence. The tarsal claws are pectinate. The strongly convex pronotum is broader than the head and rather sparsely punctate. The 11-segmented antennae are filiform with the first segment elongate.

Adults are found on the trunks and branches of beech trees.

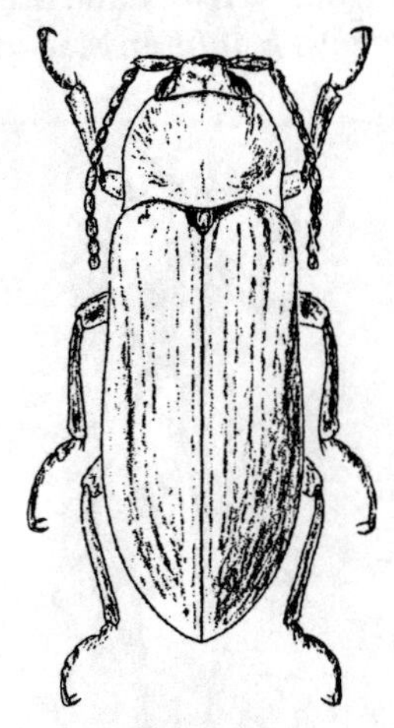

Figure 492

Figure 492 *Perothops mucida* (Gyll.)

LENGTH: 11-18 mm. RANGE: Ind. to Pa. south to Fl. Males are piceous-black and the females brownish and larger; pubescence pale and rather dense.

FAMILY 52. EUCNEMIDAE
The False Click Beetles

This small family was formerly a part of the family Elateridae. The larvae are often more like the Buprestids than like wireworms. Members of this family are not very common, but are found in maple, beech and similar wood.

1a Last segment of maxillary palpi pointed; antennae somewhat separated at their base. Fig. 493. .... *Isorhipis ruficornis* (Say)

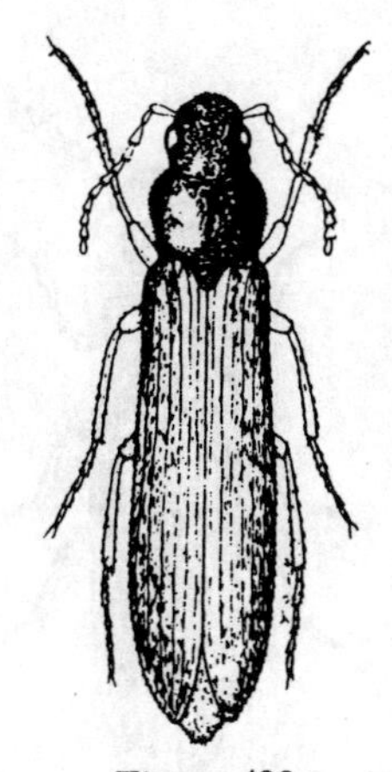

Figure 493

LENGTH: 4-7 mm. RANGE: much of the United States and Canada.

Black; elytra marked with yellow as pictured. Antennae and legs reddish-brown. Striae densely and roughly punctured.

Melasis pectinicornis Melsh. dull black and thinly covered with short gray pubescence ranges from Connecticut to Florida and west to Texas. It measures 6-8 mm.

M. rufipennis Horn measuring 8-12 mm, black with reddish elytra and antenna is destructive to trees in the Pacific Coast states.

1b Last segment of maxillary palpi dilated; antennae close together at their base. Fig. 494. .... *Sarpedon scabrosus* Bonv.

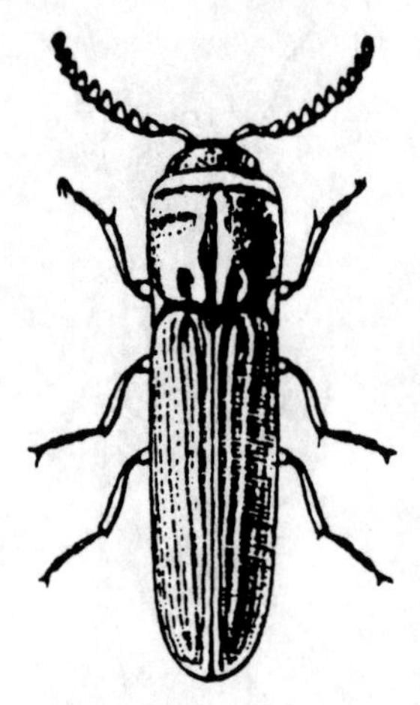

Figure 494

LENGTH: 6-8 mm. RANGE: Eastern and Southern United States and Canada.

Black; front margins of thorax reddish; antennae of male bipectinate, of female biserrate.

Deltometopus amoenicornis Say piceous with scant gray pubescence and measuring 3-5 mm is found in much of the eastern half of the United States.

Dirhagus triangularis (Say) differs in having an antennal groove on the underside of the thorax, which *Sarpedon* lacks, and in being much smaller (3-4 mm). It is black with brownish legs and antennae and is found from Georgia north into Canada.

FAMILY 53. TELEGEUSIDAE
The Long-Lipped Beetles

This small North American family is composed of 2 genera and 3 species. The body is elongate with shortened elytra and fan-shaped non-folding hind wings. The body size ranges from 5 to 35 mm. The beetles are testaceous and covered with vestiture that varies from sparse to dense. The 11-segmented antennae are stout, short, and filiform. The maxillary palpi are elongate and fan-shaped in both sexes.

These beetles are rare and little is known about their habits. The few that have been taken come from Arizona and California. One species, *Telegeusis nubifer* Martin, is shown in Figure 495.

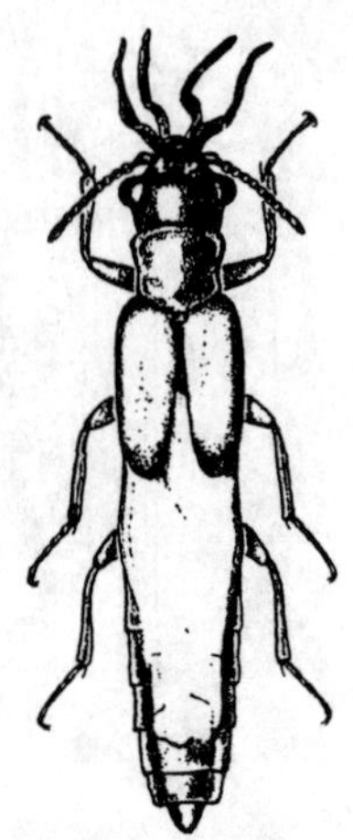

Figure 495

FAMILY 54. PHENGODIDAE
The Glowworms

The male members of this family tend to have short elytra and 12-segmented doubly plumose antennae. The females are larviform and differ from the larvae in the possession of adult genitalia. Body size ranges from 5 to 30 mm. Many species are testaceous in color, but some are piceous to black with yellow or red markings.

The male pronotum is broad, short, partly convex, and with or without explanate margins. In most species the elytra are short and narrowed greatly behind. In a few species, *Zarhipis*, they are entire. The larvae are luminescent and predaceous on other insects and small vertebrates found in the leaf litter where they live.

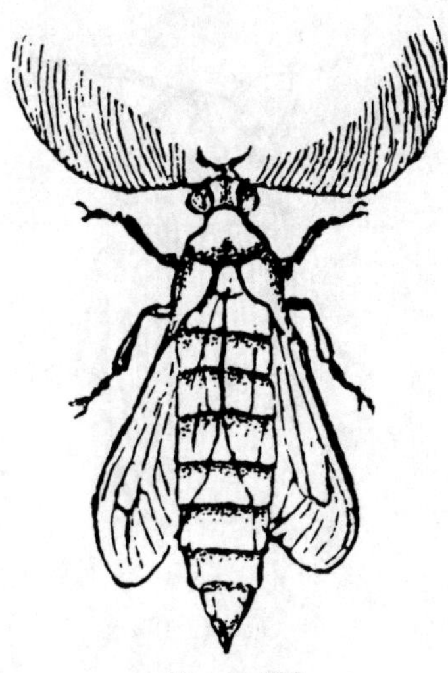

Figure 496

Figure 496 *Phengodes plumosa* (Oliv.)

LENGTH: Male; 11-12 mm. RANGE: Eastern half of U. S.

Dull yellow with fine pubescence. Antennae pulmose; elytra small and diverging. Female; elongate, wormlike.

Zarhipis piciventris Lec. is a California species. The yellowish brown larviform females measure 30 to 50 mm while the winged males are but 10 mm long. They are piceous with legs, pronotum and mandibles reddish.

Z. riversi Horn is somewhat larger. The male is black with yellowish red on head, thorax and part of abdomen. The female is yellowish-brown.

FAMILY 55. LAMPYRIDAE
The Fireflies

These insects are known by everyone because of the display they make on summer nights. They prefer damp places and are inactive by day. Over 2000 species have been described.

1a **Thorax completely covering head; antennae with second joint small, usually transverse. 2**

1b **Head not fully covered by prothorax; second antennal joint not transverse. Fig. 497. .... *Photuris pennsylvanicus* (DeG.)**

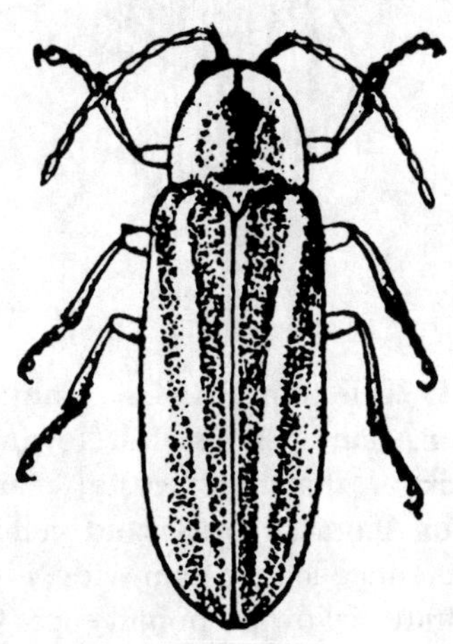

Figure 497

LENGTH: 11-15 mm. RANGE: eastern half of U. S.

Head and thorax dull yellow; thorax with median dark stripe on the red disk; elytra brownish with pale markings; labrum with three teeth. This is one of the most common fireflies.

Phausis splendidula (L.) is a European species 8-9 mm in length which has been found in the eastern United States. It and other members of its genus have a jointed needle-like appendage on the last antennal segment.

2a **(1) Eyes large, larger in male than in female; light organs well developed. 3**

2b **Eyes small; light organs feeble; antennae much compressed, not serrate; second segment short and transverse. Fig. 498. .... *Ellychnia corrusca* (L.)**

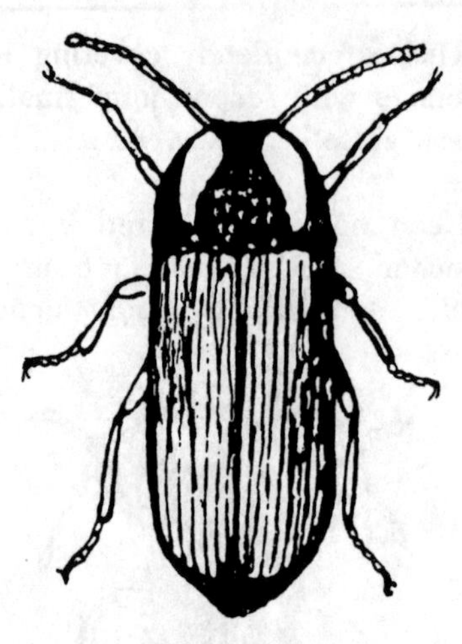
Figure 498

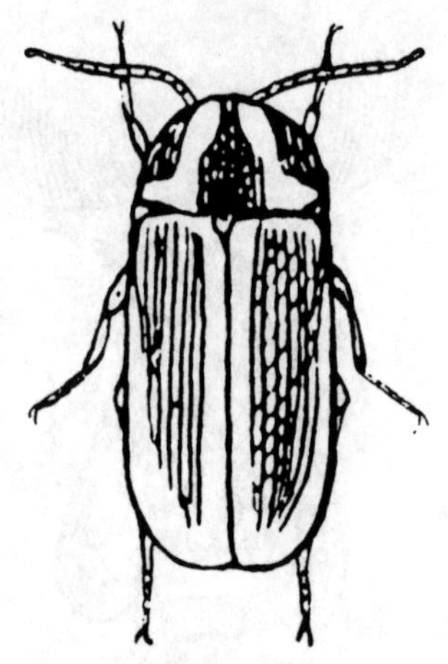
Figure 499

LENGTH: 10-14 mm. RANGE: California and Arizona to Atlantic states and Newfoundland.

Black or sometimes rusty; markings as pictured on thorax, reddish and yellow; third joint of antennae longer than wide; elytra with fine prostrate yellowish pubescence. Common throughout much of its range.

Pyropyga nigricans Say very common in California, ranges entirely across the United States. It measures 7-12 mm and is black with a reddish-yellow triangle on each side of the thorax.

3a (2) Thorax having somewhat of a ridge; light organs of female on the side of abdomen. Fig. 499. .... *Pyractomena angulata* (Say)

LENGTH: 8-15 mm. RANGE: eastern third of U. S. and Canada.

Depressed; blackish-brown with yellow and reddish markings on thorax as pictured; elytra finely granulate and punctured, each with two or three plainly evident ridges; ventral segment of female dull yellow with dusky spots.

Pollaclasis bifaria (Say) is an interesting species distinguished by the antennae which has comb-like projections on each side. It is black, with front and side of the thorax reddish yellow. It measures 9-10 mm and is known from Ohio and Indiana.

3b Thorax without a ridge, though often grooved; light organs of female at middle of abdomen. Fig. 500. .... *Photinus scintillans* Say

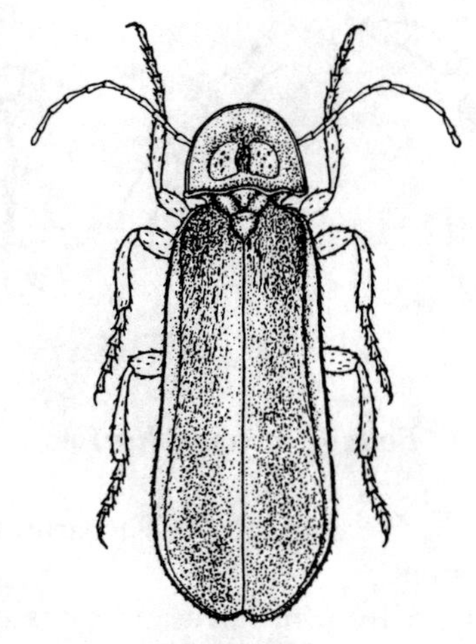

Figure 500

LENGTH: 5.5-8 mm. RANGE: Kansas to Massachusetts.

Slender; dusky brown; thorax yellowish and reddish with black center as pictured; suture and side margins of elytra pale; antennae dusky.

In some members of this family the female is always wormlike, and glows with a continuous light. The Pink Glowworm *Microphotus angustus* Lec. found in California, Colorado, Oregon and Florida is like this. The female measures 10-15 mm while the male which is winged and has a pinkish body is about 10 mm long.

FAMILY 56. CANTHARIDAE
The Soldier Beetles

These beetles have much more pliable elytra than most beetles. The head is uncovered. They are predacious. Some 1500 species are known but less than 200 in the United States.

1a Mentum small, quadrate. 2

1b Mentum very long, wider in front. Fig. 501. .... *Chauliognathus pennsylvanicus* DeG.

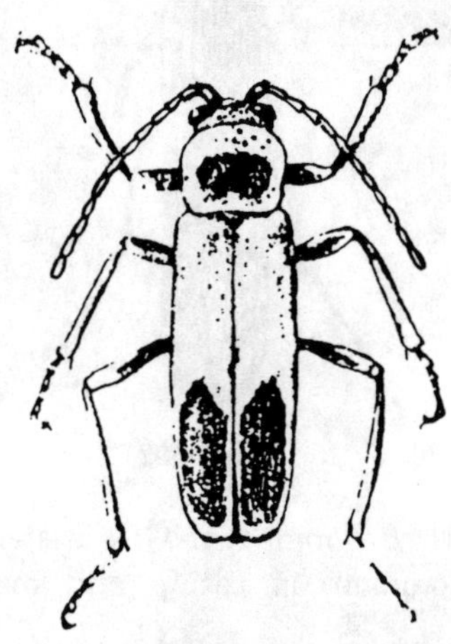

Figure 501

LENGTH: 9-12 mm. RANGE: eastern half of U. S.

Head and underparts black; thorax and elytra yellow with black markings as pictured. Very common especially on goldenrod in the fall.

C. marginatus Fab. is smaller and narrower. Its head is mostly yellow; the thorax has a mid-dorsal black stripe and a rather long but highly variable blackish spot on each elytron. It too is often very common, appearing earlier than the above. There are 16 members of this colorful genus in our fauna.

2a (1) Elytra wholly covering the wings. .. 3

2b Elytra short (a) leaving the wings exposed; last segment of maxillary palpi hatchet shaped. Fig. 502. .... *Trypherus latipennis* (Germ.)

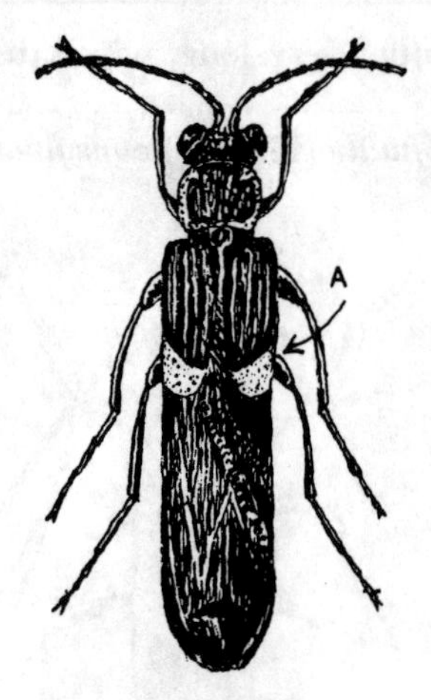

Figure 502

LENGTH: 6-7 mm. RANGE: eastern half of U. S. Common on catnip and many other plants.

Slender. Elytra very short as in Staphylinids, but the wings not folded under the elytra; eyes large and prominent in the male. Dull yellow beneath, piceous above, the tips of elytra and margins of thorax dull yellow.

Malthodes concavus Lec. in common with the several other members of its genus, has elytra shortened but not so much so as *Trypherus.* The general color is piceous with the base of the antennae and part of abdomen yellowish. It is only 2-3 mm long and ranges through the eastern third of our country.

3a (2) Thorax rounded in front, partly covering the head. 4

3b Thorax cut squarely in front, not at all covering the head. Fig. 503.

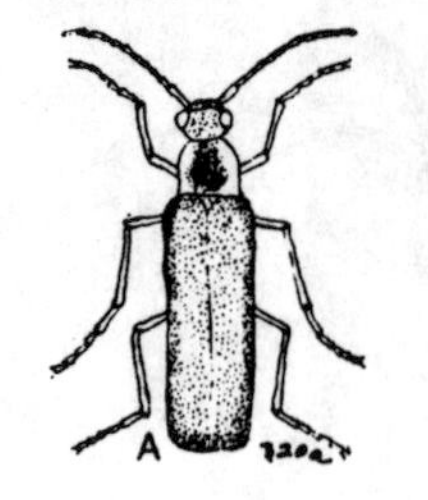

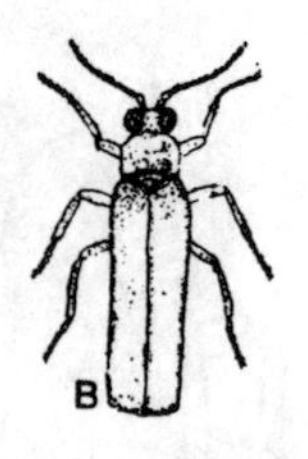

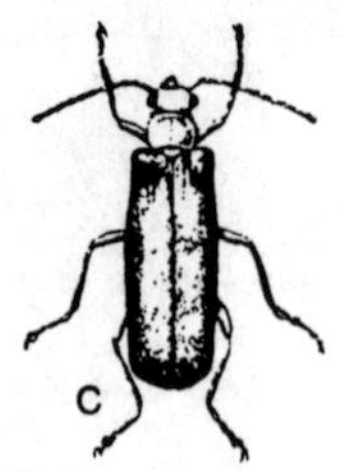

Figure 503

(a) *Podabrus rugosulus* Lec.

LENGTH: 7-8 mm. RANGE: eastern half of U. S. Common.

Black; front of head and sides of thorax yellow; head coarsely punctured, thorax less so.

(b) *P. modestus* (Say)

LENGTH: 9-13 mm. RANGE: Eastern U. S. and Canada.

Blackish front of head, margins of thorax and femora yellow.

(c) *P. tomentosus* (Say)

LENGTH: 9-12 mm. RANGE: Much of U. S.

Elytra black; head and thorax reddish-yellow. All three of these species are valued for destroying aphids.

Podabrus is a large genus of some 100 species in the U. S. and Can.

4a (3) Head short; hind angles of thorax incised. Fig. 504. .... *Silis latiloba* Blatch.

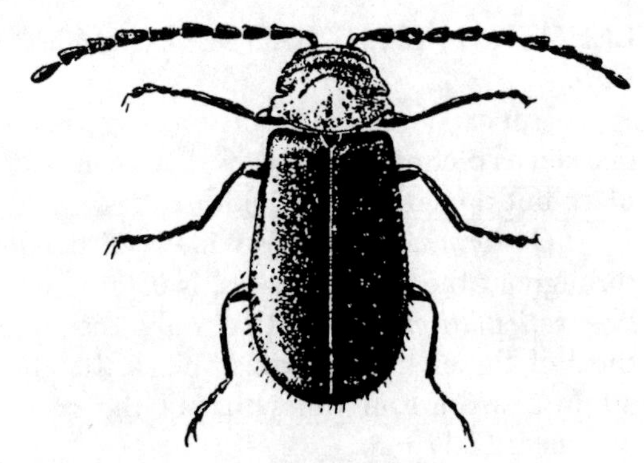

Figure 504

LENGTH: 4.5-5.5 mm. RANGE: North Central States.

Black or piceous, feebly shining; thorax reddish yellow; second segment of antennae rounded, not pubescent, less than one fourth the length of third; the third equal to the fourth.

S. *percomis* (Say) is a bluish black species, about 5 mm long which ranges from Canada and Massachusetts to Texas. The thorax which is wider than long is colored reddish-yellow. There are 25 widely distributed members of *Silis* in our territory.

4b Head fairly long; hind angles of thorax rounded. Fig. 505. .... *Cantharis bilineatas* Say

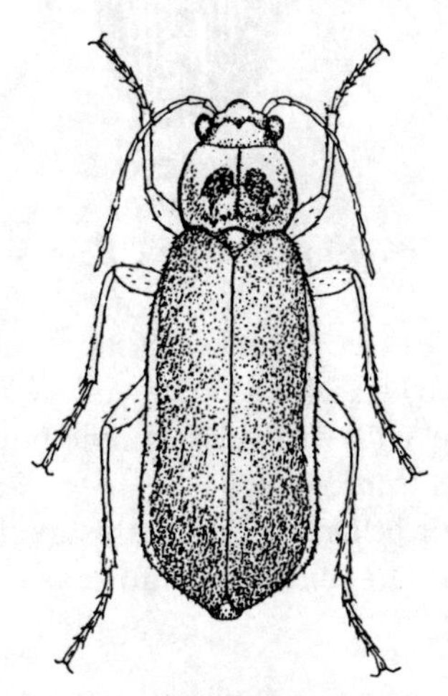

Figure 505

LENGTH: 6-8 mm. RANGE: eastern half of U. S.

Dull reddish-yellow; back of head, elytra, spots on thorax, all but two basal segments of antennae and tibia and tarsi black. Thorax widely margined; sparsely punctured.

C. divisa Lec. is a common western species ranging from Alaska to California. It measures only 6-8 mm and is black with thorax yellow with two large black spots on the disk. The genus *Cantharis* contains about 60 species in the U. S. and Can. They are difficult to separate.

FAMILY 57. LYCIDAE
The Net-winged Beetles

These often brightly colored and attractive insects, are not luminous like their near relatives the fireflies, and are active by day. About 80 species are known in our region.

1a Under side of thorax with a prominent tubular spiracle behind and at the outer extremity of the front coxae. 2

1b Under side without such spiracle as in 1a. 3

2a (1) Front prolonged into a beak; antennae with third segment scarcely as long as fourth. Fig. 506. .... *Lycus lateralis* (Melsh.)

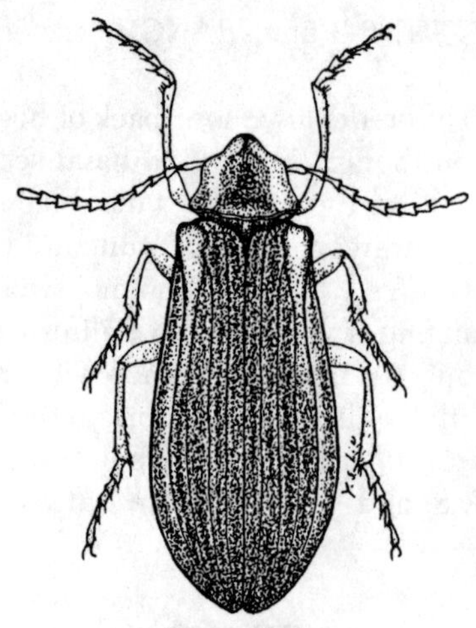

Figure 506

LENGTH: 8-10 mm. RANGE: eastern half of the U. S.

Black, apex and sides of thorax and sides of elytra to middle dull yellow.

Caeniella dimidiata (Fab.) measuring 10 mm and ranging through the Atlantic States resembles *Calopteron terminale* but has comb-like antennae and a black scutellar spot.

2b Front swollen between the antennae; without a beak; antennae much flattened. Fig. 507. .... *Calopteron reticulatum* (Fab.)

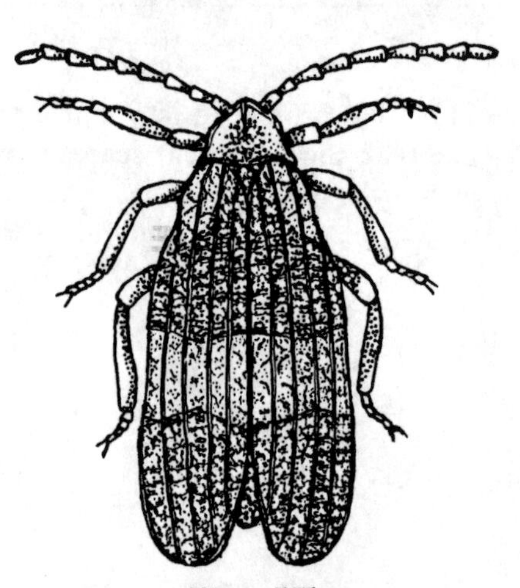

Figure 507

LENGTH: 11-19 mm. RANGE: eastern half of U. S.

Upper surface dull orange-yellow, marked as pictured with black. Elytra at widest place but three times the width at base.

C. terminale (Say) widely distributed throughout the United States, is colored much like *reticulatum* except that only the apical third of the elytra is purplish-black. Its elytra attain a width four times that of the base. It measures 11-17 mm.

3a (1) Elytra ribbed, cross-barred or netted. 4

3b Elytra with light striae (ridges) but not as in 3a. Fig. 508. .... *Calochromus perfacetus* (Say)

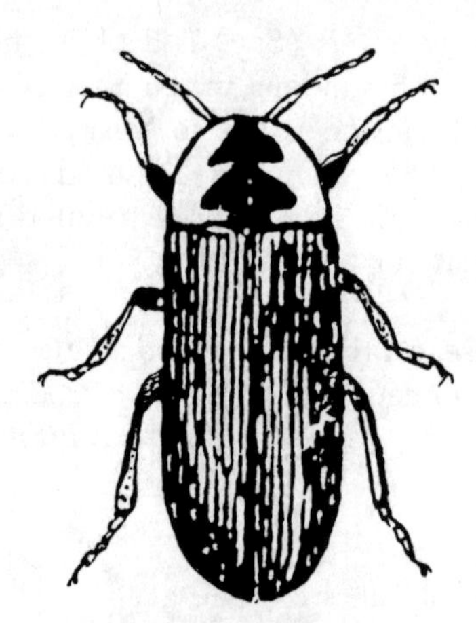

Figure 508

LENGTH: 7-10 mm. RANGE: eastern half of U. S.

Depressed. Black, with fine pubescence; thorax marked with reddish yellow as pictured. Antennae with second segment but one third as long as third.

Four other species of this genus occur in Colorado, Oregon and California.

4a (3) Thorax divided into cells. Fig. 509.

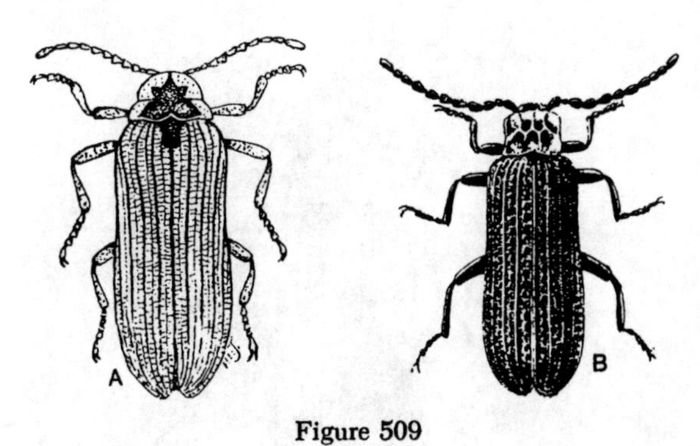

Figure 509

(a) *Dictyopterus aurora* (Hbst.)

LENGTH: 7-11 mm. RANGE: much of N. A. (also Europe and Siberia).

Depressed. Head antennae and under surface black; thorax and elytra bright scarlet; scutellum and depressions on thorax dusky; thorax with 5 cells, a center rhomboid one with two on each side.

(b) *Platycis sculptilis* (Say)

LENGTH: 5-7 mm. RANGE: Eastern half of the U. S. and Can.

Black or piceous; thorax reddish-yellow often with black spot at center. Elytra with four prominent ridges running lengthwise.

Eropterus trilineatus Melsh. ranging through the eastern one third of our country is wholly black except for dull yellow side margins on the thorax. The thorax has six rather weakly defined cells.

4b Thorax without distinct cells. Fig. 510.
.... *Plateros sollicitus* (Lec.)

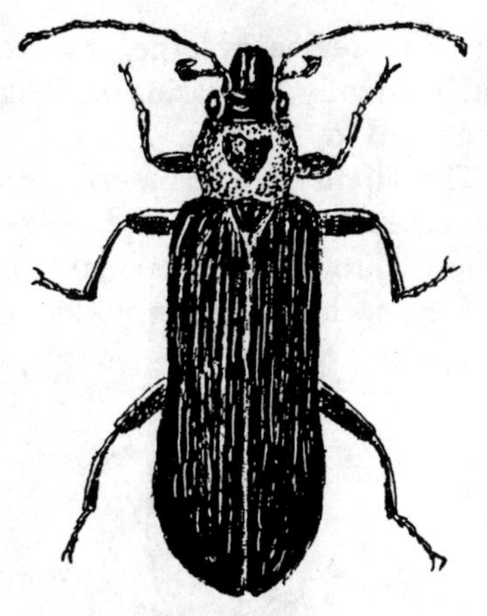
Figure 510

LENGTH: 6-7.5 mm. RANGE: eastern one third of U. S.

Dull black; thorax reddish yellow with center spot shining black; antennae of male long, the fourth segment twice as long as the third, fifth twice as long as wide. Antennae of female shorter, the seventh joint twice as long as wide.

P. modestus (Say) is the same size and has much the same color and range. It can be readily separated from the above by the intervals which alternate high and low. The genus *Plateros* contains over 30 species in our area. They are all black with fulvous or reddish markings on the pronotum. They can be separated into species only by using the male genitalia.

FAMILY 58. MICROMALTHIDAE
The Telephone-Pole Beetles

This family consists of a single species, *Micromalthinus debilis* LeC. These beetles are elongate, somewhat depressed forms, ranging in size from 1.5 to 2.5 mm. In color they vary from dark brown to piceous with yellow appendages. The 11-segmented antennae have the first two segments large, the others smaller, but gradually increasing in length as one goes outward to the apex.

The larvae live in moist, decaying oak logs. It is reported that they damage poles and buildings. Through commerce, the single North American species has been widely distributed in Europe and Africa.

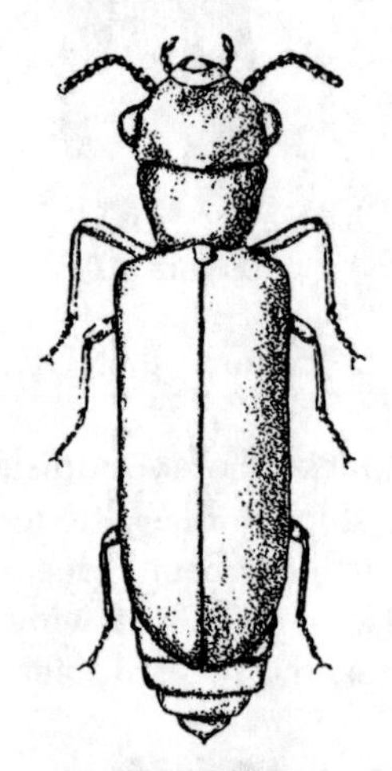

Figure 511

Figure 511 *Micromalthus debilis* LeC.

FAMILY 59. DEDRODONTIDAE
The Tooth-Neck Fungus Beetles

This family is composed of elongate, oval, convex beetles varying in size from 3 to 6 mm. Usually they are brownish in color and often bear darker spots. They have no vestiture. The 11-segmented antennae have a 3-segmented club. Also there is a pair of dorsal ocelli close to the inner margin of the eyes. The pronotum is wider than the head, narrowed in front with strong lateral angles, dentate margins, or with margins that are broadly flattened. The elytra striae are composed of punctures or polished rugose spots.

Adults are found in various types of fungi along with the larvae. Actually very little is known about the habits of these beetles. Six species in three genera are found in the U. S. and Canada. One of these is shown in Fig. 512.

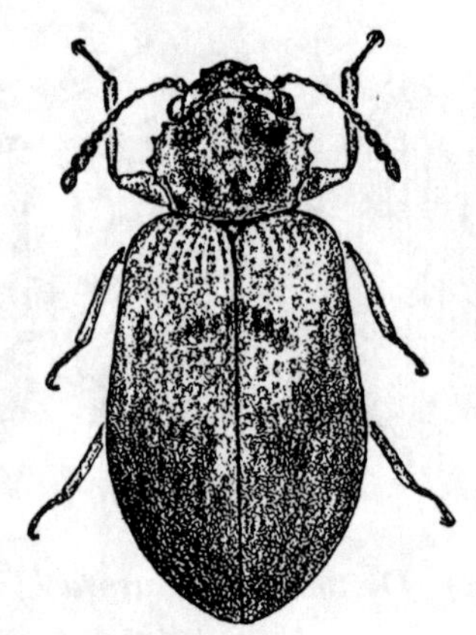

Figure 512

Figure 512 *Derodontus maculatus* (Melsh.)

LENGTH: 2.5-3 mm. RANGE: Eastern U. S. and Canada. Dull brownish-yellow with black spots. The pronotum and the appendages somewhat darker.

FAMILY 60. NOSODENDRONIDAE
The Wounded-tree Beetles

This family consists of two American species. These beetles are oval, convex, black, with scattered tufts of pale hairs. In size they vary from 4 to 6 mm. The head is densely punctate dorsally and the 11-segmented antennae terminate in a 3-segmented club. Beneath the mentum is enlarged completely enclosing the mouth. The pronotum is short and broad and ventrally excavated to receive the antennae and the first two pairs of legs.

Adults are frequently found in the fermenting sap of the wounds of trees. The larvae may develop in hollow stumps or in tree holes. Two species are found in the U. S.

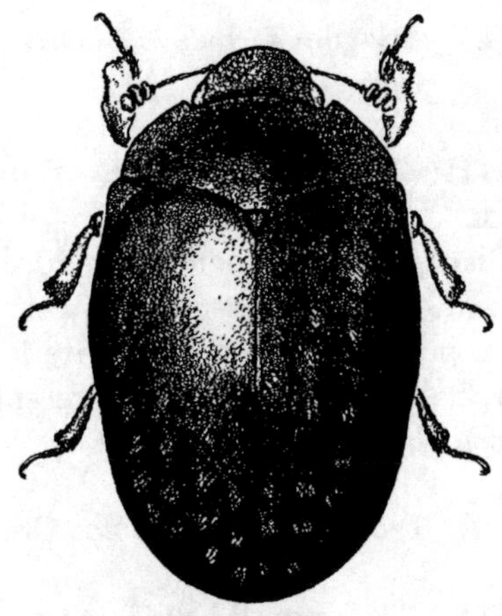

Figure 513

Figure 513 *Nosodendron californicus* Horn.

RANGE: Coastal B. C. to coastal California. Black with reddish or testaceous antennae and piceous legs.

Very similar is *N. unicolor* Say. RANGE: Kansas to Ind. The elytral punctures are finer.

FAMILY 61. DERMESTIDAE
The Skin and Larder Beetles

Some members of this family are likely to be found in any insect collection as volunteers. Unless watched closely their larvae may wreck a good collection. Over 500 species are known in all. They are small to medium sized and usually covered with dull or brilliant colored scales.

1a Length 5 mm or less; head with a distinct simple eye. 2

1b Length 6 or more mm; no simple eye. Fig. 514.

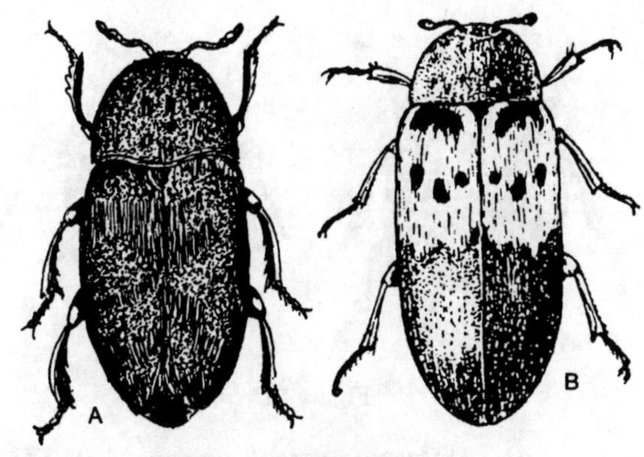

Figure 514

(a) *Dermestes caninus* Germ.

LENGTH: 7-8.5 mm. RANGE: general.

Black with mottled design on thorax and elytra in pubescence of black, gray and reddish-brown.

(b) *D. lardarius* L. The Larder Beetle

LENGTH: 6-8 mm. RANGE: cosmopolitan.

Black with gray hairs on elytra as pictured. Beneath black with scattered yellowish pubescence. It is sometimes a serious household pest and multiplies rapidly.

D. marmoratus Say 7-11 mm in length, black with white markings at base of elytra and yellowish hairs on underparts, is found in most of the Western States.

D. talpinus Mann. 5-7 mm, black above, clothed with bluish-gray, clay-yellow and black hairs and underparts mostly white, is principally western in its range.

This genus contains 15 species in our area.

2a (1) With distinct antennal grooves or pits; body covered with small colored scales. Fig. 515.

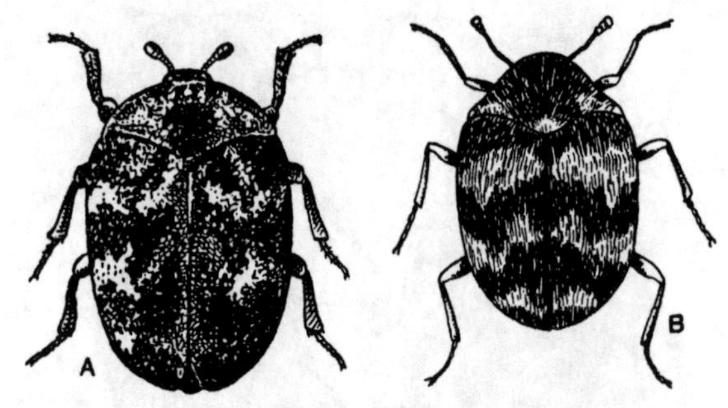

Figure 515

(a) ***Anthrenus scrophulariae*** **(L.) The Carpet Beetle**

LENGTH: 2.5-4 mm. RANGE: cosmopolitan.

Sutural stripe and apical spot on elytra, red. Remainder of pattern as pictured, black and white. Scales on under side orange and white. The larvae are known as Buffalo moths. It and the following are serious household pests.

(b) ***A. verbasci*** **(L.) The Varied Carpet Beetle**

LENGTH: 2-3 mm. RANGE: cosmopolitan.

Body black but thickly covered with yellowish-brown and white scale as pictured. Under parts with grayish-yellow scales.

A dozen more or less similar species are included in this genus in U. S. and Can.

2b Without antennal grooves; antennae 11 joined (a); 9 jointed (b). Fig. 516.

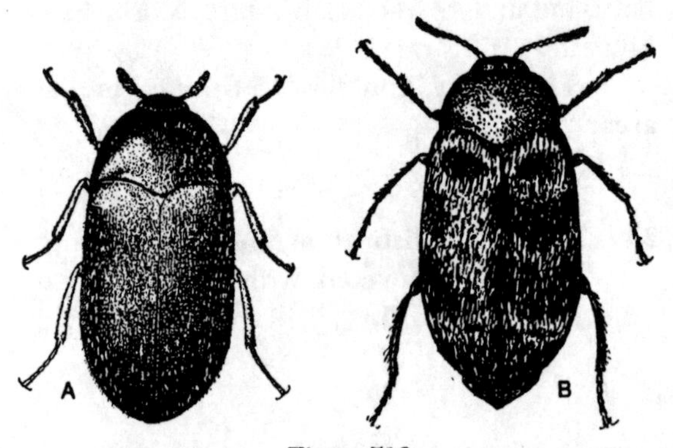

Figure 516

(a) ***Attagenus piceus*** **(Oliv.) Black Carpet Beetle**

LENGTH: 3.5-5 mm. RANGE: Europe and America.

Black, elytra sometimes reddish-brown, with scattered pubescence. A very serious museum pest. It is a very fortunate insect collection that has this species represented only by pinned specimens.

(b) ***Trogoderma versicolor*** **Creutz.**

LENGTH: 2-3.5 mm. RANGE: cosmopolitan.

Black, elytra marked with reddish and gray hairs as pictured. Eyes emarginate (notched).

Both of these genera have a dozen or more species in our fauna.

FAMILY 62. PTINIDAE
The Spider Beetles

Some 400 or more species are known. They are dull colored and feed on organic matter and often cause serious damage to food stores.

1a Erect hairs on elytra shorter; areas as shown marked with pale scales. Fig. 517. .... *Ptinus fur* L. White-Marked Spider Beetle

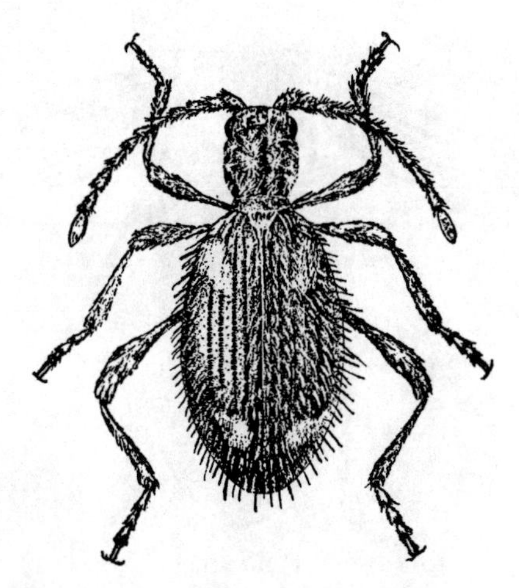

Figure 517

LENGTH: 2.5-3.5 mm. RANGE: cosmopolitan.

Male, pale brown to reddish-yellow; female, darker brown. This beetle is sometimes a serious pest of stored grain products and museum specimens.

Two other similar species are the Hairy Spider Beetle *P. villiger* (Reit) and the Storehouse Beetle, *Gibbium psylloides* (Czemp.) shining brown above and measuring 3 mm or less, and are in each case economic and widely distributed.

1b Erect hairs of elytra longer; no markings on elytra, except on humeral area of female. Fig. 518. .... *Ptinus clavipes* Panzer. Brown Spider Beetle

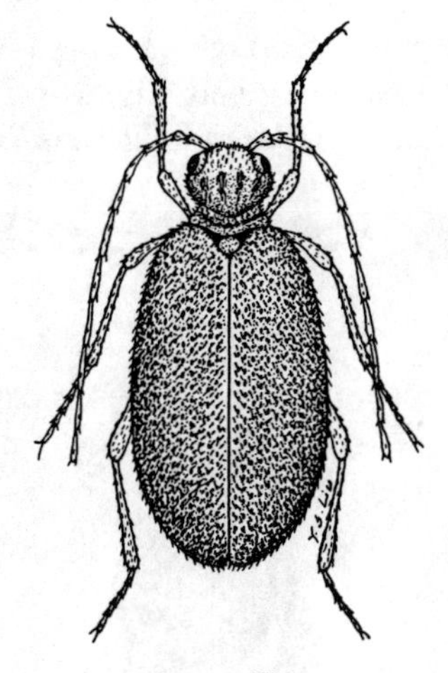

Figure 518

LENGTH: 2.7-3.5 mm. RANGE: widely distributed.

Pale brown to darker. Head, back of antennae smooth. An introduced species.

Sphaericus gibboides (Boield.) is a tiny species (1.8-2.2 mm), brown, clothed with pale brown and yellow scales. It is distributed world wide and lives in all kinds of food products.

Mezium americanum Lap. 2.5-3 mm long, dull yellow with shining black elytra, feeds upon dry animal substances and is cosmopolitan. The thorax is sub-cylindrical with a blunt tubercle at each side.

These genera are all small with 1 or 2 species. *Ptinus* with 35 species is the only one of any size.

FAMILY 63. ANOBIIDAE
The Drug-store and Death-watch Beetles

These are small dull-colored beetles which bore in wood or other organic substances. The mating call in some cases is made by bumping the head against the wall of the wooden burrow, giving rise to strange tales of haunted houses, etc. More than 900 species are known.

1a **Thorax not margined at sides; front of head not margined. Fig. 519. .... *Eucrada humeralis* (Melsh.)**

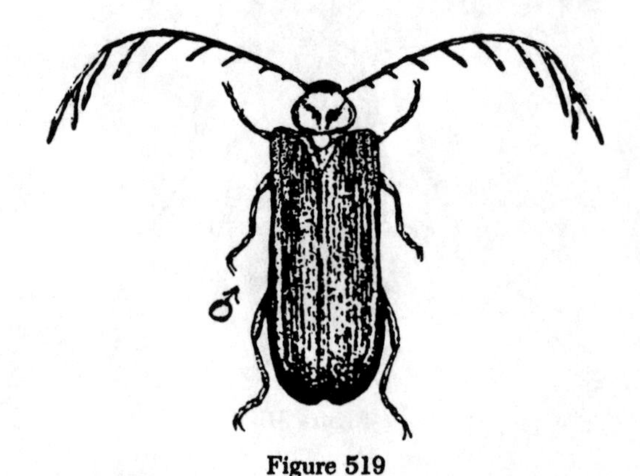

Figure 519

LENGTH: 4-6 mm. RANGE: Eastern half of United States and Canada.

Dull. Black, marked on thorax and elytra as pictured with reddish-yellow; elytra with coarse punctures in rows.

Ernobius conicola Fisher. 5-6 mm long, shining brown with yellow pubescence, breeds in the cones and dead wood of the Monterey cypress in California. This is a large genus of about 30 species all associated with coniferous trees.

1b **Thorax margined at sides and front of head with margin at apex. 2**

2a **(1) Antennae of male fan-shaped; outer margin of front tibia forming an apical tooth with margin above it finely dentate. Fig. 520. .... *Ptilinus ruficornis* Say**

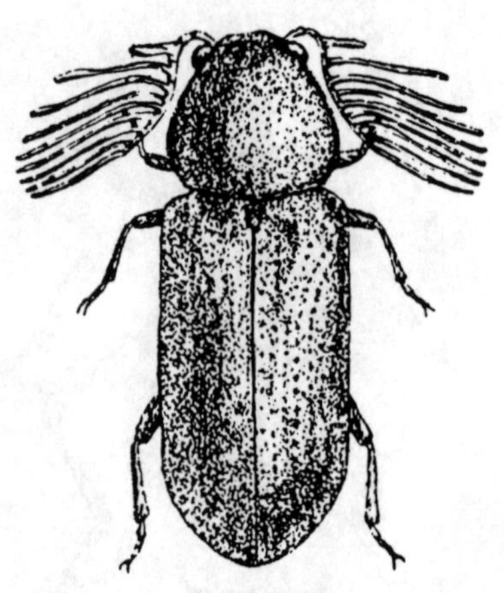

Figure 520

LENGTH: 3-5 mm. RANGE: Eastern United States and Canada.

Cylindrical. Dull. Black; antennae and legs reddish-yellow. Often common on dead branches of oak and maple in which the larvae live.

P. pruinosus Csy. is also black and dull. It measures from 3 to 3.5 mm and is known from Indiana and Ontario. It differs from *ruficornis* in that the margin of the thorax is poorly defined.

2b **Not as in 2a. 3**

3a **(2) Head when at rest received on the under surface of the thorax which is excavated for it. Fig. 521. .... *Stegobium paniceum* (L.)**

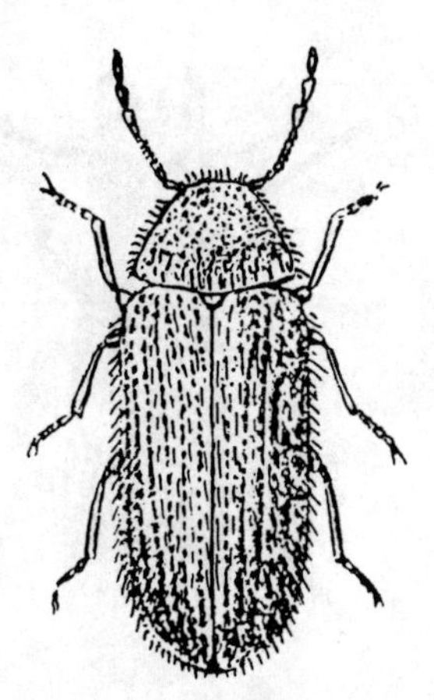

Figure 521

LENGTH: 2.5-3.5 mm. RANGE: cosmopolitan.

Fairly robust. Reddish-brown with yellowish pubescence. Elytra finely striate, the striae punctured. This "drug store beetle" has some amazing tastes in the field of drugs as well as food products.

Xeranobium desertum Fall 7-8 mm long, brown with gray hairs, breeds in the stems of greasewood in the Southwest.

Trichodesma gibbosa (Say) measuring 4.5-6.5 mm and known from Iowa, Florida, Canada, etc., is oblong, robust and densely covered with grayish-white hairs. The thorax is a bit narrower than the elytra, its disk elevated and bearing many fine erect hairs. At the crest of the thorax is a divided tuft of brown and brownish-yellow hairs with two smaller tufts in front of it. The front coxae are widely separated.

3b Head very strongly reflexed and retracted; the mandibles often reaching the metasternum. 4

4a (3) 2nd and 3rd legs received at rest in special cavities on ventral side; antennae usually received between front coxae. Fig. 522. .... *Tricorynus confusus* (Fall)

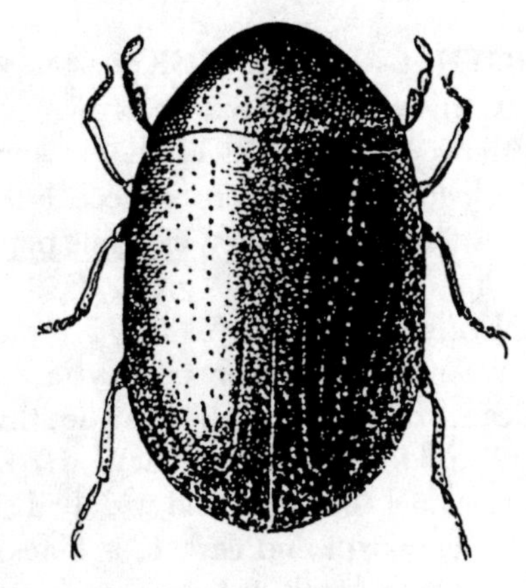

Figure 522

LENGTH: 2-3 mm. RANGE: general.

Convex. Dark piceous brown, pubescent, coarse punctures on sides of elytra and thorax.

In our territory this genus includes some 90 species mostly brown or piceous, and similar in shape.

Dorcatoma falli White. about the same size and cosmopolitan in range is shining black with gray pubescence. The segments of the antennal club are notched on their front edge. Two similar species here.

4b Without grooves to receive 2nd and 3rd legs; grooves for antennae on head. Fig. 523. .... *Lasioderma serricorne* (Fab.) Cigarette Beetle

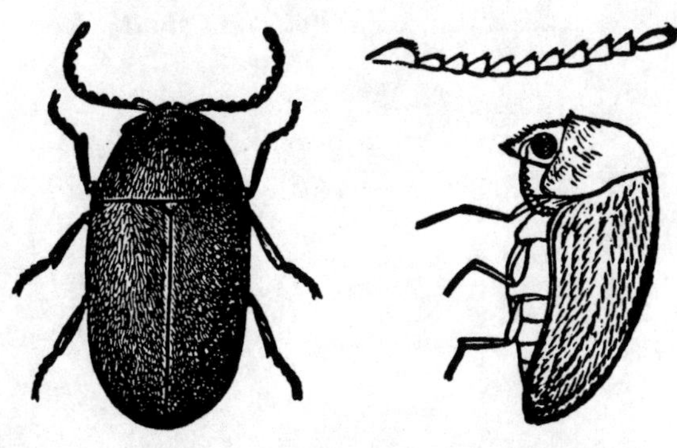

Figure 523

LENGTH: 2-3 mm. RANGE: cosmopolitan.

Convex. Dull. Reddish-brown or brownish-yellow. Front angles of thorax acute. This beetle feeds not only on tobacco but also on such tasty foods as ginger, cayenne pepper, figs and other hot stuff.

Vrilletta decorata VanD., 5-6 mm, black with yellow markings on the elytra, breeds in the dead wood of live oak in the Southwest.

Petalium bistriatum (Say) 1.7-2 mm in length, is fairly common and widely distributed from Texas north and east. It is black or dark reddish-brown, with antennae and legs usually lighter. It differs from the above species in having the front coxae expanded at the apex into horizontal plates. In our fauna 23 small, similar species are in the genus *Petalium.*

FAMILY 64. BOSTRICHIDAE
The Horned Powder-Post Beetles

Many of these are so tiny that their larvae make "pinholes" in lumber; a few species are large with a maximum length of about two inches. Reddish-brown to black are the usual colors.

1a Head not covered by the prothorax; front cover separated by a lobe. Fig. 524. .... *Polycaon stouti* (Lec.)

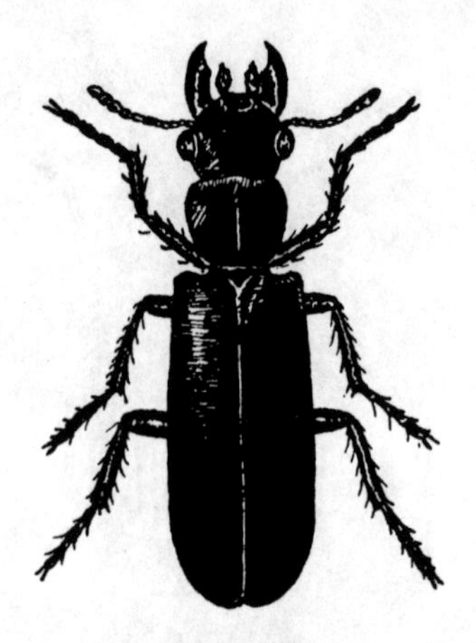

Figure 524

LENGTH: 11-21 mm. RANGE: Western States.

Wholly black; punctures on head and thorax coarse, on elytra fine. It breeds in the dead wood of many western trees and occasionally invades living wood.

Melalgus confertus (Lec.) 7-15 mm in length, elognated, cylindrical, black with brown elytra, damages many fruit trees and other plants in California and Oregon.

The Spotted Limb Borer *Psoa maculata* (Lec.) 6-8 mm is another destructive California species. It is bronze-colored with a covering of gray hairs. The elytra, blackish or bluish-green, bear several reddish, yellow or white spots.

1b Head covered by prothorax. 2

2a (1) First and second segments of antennae longer than the middle ones. Fig. 525. .... *Xylobiops basilaris* (Say)

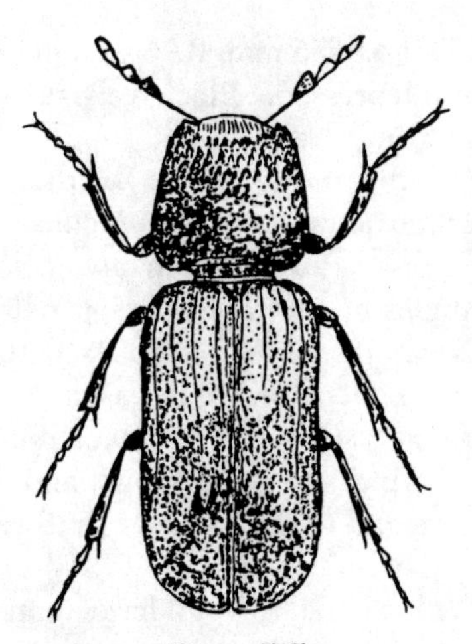

Figure 525

LENGTH: 5-6 mm. RANGE: United States and Canada.

Cylindrical. Black, marked with dull reddish-yellow as pictured; antennae with 10 segments.

The "Lead Cable Borer" *Scobicia declivis* (Lec.) is an interesting near relative, known from the West. This cylindrical black or dark brown beetle (5-6 mm) feeds on dry wood of several species as a steady diet, but tops off on lead telephone cables to the consternation of the linemen, who have dubbed it the "Short Circuit Beetle."

2b **First and second antennal segments shorter than the middle ones; front without margin. Fig. 526. .... *Amphicerus bicaudatus* (Say)**

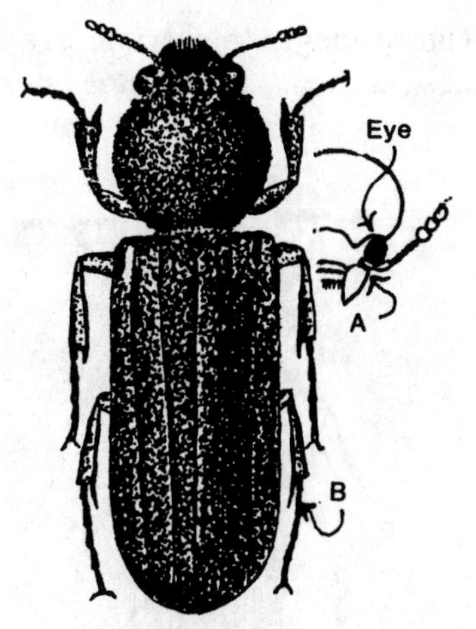

Figure 526

LENGTH: 6-9 mm. RANGE: East of the Rockies.

Cylindrical. Brownish-black, with scattered recumbent pubescence. Its official common name is "Apple Twig Borer."

The "big boy" of the family is *Dinapate wrighti* Horn which measures 30-52 mm, dark brown to black, and bores in palms in Southern California.

Lichenophanes bicornis (Web.) 7-12 mm in length, ranges from Iowa to Florida and Connecticut. It is dull, piceous and ornamented with irregular patches of yellow scales. Each elytron has two longitudinal ridges which distinguish it from *Lichenophanes armiger* (Lec.) which has but one short ridge on each elytron and is somewhat smaller.

Some 60 odd species are found in the U. S. These are spread over 31 genera, many genera having then only one or two species.

FAMILY 65. LYCTIDAE
The Powder-post Beetles

Less than 100 species of tiny but highly destructive woodborers belong to this family. They range in color from yellow to brown, reddish and black.

1a **Thorax longer than wide. Fig. 527. .... *Lyctus opaculus* Lec.**

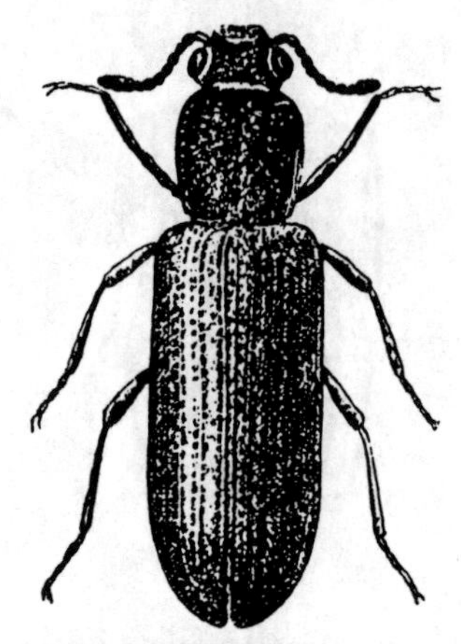

Figure 527

LENGTH: 3-5 mm. RANGE: Eastern half United States.

Dark reddish-brown or blackish, with scattered yellow hairs, head and thorax usually darker than elytra. Feeds in dead oak and other wood.

L. brunneus (Steph.) is an Oriental species introduced in bamboo goods. The head and thorax are black and the other parts brown. (4-5 mm)

1b **Thorax as wide as long. Fig. 528. .... *Lyctus planicollis* Lec. Southern Lyctus Beetle**

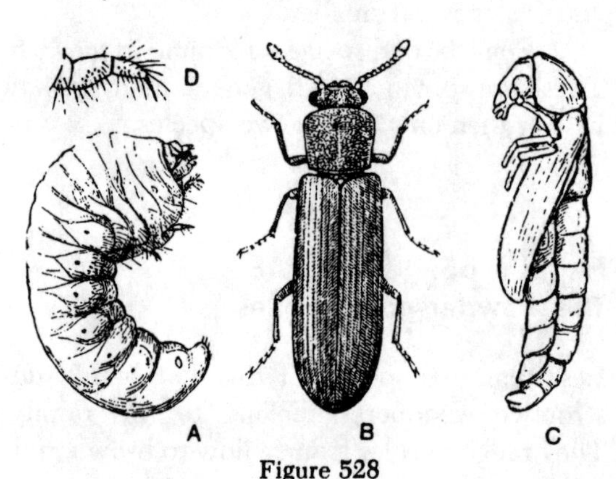

Figure 528

LENGTH: 4.5-5.5 mm. RANGE: general.

Subdepressed. Black, elytra with fine punctures in rows.

A, larva and leg (D); B, adult; C, pupa.

Trogoxylon parallelopipedus (Melsh.) may be distinguished from *planicollis* by the front angles of the thorax being not rounded. The elytra are only a bit wider than the thorax, but about three times as long. It has a coat of fine yellowish hairs over dull reddish-brown. It measures 3.5-4 mm and is known from Iowa and to the east and southeast.

1c **With single row of large punctures on elytra. Fig. 529. .... *Lyctus linearis* (Goeze.)**

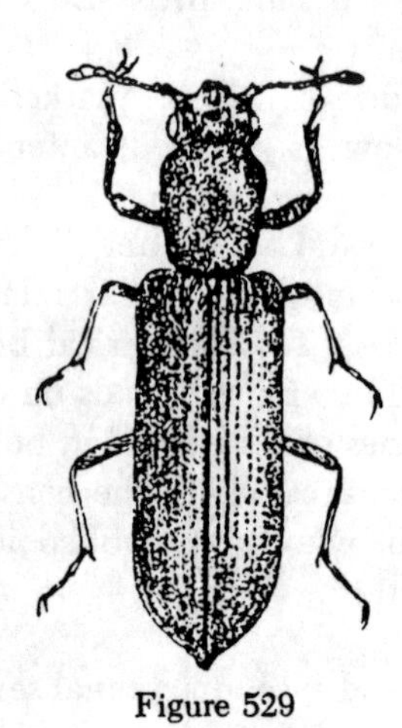

Figure 529

LENGTH: 2.5-3.5 mm. RANGE: cosmopolitan.

Subdepressed. Reddish-brown. Introduced from Europe.

The Western Lyctus, *L. cavicollis* Lec., rusty-brown, slender and measuring 2.5 to 3 mm bores in hickory, oak, orange and eucalyptus on the West Coast area.

FAMILY 66. TROGOSSITIDAE
The Bark-Gnawing Beetles

These cylindrical or often flattened beetles range from small to medium sizes. Many are black or brown; some have brilliant metallic shades. Almost 700 species are known.

1a Form elongate, thorax emarginate at apex; head large. Fig. 530.

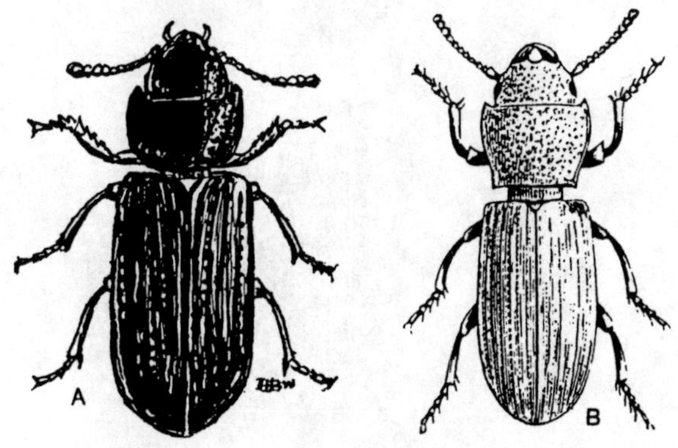

Figure 530

(a) *Tenebroides mauritanicus* (L.) The Cadelle

LENGTH: 9-10.5 mm. RANGE: cosmopolitan.

Blackish throughout, antennae a gradually widening club. Eighth joint of antennae equal to ninth.

(b) *Tenebroides corticalis* (Melsh.)

LENGTH: 7-8 mm. RANGE: Alaska, Canada, and the U. S.

Black, rather dull, antennae and legs lighter. Eighth joint of antennae smaller than the ninth.

There are about 20 members of this genus in our area. All are brownish and very similar in appearance.

1b Form oval, with flattened margins. Front coxal cavities closed (a); open behind (b). Fig. 531.

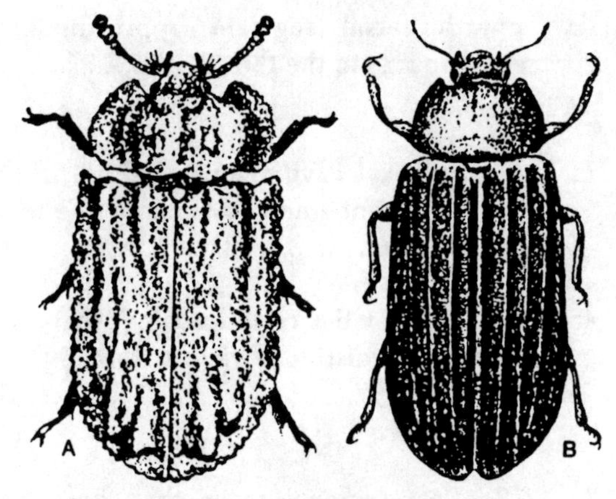

Figure 531

(a) *Calitys scabra* Thunb.

LENGTH: 9-10 mm. RANGE: Europe and Southern Can. and northern U. S.

Reddish. Thoracic margins wide. Lives on fungi on logs.

(b) *Grynocharis quadrilineata* (Melsh.)

LENGTH: 5-6 mm. RANGE: Wis. to Me. and Manitoba to Que.

Black, feebly shining, antennae and legs lighter. Four narrow ridges on each elytron with 3-4 rows of punctures on each interval. Under bark.

FAMILY 67. CLERIDAE
The Checkered Beetles

These beetles are often highly colored with intricate patterns. They are small to medium in size, often cylindrical and usually hairy. The adults are found under bark and on flowers; they are predacious. About 300 species are known from our region and some 3500 for the whole world.

1a **Fourth tarsal segment very small, frequently embedded within the lobes of the third. 9**

1b **Fourth tarsal segment approximately equal in size to the third. 2**

2a **(1) Procoxal cavities open behind; first tarsal segment small covered by the second segment. 4**

2b **Procoxal cavities closed behind; first tarsal segment distinctly visible from above. 3**

3a **(2) Last segment more than half the length of the entire antennae; punctures on elytra scattered. Fig. 532. .... *Monophylla terminata* (Say)**

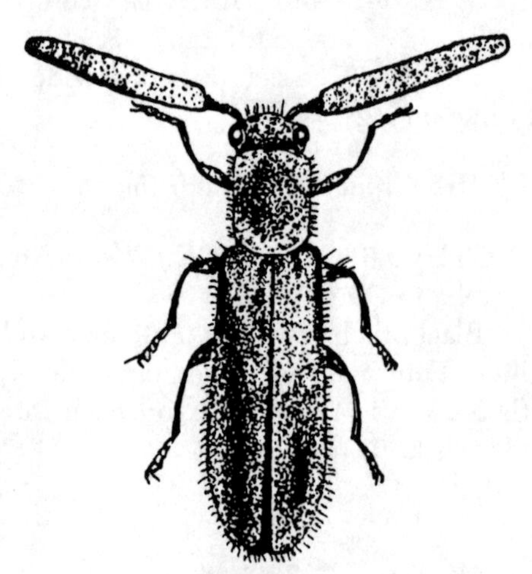

Figure 532

LENGTH: 4-7 mm. RANGE: Texas and Arizona east to Atlantic coast.

Black, somewhat shining; thorax and elytra marked with yellow as pictured; abdomen of male wholly yellow, the last segment black in the female.

Cymatoderella collaris (Spin.) 5-6 mm long is black with a red thorax; the latter sometimes margined with black. It ranges through the Southern States. Its last anennal joint is not abnormally long.

3b **Last segment of antennae never longer than the preceding four taken together; punctures of elytra in rows. Fig. 533. .... *Cymatodera bicolor* Say**

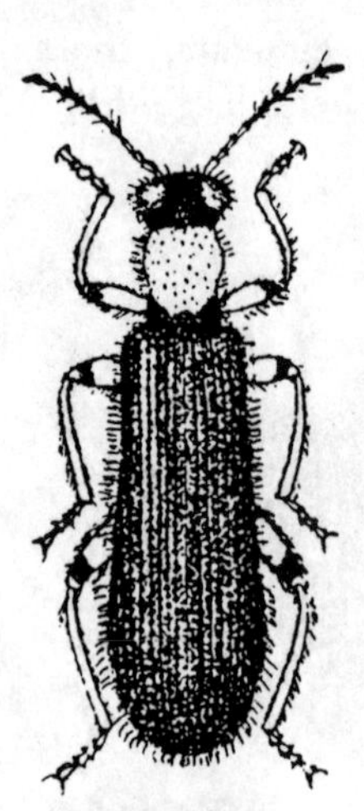

Figure 533

LENGTH: 6-10 mm. RANGE: eastern half of U. S.

Black, somewhat dull; thorax, legs and first two joints of antennae reddish-yellow marked with black as pictured.

C. ovipennis Lec., 8-11 mm long and light brown with a paler line on elytra is native of the Southwestern States.

C. inornata Say is 7-9 mm long, wholly dark brown with a coat of yellowish hairs. The antennae and the base of the femora are reddish-brown. It ranges across the eastern half of the United States and into Canada. There are about 70 members of this genus in the U. S., many in the Southwest.

4a **(2) Eyes with a notch in front (a). Fig. 534. .... 5**

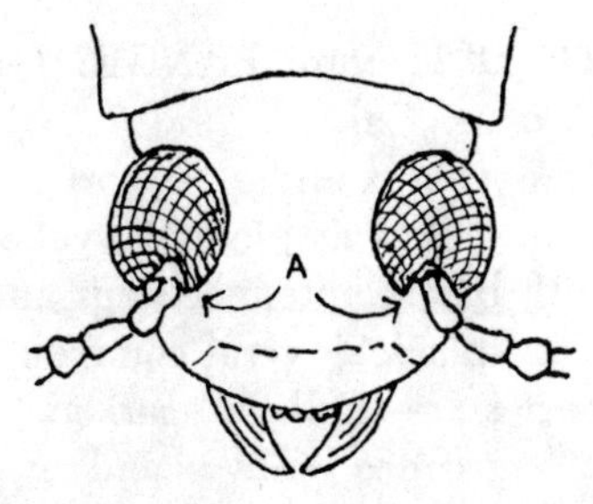

Figure 534

4b **Eyes without a notch. 8**

5a **(4) Last three joints of antennae forming a distinct club; eyes finely granulate. 6**

5b **Antennae serrate, but not clubbed; eyes coarsely granulate. Fig. 535. .... *Priocera castanea* Newn.**

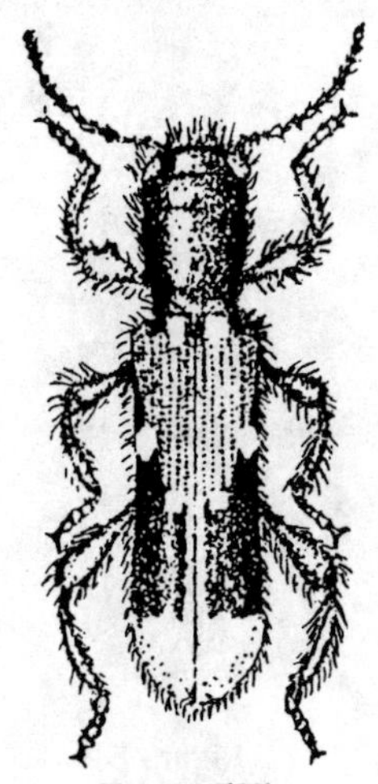

Figure 535

LENGTH: 6.5-10 mm. RANGE: eastern half of the U. S. and Canada.

Head, thorax and legs dark chestnut-brown shining; elytra reddish brown marked as pictured with black and yellow.

Colyphus ornaticollis (Lec.) with a length of 8 mm, colored black with pinkish marked thorax is known from Ohio and Indiana.

6a **(5) Last segment of maxillary palpi slender. 7**

6b **Last segment of maxillary palpi slightly broader than next to last segment; antennal club triangular. Fig. 536. .... *Trichodes nutalli* Kby.**

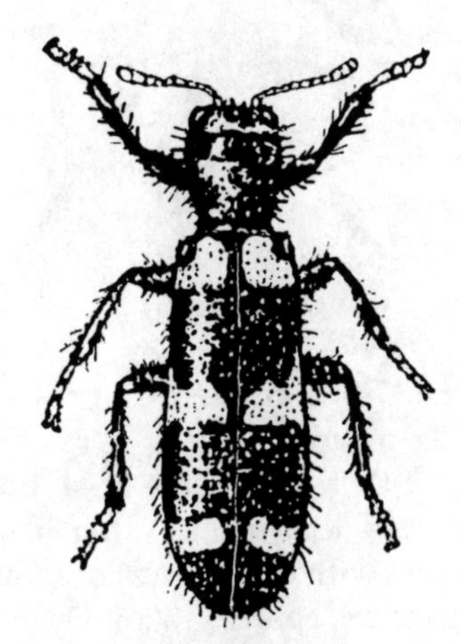

Figure 536

LENGTH: 8-11 mm. RANGE: Eastern half of U. S. and Canada and Colorado.

Dark blue, purplish or greenish blue; elytra darker blue, marked as shown with reddish-yellow. A very attractive beetle.

T. ornatus Say, a somewhat similarly marked and colored beetle, ranges through most of the Western States. It is 6-7 mm long and metallic blue with 3 yellow bars across the elytra. There are 11 members of this colorful genus in our fauna.

7a **(6) Hind tarsus slender and elongated; thorax with impressed line at middle and with a deep subapical groove. Fig. 537. .. .. *Thanasimus dubius* (Fab.)**

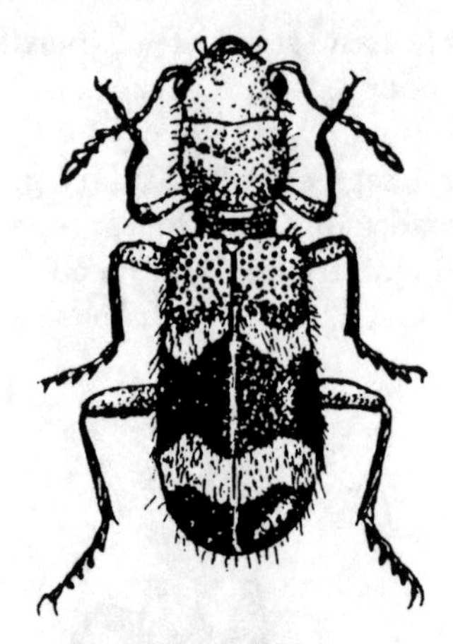

Figure 537

LENGTH: 7-9 mm. RANGE: Alas. to Me., south to Tenn. and N. Mex.

Head, thorax and base of elytra dull red. Antennae and legs red to pitch black; elytra mostly black with cross bands of whitish hairs.

Enoclerus cupressi Van Dyke living in California serves a useful purpose in destroying the Cypress Barkbeetle. It is black with triangular orange spots on the elytra and is 7 mm long.

7b Hind tarsi broadly dilated; thorax usually smooth. Fig. 538. .... *Enoclerus ichneumoneus* (Fab.)

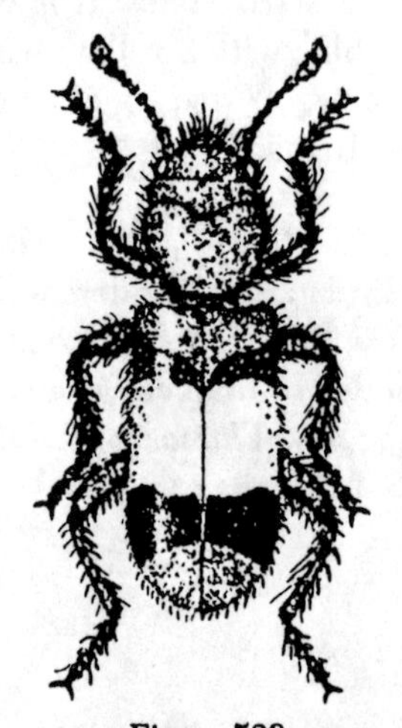

Figure 538

LENGTH: 8-11 mm. RANGE: Can. to Fl., west to Tx. and Cal.

Head, thorax and elytra brick red; median band on elytra as pictured, yellowish margined with black; antennae, legs and much of under parts black. Several common closely related species are similarly marked.

E. nigrifrons (Say) and *E. rosmarus* (Say) have the abdomen black. In the first the elytral punctures are finer than in *rosmarus* which has red antennae. Both range over the Eastern half of the U. S. *Enoclerus* is a large genus with over 40 species in the U. S. and Can.

8a (4) Elytra bright red; antennae with 3 segments in club. Fig. 539. .... *Zenodosus sanguineus* (Say)

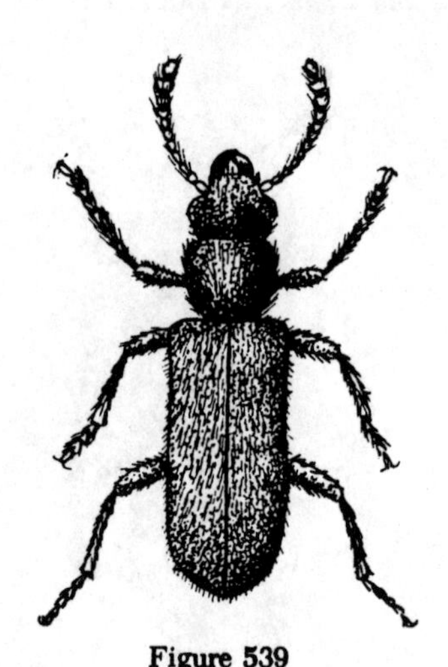

Figure 539

LENGTH: 4.5-6.5 mm. RANGE: Ont. to N. C. west to Col.

Head, thorax, and under parts brown, elytra crimson or blood-red, rather dull. There is but this one species in the genus. These beautiful beetles are found under bark. They are often found in the crass and debris at the base of old, dead beech trees.

8b **Elytra black, sometimes marked in yellow; antennal club small, spherical and with but two segments. Fig. 540.**

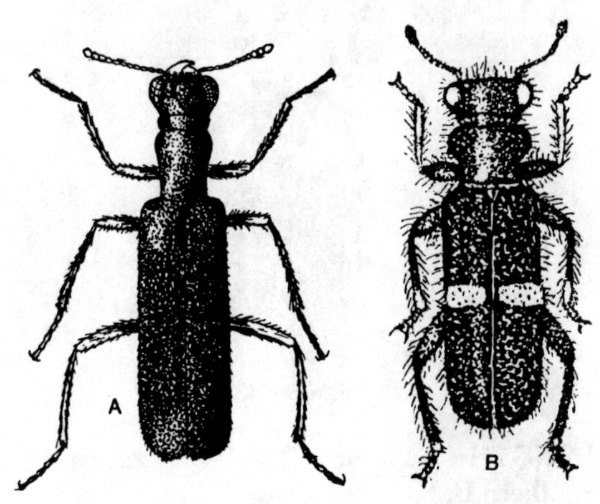

Figure 540

(a) ***Isohydnocera tabida*** **(Lec.)**

LENGTH: 5-7 mm. RANGE: Ont. to Ala. west to Kan.

Bluish black, antennae and legs pale reddish-yellow.

(b) ***Phyllobaenus unifasciata*** **(Say)**

LENGTH: 3.5-4.5 mm. RANGE: East. Can. to Ala. west to Col. and Ariz.

Bluish-black; elytra with bar of dense white hairs as pictured.

P. scaber (Lec)., dull brownish gray and 4-5 mm long; inhabits the Pacific coast. About 60 members of this genus occur in our fauna.

9a **(1) Last three segments of antennae large, flat and dilated; six segments of abdomen visible. See (a), Fig. 541. .... 10**

Figure 541

9b **Antennae ending in a small compact club; but five segments of abdomen visible. Fig. 541. .... *Necrobia ruficollis* (Fab.)**

LENGTH: 4-5 mm. RANGE: it and the other members of its genus mentioned here seem to be scattered world wide.

Red; head, antennae, and abdomen black, elytra metallic blue or green marked at base as shown with red.

Necrobia rufipes (DeG.) and *N. violacea* (L.) differ from *ruficollis* in being wholly blue or green above. In *rufipes* the base of antennae and legs are red but in *violacea* they are dark. These are known as "Ham Beetles." When ham is scarce they gladly congregate in other meat supplies, cheese, spoiled fish or dead animals in the field.

All three species have coarsely granulated eyes; only the last three of the 11 segments forming the antennae are in the club; the claws are toothed at their base. These three species likely constitute the only serious offenders within the family.

10a **(9) Notch of eye in lower front. 11**

10b Notch of eye on inner side. Fig. 542. ....
.... *Phlogistosternus dislocatus* (Say)

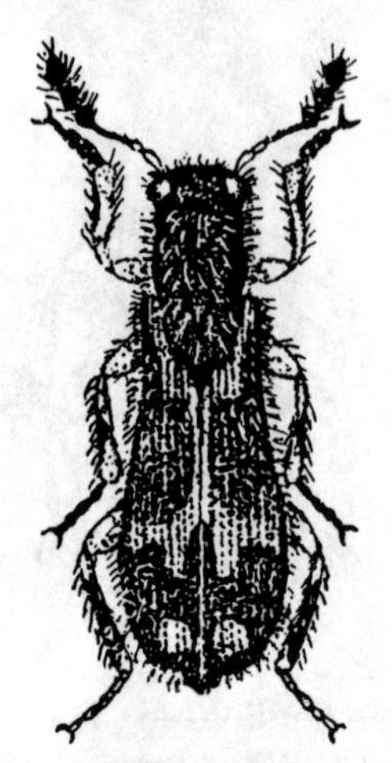

Figure 542

LENGTH: 4-6 mm. RANGE: eastern half of the U. S. and Canada.

Black, with yellowish legs, antennae and markings on elytra; elytra with coarse deep punctures.

Neichnea laticornis (Say) with a length of 5-7 mm; black with mid-stripe on head and sides of thorax orange yellow, is found from Canada to North Carolina and west to Illinois.

11a (10) First segment of antennae not smaller than second; visible from above. 12

11b First segment small and hidden from above by overlapping second segment. Fig. 543. ....
.... *Neorthopleura thoracica* (Say)

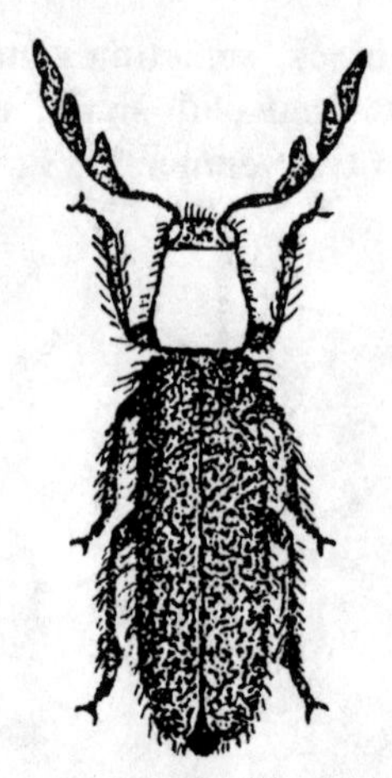

Figure 543

LENGTH: 3-12 mm. RANGE: Eastern and Southern U. S.

Black; thorax reddish yellow, often clouded at front and back. Elytra sometimes with indistinct pale cross bar at middle.

The eyes are coarsely granulate; the antennae have 11 segments, all but two of which are in the club. The elytra have many coarse punctures. Several other species are found in Tex.

12a (11) Thorax widest behind the middle then rather abruptly constricted. Fig. 544.

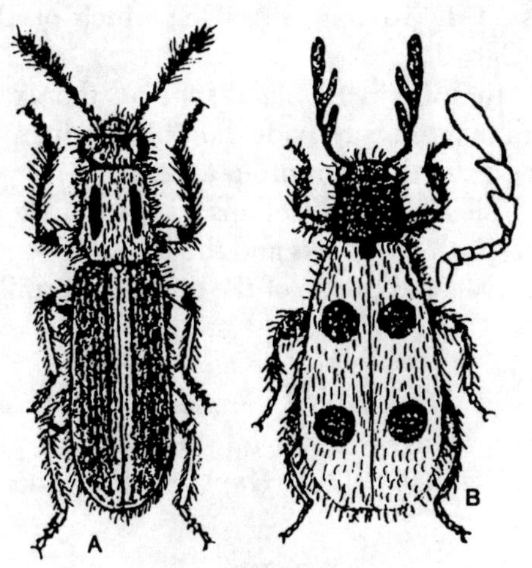

Figure 544

(a) *Cregya oculata* Say

LENGTH: 4-6.5 mm. RANGE: eastern half of the U. S.

Black, shining; thorax and elytra yellow, marked as pictured, with black; legs yellow. Several similar species are in the same range.

(b) *Pelonides quadripunctatum* Say

LENGTH: 5-7 mm. RANGE: throughout the North and South Central States.

Black, elytra bright orange with 2 black spots as pictured on each wing cover. The antennae have but 10 joints.

12b Thorax not as in 4a. Fig. 545. .... *Chariessa pilosa* Forst.

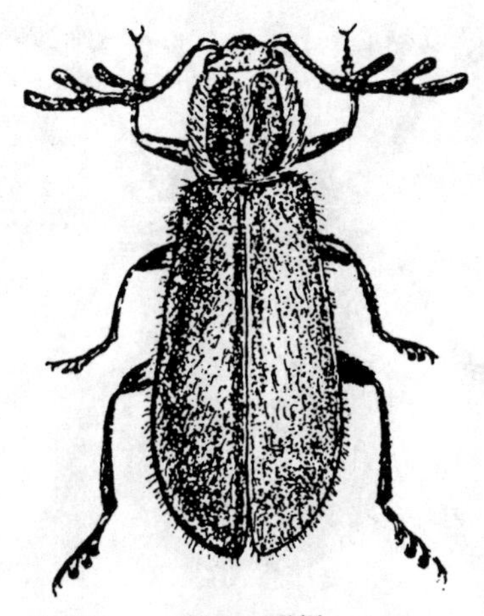
Figure 545

LENGTH: 7-13 mm. RANGE: Eastern Canada to Texas and Florida.

Black, rather dull; thorax reddish with black markings; elytra with dense fine punctures.

C. elegans Horn, is a western species 11-12 mm, pale red with bluish-black elytra. It ranges from Oregon to California and Texas.

FAMILY 68. MELYRIDAE
The Soft-winged Flower Beetles

More than 100 species have been described for our region but only a few are common in general collections.

1a Antennae with 11 segments. 3

1b Antenna with apparently 10 segments. (The second segment being small and hidden.) 2

2a (1) Elytra marked with spots. Fig. 546. .... *Collops quadrimaculatus* (Fab.)

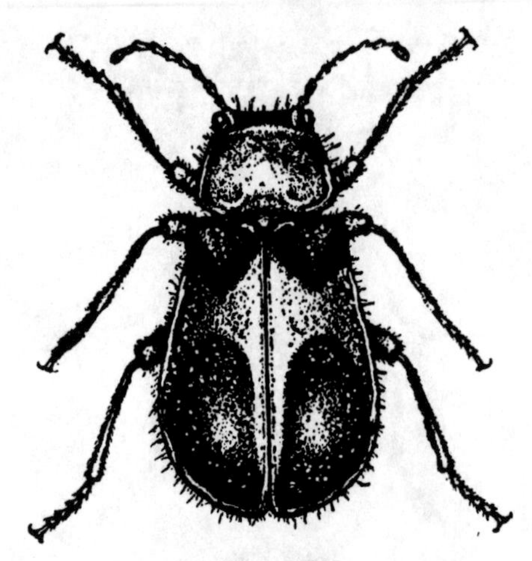

Figure 546

LENGTH: 4-6 mm. RANGE: much of the U. S. especially east of the Rockies. Common.

Head, abdomen and femora, black; thorax and elytra reddish-yellow, the marking blue or bluish-black.

C. bipunctatus Say about the same size (5-6 mm), has bluish-black elytra with a reddish thorax on which are two oblique black spots. It ranges from the Missouri river to the Pacific. There are about 30 species of *Collops* in our fauna.

2b Elytra marked with stripes. Fig. 547. .... *Collops vittatus* (Say)

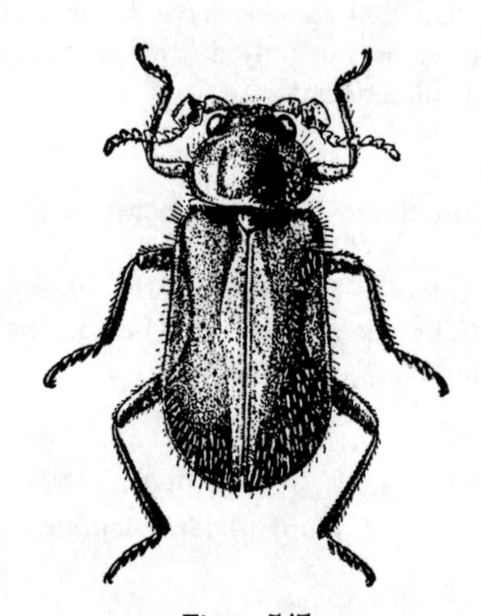

Figure 547

LENGTH: 5-7 mm. RANGE: Much of U. S. and Canada.

Reddish with black spot on thorax and elytra marked with wide dark blue stripes. The spot on the thorax is often absent.

This "Striped Collops" is a valuable predator. Both the adults and the larvae feed upon the larvae and pupae of the alfalfa caterpillar.

3a (1) Front tarsi simple; elytra of male prolonged at tip into a hook-like piece. Fig. 548. .... *Hypebaeus apicalis* Say

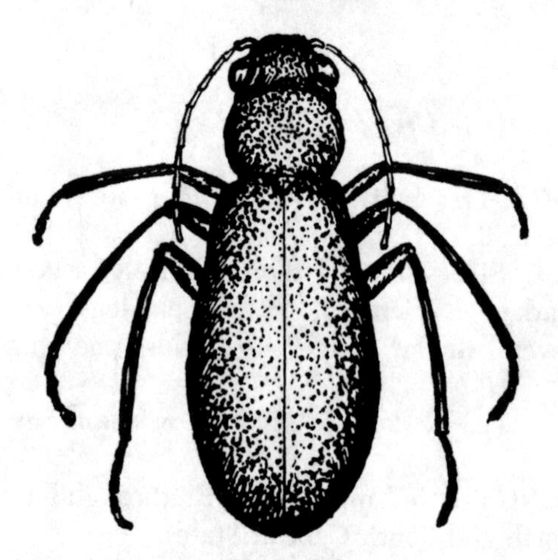

Figure 548

LENGTH: 2-2.5 mm. RANGE: Eastern Canada to Fla.

Wholly black above; legs and base of antennae pale dull yellow. A small group of 5 species.

Malachius erichsoni Lec. measuring 3.5 to 4 mm and ranging through much of the eastern half of the United States, does not have an appendage on the elytra. The head and elytra are black, the latter with a bluish tinge; the thorax is reddish yellow. A group of 35 species, mostly Western.

3b Front tarsi of male with second joint prolonged covering the third; elytra of both

sexes without an appendage. Fig. 549. .... *Attalus circumscriptus* (Say)

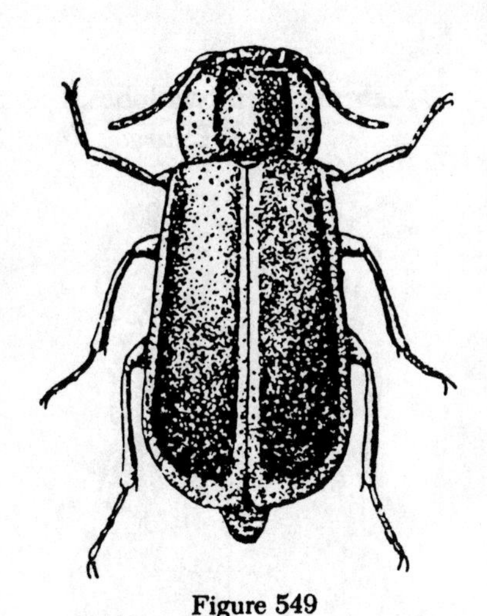

Figure 549

LENGTH: 3 mm. RANGE: South Central and Southern States west to Arizona.

Black; margins of thorax and elytra as pictured, pale dull yellow. Under surface black, the ventral segments margined behind with yellow; legs yellowish and dusky.

A. terminalis (Er.) differs from the above in having the head and thorax wholly black; the elytra of the male are narrowly margined with yellow but in the female they too are wholly black. It measures 2.5 mm and ranges over much of the eastern half of the U.S.

More than 30 members of this genus live in the U. S. and Canada.

FAMILY 69. LYMEXYLONIDAE
The Ship-Timber Beetles

This is a small family, two genera and two species, of medium sized beetles, 10-15 mm in length. These beetles are elongate, cylindrical, black or brown in color, with yellow appendages, and covered with a fine pile. The short 11-segmented antennae are serrate and the maxillary palpi are flabellate.

Adults are taken from the trunks of trees and from under bark. The larvae are wood borers in the sapwood and heartwood of hard wood trees.

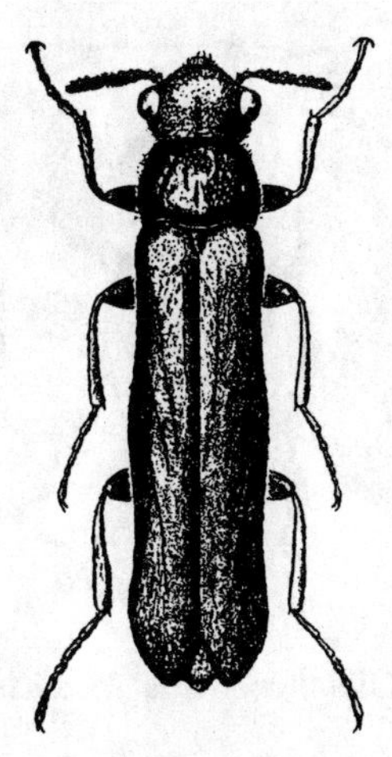

Figure 550

Figure 550 *Hylectes lugubris* Say. LENGTH: 10-12 mm. RANGE: Eastern U. S. Head, pronotum, and prosternum rufous, elytra and mesosternum black, abdomen and legs yellow, antennae black with the basal segments more or less yellow.

The second species, *Melittomma sericeum* Harris has the range the same as the above. LENGTH: 10-15 mm. Entirely dark brown with the under side and the appendages reddish. Densely pubescent with brownish hairs. Comes to lights. Also taken on oak.

FAMILY 70. NITIDULIDAE
The Sap-Feeding Beetles

These small beetles feed on decaying plant materials and are found in garbage, where sap is escaping from plants, under dead bark and sometimes in ants' nests. They are usually black, brown or grayish. This family is cosmopolitan and numbers over 2500 species.

1a **Labrum united with the front; black usually marked with yellow or reddish. Fig. 551.**

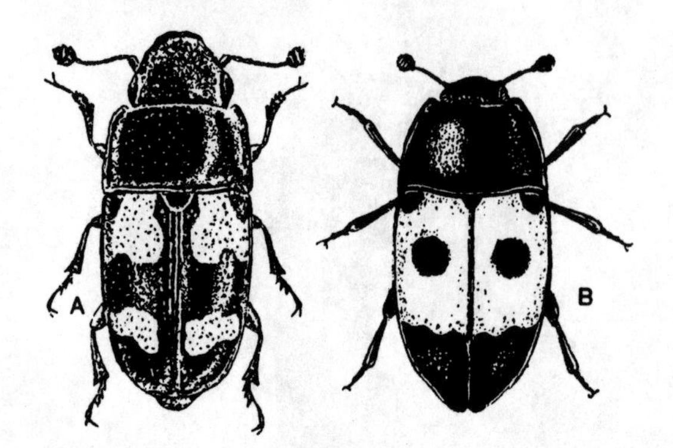

Figure 551

(a) ***Glischrochilus fasciatus*** **(Oliv.)**

LENGTH: 4-7 mm. RANGE: general.

Shining black marked with yellowish as pictured. Spots variable in size. This beetle is often abundant in and under garbage pails.

(b) ***G. sanguinolentus*** **(Oliv.)**

LENGTH: 4-6 mm. RANGE: Eastern half United States and Canada.

Shining black; elytra bright red with humeral and discal spots and tips black; abdomen and metasternum red.

1b **Labrum free; not shining black. 2**

2a **(1) Elytra wholly covering abdomen or at most exposing but one segment. 3**

2b **Elytra exposing 2 or more segments of abdomen. 6**

3a **(2) Tarsi distinctly widened; tips of elytra usually truncate. 4**

3b **Tarsi not widened; elytra entirely covering abdomen. 5**

4a **(3) Labrum with two lobes. Fig. 552. .... *Epuraea duryi* Blatch.**

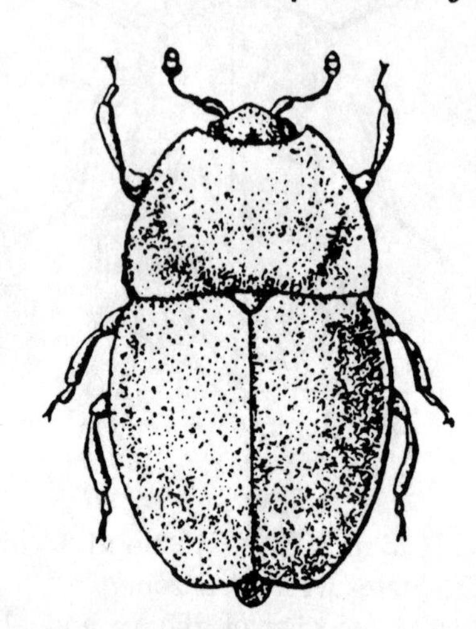

Figure 552

LENGTH: 3.5-5 mm. RANGE: Eastern United States.

Flattened. Shining pale yellow with scattered pubescence.

A smaller member (2.5-3 mm) of this genus is *Epuraea obtusicollis* Rttr. Its color is dark reddish-brown to blackish with paler margins to elytra and thorax. It is widely distributed. More than 30 species of this genus have been described for our area. They are all yellowish and difficult to separate.

4b **Labrum but feebly notched. Fig. 553. .... *Nitidula bipunctata* (L.)**

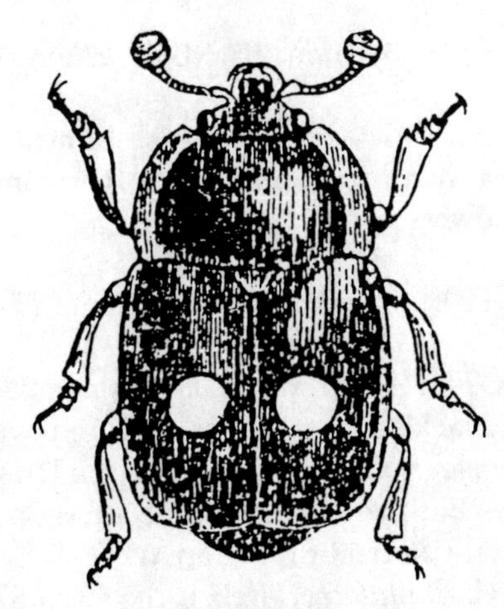

Figure 553

LENGTH: 4.5-6 mm. RANGE: Europe and North America.

Blackish, marked with a red spot on each elytron; covered with a fine pubescence. Common on old bones and skins.

Stelidota geminata (Say) 2-2.5 mm is widely distributed and common. It is dark reddish-brown with two indistinct cross-bars on the elytra. Its abdomen is entirely covered by the elytra.

5a (3) **Mentum covering base of maxilla; thorax and elytra with wide flat margins. Fig. 553b. .... *Prometopia sexmaculata* (Say)**

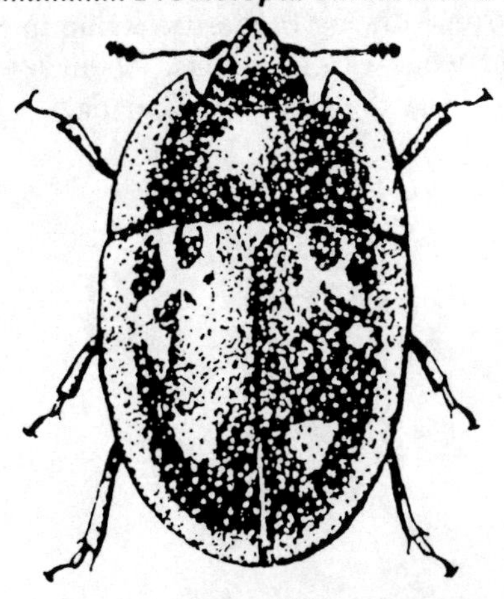

Figure 553b

LENGTH: 5-6 mm. RANGE: East of Rockies and Cal.

Piceous, marked on elytra and margins with pale reddish-brown as shown. Feeds on sap in spring.

Another species common to the two continents is *Omosita colon* (L.) It is blackish with dull yellow margins on elytra and thorax and 4 or 5 dull yellowish spots on elytra. It measures 2-3 mm.

5b **Mentum not covering the maxilla. Fairly large. Fig. 554. .. *Phenolia grossa* (Fab.)**

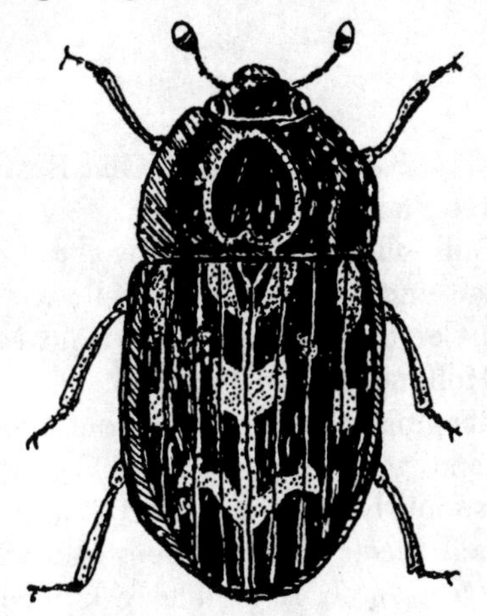

Figure 554

LENGTH: 6.5-8 mm. RANGE: Eastern half United States and Canada.

Piceous, somewhat dull; elytra with indistinct reddish markings as shown. Common beneath bark and in fungi.

Lobiopa undulata (Say) with perhaps much the same range, differs from the above in the front extending into lobes over the base of the antenna. It is piceous with dull yellowish margins and spots on elytra. It measures 4-5 mm. There are a half dozen similar species here.

6a (2) **Form elongate, resembling the rove beetles. Fig. 555. .... *Conotelus obscurus* Er.**

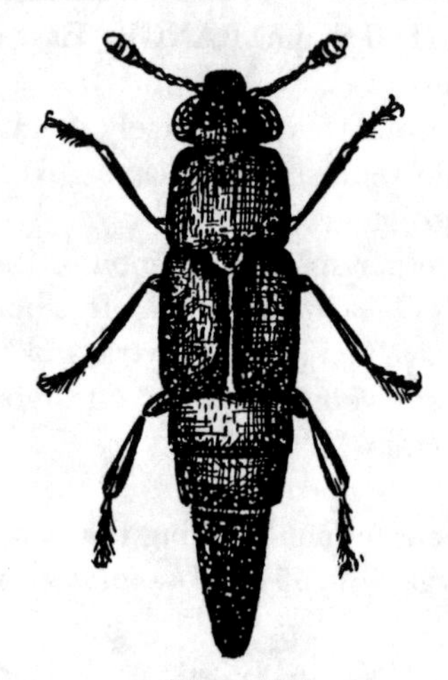

Figure 555

LENGTH: 3.5-4.5 mm. RANGE: Eastern half of United States.

Dull black, with scattered pubescence; antennae and legs brownish-yellow, the club darker. Common in flowers of Wild Morning-glory, Hollyhock, and Dogwood.

Ranging through the Southwest, also black and measuring 3 mm. *C. mexicanus* Murr. seriously attacks the buds and flowers of several species of fruit trees and shrubs as well as cotton. It is popularly known as the Fruit Bud Beetle.

6b Not elongate. Fig. 556.

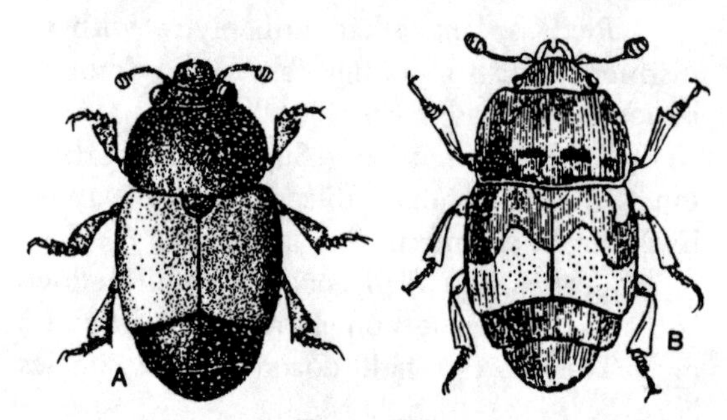

Figure 556

(a) *Carpophilus pallipennis* (Say)

LENGTH: 3-4 mm. RANGE: Wherever cacti grow.

Reddish or yellowish-brown; elytra lighter than head and thorax. It is common in the flowers of cacti.

(b) *Carpophilus hemipterus* (L.)

LENGTH: 4 mm. RANGE: cosmopolitan.

Blackish, marked as pictured with dull yellow, legs pale. It is known as the Dried Fruit Beetle. Both adults and larvae are serious pests especially in fruit drying areas.

C. dimidiatus (Fab.), the Corn Sap Beetle, differs from *hemipterus* in having a squarish thorax with distinct front angles. It is lighter in color, a bit smaller and is cosmopolitan. Over 30 members of this genus are found in the U. S. and Can.

FAMILY 71. RHIZOPHAGIDAE
The Root-Eating Beetles

Members of this family are depressed. elongate beetles ranging size from 1.5 to 3 mm. The 10-segmented antennae usually have a 2-segmented club. In color they vary from brown to piceous.

These beetles are usually found under bark, some having the larvae living in the tunnels of wood-eating insects. About 60 species are found in the U. S. and Canada.

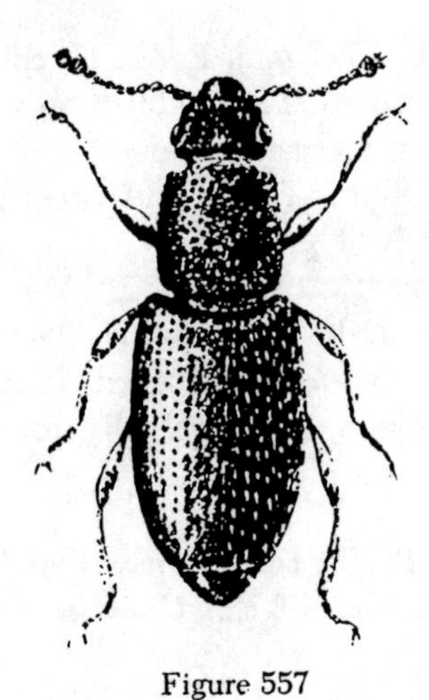

Figure 557

Figure 557 *Monotoma picipes* Hbst.

LENGTH: 2-2.3 mm. RANGE: Europe and much of the United States.

Dull black or brownish; antenna and legs reddish-brown; thorax with coarse dense punctures. Elytral striae with coarse punctures and yellowish hairs.

Bactridium ephippigerum (Guer.) is about the same size and color but is shining and its head does not narrow behind the eyes. The punctures and striae are fewer. It inhabits the eastern half of the United States.

Rhizophagus bipunctatus Say. Length: 2.5-3 mm. Range: Eastern U. S. and Canada. Black or brown, shining, antennae and legs reddish, elytra usually with two red spots, one apical, one basal; elytra with rows of punctures.

FAMILY 72. CUCUJIDAE
The Flat Bark Beetles

The some 1000 species of this family are highly variable in size, shape, markings and habits. They are often flattened to live under bark but some are serious pests in grain products.

1a Front coxal cavities open behind. Fig. 558a. 4

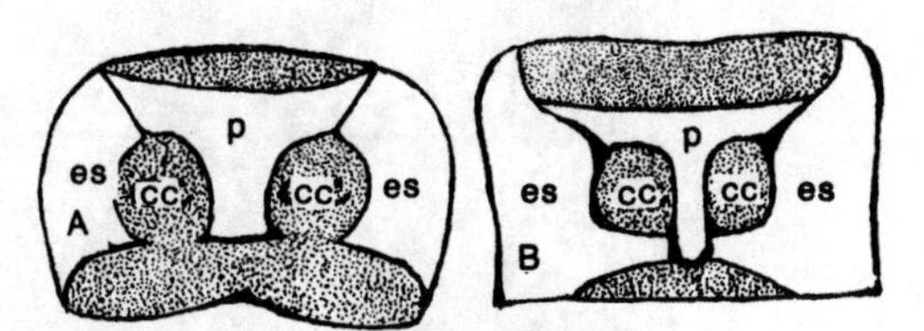

Figure 558

1b Front coxal cavities closed behind. Fig. 558b. 2

2a (1) Third joint of tarsi lobed beneath, antennae long, slender, filiform. Fig. 559. .... *Telephanus velox* Hald.

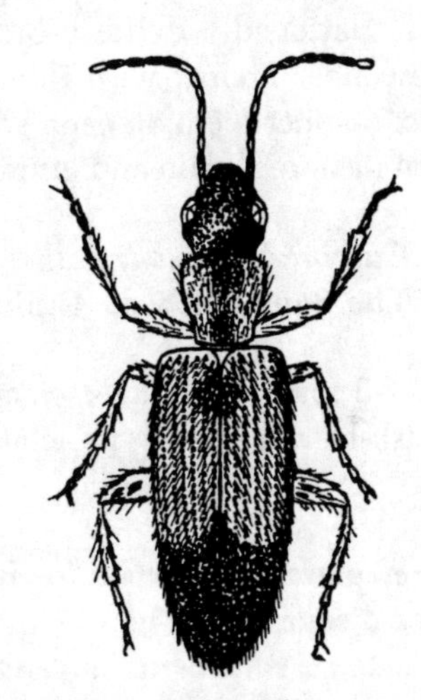

Figure 559

LENGTH: 4-4.5 mm. RANGE: Eastern half of United States.

Pale brownish-yellow with darker brown markings as pictured. Rather thickly clothed with fairly long pubescence. Common in debris and under stones. Slow to get started but a swift runner.

2b Antennae clubbed; tarsi not lobed. 3

3a (2) **Last three joints of antennae enlarging abruptly into the club. Fig. 560.**

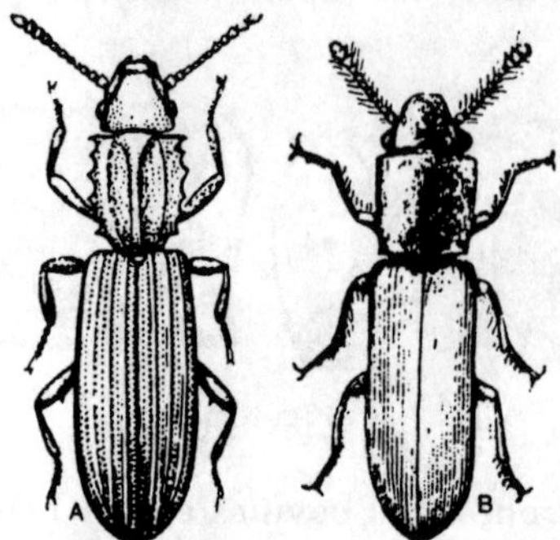

Figure 560

(a) ***Oryzaephilus surinamensis*** **(L.)**
The Saw-toothed Grain Beetle

LENGTH: 2.5 mm. RANGE: cosmopolitan.

Form flattened. Reddish-brown with paler pubescence. Thorax with three elevated lines and six distinct teeth on each edge. Much too common in stored grain and grain products.

(b) ***Cathartus quadricollis*** **(Guer.)**
The Square-necked Grain Beetle

LENGTH: 2-3 mm. RANGE: cosmopolitan.

Reddish-brown, feebly punctate. Another common pest of grain products.

3b **Antennae gradually widening into a club of 3 or 4 segments. Fig. 561. .... *Ahasverus advena* (Waltl)**

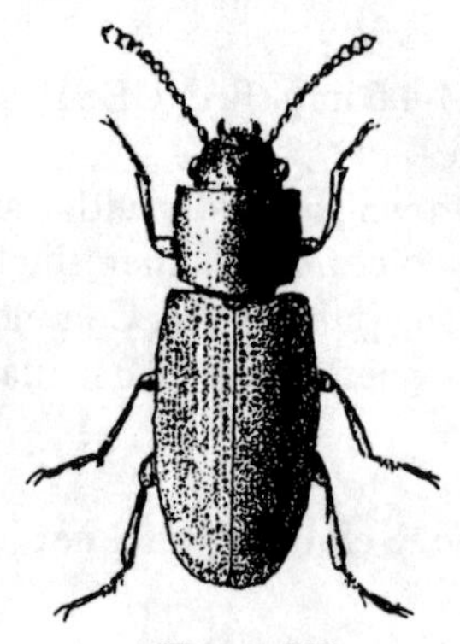

Figure 561

LENGTH: 1.7-2 mm. RANGE: cosmopolitan.

Subconvex. Pale reddish-brown, shining, with fine pubescence. Rows of coarse punctures on elytra. Feeds and breeds in stored grain and fruit products.

Cathartus rectus Lec. may be distinguished from the above in that the front angles of the thorax are not toothed. It is similar in size and color and is widely distributed.

4a (1) **Bright red; head widest behind the eyes. Fig. 562. ...... *Cucujus clavipes* Fab.**

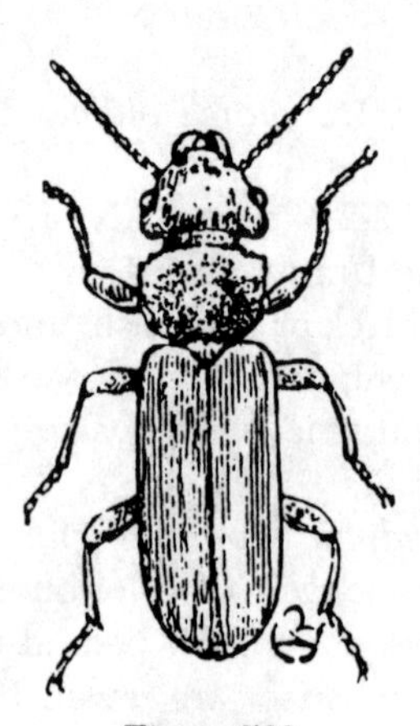

Figure 562

LENGTH: 10-14 mm. RANGE: throughout much of North America.

Much flattened. Scarlet red above, darker beneath; antennae and eyes black. An attractive beetle and beneficial in that both adult and larva feed on bark beetles and wood borers, the tunnels of which they invade.

The western form is variety *puniceus* Mann.

4b **Head widest across the eyes. 5**

5a (4) **Thorax toothed at sides; first antennal segment about as long as the head. Fig. 563. .... *Uleiota dubius* Fab.**

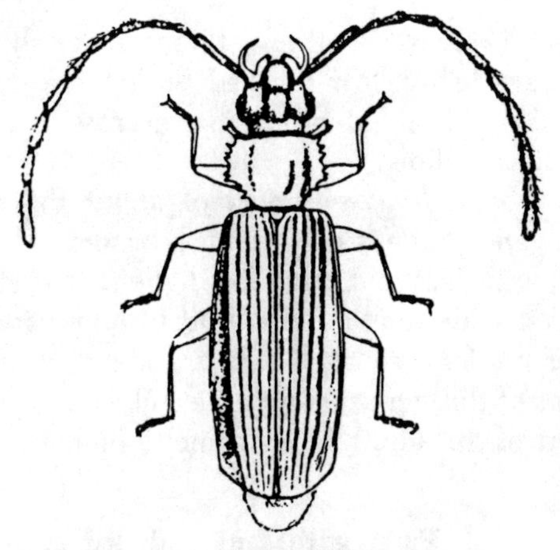

Figure 563

LENGTH: 4-6 mm. RANGE: much of United States.

Much flattened. Brownish-black, sometimes quite pale brown. Sides of elytra and the head and thorax usually paler. Antennae thickly pubescent. Very common under bark.

U. debilis Lec. differs from *dubius* in being darker in color and in being more slender with less serrate thorax.

5b Thorax without teeth at sides; first joint of antennae much shorter than head. Fig. 564. ...... *Laemophloeus minutus* (Oliv.)

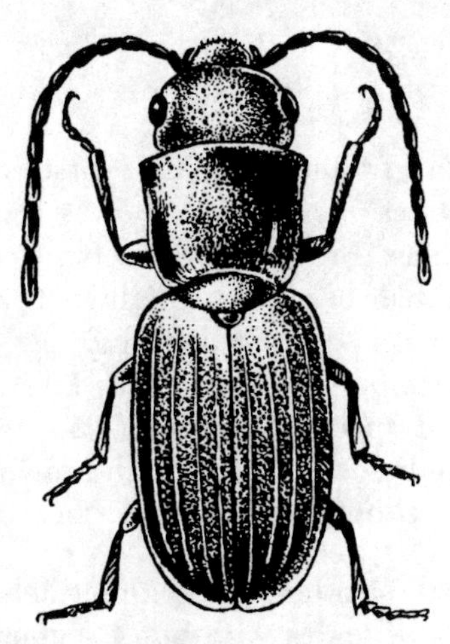

Figure 564

LENGTH: 1.5-2 mm. RANGE: cosmopolitan.

Flattened. Reddish-brown. Popularly known as the Flat Grain Beetle, it may be recognized by its small size and long antennae. It infests both dried fruit and cereals.

A related species, the Rust-red Grain Beetle *L. ferrugineus* (Steph.) with cosmopolitan distribution, is often found with the above which it resembles closely. Common in the East are two other species found under bark. Both are larger, 3 to 4.5 mm and have on the elytra a pale spot before the middle. *L. biguttatus* Say has the elytra densely punctate, whereas *L. fasciatus* Melsh. is finely and sparsely punctate.

FAMILY 73. SPHINDIDAE
The Dry-Fungus Beetles

This small family is made up of six N. American species. These beetles are convex, oblong, and range in sizes from 1.5 to 3 mm. In color they are reddish-brown or piceous. The 10-segmented antennae have a 3-segmented club made up of a small eighth segment, a wider, quadrate ninth segment, and a larger and wider tenth segment. The vestiture varies from a coating of dense semi-erect hairs to almost none.

All species are found living on dry fungi on the trunks of trees and on logs.

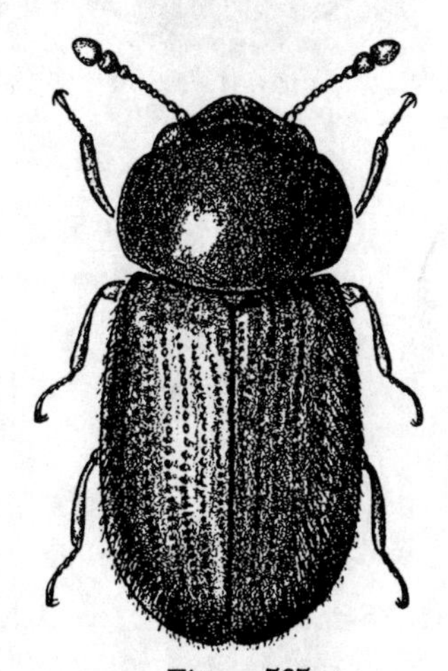

Figure 565

Figure 565 *Sphindus trinifer* Csy. Length: 1.7 mm. Range: Ontario. Shining, piceous-black; elytra and appendages with testaceous pile; pronotum minutely and not closely punctate.

A more frequently encountered species is *S. americanus* LeC. RANGE: Eastern U. S. and Canada. LENGTH: 2-2.5 mm. More or less rufo-piceous with humeral angles and apices more reddish; appendages reddish-brown. Pronotum finely and closely punctate.

FAMILY 74. CRYPTOPHAGIDAE
The Silken Fungus Beetles

These also live on fungi, decaying organic matter and sometimes with the ants. They are small. Almost 1000 species are known.

1a Antennae inserted under sides of the front and widely separated at base. 2

1b Antennae inserted on the front and close together at base. Fig. 566. .... *Anchicera ephippiata* Zimm.

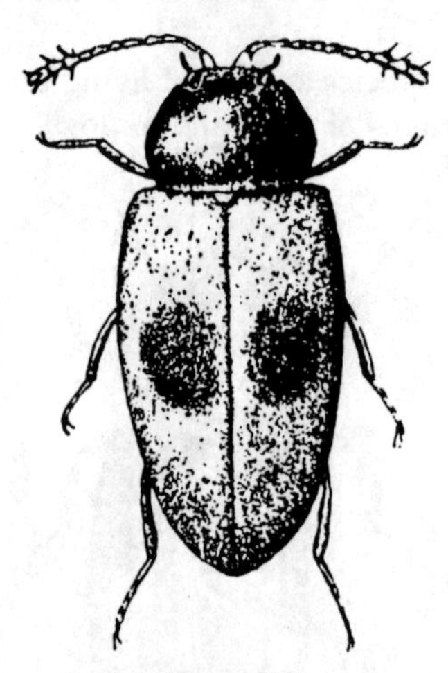

Figure 566

LENGTH: 1.5 mm. RANGE: widely distributed.

Convex. Head and thorax blackish; elytra reddish-yellow marked with blackish dot or disk, often extending into a cross-bar; legs reddish-yellow.

A. ochracea Zimm., of about the same size and shape is dark reddish-brown to blackish with legs and antennae lighter. It is shiny, has a wide range and is said to hibernate under mullein leaves. There are about 30 members of this genus in our area. Like many members of this family, they come to lights.

2a (1) Tarsi with 2nd and 3rd segments lobed beneath, the latter the larger; 4th segment very small. Fig. 567. .... *Toramus pulchellus* Lec.*

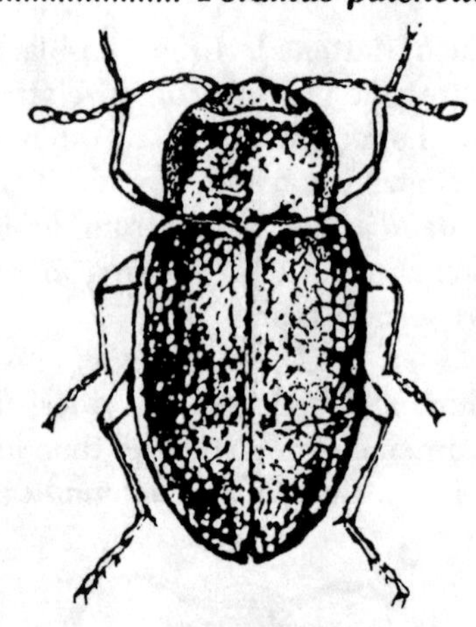

Figure 567

LENGTH: 1.5 mm. RANGE: Eastern half of United States.

Convex. Brownish-yellow to piceous; legs and basal half of antennae lighter. Elytra with two yellowish bands as pictured.

Telmatophilus americanus Lec. a widely distributed species is larger (2.5-3 mm) and densely pubescent. It is reddish-brown to piceous with yellowish to ash-gray pubescence.

2b Tarsi slender and without lobes, hind tarsi of males with but 4 segments. Fig. 568.

*These beetles are treated as Languriidae by some.

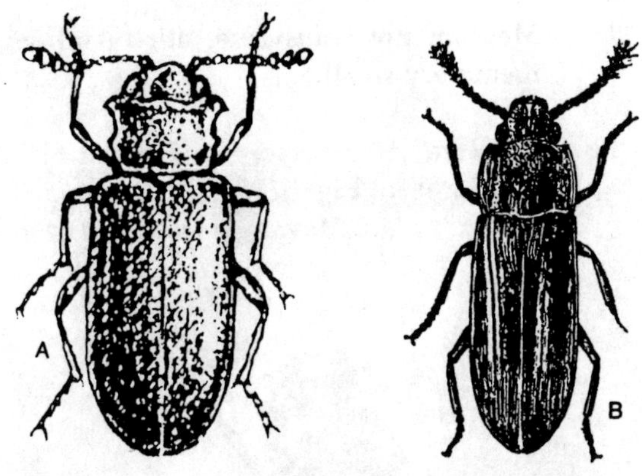

Figure 568

(a) ***Cryptophagus acutangulus Gyll.***

LENGTH: 2.5 mm. RANGE: Europe and North America.

Subdepressed. Pale brownish-yellow, head and thorax darker. Abundant; often found in cereals and sacked sugar. This is a large (40 spp.) and difficult group.

(b) ***Pharaxonotha kirschi* Reit.* (Mexican Grain Beetle)**

LENGTH: 4.5-5 mm. RANGE: Mexico and occasionally in the United States.

Shining, deep brown; antennae long. A pest of stored grain.

Henoticus serratus (Gyll.) is another cosmopolitan species. It measures about 2 mm and is dark reddish-brown with scattered yellowish hairs. The apical angles of the thorax are not thickened as in *Cryptophagus.*

FAMILY 75. LANGURIIDAE
The Lizard Beetles

Lizard beetles are narrow, elongate, shiny, parallel-sided beetles. The 11-segmented antennae possess a club varying from 3 to 5 segments. They are dark, sometimes metallic, and often marked with orange. (Some beetles traditionally assigned to the Cryptophagidae actually belong here. Refer to note in that section.)

The larvae are stem borers and the adults feed on the pollen and leaves of the host plant.

1a **Ocular stria extending from the antennal sockets to the base of the eye present. Fig. 569. .... *Acropteroxys gracilis* Newn.**

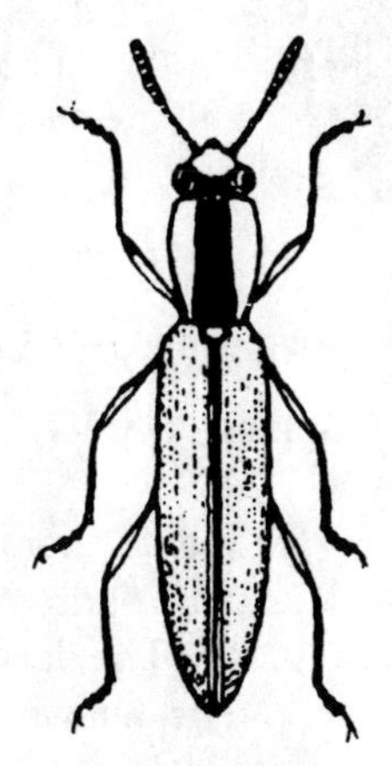

Figure 569

LENGTH: 8-10 mm. RANGE: Eastrn half United States.

Head in part or wholly red; thorax yellowish-red with broad greenish-black stripe extending its entire length as pictured. Feeds on ragweed and related plants.

Longer and wider and with much the same range. *A. lecontei* (Cr.) has the head wholly black and only a large spot at center of thorax.

1b **Ocular stria absent. Fig. 570.**

*These beetles are treated as Languriidae by some.

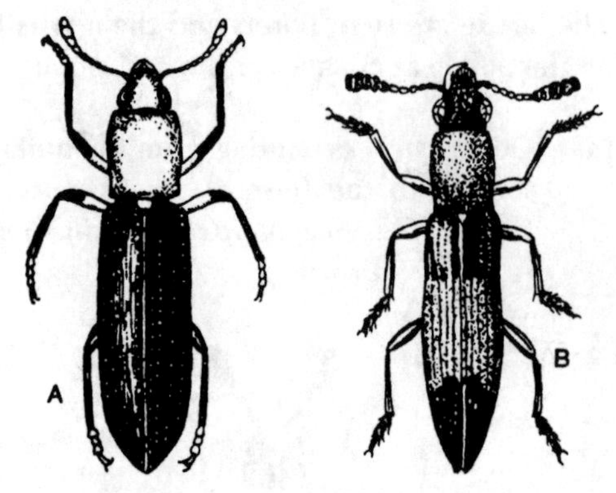

Figure 570

(a) *Languria mozardi* Latr.

LENGTH: 5-8 mm. RANGE: Eastern half United States.

Shining. Head and thorax wholly red; elytra bluish-black; tarsi, antennae and apical half of femora black. The larvae infests the stems of clover and is known as the Clover Stem Borer.

(b) *Languria trifasciata* Say

LENGTH: 6-8 mm. RANGE: Eastern half United States.

Shining. Marked with red and bluish-black as pictured; antennae in part reddish-yellow.

There are 12 other species of these colorful beetles in our fauna.

FAMILY 76. EROTYLIDAE
The Pleasing Fungus Beetles

This is a large family, almost 1500 species, of small to large beetles ranging in size from 3 to 20 or more mm. Their shape varies from subrotundate to almost elliptical. Most frequently they are black marked with orange, but some species are testaceous with black markings. The 11-segmented antennae usually have a 3-segmented club, rarely 4 or 5 segmented.

Both adults and larvae are found on or in fleshy fungi where they feed.

1a **Mentum not transverse, 4th tarsal segment very small. 2**

1b **Mentum transverse, 4th tarsal joint readily seen. Fig. 571. .... *Megalodacne heros*** (Say)

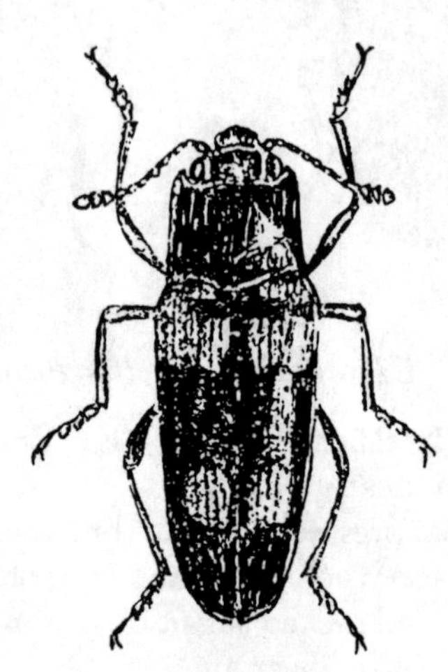

Figure 571

LENGTH: 18-22 mm. RANGE: Eastern U. S. and Canada.

Shining. Black, with red markings.

M. fasciata (Fab.) which is colored and marked much like *heros* is readily distinguished by its smaller size (9-15 mm) and by rows of fine punctures on elytra which the former lacks. Same range as above.

Both of these very attractive beetles hibernate gregariously under bark.

2a **(1) Body oval. Fig. 572. .... *Tritoma humeralis* Fab.**

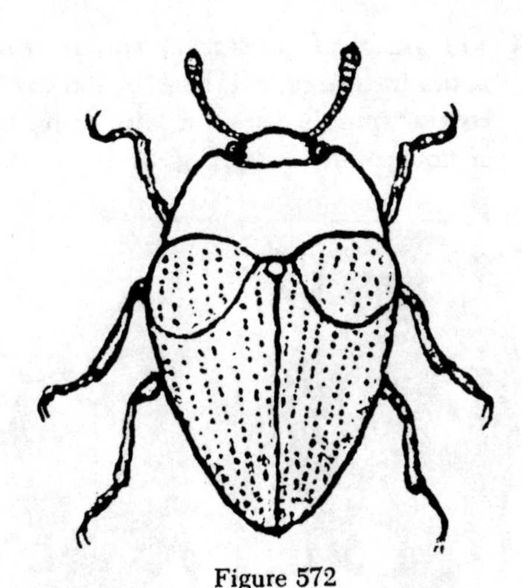

Figure 572

LENGTH: 3-4 mm. RANGE: Eastern Can. and U. S. south to S. Car., west to Kan.

Black with reddish-yellow spot on humerus of each elytron.

T. sanguipennis (Say) about 4 mm long has its elytra wholly dark red with thorax black while *T. angulata* Say has thorax and elytra wholly black. Both have ranges similar to *T. humeralis*. There are 10 more members of this genus in our area.

2b Body more elongate. 3

3a (2) Eyes with large facets. Fig. 573. .... ***Ischyrus quadripunctatus* (Oliv.)**

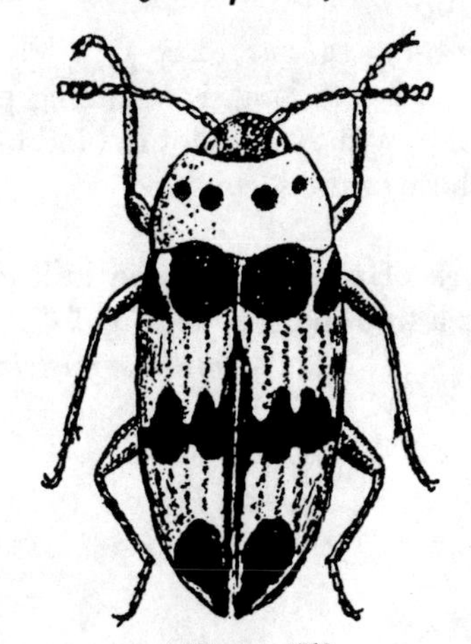

Figure 573

LENGTH: 7-8 mm. RANGE: Eastern half of United States.

Reddish-yellow with black markings; underparts black, with sides of abdomen yellow. Elytra with rows of distinct punctures. Feeds on fungi. Like several other species of the family it gathers in large numbers to hibernate under bark or beneath logs. If you happen to hit the "jack pot," collecting is simple.

3b Eyes with fine facets. Fig. 574. .... ***Triplax thoracica* Say**

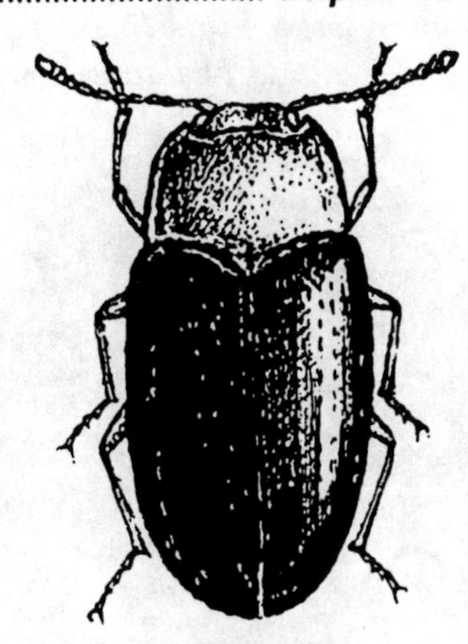

Figure 574

LENGTH: 3.5-5 mm. RANGE: From the Rockies eastward.

Head and thorax yellow; elytra and apical half of antennae black. Underparts reddish-yellow. Beneath bark and in fleshly fungi.

T. festiva Lac. a widely distributed species especially in the South is a bit larger than the preceding. The ground color is black, but attractively marked with thorax, scutellum and broad median band reddish-yellow. About 20 members of this genus are widely distributed over the U. S. and Can.

FAMILY 77. PHALACRIDAE
The Shining Flower Beetles

The little over 100 American species of this family are tiny, highly convex, shining beetles. They and their larvae live on or in flowers. Sweeping is likely the surest way to get them.

1a Antennae inserted under the sides of the front, their base concealed from above; front and hind tarsi equal in length; scutellum large. Fig. 575. .... *Phalacrus simplex* Lec.

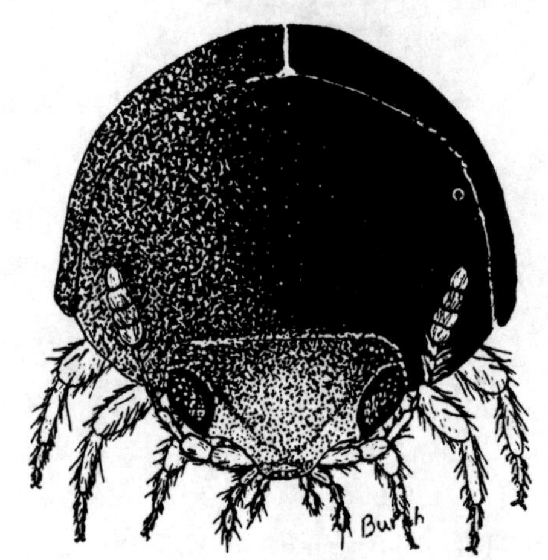

Figure 575

LENGTH: 2-2.3 mm. RANGE: Central and Western United States.

Red-brown or blackish; legs and antennae paler.

P. politus Melsh., measuring around 2 mm is somewhat shining and black with paler legs and antennae. The antenna is slender, its third joint equalling the fourth and fifth combined. Eastern half of the U. S. About 25 members of this genus are found in our area.

1b Antennae inserted at sides of front with base visible from above; hind tarsi longer than others; scutellum not large. 2

2a (1) Apex of prosternal spine with an acute free edge, and bearing a transverse row of spinules; second joint of hind tarsi not usually long. Fig. 576.

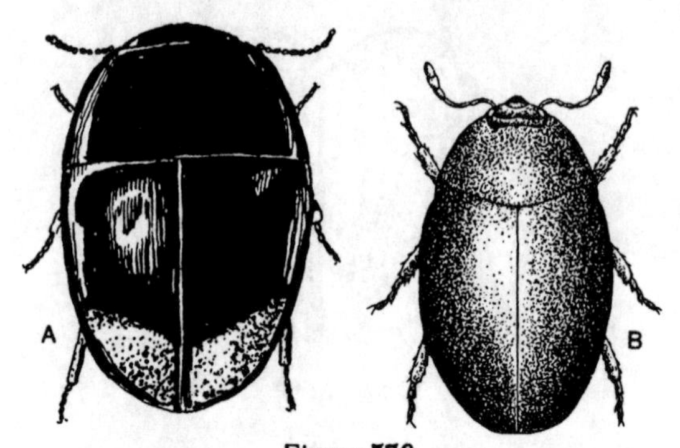

Figure 576

(a) *Stilbus apicalis* (Melsh.)

LENGTH: 2-2.3 mm. RANGE: U. S.

Strongly convex. Shining. Dark chestnut-brown to blackish. Under parts reddish-brown; legs and antennae paler.

(b) *S. nitidus* (Melsh.)

LENGTH: 1.3-1.5 mm. RANGE: Eastern half United States.

Very shining. Pale reddish-yellow; elytra rounded at apex. *Stilbus* is the largest genus with 35 spp.

Gorginus rubens (Lec.) widely distributed, about 2 mm long, is a shining pale reddish species, with basal joint of hind tarsi longer than those of the second leg.

2b Apex of prosternal spine inflexed, without a free sharp edge. Fig. 577. .... *Olibrus semistriatus* Lec.

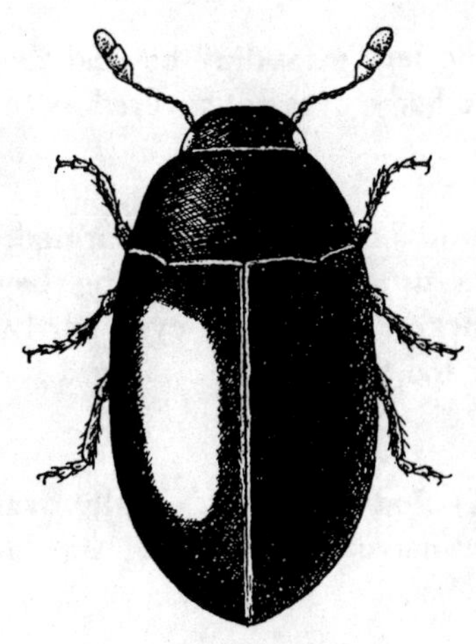

Figure 577

LENGTH: 1.8-2.2 mm. RANGE: Eastern half United States and Canada.

Shining. Dark chestnut-brown above; under parts, including legs and antennae, pale reddish-yellow.

Acylomus ergoti Csy., measuring around 2 mm and shining black or chestnut brown, ranges through the Central and Eastern States.

A. piceous Csy. is much like *ergoti* in size, but differs in being distinctly narrowed behind the base of the elytra. It is piceous-brown and is eastern in its range.

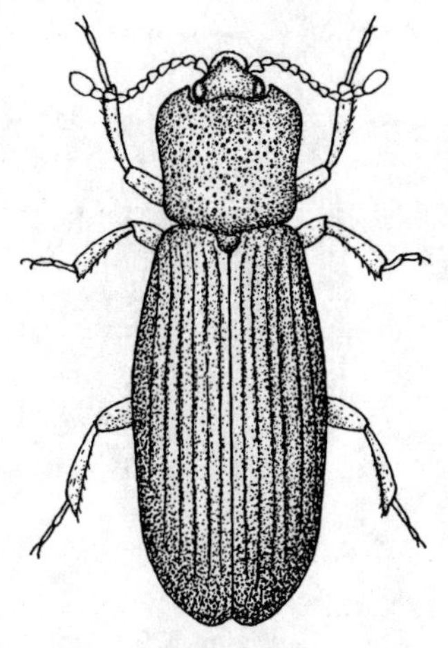

Figure 578

Figure 578 *Cerylon castaneum* Say.

LENGTH: 2-3 mm. RANGE: Eastern half of U. S.

Depressed. Shining. Dark reddish-brown; elytra with punctured striae.

Philothermus glabriculus Lec. is a common species throughout the eastern half of the United States. In size and color it is much the same as the preceding but its antennae have 11 segments 2 of which make the club instead of 10 and 1.

FAMILY 78. CERYLONIDAE
The Minute Bark Beetles

This small family of beetles is made up of about 10 N. American species. These beetles are oblong, flat and range in size from 2 to 3 mm. They are reddish or dark brown in color with no vestiture. Both adults and larvae are found under bark.

FAMILY 79. CORYLOPHIDAE
The Minute Fungus Beetles

This family consists of small beetles ranging in size from .5 to 5 mm in length. An outstanding characteristic is the expanded pronotum covering the head so that the head is not visible from above. They vary in shape from round to elongate. The antennae have from 9 to 11 segments with the two basal segments enlarged and with an apical club. Adults are found under bark and in various kinds of decaying vegetative matter. About 60 species are found in the U. S. and Canada.

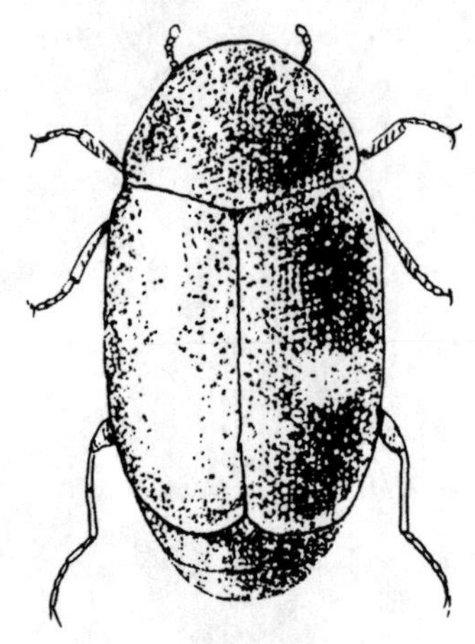

Figure 579

Figure 579 *Molamba ornata* Csy.

LENGTH: 1.3-1.5 mm. RANGE: Central United States.

Piceous, antennae and legs yellowish; thorax reddish-yellow with dark discal spot as pictured, elytra with two spots and edge of tips, dull yellow.

Corylophodes marginicollis (Lec.) measures less than 1 mm. It is shining blackish with paler legs and antennae and is fairly abundant. Eastern third of U. S.

FAMILY 80. COCCINELLIDAE
The Ladybird Beetles

There are some 3000 known species of Lady Beetles. Some are very small; not any are large. Many are marked with spots on a colored background, but some possess a solid color. A few species eat plants, but for the most part they are preditors and are valued for the help they thus give. They are noticed by everyone and are favorites with many collectors.

1a **Middle coxae narrowly separated; body glabrous, elongate, oval; femora of the long legs extending beyond the sides of the body; eyes not covered by thorax. 15**

1b **Middle coxae widely separated; the femora usually not extending beyond the sides of the body; eyes partly covered by the thorax. 2**

2a **(1) Body compact, usually oval; epipleura narrow; usually flat and horizontal. 3**

2b **Body loosely jointed, usually rounded in form; epipleura wide and concave. 8**

3a **(2) Body pubescent. 6**

3b **Body glabrous. 4**

4a **(3) Front tibiae with a strong spine on outer edge near middle; eyes with a small notch in front margin. Fig. 580. .... *Brachyacantha ursina* (Fab.)**

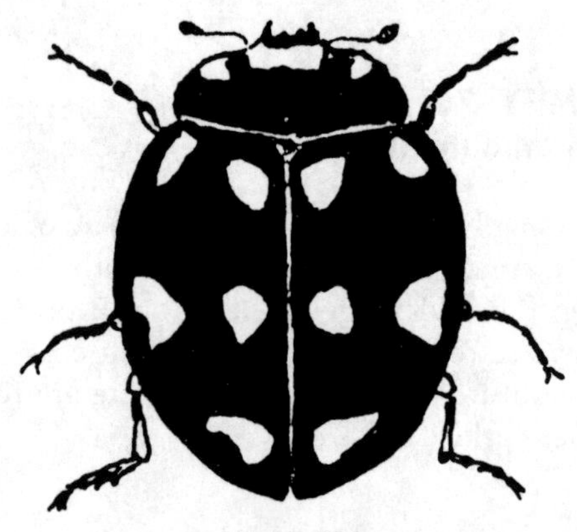

Figure 580

LENGTH: 3-4 mm. RANGE: Across the U. S. and Canada; often common.

Black shining; head yellow; male with

front margin of thorax yellow; in female, only the front angles so colored; other markings as pictured, yellow; underparts black, legs light.

B. quadripunctata, Melsh., measuring 3 to 3.5 mm is shining, black with yellow head in male, but in female, there is only a yellow V on an otherwise black head. It ranges from Iowa to Florida. There are about 20 members of this genus in the U. S. and Can.

4b No spines on front tibiae; eyes without notch. 5

5a (4) Side margins of elytra marked. Fig. 581. .... *Hyperaspis undulata* (Say)

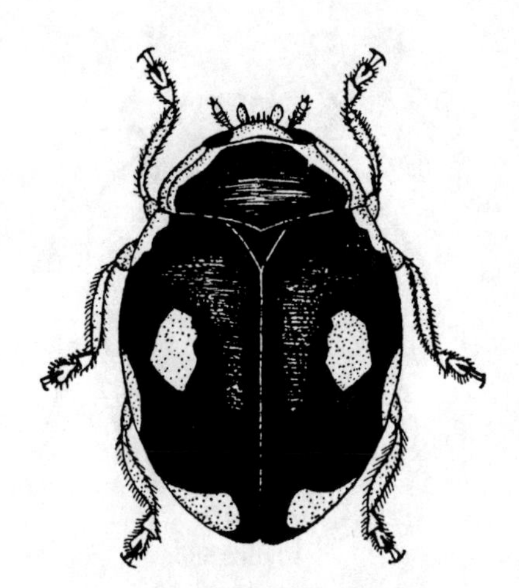

Figure 581

LENGTH: 2-2.5 mm. RANGE: most of the United States, also in Canada.

Black, shining; male marked as pictured, with yellow; female with head and front margin of thorax, black.

H. lateralis Muls. The Lateral Lady Beetle, 3 mm, is shining black with front margin of thorax, humeral margin and an apical and larger median spot on elytra, yellow to red. It ranges from Montana to Mexico.

5b Side margins of elytra wholly black. Fig. 582. .... *Hyperaspis signata* (Oliv.)

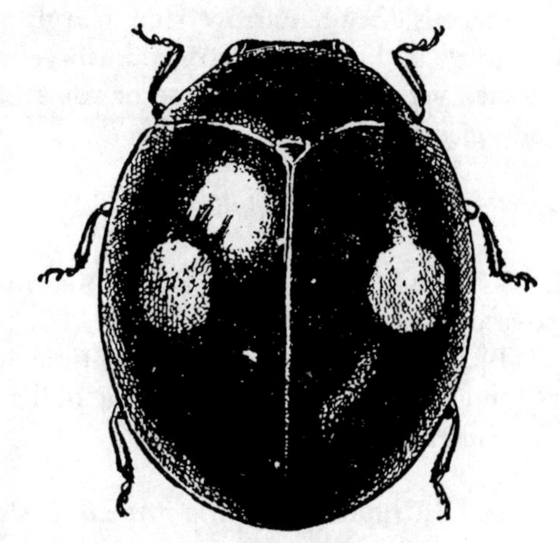

Figure 582

LENGTH: 2.3-3.5 mm. RANGE: Eastern half of United States.

Black, shining with many punctures; male with head and narrow apical and side margins of thorax yellow; female with head and thorax wholly black. Spot on disk and sometimes a smaller one near apex of elytra yellow.

H. quadrivittata Lec., has two yellow stripes on margins of the black elytra. It ranges widely, especially in the West, and measures 2.5 mm. There are over 70 species in this genus.

6a (3) Elytra and thorax marked with yellow or red. Fig. 583.

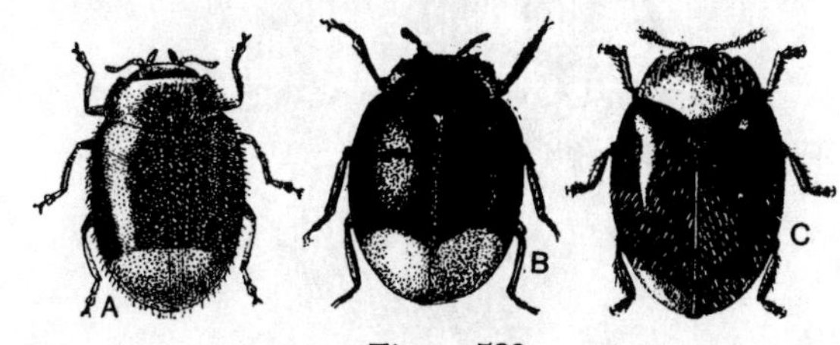

Figure 583

(a) *Diomus terminatus* Say

LENGTH: 1.5-1.8 mm. RANGE: Eastern half of United States.

Piceous; head, narrow side margins of thorax, legs and apex of elytra reddish-yellow. Abdomen yellow except at base or sometimes wholly piceous.

(b) *Scymnus caudalis* **LeC.**

LENGTH: 2-2.3 mm. RANGE: Colorado and eastward.

Black; head, sides of thorax, tibia and tarsi and often spot at tip of elytra dull reddish; femora piceous.

(c) *Cryptolaemus montrouzieri* **Muls.**

LENGTH: 3-3.5 mm. RANGE: California.

Shining. Black; head, thorax, tips of elytra, and abdomen reddish. Introduced from Australia to California to combat mealy bugs.

6b Elytra and thorax wholly black. 7

7a (6) Pubescence distinctly apparent; tarsal claws bifid. Fig. 584. .... *Stethorus punctum* (Lec.)

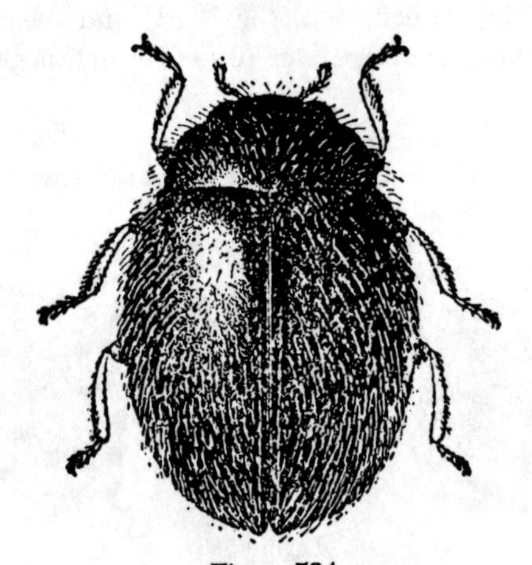

Figure 584

LENGTH: 1.2-1.5 mm. RANGE: Eastern half of the United States.

Shining. Black; legs and antennae, except base of femora yellow. Distinguished from *Scymnus* by much shorter clypeus.

Scymnus tenebrosus Muls. with a length of 1.8-2.5 mm and ranging through the eastern third of the United States is uniformly black, shining, with legs red or reddish-brown.

Didion nanus Lec., also wholly black with reddish legs is much smaller, measuring only 1.3-1.5 mm. It seems to range from coast to coast.*

7b Apparently without pubescence unless examined with high magnification. Tarsal claws single pointed. Fig. 585. .... *Microweisea misella* (Lec.)

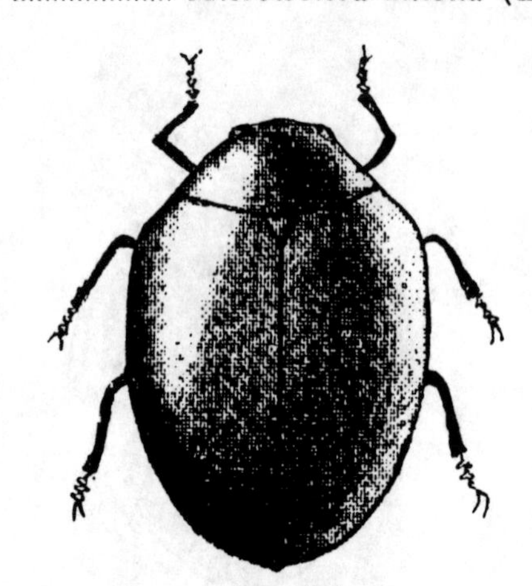

Figure 585

LENGTH: .8-1 mm. RANGE: much of the United States.

Wholly black. It is valued in the West as a predator of the San Jose Scale.

M. suturalis Sz. less than 1 mm long is shining black with brownish-red elytra, though black at the suture. It is valued in Southern California for feeding on San Jose Scale and red spiders.

8a (2) Upper surface pubescent. 9

**Scymnus* and several related genera contain about 150 small, round, black beetles. Identification is difficult.

8b **Upper surface glabrous.** **10**

9a **(8) Facets of eyes coarse; wholly black above. Fig. 586.** .... ***Rhyzobius ventralis*** **(Er.)**

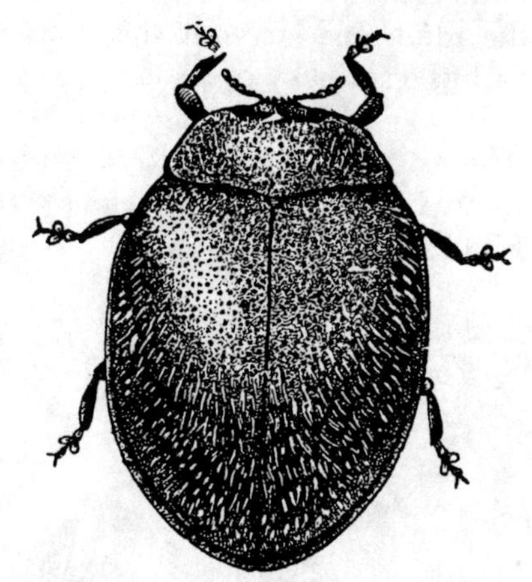

Figure 586

LENGTH: 2-2.5 mm. RANGE: California.

Shining or velvety black; abdomen reddish.

This Black Lady Beetle was introduced from Australia in 1892 and plays an important role in destroying mealybugs and soft scales in the orchards of California.

9b **Facets of eyes fine; upper surface red and black. Fig. 587.** .... ***Rodolia cardinalis*** **(Muls.)**

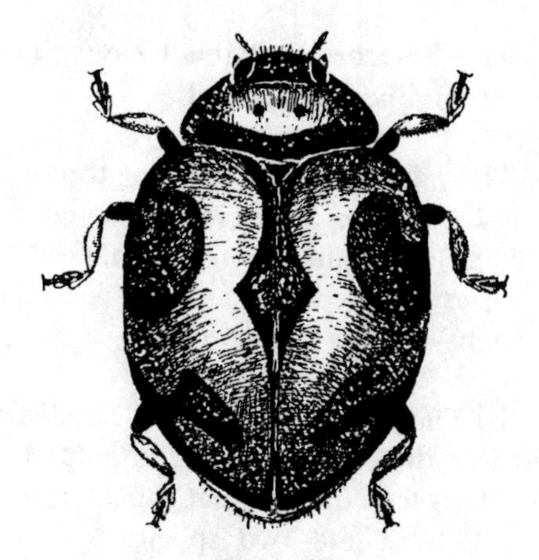

Figure 587

LENGTH: 2.5-3.5 mm. RANGE: California and Florida.

Red with variable black markings; (the males usually have more black than red while the opposite condition is true of the females). The thick pubescence often grays and obscures the pattern.

This is the widely publicized Vedalia Lady Beetle, which has figured so favorably in keeping the Cottony Cushion Scale in check. It was first brought from its native Australia in 1888.

10a **(8) Frontal plate not covering the base of the antennae.** **11**

10b **Frontal plate broadly dilated, covering the base of the antennae and subdividing the eyes; upper surface glabrous; front tibia with a small tooth, on outer margin near base. Fig. 588.**

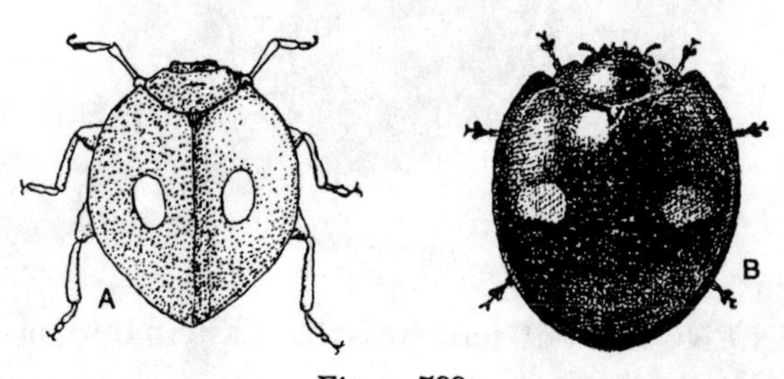

Figure 588

(a) ***Chilocorus stigma*** **Say Twice-stabbed Lady Beetle.**

LENGTH: 4-5 mm. RANGE: throughout much of the United States and Canada.

Black and shining; elytra marked with two spots as pictured with red; the ventral segments red.

(b) The Asiatic Ladybird, *C. similis Rossi,* was introduced from China to help control the San Jose scale but did not satisfactorily establish itself. These two species are nearly identical.

Axion plagiatum (Oliv.) 6-7 mm, is marked like the preceding species but the red spots are much larger especially in the female. Ranges from Ind. to Texas west to Cal.

11a (10) Upper surface pubescent. 12

11b Upper surface glabrous. 13

12a (11) Each elytron with seven black spots; thorax with 5 spots. Fig. 589. .... *Epilachna borealis* Fabr. Squash Beetle

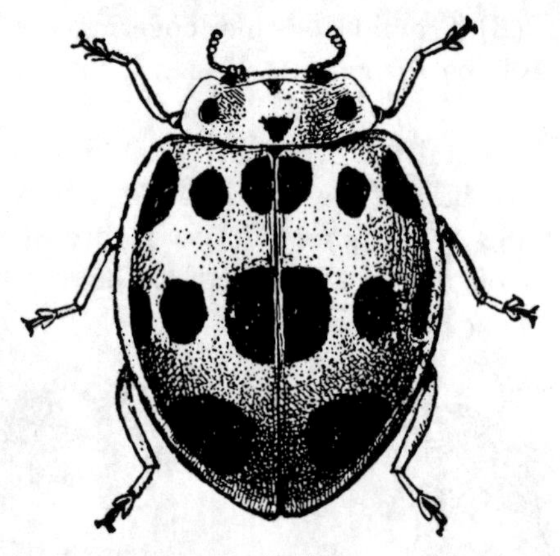

Figure 589

LENGTH: 7-10 mm. RANGE: Eastern third of United States.

Head, legs and upper surface yellow; markings black; under surface yellow to dark brown.

This genus has 4 known North American species. They are the "black sheep" of the Lady Beetle family as they are plant feeders while all the others are predacious on injurious pests. Both the adults and larvae of this species feed on cucurbits, especially squashes.

12b Thorax plain, each elytron with eight spots. Fig. 590. .... *Epilachna varivestis* Muls. Mexican Bean Beetle

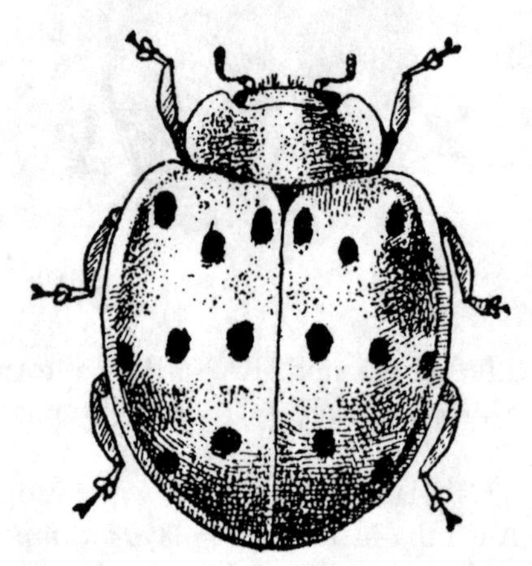

Figure 590

LENGTH: 6-9 mm. RANGE: originally Mexico but now covering much of the United States and Ontario.

Yellow to coppery brown, marked as pictured with black.

For many years this serious enemy of beans was found only in a few of our Western States. In 1920 it was found in Alabama from which point of infestation it has steadily spread. Both larvae and adults are leaf eaters. Another name is Bean Lady Beetle.

13a (11b) Antennae but slightly longer than the head; epipleura not reaching the tip of elytron. 14

13b Antennae extending to or beyond the middle of thorax; epipleura entire. Fig. 591a. .... *Anatis labiculata* Say
Fifteen-Spotted Lady Beetle

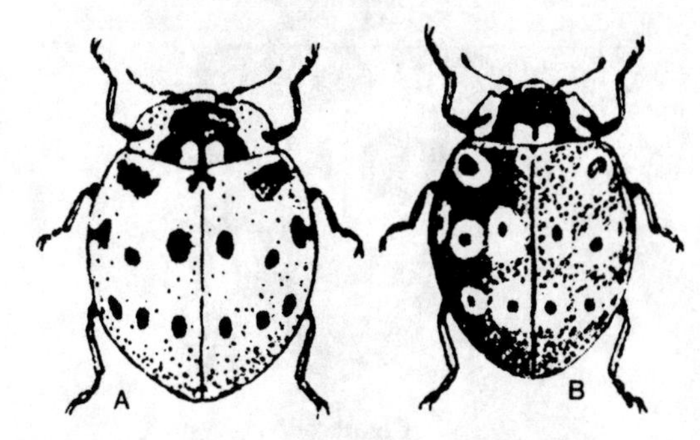

Figure 591

LENGTH: 7-10 mm. RANGE: Eastern third of U. S.

Head black and yellow; thorax and elytra yellowish to dark reddish-brown marked as pictured with black; the black spots almost obscure on the darker specimens.

In *A. mali* (Say) (Fig. 591b) each elytral spot is surrounded by a pale border. Ranges from B. C. and Wash. to Ind. to the Northeast.

14a (13) (a, b, c) Elytra plain red without any black marks. Fig. 592. .... *Cycloneda munda* (Say)
Red Lady Beetle

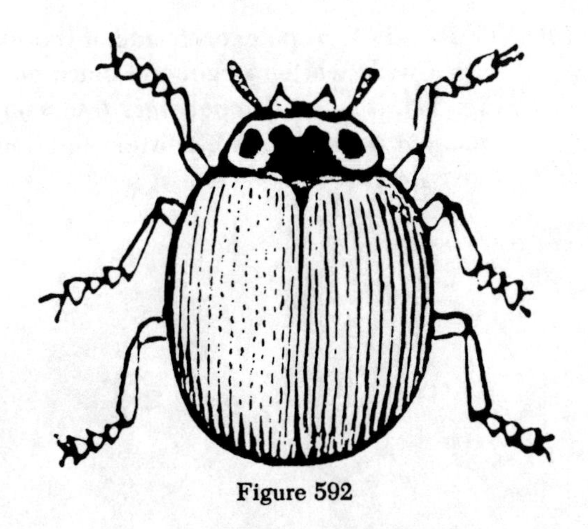
Figure 592

LENGTH: 4-5.5 mm. RANGE: United States and Canada.

Head black with two white spots in female and front margin white in male. Thorax black and white as pictured; elytra with varying intensities of reddish yellow sometimes very bright red.

C. sanguinea (L.) is highly similar and ranges throughout much of North America and South America.

14b Each red elytron with one round black spot. Fig. 593. .... *Adalia bipunctata* (L.)
Two-Spotted Lady Beetle

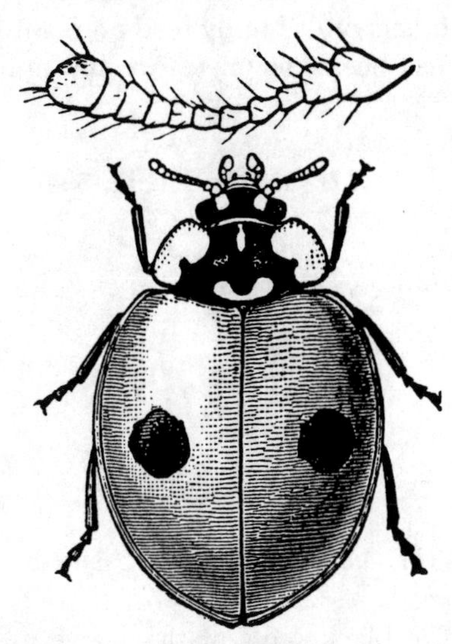
Figure 593

LENGTH: 4-5 mm. RANGE: Europe and North America.

Head and thorax marked with black on a yellowish background; elytra yellowish red, with black spot as pictured; under parts black or reddish brown.

This very common species hibernates in houses and is seen at the windows in early spring wanting out.

14c Elytra red with nine black spots. Fig. 594a.

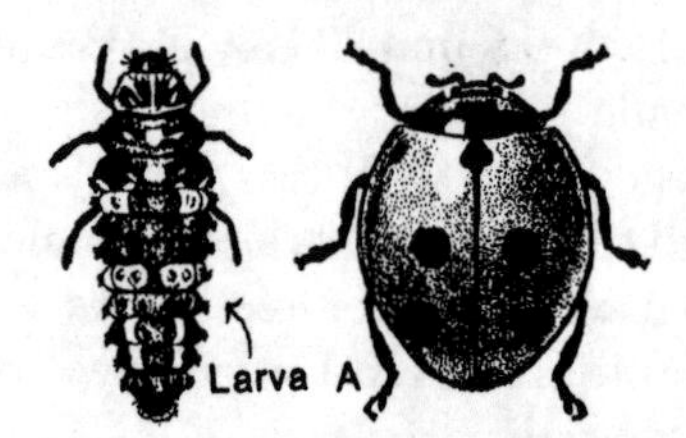

Figure 594a

(a) *Coccinella novemnotata* Hbst.

LENGTH: 5.5-7 mm. RANGE: much of U. S.

Head and thorax black and pale yellowish; elytra yellowish-red marked with black as shown; legs and under surface black.

Both adults and larvae of this species, like many others of its family feed on plant lice. An adult has been known to eat 100 aphids per day.

(b) *C. trifasciata* L. Fig. 594b.

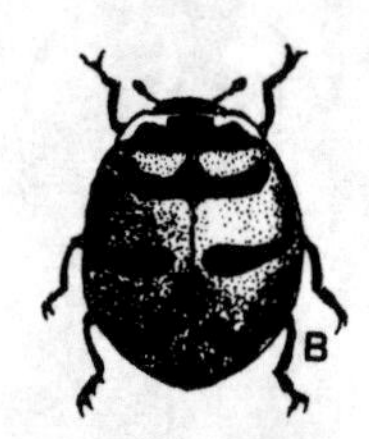

Figure 594b

LENGTH: 4.5-5 mm. RANGE: Northern U. S. and Canada.

The black spots of the above are fused into bars for this species, as pictured. *Coccinella* has 13 species and many varieties in our range.

15a (1) Claws divided into 2 sharp points, unequal in length; thorax without a margin. 16

15b Claws with a large quadrate basal tooth (a); thorax with a distinct narrow margin at base. Fig. 595. .... *Coleomegilla maculata* DeG. Spotted Lady Beetle

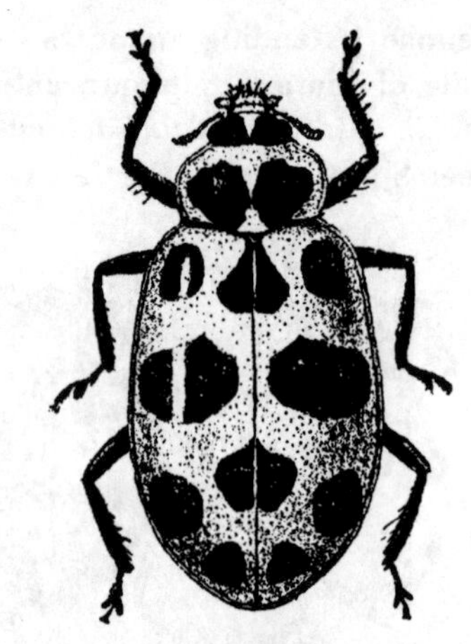

Figure 595

LENGTH: 5-7 mm. RANGE: most of the United States and Canada.

Bright red or pink (not orange-red as in many lady beetles) marked with black; undersurface black.

This very common species is destructive to several important aphid pests. The adults congregate in rather large numbers for hibernation and are thus found from late fall to early spring.

16a (15) Legs wholly black; thorax with narrow white margin which does not have a black dot. 17

16b Tibiae and tarsi pale; each side of thorax with a wide white margin and black dot. Fig. 596. .... *Hippodamia tredecimpunctata tribalis* (Say) Thirteen-Spotted Lady Beetle

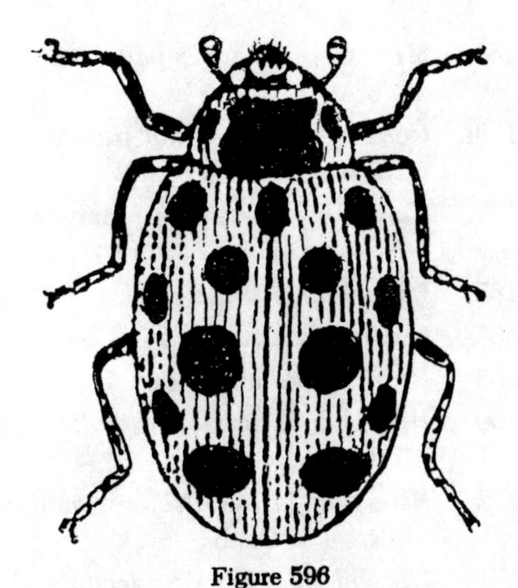

Figure 596

LENGTH: 4.5-5.5 mm. RANGE: widely distributed throughout North America.

Head black at base, elytra yellowish-red with black, markings which are variable in size in different specimens; femora and under surface of body black.

17a (16) Black disk of thorax deeply indented at front and back with white. Fig. 597. .... *Hippodamia parenthesis* (Say) Parenthesis Lady Beetle

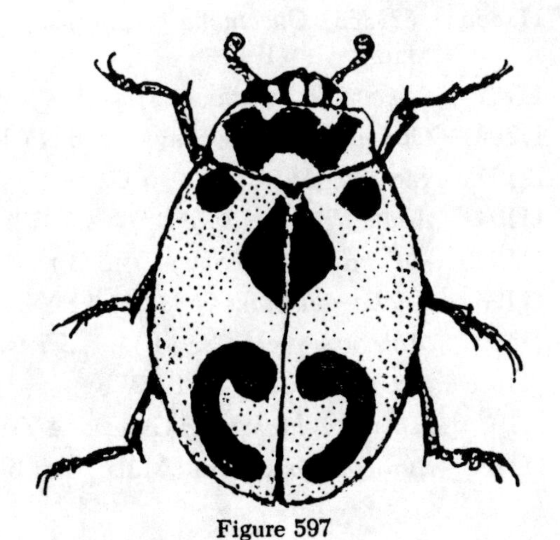

Figure 597

LENGTH: 4-5 mm. RANGE: most of the United States.

Elytra yellowish-red marked as pictured with black; legs and undersurface black.

We have known novice gardners to carefully pick the lady beetles from some aphid infested plant, supposing them to be the parents of the plant lice. If they had looked a bit closer they could have seen the lady beetles ravenously devouring the aphids and would not have made this costly mistake. It's a smart man who can correctly separate all his friends and enemies.

Asa Fitch, back in 1861, named twelve new species all from New York (maybe all from one garden). All twelve of these minor variants were presently made synonyms of the above species. Thus fame is sometimes short-lived with beetles as well as with politicians.

17b Black disk of thorax marked with two converging yellowish-white dashes. Fig. 598. .... *Hippodamia convergens* Guer. Convergent Lady Beetle; a, adult; b, pupa; c. larva.

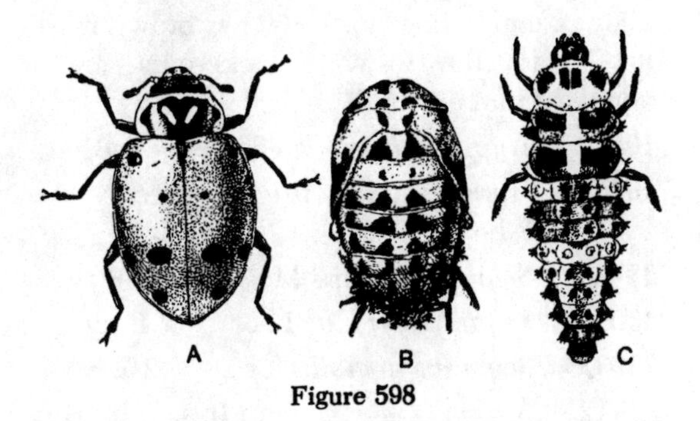

Figure 598

LENGTH: 6-7 mm. RANGE: throughout the United States and Canada.

Elytra from light to darker yellowish red marked with six or less black spots on each elytron. Legs and undersurface are wholly black.

Hippodamia quindecimmaculata Muls. is distinguished by the much larger spot and the front one along the suture somewhat double.

The species *glacialis* (Fab.) lacks the three front spots or sometimes retains the humeral one. The first of these is Midwestern and the second ranges across the top of the U. S.

The genus *Hippodamia* is a highly important one from the economic standpoint. It numbers less than 20 species but almost every one plays an important role.

Other Lady Beetles

The Coccinellidae is another family, many species of which can be determined fairly accurately by the markings on the upper surface. A number of these as shown by Charles W. Leng and others, are reproduced here. Fig. 599.

Leng's catalog* numbers appear over the drawings and opposite the names. The figures and letters following the names indicate the average size and color of the species. The main or ground color is indicated by capital letters. (B, black; N, brown; O, orange; R, red; W, white and Y, yellow.) Lower case letters following the capitals show the color of the markings. Thus No. 11028 as seen below is black with reddish-yellow markings and measures about 2 mm; while No. 11150 has brown marking on a yellowish-white background and is about 2.5 mm in length.

10880 *Hyperaspis proba* (Say) 3 B y-r
10954 *Hyperaspidius vittigera* (Lec.) 2 N yw
11028 *Scymnus collaris* Melsh. 2 B ry
11094 *Scymnus ornatus* Lec. 2 B oy
11097 *Scymnus amabilis* Lec. 2 B y
11112 *Scymnus myrimidon* Muls. 2 B y
11147 *Coccidula lepida* Lec. 3 YR b
11150 *Psyllobora vigintia-maculata* (Say) 2.5 YW n
11150a var. *parvinotata* Csy. 2 YW n
11150c var. *renifer* Csy. 1.6 YW n
11150d var. *taedata* Lec. 2.5 YW n
11154 *Anisostricta strigata* Thunb. 3 B y
11155 *Naemia seriata* (Melsh.) 6 RY b
11156 *Macronaemia episcopalis* (Kby.) 4 Y b
11158a *Coleomegila maculata floridana* (Leng) 5 R b
11159 *Coleomegila vittigera* (Mann.) 5 R b
11165 *Hippodamia sinuata* Muls. 5 R b
11165a var. *spuria* Lec. 5 OR b
11168 *Hippodamia americana* Cr. 5 R b
11169 *Hippodamia oregonensis dispar* Csy. 4.5 R b
11172 *Hippodamia quindecim-maculata* Muls. 6 OR b
11174 *Hippodamia glacialis lecontei* Muls. 6 R b
11175 *Hippodamia quinquesignata* (Kby.) 6 R b
11176 *Hippodamia glacialis extensa* Muls. 5 R b
11178 *Neoharmonia venusta* (Melsh.) 6 Y b
11179 *Neoharmonia notulata* (Muls.) 5 B y
11181 *Coccinella trifasciata* L. 5 R b
11183 *Coccinella tricuspis* Kby. 4 R b
11185 *Coccinella transversoguttata* Fald. 7 O b
11185b var. *nugatoria* Muls. 7 O b
11185c (21863) *Coccinella californica* Mann. 6 R b
11187 *Coccinella monticola* Muls. 6 R b
11192 *Olla abdominalis* (Say) 5 NY b
11192a var. *plagiata* Csy. 5 B r-w
11194 *Adalia frigida* (Schn.) 4.5 R b
11194b var. *humeralis* (Say) 5 B r
11195 *Adalia annectans* Cr. 4.5 RY b
11196 *Mulsantia picta* (Rand.) 5 Y-ish b
11196a var. *minor* Csy. 4 YW b
11197 *Mulsantia hudsonica* Csy. 4 YW b
11198 *Agrabia cyanoptera* (Muls.) 5 Blue

*An obsolete catalog whose numbers are used here only for convenience.

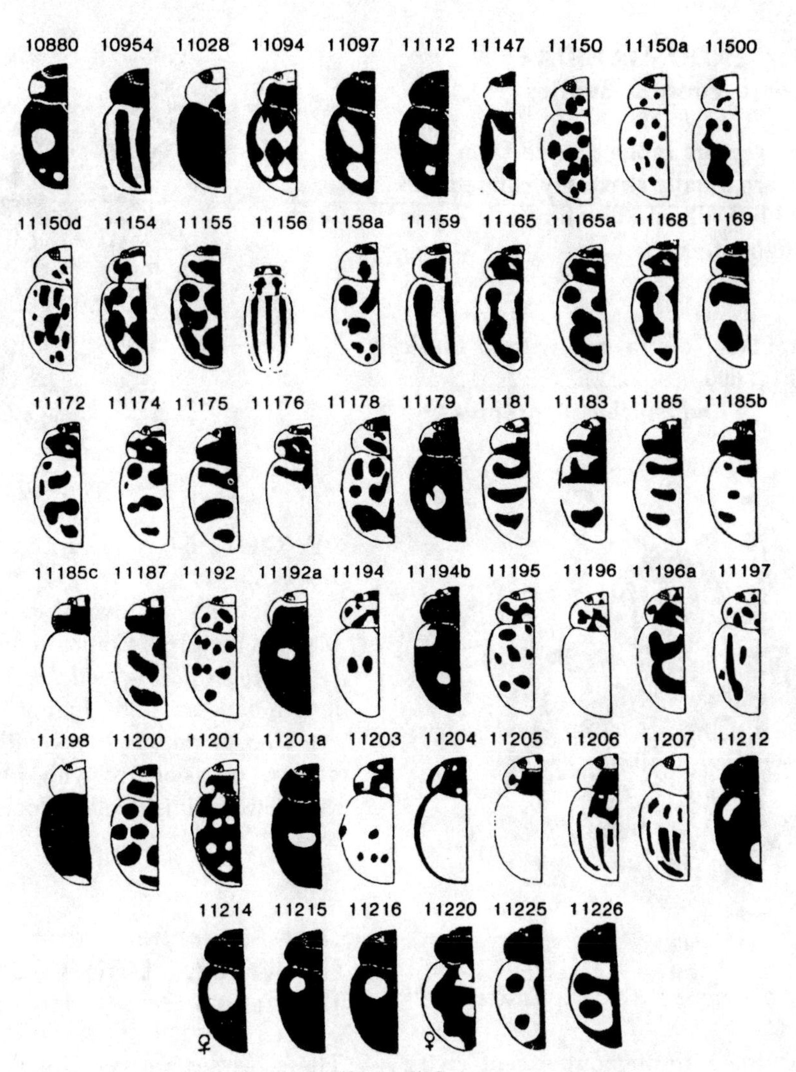

Figure 599

11200 *Anisocalvia duodecim-maculata* (Gebl.) 5 Y b

11201 *Anisocalvia quatuordecimguttata* (L.) 5 B yw

11201a var. *similis* (Rand.) 5 B r

11203 *Anatis rathvoni* Lec. 8 R b

11204 *Anatis rathvoni s. lecontei* Csy. 10 NR b

11205 *Neomysia pullata* (Say) 7 YN b

11206 *Neomysia subvittata* (Muls.) 6 YN b

11207 *Neomysia horni* (Cr.) 7 YN b

11212 *Axion tripustulatum* (DeG.) 6.5 B r

11214 *Axion plagiatum* (Oliv.) 6.5 B r

11215 *Chilocorus tumidus* Leng 5 B r

11216 *Chilocorus cacti* L. 5 B r

11220 *Exochomus marginipennis* Lec. 3 B yn

11225 *Exochomus septentrionis* Weise 4 Y b

11226 *Exochomus davisi* Leng 5 Y b

FAMILY 81. ENDOMYCHIDAE
The Handsome Fungus Beetles

These fungus feeders range in size from 1 to 25 mm. They are usually strikingly colored and some are highly brilliant. The tarsi are 4 segmented throughout.

1a **Thorax black; elytra red with 2 black spots. Fig. 600. .... *Endomychus biguttatus* Say**

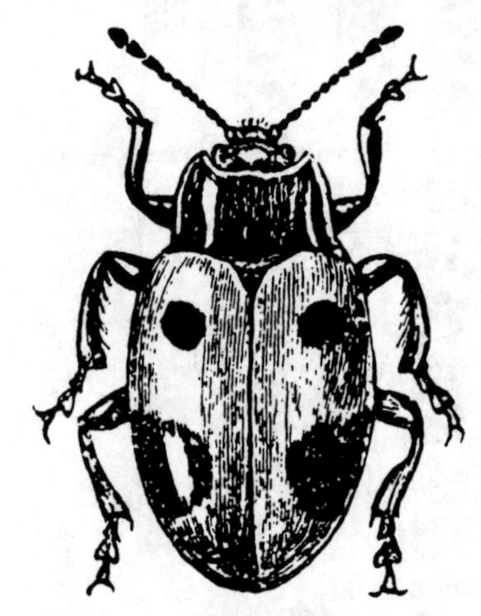

Figure 600

LENGTH: 3.5-4 mm. RANGE: Eastern half of U. S. and Can.

Shining. Black throughout except elytra which are red with black spots as pictured. Hibernates in adult stage. Common.

Bystus ulkei (Crotch) likely scattered over much of the eastern half of the U. S. measures less than 2 mm and has reddish-brown thorax and elytra. Its rounded body, which is covered with pubescence distinguishes this genus from *Endomychus.*

1b **Not colored as in 1a. Fig. 601. .... *Aphorista vittata* (Fab.)**

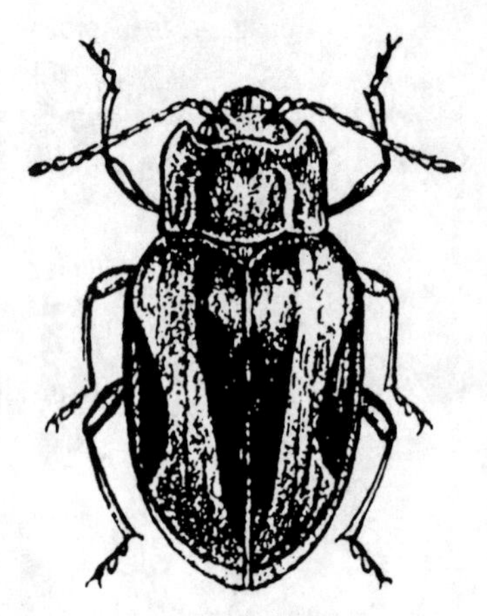

Figure 601

LENGTH: 5-6 mm. RANGE: East of the Rockies.

Shining. Brownish-red marked with black as pictured. Common in fungi.

Mycetina perpulchra Newn. with a length of 4 mm and ranging through the Eastern U. S. is shining black with thorax reddish-yellow, occasionally with blackish stripe on disk. Two reddish-yellow spots are on each elytron.

FAMILY 82. LATHRIDIIDAE
The Minute Brown Scavenger Beetles

These beetles are very small, usually reddish-yellow, or occasionally black. They are found in decaying leaves and under stones.

1a **Glabrous; front coxae separated. Fig. 602. .... *Enicmus minutus* (L.)**

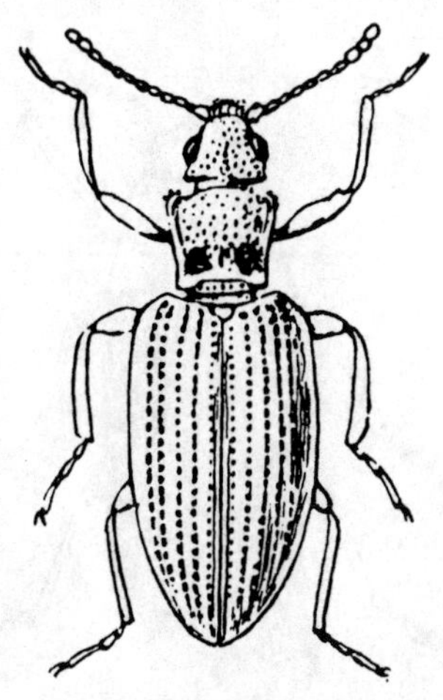

Figure 602

LENGTH: 1.5-2 mm. RANGE: cosmopolitan.

Usually brown; sometimes yellowish or blackish. Disk of thorax granulate.

Cartodere constrictus Gyll., 1.3-1.8 mm has elytra without tubercles and the antennal club but 2 segments. It is cosmopolitan.

Holoparamecus kunzei Aube dull reddish-yellow and around 1 mm in length, is known entirely across our continent. It is unlike the above two species in having the front coxal cavities open behind.

1b Pubescent; front coxae touching. Fig. 603. ....
...... *Melanophthalma distinguenda* Com.

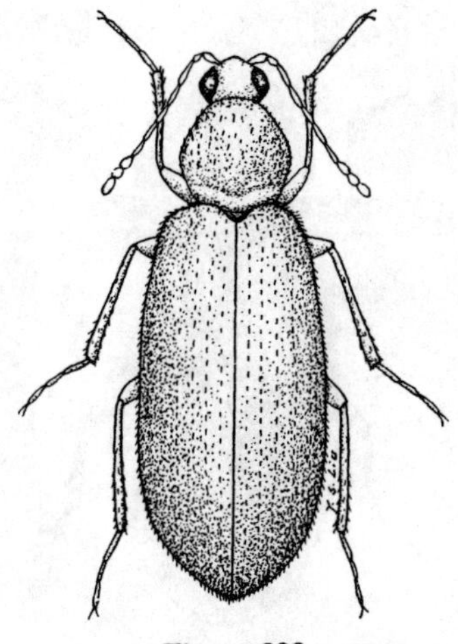

Figure 603

LENGTH: 1.5-1.8 mm. RANGE: cosmopolitan.

Dull brownish-yellow the elytra usually darker; long conspicuous pubescence yellowish.

Corticaria serrata (Payk.) is more elongate than the above and has but five ventral segments in the male. It is dull reddish-brown or reddish-yellow, measures about 2 mm and is cosmopolitan.

C. elongata Gyll. 1.4-1.8 mm, is also common and cosmopolitan. It is light brownish- or reddish-yellow with rather long pale pubescence; the thorax is squarish and wider than long.

FAMILY 83. BIPHYLLIDAE
The False Skin Beetles

There are only four North American species of this family. These beetles are small, oval, somewhat convex beetles, between 2.5 and 3 mm in length. Their color ranges from testaceous to piceous and they are densely pubescent. The 11-segmented antennae have a 3-segmented club. In our species there are fine raised lines on the sides of the pronotal disc. These beetles are taken from fungi or from under bark.

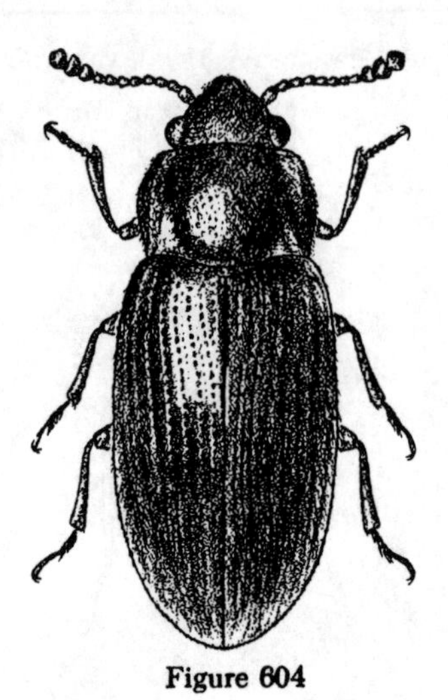
Figure 604

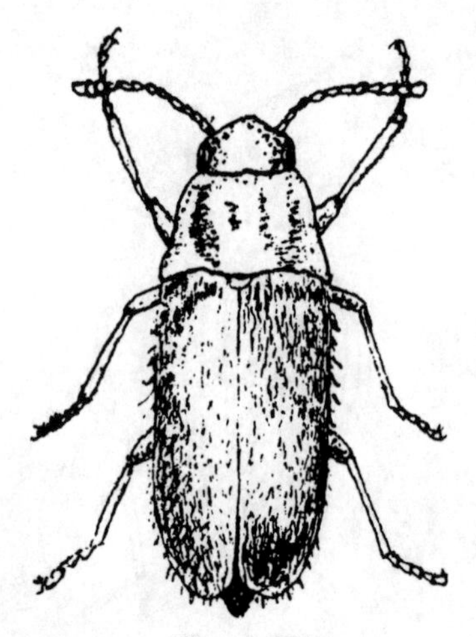
Figure 605

Figure 604 *Diplocoelus brunneus* LeC. LENGTH: 3.5 mm. RANGE: Ind. eastward. Dark reddish-brown to piceous, the undersides and appendages paler; body covered with dense pubescence. Pronotum with two distinct lines on each side and the vestiges of a third line.

FAMILY 84. BYTURIDAE
The Fruitworms

A distinctive feature of the members of this family is the lobed second and third tarsal segments. These beetles are small, ranging between 3 and 4.5 mm, oval, covered with long pubescence, and varying in color from a pale brown to piceous. The 11-segmented antennae have a 3-segmented club.

The larvae are small white worms found in the fruit heads of raspberries and blackberries. Adults are taken from the berry flowers and others.

Figure 605 *Byturus unicolor* Say. RANGE: east of the Rockies. A robust orange-yellow in color, abdominal segments darker. *B. sordidus Barber* entirely piceous or dark castaneous with white vestiture and *B. rubi* Barber ochraceous with yellow vestiture above and whitish beneath have the same size, range, and habits as *unicolor*. However, they differ by having a well-developed tooth on the apical third of the front tibia of the male. *B. bakeri* Barber, testaceous with the pronotum and head somewhat darker, pubescence grayish is the western raspberry fruit worm. A second West Coast form is *Byturellus grisescens* Jayne. It differs by having the antennae longer than the width of the head; brownish with the elytra mottled with black; pubescence a mixture of white and brown.

FAMILY 85. MONOMMATIDAE
The Opossum Beetles

This family is made up of compressed, smooth, oval, dorsally convex, ventrally flattened beetles ranging in size from 5 to 12 mm. They are piceous to black with no dorsal vestiture. The 11-segmented antennae have a 2-3 segmented club.

The larvae live in rotten wood. Adults are taken in decaying cacti, yucca, on dead twigs, and in other vegetative debris. Adults also come to lights. Five species are taken in the southern U. S. ranging from Fla. to So. Cal.

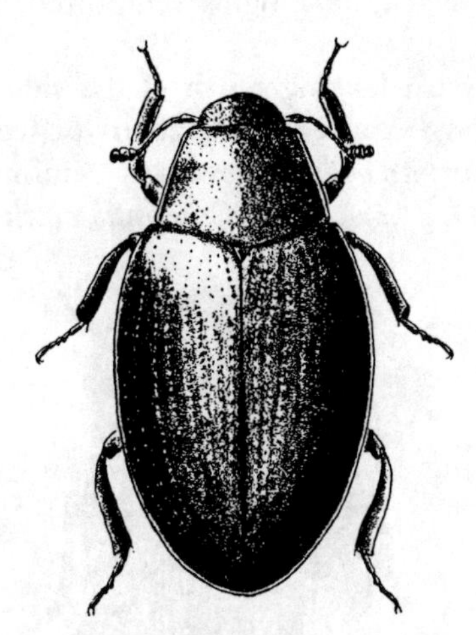

Figure 606

Figure 606 *Hyporhagus opuntiae* Horn.

LENGTH: 5-7 mm. RANGE: Ariz. Piceous-black the head, margins of the pronotum, lateral margins of the elytra, legs, and ventral surface reddish.

H. punctulatus Thom. LENGTH: 3-5 mm. This similar species is frequently taken in Florida.

FAMILY 86. CEPHALOIDAE
The False Longhorn Beetles

These beetles are elongate, convex, somewhat fusiform, long-legged, and resembling long-horn beetles. In size they range from 8 to 20 mm and are colored testaceous to black with all sorts of dark and pale markings or variations. The vestiture is fine and short and the 11-segmented antennae are filiform with the apical segment enlarged. The tarsal claws are pectinate.

Adults are taken from flowers and by beating various types of foliage. Ten species are found in the U. S. and Canada.

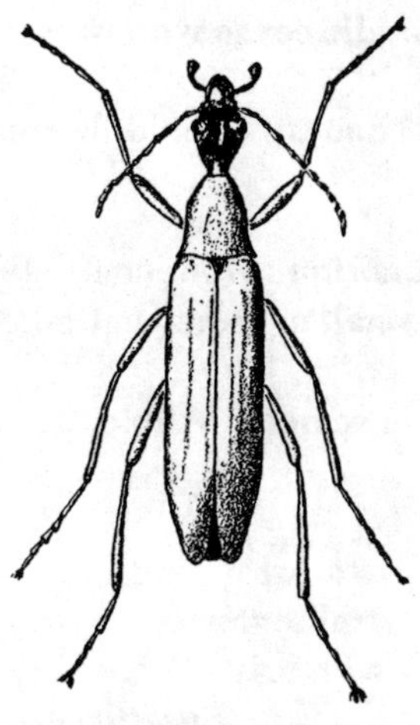

Figure 607

Figure 607 *Cephaloon tenuicorne* LeC.

LENGTH: 8-13 mm. RANGE: Pacific Coast from B. C. and Western Mont. to Cal.

Variously colored as described above, the males tending to be piceous and the females testaceous.

C. lepturides New. RANGE: Ind. north and eastward. Size and color of the above and is a common Eastern form.

FAMILY 87. TENEBRIONIDAE
The Darkling Beetles

These beetles are usually black or brown though occasionally more brightly marked. Some are very small; others are medium sized or large. The family is a large one. Most of our American species are western in their range.

1a **Ventral segments wholly corneous (stiff and heavily chitinized). 2**

1b Hind margins of third and fourth ventral segments corneous. 9

2a (1) Middle coxae with trochantin. 3

2b No trochantins on middle coxae. 6

3a (2) Labrum prominent; the mentum either small or somewhat notched. 4

3b Labrum scarcely visible. 5

4a (3) Elytra wrapping well around the body; basal segments of front tarsi long; femora with bristly hairs. Fig. 608. .... *Eusattus erosus* (Horn)

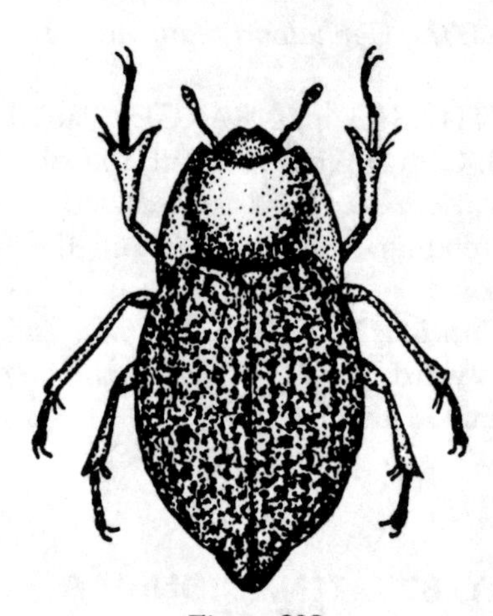

Figure 608

LENGTH: 17-19 mm. RANGE: Southwestern states.

Moderately shining. Black; elytra with coarse erosions and elevated smooth areas, but without definite ridges.

E. costatus (Horn), having much the same range, size and color of the preceding is convex, and rather dull. The elytra is subcostate with the intervening spaces closely punctured.

Eusattus reticulatus (Say) 12-15 mm in length, is common in Arizona and New Mexico and may be found in some of the adjoining states. Similar to *E. erosus,* it differs in having the epipleura and the elytral fold nearly smooth instead of roughly sculptured.

4b Elytra leaving much of the sides of the body exposed; basal joints of front tarsi shorter; eyes transversely reniform. Fig. 609. .... *Coniontis viatica* Esch.

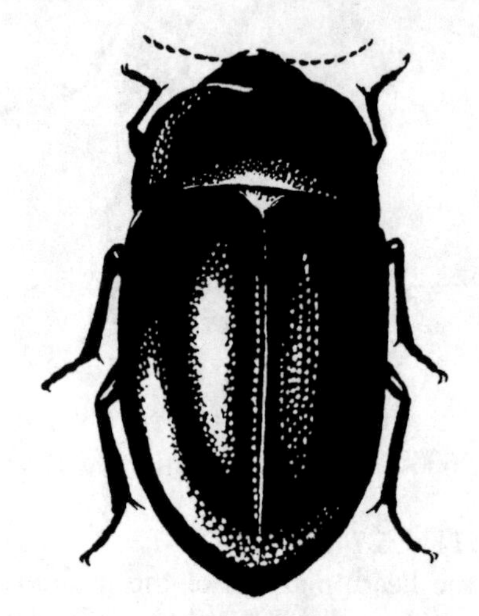

Figure 609

LENGTH: 13-15 mm. RANGE: California.

Smooth, shining. Black; elongate, thorax wider than the base of elytra; legs wholly black, the under surface of front tibia being covered with spines. Lives along the ocean shore.

C. *robusta* Horn, more obtuse than the above and having heavier legs is variably punctured although sometimes wholly smooth. Its size and range is similar to that *viatica.*

Over 100 species of this genus in the West.

5a (3) Mentum covering all the mouth parts without an attaching stalk. Fig. 610.

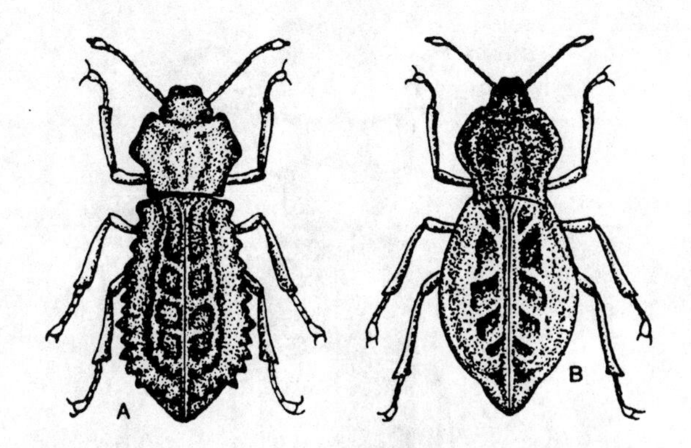

Figure 610

(a) ***Astrotus contortus*** **(Lec.)**

LENGTH: 10-12 mm. RANGE: Texas.

Dark brown with scattered scale-like hairs usually covered with a coating of gray matter.

(b) ***Astrotus regularis*** **Horn**

LENGTH: 10-12 mm. RANGE: Texas.

Color and covering substantially the same as *contortus*.

Stenosides anastomosis (Say) 12-15 mm in length and known from Kansas, Colorado and Arizona has the coxal cavities nearly rounded, the antennae slender with the segments longer than broad and the apex of the prosternum deflexed.

5b Mentum smaller, plainly stalked. Fig. 611.

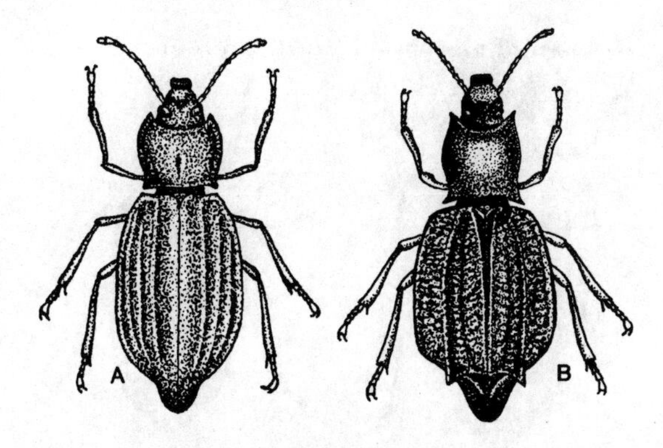

Figure 611

(a) ***Asidina semilaevis*** **(Horn)**

LENGTH: 18-23 mm. RANGE: southwestern states.

Dull black. Under surface with scattered coarse punctures. A stripe between suture and first costa.

(b) ***Philolithus aegrotus*** **Lec.**

LENGTH: 20-23 mm. RANGE: middle and southwestern states.

Black. Thorax with sides sinuate and base notched.

6a (2) Mentum unusually large, covering both the ligula and the maxilla. 7

6b Ligula and maxilla not covered by mentum. Fig. 612.

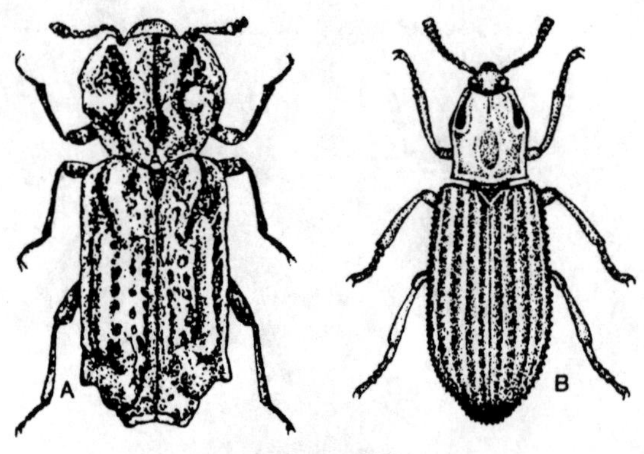

Figure 612

(a) ***Phellopsis obcordata* (Kby.)**

LENGTH: 11-14 mm. RANGE: Eastern United States.

Dull and somewhat flattened. Dark brown with rust-red effects. Thorax roughened and elevated as are also parts of the elytra. Feeds on fungi. *P. porcata* LeC. is common in the Pacific Northwest.

Closely related on our west coast *Phloeodes pustulosus* Lec. and *P. diabolicus* Lec. are known as Ironclad Beetles because of their very hard shell. The latter is wholly black while the former is grayish-black with whitish and black markings. Both are much roughened and measure 15-25 mm. They are found in California.

(b) ***Usechus lacerta* Mots.**

LENGTH: 5 mm. RANGE: Southwestern United States.

Differs from (a) in the eyes having large facets instead of small fine ones. The coxae are widely separated.

7a (6) Front tibae with but one terminal spur; mandibles with distinct external groove. Fig. 613. .... *Cnemodinus testaceous* Horn

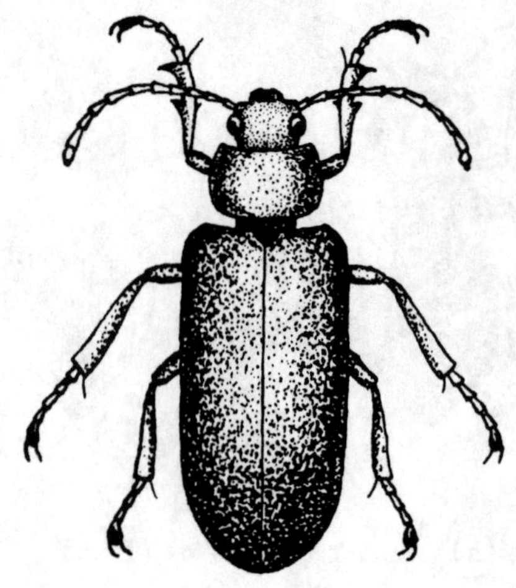
Figure 613

LENGTH: 7-8 mm. RANGE: Southwestern United States.

Yellowish-brown, pubescent with scattered yellow hairs. Scutellum triangular, longer than wide.

Metopoloba pruinosa (Horn) 12 mm long, is shining, rufopiceous, with the thorax broader than long. The body is winged as in the above but differs from it in having two spurs on the front tibiae.

7b Two terminal spurs on front tibae; mandibles not grooved on outer surface. 8

8a (7) Hind coxae small and widely separated; eyes with fine facets; legs long and slender. Fig. 614. .... *Craniotus pubescens* Lec.

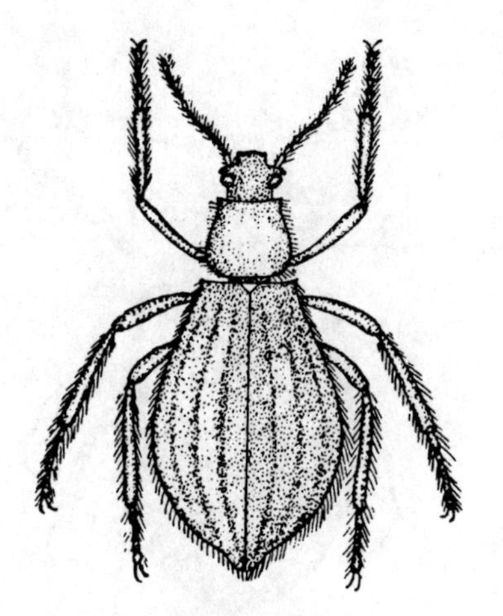

Figure 614

LENGTH: 11-13 mm. RANGE: So. Cal.

Black. Antennae with 11 segments but appearing as 10 because of the very small conical first segment.

Bothrotes canaliculatus (Say) ranging in the Central and Eastern States, has its thorax as wide as elytra. It measures 15-18 mm.

8b Hind coxae not widely separated; scutellum well developed; antennae filiform. Fig. 615. .... *Trimytis pruinosa* Lec.

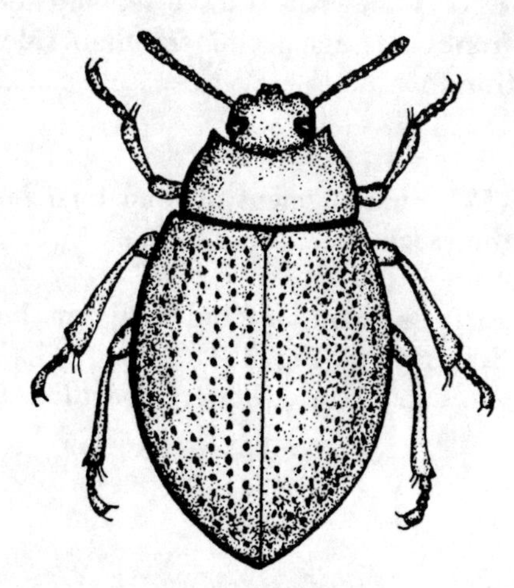

Figure 615

LENGTH: 5-7 mm. RANGE: Western States.

Shining. Black; thorax below with scattered coarse punctures.

Metoponium bicolor Horn known from California and Arizona, (6-7 mm) is black except head and thorax which are rusty-brown.

9a (1) Front wholly corneous. 10

9b Front with a coriaceous margin or band between it and the labrum, metasternum very short. No inner wings. Fig. 616. .... *Meracantha contracta* (Beauv.)

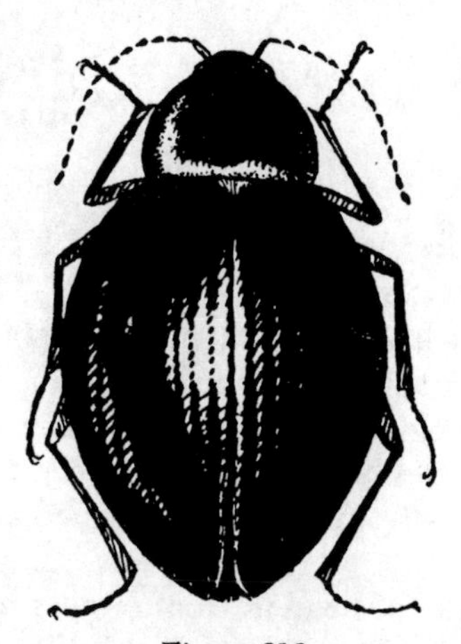

Figure 616

LENGTH: 11-13 mm. RANGE: Eastern United States and Canada.

Very convex, robust. Black with greenish bronze sheen; antennae and legs piceous; elytral striae but feebly impressed with intervals thickly and finely punctured.

This very interesting beetle found under bark does not seem much like others of its family. There is but this one species of the genus, although it has been renamed several times to make 4 or 5 synonyms.

10a (9) Tarsi compressed; first joint very short. 11

10b **Tarsi not compressed, first joint not especially short.** **12**

11a **(10) Sides of head in front of eyes prominent; antennae 10 jointed. Fig. 617.** .... ***Bolitotherus cornutus*** **(Panz.)**

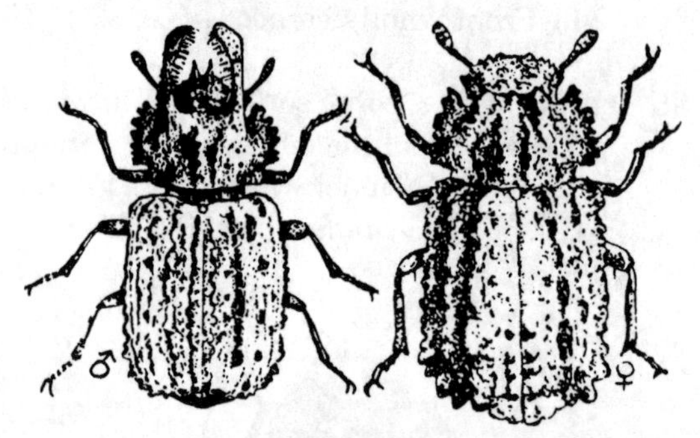

Figure 617

LENGTH: 10-12 mm. RANGE: Eastern half United States.

Black or brownish, much roughened. The male (a) with two horns as shown. They live in old fungi, on old logs and when disturbed lie motionless so that they are very difficult to distinguish from pieces of the wood and fungi.

11b **Sides of head in front of eyes not prominent; antennae pectinate. Fig. 618.** .... ***Rhipidandrus flabellicornis*** **(Sturm.)**

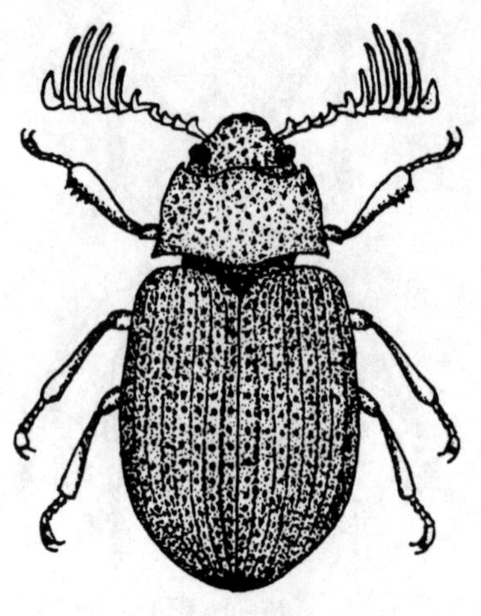

Figure 618

LENGTH: 2-3 mm. RANGE: Eastern half of U. S. and Can.

Convex. Black, legs and antennae reddish-brown. Eyes large, convex, coarsely granulate. Antennae pectinate beginning at fifth joint.

12a **(10) Eyes usually rounded, more prominent than sides of front.** **13**

12b **Eyes somewhat transverse, notched in front and less prominent than sides of front.** **17**

13a **(12) First segment of hind tarsi longer than second.** **14**

13b **First segment of hind tarsi not longer than second. Fig. 619.** .... ***Diaperis maculata*** **Oliv.**

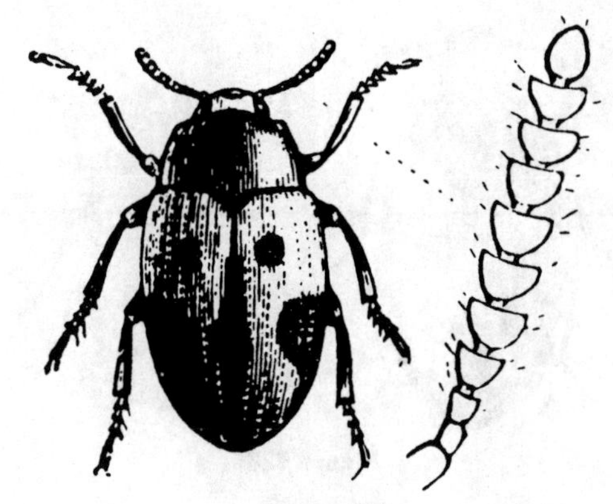

Figure 619

LENGTH: 6-7 mm. RANGE: Eastern U. S. and Can.

Convex. Black. Head and elytra marked with orange-red as pictured; legs wholly black. Commonly found living gregariously under bark and in fungi.

Out in the southwest *D. rufipes* Horn, about the same size as the above but with pale orange markings on the legs and 2 cross bands and an apical spot orange on each elytron takes over.

14a (13) First joint of hind tarsi longer than 2nd and 3rd together. 15

14b First joint of hind tarsi longer than the second only. Fig. 620. .... *Neomida bicornis* (Fab.)

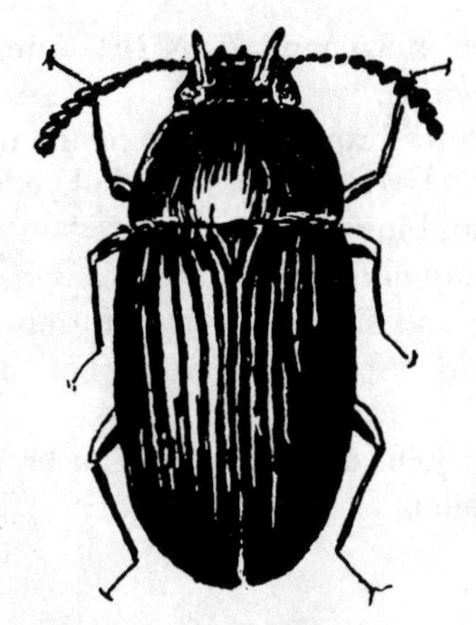

Figure 620

LENGTH: 3-4 mm. RANGE: Eastern half of U. S. and Can.

Metallic green or blue, often with head, pronotum, and scutellum red; strongly shining; males with two prominent horns between the eyes and two smaller ones on the clypeus; females hornless; common on leathery fungi.

15a (14) Last joint of maxillary palpi elongate, triangular. Fig. 621. .... *Alphitophagus bifasciatus* (Say)

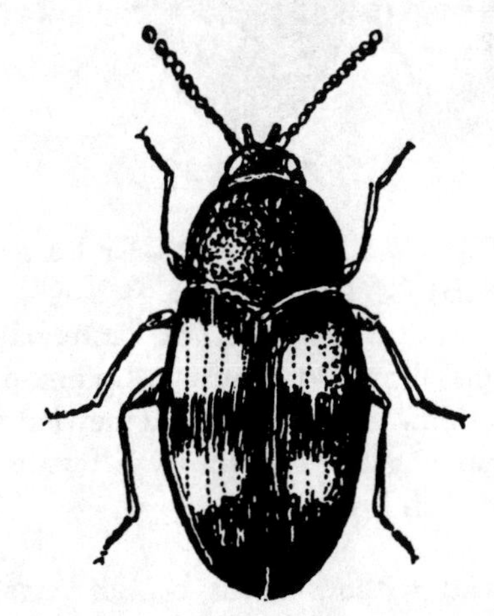

Figure 621

LENGTH: 2-2.5 mm. RANGE: Europe and North America.

Reddish-brown to blackish; antennae and legs paler; elytra marked with dull yellow. Under bark and in fungi.

Pentaphyllus pallidus Lec. is much the same size and shape but is uniformly colored a pale reddish brown, without spots or striae.

15b **Last joint of maxillary palpi broadly triangular. 16**

16a **(15) Head with horns if male, or tubercules if female on concave front. Fig. 622. .... *Platydema excavatum* (Say)**

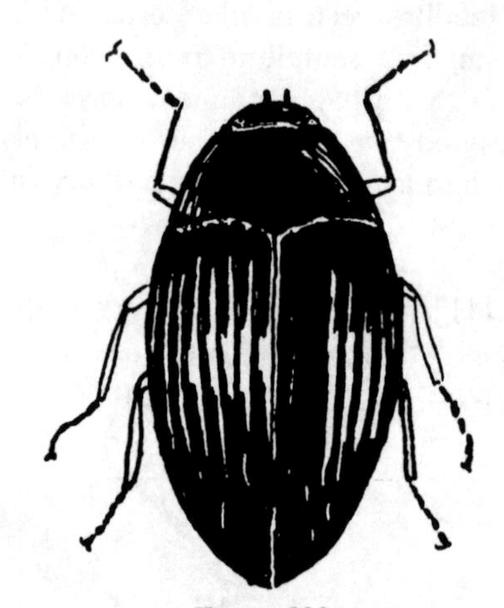

Figure 622

LENGTH: 4.5-5.5 mm. RANGE: East of the Rockies and Central America.

Convex. Shining. Black; antennae and legs reddish-brown or blackish. Common.

P. erythrocerum C. & B. somewhat smaller and ranging in our southeast differs in being a dull reddish brown.

16b **Head without horns or tubercles. Fig. 623.**

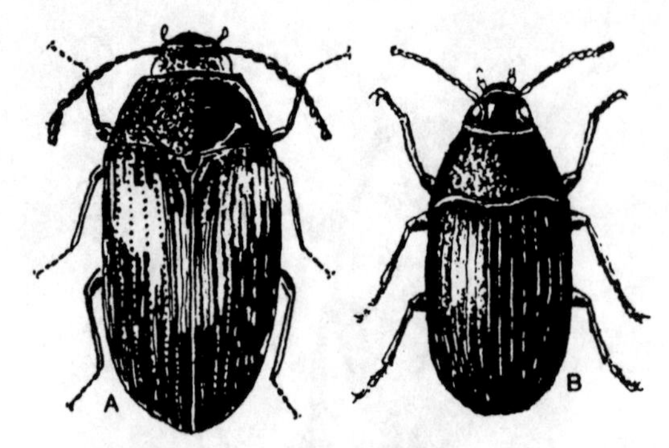

Figure 623

(a) *Platydema ellipticum* (Fab.)

LENGTH: 5.5-7 mm. RANGE: Eastern half of United States.

Dull black with irregular reddish spot near base of each elytron.

(b) *Platydema subcostatum* C. & B.

LENGTH: 5.5-6.5 mm. RANGE: Eastern half United States.

Shining. Black. Antennae reddish-brown; legs piceous. Common under bark and in fungi.

P. ruficorne Sturm measuring 4-6 mm and likely ranging through much of the eastern third of the United States, is dull black with a purplish tinge; the legs and under surface are brownish yellow; antennae pale reddish-yellow. The elytra have fine striae. There are 19 species of this genus in our territory.

17a **(12) Front coxae rounded; middle coxae with trochantin; third segment of antenna usually longer than 4th. 20**

17b **Front coxae somewhat transverse; no trochantin on middle coxae; 3rd segment of antennae short. 18**

18a **(17) Last 2 or 3 segments of antennae abruptly broadened. Fig. 624.**

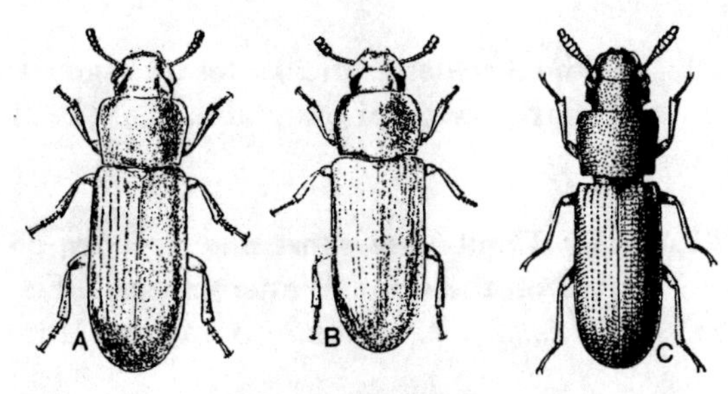

Figure 624

(a) ***Tribolium confusum* Duv. Confused Flour Beetle**

LENGTH: 4.5-5 mm. RANGE: Cosmopolitan.

Depressed. Reddish-brown. A very common pest of stored food products. Is distinguished from the Red Flour Beetle *T. castaneum* (Hbst.) (Fig. b.) which is also a universal food pest, by its gradually broadening antennae in contrast to the third abruptly widening segments at apex in the latter.

(c) ***Latheticus oryzae* Wat.**

LENGTH: 3 m. RANGE: cosmopolitan.

Pale yellow; antennae enlarging uniformly from base to beyond middle. A pest in cereal products and mills.

18b Antennae widening gradually from base to apex. 19

19a (18) Scutellum scarcely wider than long; eyes but slightly emarginate; transverse. .... *Gnatocerus maxillosus* (Fab.) Fig. 625. ....

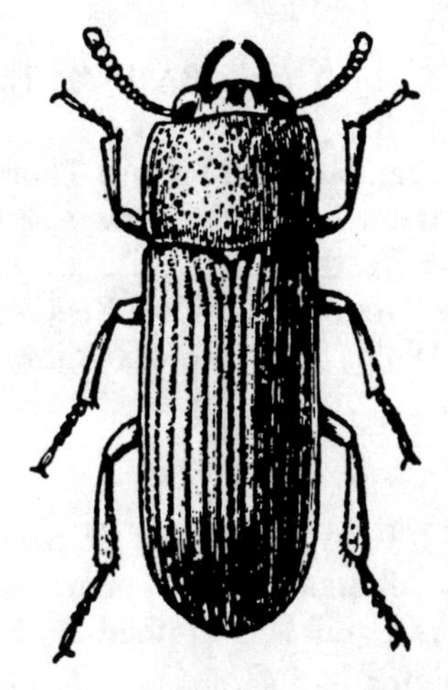

Figure 625

LENGTH: 3-4 mm. RANGE: cosmopolitan.

Reddish-brown. First segment of hind tarsus not longer than second and third. This introduced species thrives best in the Southern States.

The Broad-horned Flour Beetle *G. cornutus* (Fab.) is an introduced cosmopolitan pest of cereal products. It is shining brown, slender and 3-4 mm long. Its name comes from the broad horns on the mandibles of the male.

19b Scutellum plainly wider than long; eyes rounded, not divided. Fig. 626. .... *Palorus ratzeburgi* Wissm.

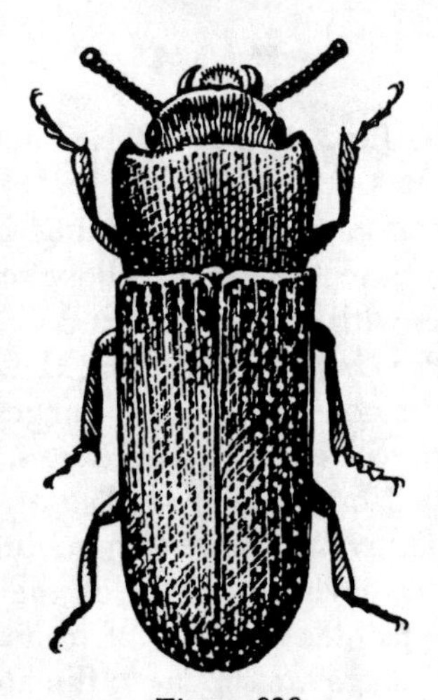

Figure 626

LENGTH: 3-3.5 mm. RANGE: Europe and North America.

Shining. Reddish-brown. Thorax convex; nearly square. Popularly known as the Small-eyed Flour Beetle.

The Depressed Flour Beetle *P. subdepressus* (Woll.) is a somewhat similar economic species.

20a (17) Fourth segment of maxillary palpus, elongate, oval, acuminate. Head, thorax and legs clothed with yellowish scales. Fig. 627. .... *Alaudes singularis* Horn

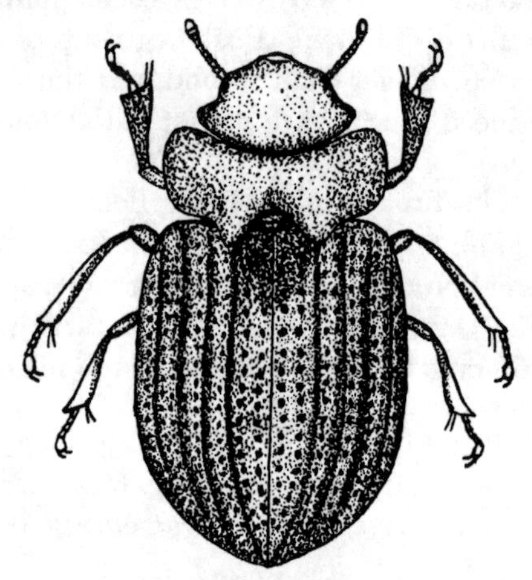

Figure 627

LENGTH: 1.5-2 mm. RANGE: Southwestern United States.

Brownish moderately shining. Head, thorax and legs clothed with yellow scales. This insect lives with ants and is blind.

Uloma impressa Melsh. 11-12 mm in length, common under bark, ranges through the eastern part of the United States. It is shining, chestnut-brown with reddish-brown legs.

U. imberbis Lec. 8-9 mm differs from *impressa* in having the end of the antennae obliquely pointed instead of rounded. The colors and range are similar to the above.

20b Without scales as in 20a; fourth segment of maxillary palpi triangular. 21

21a (20) Front very short and broadly dilated on the sides. Western species. Fig. 628. .... *Conibius gagates* Horn

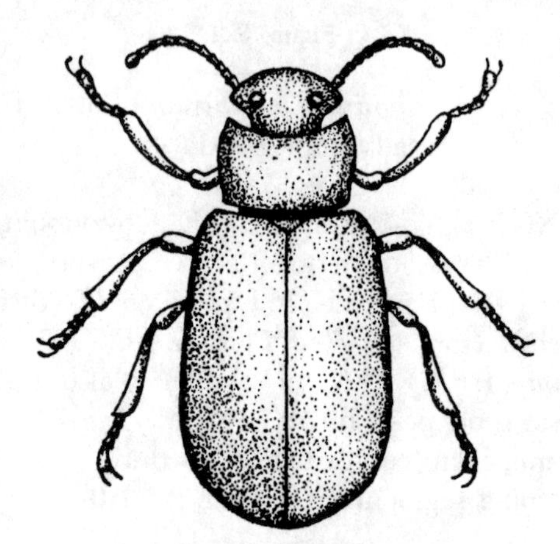

Figure 628

LENGTH: 6-7 mm. RANGE: Southwest.

Rather dull. Black. Head, thorax and elytra densely set with fine punctures. Under parts black, shining with whitish bloom when recently emerged.

In the genus *Blapstinus,* the antennal segments 4-8 are longer than broad instead of just the opposite, as in *Conibius.* Some 50 species are known.

B. dilatus Lec. 9-10 mm is abundant under logs in Arizona and California. It is black or dark brown; the thorax is feebly convex and is strongly sinuate at its base. It has a thin cover of brownish hairs.

B. pulverulentus Mann. 5-6 mm is abundant throughout California. It is shining black. The thorax is less sinuate at its base than in *dilatatus.* More than 50 *Blapstinus* occur in our area.

21b Not as in 21a. 22

22a **(21) Tarsi with a dense coating of fine silken hairs beneath. Fig. 629. .... *Merinus laevis* (Oliv.)**

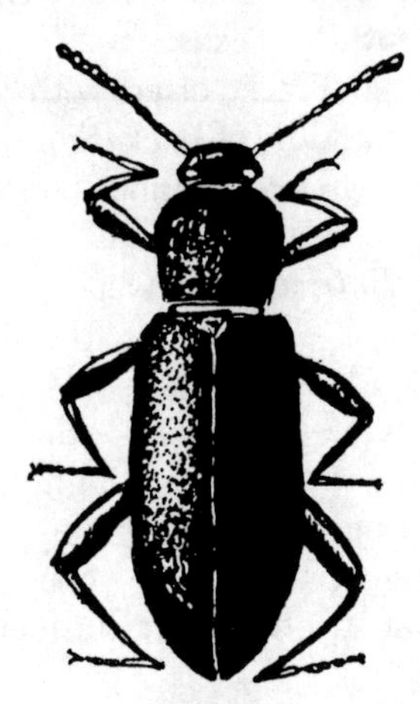

Figure 629

LENGTH: 18-26 mm. RANGE: Eastern United States.

Feebly shining. Black. Femora strongly club-shaped; epipleura not reaching tip of elytra.

A very common and widely scattered species, *Alobates pennsylvanica* (DeG.) resembles the former but the femora are more slender and the epipleura reaches to tip of elytra. It measures 20-23 mm. Hibernates gregariously under bark. Associated with it but much less common. *A. morio* (Fab.) may be readily distinguished by a tuft of long yellowish hair on the mentum.

22b **Tarsi with scattered coarse pubescence or spines beneath. 23**

23a **(22) Elongate sub-depressed, cosmopolitan beetles; elytra not extending over most of the sides of the body. Fig. 630.**

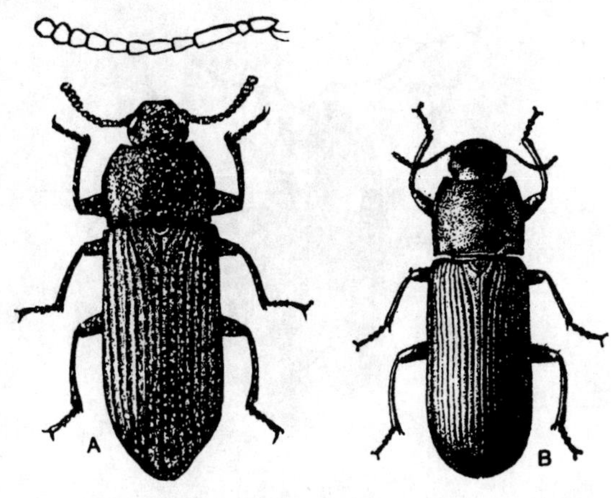

Figure 630

(a) *Tenebrio obscurus* Fab. The Dark Mealworm

LENGTH: 14-17 mm. RANGE: cosmopolitan.

Dull. Blackish on dark reddish-brown. Intervals granulate.

(b) *Tenebrio molitor* L. The Yellow Mealworm

LENGTH: 13-16 mm. RANGE: cosmopolitan.

Shining. Blackish. Intervals convex densely punctured.

Both of these species have been introduced from Europe. Their larvae, known as meal worms, resemble wireworms. They are abundant in the grain debris at mills and storage places. A similar introduced beetle, *Neatus tenebrioides* Beauv., is found under bark. Rufopiceous, shining. 10-12 mm.

23b **Elytra extended to cover much of the sides of the body; segments at apex of antennae broader. 24**

24a **(23) Epipleura very narrow not reaching the humeral angle. Fig. 631. .... *Embaphion muricatum* Say**

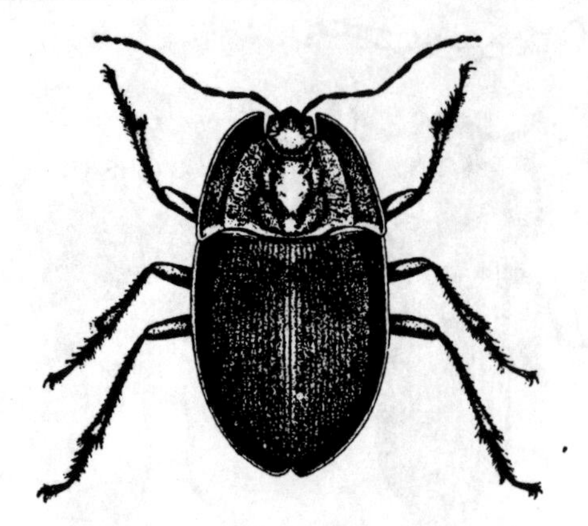

Figure 631

LENGTH: 15-18 mm. RANGE: Western half of United States.

Shining. Black, occasionally brownish. The sides of the thorax and abdoman extend into thin lateral plates which are inclined at an angle and often tinted bluish. The wing covers are attached to the back. The larvae along with those of the following genus are known as false wireworms.

24b Epipleura broader at the base and reaching the humeral angles. 25

25a (24) Front tarsi of both sexes wholly spinous beneath; elytra flattened. Fig. 632.

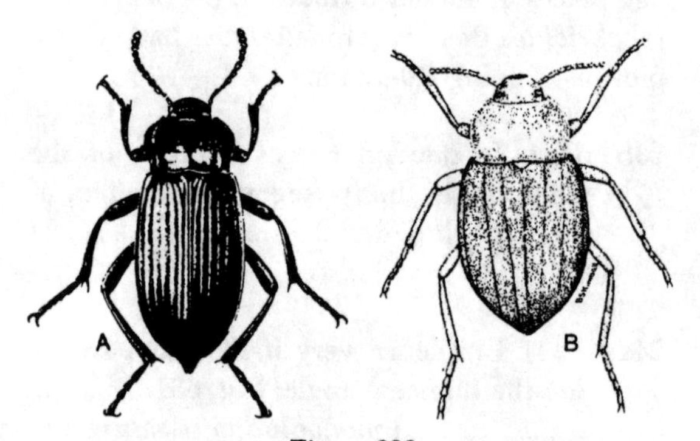

Figure 632

(a) ***Eleodes suturalis* (Say)**

LENGTH: 23-26 mm. RANGE: From the Dakotas south to Texas.

Shining. Black; often with a broad dull red stripe at middle of back; thorax flat. Flightless since wing covers cannot be raised.

(b) ***E. tricostatus* (Say)**

LENGTH: 16-20 mm. RANGE: Mexico to Canada in western half of continent.

Black with tinge of ash gray. Elytra with heavy ridges giving it its name.

This genus is a large one and contains many of the beetles most frequently seen in the West.

25b Anterior tarsi of males with first two or three segments only, pubescent beneath (not spinous). Fig. 633.

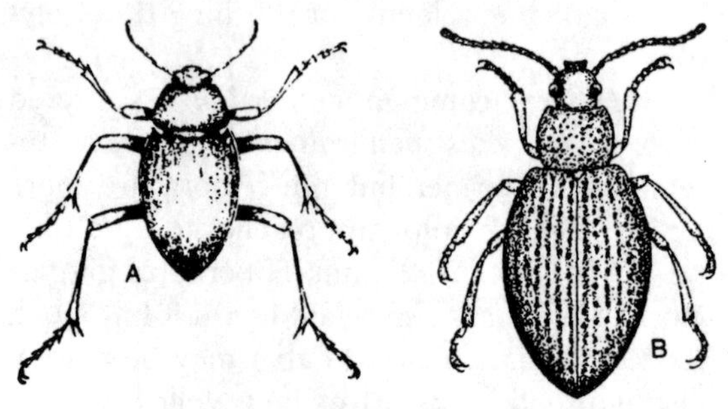

Figure 633

(a) ***Eleodes opacus* (Say)**

LENGTH: 11-14 mm. RANGE: West of the Missouri river.

Dull. Black, somewhat grayed with very fine white pubescence. Each elytron with 5 ridges. The larvae is known as the Plains False Wireworm and is a serious pest of wheat.

(b) ***Eleodes granosus* Lec.**

LENGTH: 16-19 mm. RANGE: California and Nevada.

Black. Elytra covered with rows of elevated rounded tubercles, not hairy.

A very large species (30-35 mm) is *E. giganteus* Mann. from our West Coast. Its supper surface is smooth and shiny. Over 100 species of *Eleodes* are found in our fauna, all western.

FAMILY 88. ALLECULIDAE
The Comb-clawed Bark Beetles

These are for the most part medium-sized plain brownish beetles, which may be found on leaves and flowers or under bark. The comb-cut tarsal claws are characteristic. The larvae, like those of the next family, resemble wireworms.

1a Tarsi with lobes on under side; thorax rounded in front. Fig. 634. .... *Hymenorus pilosus* (Melsh.)

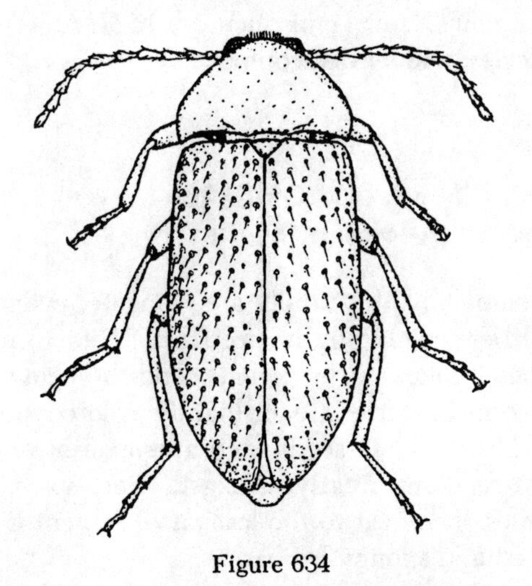

Figure 634

LENGTH: 7-8 mm. RANGE: Central and Eastern States.

Shining. Dark reddish-brown to piceous; antennae, tibiae and tarsi lighter. Eyes separated by more than their own width.

Lobopoda punctulata (Melsh.) distinguished from the above genus by the head resting upon the prosternum and coxae, is dark chestnut with scattered fine yellowish pubescence. It has a wide range and measures 9-10 mm.

The genus *Hymenorus* includes more than 100 species mostly found in a band stretching across the southern U. S. from Fla. to Cal.

1b Without lobes on under side of tarsi; apical segment of maxillary palpi broadly triangular. 2

2a (1) Third segment of the long filiform antennae distinctly shorter than fourth; mandibles sharp pointed at tip. Fig. 635. .... *Androchirus erythropus* (Kby.)

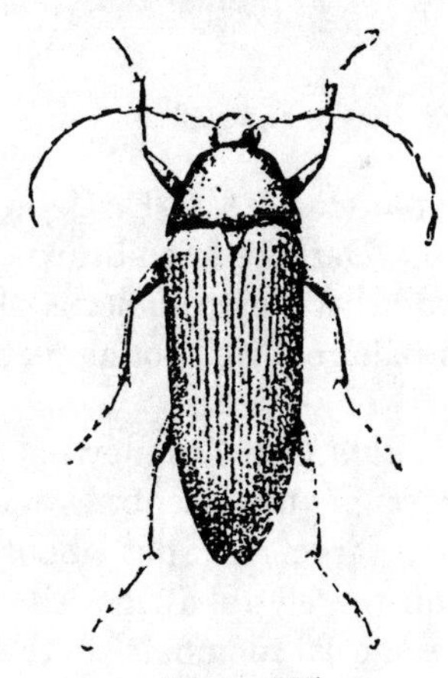

Figure 635

LENGTH: 9-10 mm. RANGE: Eastern half of United States and Canada.

Dull. Grayish black; legs and antennae pale reddish-yellow. Third segment of antennae twice as long as second; eyes small and rather widely separated.

A. femoralis (Oliv.) ranging through the Southeast, is considerably darker in color.

2b **Third segment of antennae wholly (or rarely, nearly) equal to the fourth. 3**

3a **(2) Form elongate, sides somewhat parallel; antennae short and heavy; front tarsi not as long as tibiae. Fig. 636. .... *Mycetochara foveata* Lec.**

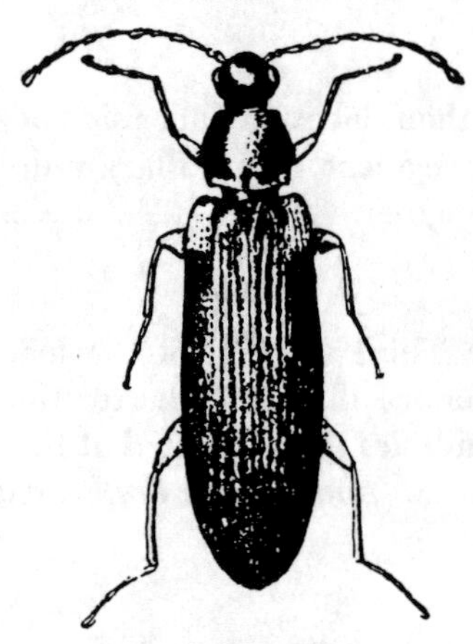

Figure 636

LENGTH: 5-6 mm. RANGE: Central States.

Shining Dark reddish-brown; antennae and legs reddish-yellow; humerus of each elytron with a pale reddish spot as pictured. (After Blatchley)

M. binotata (Say), somewhat larger and ranging through the Central and Eastern States; has a large red spot about one-third way back on the elytra. It is a shining black. There are some 20 members of this genus in our area.

3b **Form more oval, with sides rounded; antennae slender. Fig. 637. .... *Isomira quadristriata* Couper**

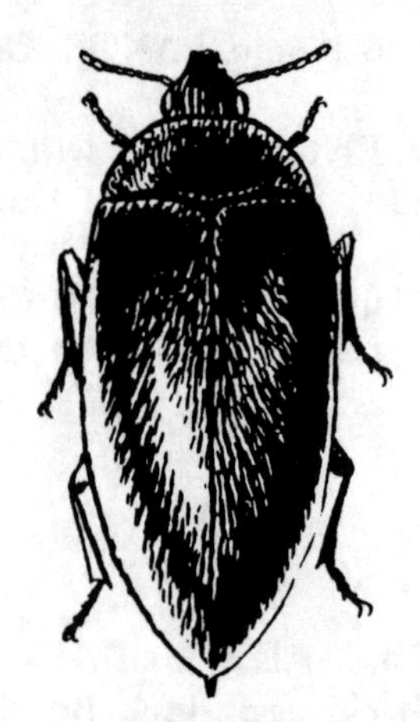

Figure 637

LENGTH: 5-6.5 mm. RANGE: Eastern half United States and Canada.

Dark reddish-brown to piceous; the head and thorax nearly black; antennae and tibiae paler.

I. sericea (Say), with somewhat the same range and a bit smaller; is uniformly pale brownish-yellow and heavily covered with short pubescence. Its fourth segment of the maxillary palpi is long and slender instead of heavy as with most of the other members of the genus. This genus includes 15 N. American species, widely distributed.

FAMILY 89. LAGRIIDAE
The Long-jointed Beetles

Members of this family are elongate, cylindrical beetles ranging in size from 6.5 to 15 mm. Their color varies from reddish-yellow to brownish. Some are brilliantly colored green or blue. The 11-segmented antennae have the last segment greatly enlarged, more so in the males. The next to the last tarsal sgment is dilated and spongy beneath.

Adults are beaten from foliage and many species are attracted to lights. The larvae breed in vegetable debris in stumps and under bark.

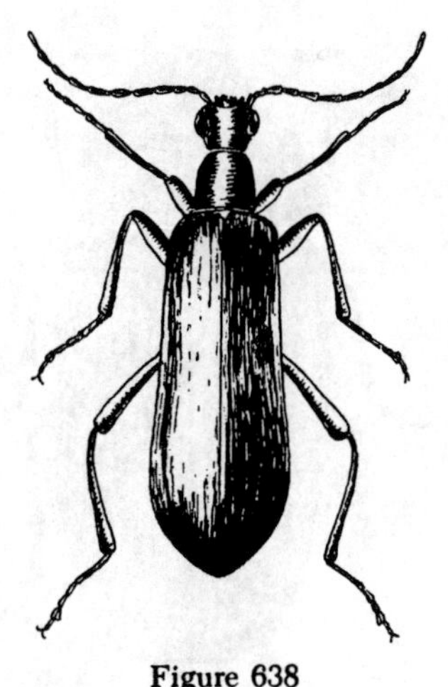

Figure 638

Fig. 638. .... *Arthromacra aenea* (Say)

LENGTH: 9-14 mm. RANGE: Eastern United States and Canada.

Convex, slender. Above brilliant green, blue coppery or dark brown with metallic sheen; antennae reddish-brown; under parts dark bronze.

The genus *Statira* differs in having large prominent eyes, distinct striations on elytra and head constricted at the base. *Statira gagatina* Melsh 6.5-8 mm is piceous and rather widely distributed. There are 18 members of this genus in our fauna.

FAMILY 90. SALPINGIDAE
The Narrow-Waisted Bark Beetles

These beetles are elongate to robust or subdepressed, ranging in size from 2.5 to 15 mm. The pronotum is not or poorly margined and subquadrate in shape. The color is dark, sometimes metallic, shining, and with no or with poorly developed vestiture. The 11-segmented antennae are clavate, sometimes serrate.

Adults are found on flowers, under bark, under rocks, and in detritus. The family is widely distributed with 34 N. American species.

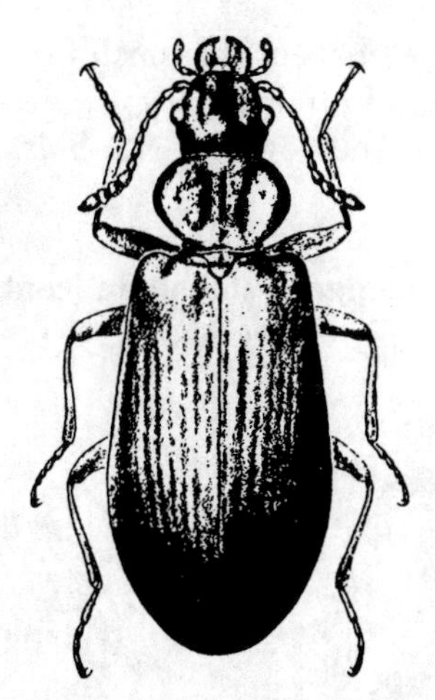

Figure 639

Fig. 639 *Pytho niger* Kby. LENGTH: 10-15 mm. Range: Eastern Can. and Northeastern U. S. and B. C. Shining piceous-black with reddish-brown appendages; pronotum with a deep lateral impression each side; elytra with rounded costae. Members of this genus are taken under the bark of dead pine and other conifers.

Rhinosimus viridiaenus Rand. LENGTH: 2.5-4 mm. RANGE: B. C. and Wash. through Ind. to N. Eng. and Que. Black with a aeneous or greenish metallic reflection; legs yellowish; front of head extended appearing as a beak as in the weevils. Found on dead basswood and alder limbs, sometimes very common.

Mycterus scaber Hald. measuring 4.5 mm, black with yellow legs and antennae, ranges through the Eastern half of the United States.

FAMILY 91. MYCETOPHAGIDAE
The Hairy Fungus Beetles

The tiny members of this small family live under bark and in fungi. They are densely punctured above and covered with hair.

1a Eyes elongate, sinuate in front, antennae gradually enlarging. Fig. 640.

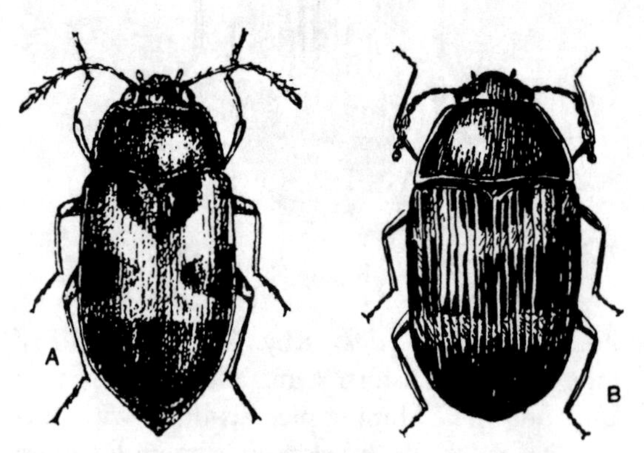

Figure 640

(a) *Mycetophagus punctatus* Say

LENGTH: 4-5.5 mm. RANGE: Eastern half United States, west to Idaho.

Subdepressed. Blackish above with reddish-yellow markings. Antennae reddish becoming blackish at apex.

(b) *Mycetophagus flexuosus* Say

LENGTH: 3-4 mm. RANGE: Eastern half of U. S. and Can. west to Mont. and Manit.

Above black with reddish-yellow markings. (Size and arrangement of markings are variable). Feeds on sap and fungi. There are 15 *Mycetophagus* species in the U. S. and Can.

1b Eyes rounded, not sinuate; segments 9-11 of antennae enlarged abruptly. Fig. 641. .... *Typhaea stercorea* L.

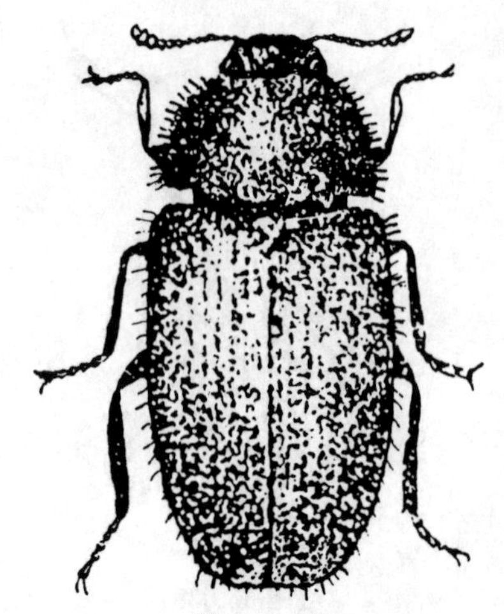

Figure 641

LENGTH: 2.5-3 mm. RANGE: cosmopolitan.

Dull reddish-yellow, elytra occasionally blackish. Antennae shorter than head and thorax, the club 3 jointed. A pest of cereal products.

Lithargus sexpunctatus Say about the same size and widely distributed is characterized by three yellow spots on each elytron. The ground color is piceous.

L. didesmus Say 2-2.2 mm in length, is shining, piceous; each elytron with yellow crossband back of the middle and a yellow humeral spot. It differs from *sexpunctatus* in lacking the basal impressions. The pubescence stands in irregular patches. It is found in many of the states east of the Missouri river. Four other species are found in our area.

Thrimolus minutus Csy. is a very little, dark brown or clay-yellow species measuring only 1-1.2 mm. The head is large and transverse; the thorax much wider than long, has its hind angles broadly rounded in contrast with the other genera of the family. RANGE: Ind. to Pa., south to Fla. and Tex.

FAMILY 92. CIIDAE
The Minute Tree-fungus Beetles

These black or brownish, somewhat-elongate beetles seldom measure over 3 mm. They live under bark and in bracket fungi.

1a Clypeus with 2 triangular teeth; thorax of male with a lobe in front. Fig. 642. .... *Cis cornutus* Blatch.

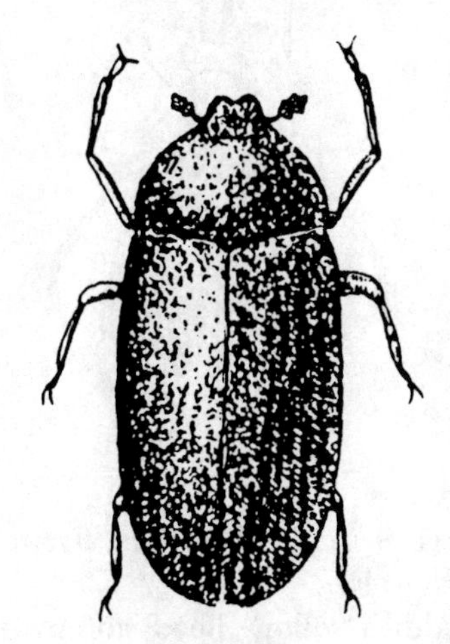

Figure 642

LENGTH: 2-2.5 mm. RANGE: Eastern United States and Canada.

Robust. Dark reddish-brown, clothed with erect yellowish hairs; the legs and antennae paler.

Ceracis is another important genus being characterized by antennae of nine segments instead of 10 as in *Cis*. *C. thoracicorne* Ziegl. which ranges widely throughout the United States and Canada is but 1.5 mm long. It is dark reddish brown and highly polished.

1b Clypeus without teeth; third segment of antennae longer than fourth. Fig. 643. .... *Cis fuscipes* Mellie

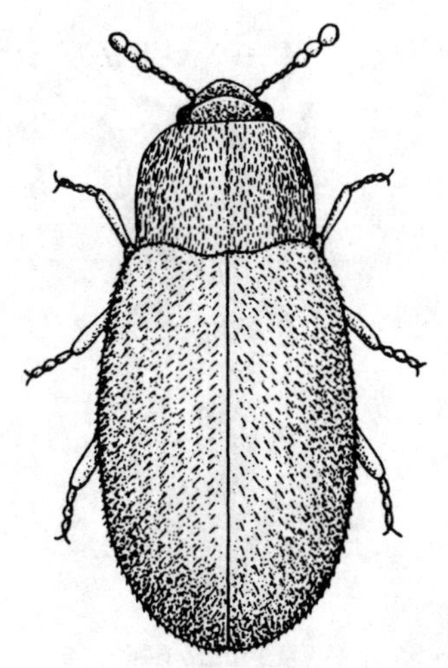

Figure 643

LENGTH: 2.5-3 mm. RANGE: Eastern half of United States.

Convex. Black to dark reddish-brown; antennae and legs paler. Common beneath bark and in fungi.

The members of the genus *Ceracis* have only 8 antennal segments. *C. sallei* Mell. dull reddish-yellow with elytra blackish at base and ranging widely, is but 1.3 mm in length.

FAMILY 93. COLYDIIDAE
The Cylindrical Bark Beetles

These long slim beetles seem to be related to the Cucujids. They live under the bark of trees. Many of them are predacious. Around 1000 species have been described. Fig. 644. .... *Colydium lineola* Say

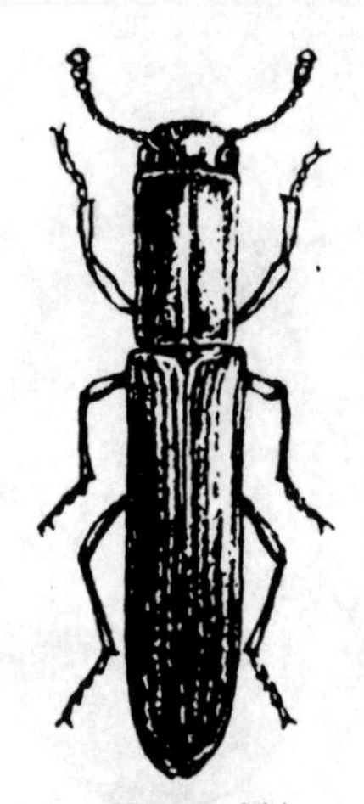

Figure 644

LENGTH: 4-7 mm. RANGE: United States and Canada.

Black moderately shining; legs and antennae paler. Thorax with deeply impressed center line; alternate intervals of elytra finely ribbed.

Bitoma quadriguttata Say is widely distributed, about 3 mm long, blackish-brown with 3 dull reddish spots on each elytron, and found beneath bark and under logs. The first tarsal joint is much shorter than in the preceding species.

Namunaria guttulatus (Lec.) 4-5 mm long, black with reddish-brown legs, antennae and margins of thorax and elytra ranges from Indiana eastward. It differs from *Colydium* in having the first joint of the tarsi short. There are but two segments in the antennal club.

Bothrideres geminatus (Say) 3-4.5 mm, dark reddish-brown, is common and ranges from Texas to our east coast.

FAMILY 94. PYROCHROIDAE
The Fire-colored Beetles

These medium sized soft bodied beetles usually have some "fire color" on them, hence the name. They are often found under bark. The family is small with less than a score of species known for our region.

1a Eyes very large, almost touching. Fig. 645. .... *Dendroides canadensis* Latr.

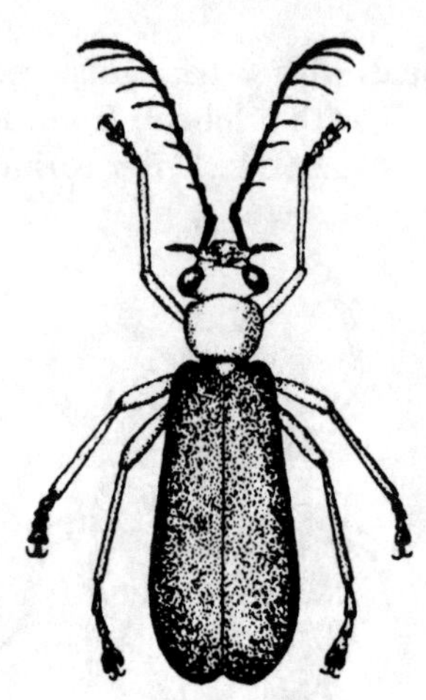

Figure 645

LENGTH: 9-13 mm. RANGE: From B. C. east and south to Fla.

Reddish yellow; head, antennae and elytra, piceous; the antennal branches are much longer in the male.

D. concolor Newn. ranging throughout the eastern half of the United States and Canada and about the same size, is pale brownish-yellow throughout.

1b Eyes normal in size and well separated. Fig. 646. ....
.... *Neopyrochroa femoralis* (Lec.)

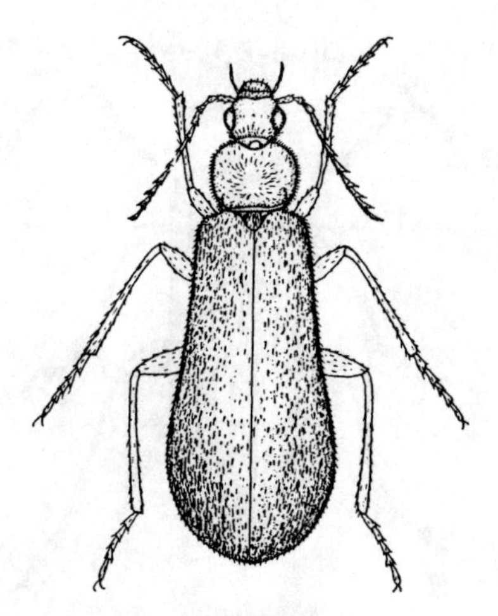

Figure 646

LENGTH: 14-17 mm. RANGE: Both this and the next species are found east of the Rockies.

Head and thorax yellowish-red, elytra bluish-black with scattered sub-erect pubescence, shining; under surface, tibiae, tarsi and palpi blackish.

N. flabellata (Fab.) attains the same size. It is reddish throughout except the elytra and most of the antennae which are black. The thorax is wider than in the above.

Two other genera belong to this small family. *Ischalia costata* (Lec.) 5 mm, is dull brownish-yellow; the elytra have a pale stripe on the sides and the tips are pale. The antennae are simple. Ranges from Ind. to Ga. and north.

Schizotus cervicalis Newn. 7 mm, is blackish; front of head, the thorax and sutural and marginal lines on the elytra reddish. The antennae bear branches. Ranges from the upper Midwest to N. Eng. and Eastern Can.

FAMILY 95. OTHNIIDAE
The False Tiger Beetles

These moderate sized beetles, 5-7 mm in length, resemble tiger beetles in their general appearance. In color they range from brown to piceous and are often mottled with rather dense pubescence. The 11-segmented antennae end in a 3-segmented club.

Adults are found on the leaves and bark of trees, on the tops and sides of stumps, and under bark. Worldwide there is only one genus containing 16 species, five of which are found in the U. S. and Can.

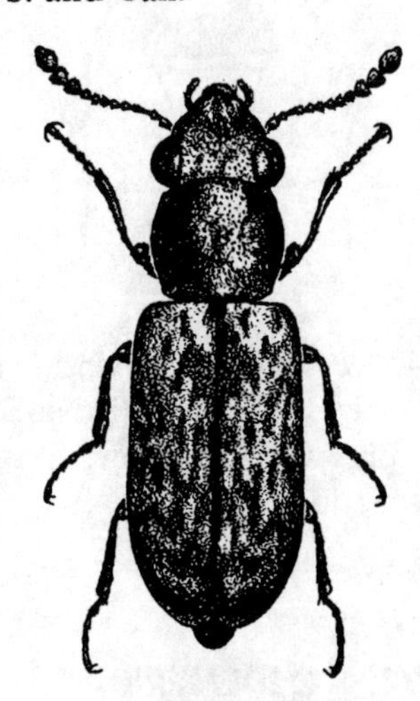

Figure 647

Figure 647 *Elacatis umbrosa* (LeC.) LENGTH: 5 mm. RANGE: Neb. Piceous black, mottled with pale. *E. lugubris* (Horn). Dark blackish, bronzed; legs black, tarsi paler; elytra coarsely, densely punctate, with white hair clusters on the disc. Taken in northern Id., eastern Or., and SE B. C. on ponderosa pine.

FAMILY 96. PROSTOMIDAE
The Jugular-horned Beetles

This family consists of the single species described below. These are moderate sized beetles, 5-6 mm in length, elongate, subdepressed, brownish in color, and shining. The 11-segmented antennae are monilliform and end in a loose 3-segmented club. The head across the eyes is wider than the pronotum and elytra.

The pronotum is quadrate, and the elytra are subparallel with series of punctures. The mandibles are longer than the head and internally serrate.

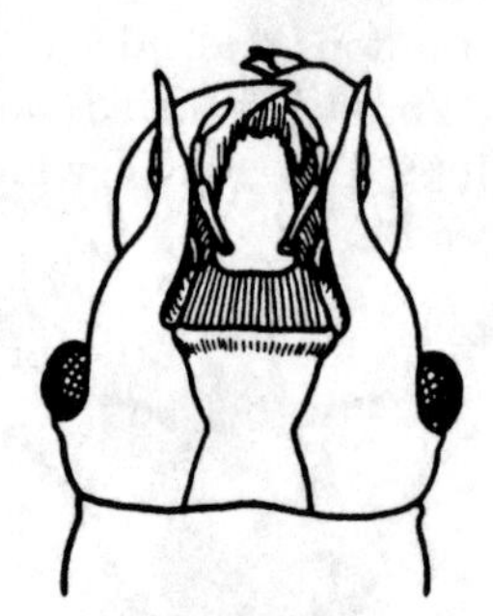

Figure 648

Figure 648 *Prostomis mandibularis* F. LENGTH: 4.8-6.2 mm. RANGE: Pacific Northwest plus Cal. and Nev. Testaceous to rufotestaceous in color.

FAMILY 97. OEDEMERIDAE
The Pollen Feeding Beetles

Some sixty species of this family are known in our region; 800 world wide. They are elongate and soft bodied. They often visit flowers, flying in the twilight or at night. Some may be also found under debris or on decaying stumps and logs.

1a Front tibiae with one spur; brownish yellow with black at tip of elytron. Fig. 649. .... *Nacerdes melanura* (L)

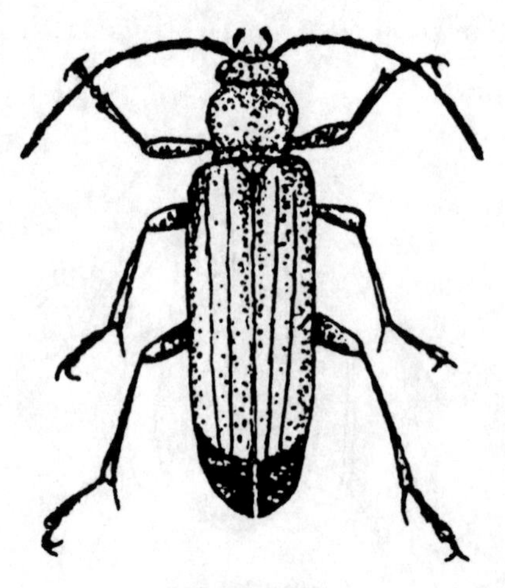

Figure 649

LENGTH: 8-12 mm. RANGE: throughout our country (also Europe and Asia, its original home).

Dull yellow above marked on elytron as pictured, with blue-black; under parts dusky; four elevated lines on each elytron. It breeds in coniferous timber and causes considerable damage.

1b Front tibiae with two spurs; claws toothed at base. Fig. 650. .... *Asclera puncticollis* (Say)

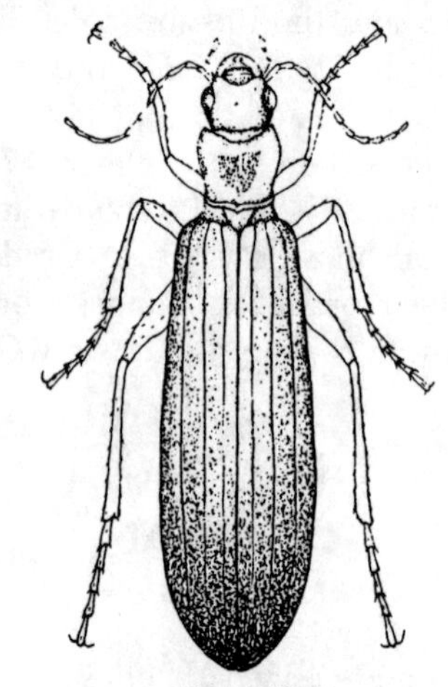

Figure 650

LENGTH: 6.5-8 mm. RANGE: Eastern and central U. S.

Dull black; thorax entirely red and smooth except 3 fovea and a row of punctures near the base. Elytra each with three longitudinal ridges.

Oxycopis thoracica (Fab.) 5-7 mm with thorax reddish-yellow; elytra purplish or blue and other parts piceous, ranges through the Central and Southern States.

A. ruficollis (Say), pronotum entirely red, is more common. Eastern U. S. and Can.

FAMILY 98. MELANDRYIDAE
The False Darkling Beetles

These beetles live under bark or in fungi and vary considerably in size. Black or brown is their usual color. They are often covered with a heavy coat of pubescence.

1a **The head at most gradually narrowed or rounded behind the eyes. 2**

1b **Head very strongly and abruptly constricted just behind the eyes. Fig. 651. .... *Anaspis rufa* Say**

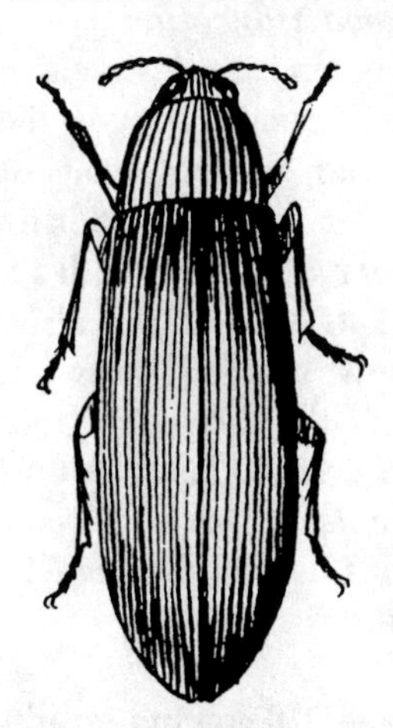

Figure 651

LENGTH: 3-4 mm. RANGE: widely distributed.

Thorax one-half wider than long; head yellow or sometimes dull; abdomen and antennae fuscous or dull yellow. Nine other species in our area.

Pentaria trifasciatus Melsh., has its fourth tarsal segment of front and middle legs as long as third instead of very small as in the above. The elytra are yellow with base, tips and wide middle cross-band, black. Range: Most of the U. S. except N. Eng.

2a **(1) Tarsal claws undivided. 3**

2b **Tarsal claws with an appendage at base. Fig. 652. .... *Osphya varians* (Lec.)**

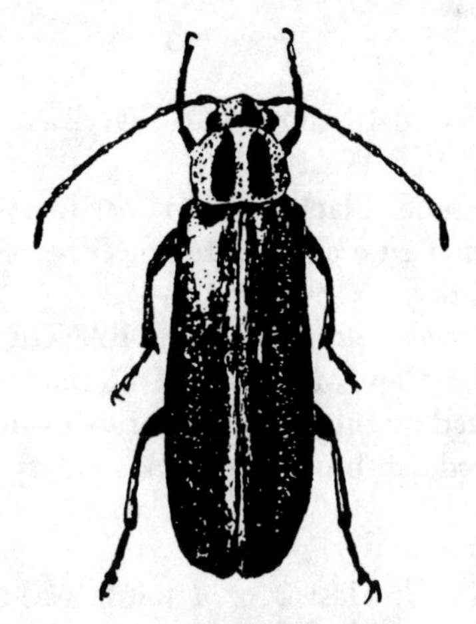

Figure 652

LENGTH: 5-8 mm. RANGE: Que., N. Y., Ind. to Tex.

Blackish. With scattered fine gray prostrate hairs. Thorax reddish-yellow marked with black as pictured, the marks often fused into one or variable.

3a **(2) Next to last segment of tarsi simple. 4**

3b **Next to last segment of tarsi excavate and notched, somewhat lobed beneath. Fig. 653. .... *Melandrya striata* Say**

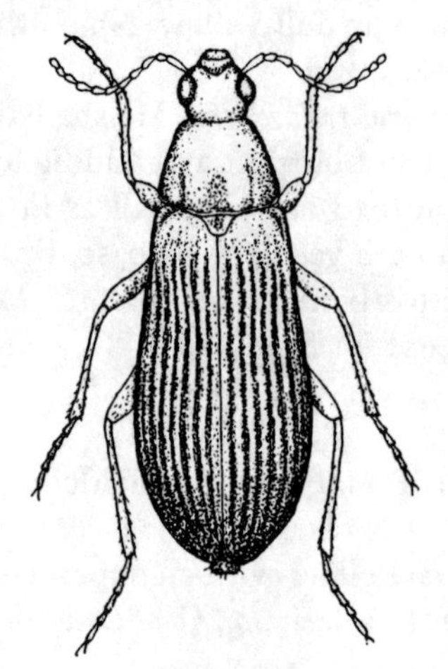

Figure 653

LENGTH: 8-15 mm. RANGE: Eastern half United States, West B. C.

Shining. Black. Antennae reddish-brown, pubescence fine and scattered. Often common under bark.

Dircaea liturata LeC. RANGE: East. U. S. and Can. Measuring 7-11 mm, may be recognized by the two irregular yellowish spots on the reddish-brown to blackish elytra.

4a **(3) The last 3 or 4 joints of antennae abruptly thickening into a club. Fig. 654. .... *Pisenus humeralis* (Kby.)**

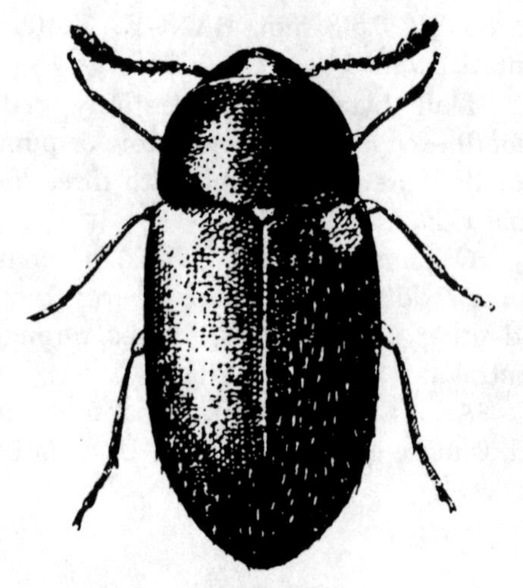

Figure 654

LENGTH: 3-4 mm. RANGE: Central and Eastern United States.

Convex. Shining. Blackish with scattered fine short yellowish hairs; legs and antennae dark reddish-brown. Elytra marked with a reddish spot on each humeri as shown. Often common on fungi.

Abstrulia tesselata (Melsh.), with a length of 3.5 mm and ranging from New York to Virginia westward, differs from the above in having four segments in the antennal club instead of but three. It is piceous with dull yellow legs and antenna.

Tetratoma truncorum Lec. 4.5-6 mm in length and known from Iowa eastward, is elongate-oval and strongly convex. The elytra are steel blue; head and antennae black; thorax, legs and under surface reddish-yellow. The thorax is wider than long with its sides much rounded and its hind angles obtuse. It is told from the above species by its larger size and the margins of the thorax not turning upward as in the other two. Common on fungi in late fall. The second species, *T. concolor,* LeC., mostly piceous, is found in the Pacific Northwest and Col.

4b **Antennae thickening gradually or filiform. 5**

5a **(4) Front coxal cavities with opening at side. Fig. 655. .. *Penthe obliquata* (Fab.)**

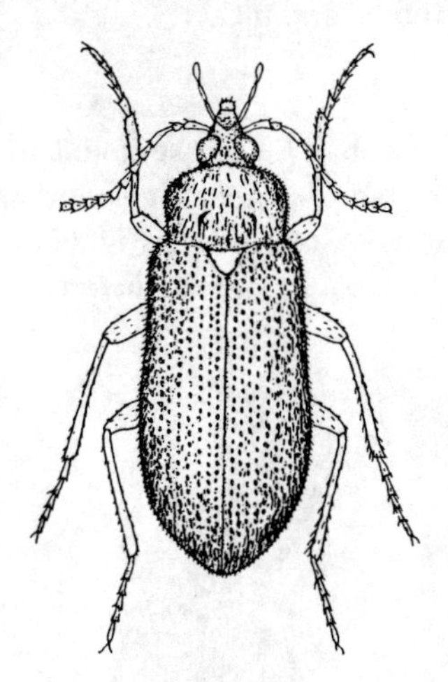

Figure 655

LENGTH: 11-14 mm. RANGE: Eastern half United States and Canada.

Somewhat flattened. Velvety black throughout except scutellum which is densely covered with long yellow-orange hairs.

P. pimelia (Fab.) about the same size and range is almost identical with the above except that the scutellum is black (lacking the orange hairs). Both species hibernate under bark gregariously.

5b **Front coxal cavity without a fissure at side. Fig. 656. .... *Eustrophinus bicolor* (Fab.)**

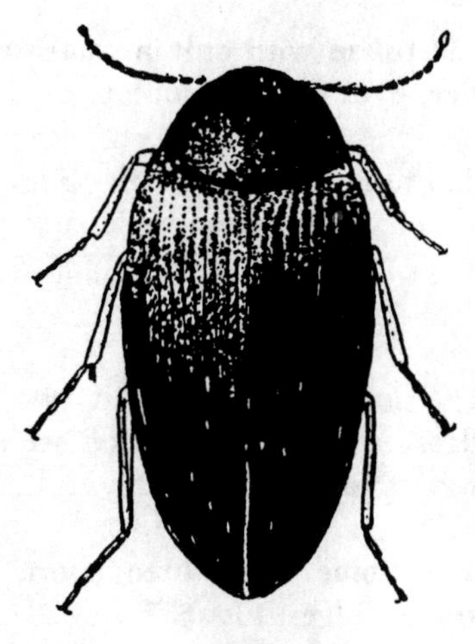

Figure 656

LENGTH: 5-6 mm. RANGE: Across U. S. and S. Canada.

Convex. Shining. Black, sparsely pubescent; abdomen and legs reddish-brown; last segment of antennae reddish-yellow.

Holostrophus bifasciatus (Say) ranging through the eastern half of our country and measuring 4-6 mm, has blackish elytra marked with two yellowish bands.

Hallomenus scapularis (Melsh.), 4.5-5 mm in length, lives in fleshy fungi and is widely distributed. It is piceous to dark reddish-brown, with antennae, legs and base of elytra dull brownish-yellow. The thorax is one-half wider than long, with deep impressions on each side near the base. Eastern and southern U. S., west as far as N. Mex.

FAMILY 99. MORDELLIDAE
The Tumbling Flower Beetles

Few insects do a better job living up to their name than these. For the most part rather small wedge or spike shaped beetles, they live as adults feeding among flowers but when disturbed kick and tumble in all directions. We have a good 150 species; the world some 800.

1a Hind tibiae with only a small subapical ridge; eyes finely granulate. 4

1b Hind tibiae and tarsi with oblique ridges on their outer face; eyes coarsely granulate. 2

2a (1) Hind tibiae with but two oblique ridges on outer face, which are equal in length and parallel. 3

2b Hind tibiae with three short, oblique parallel ridges. Fig. 657. .... ***Mordellistena sexnotata*** **Dury**

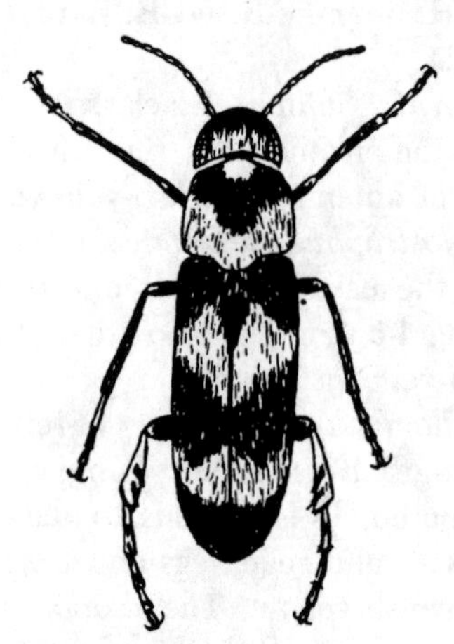

Figure 657

LENGTH: 3 mm. RANGE: Iowa to Ohio.

Brownish yellow, marked as pictured with dark brown; first segment of hind tarsi with two oblique ridges and the second segment with but one.

M. pubescence Fab. (3 mm) is characterized by four oblique ridges on the hind tibiae. It is black with a brownish pubescence. The elytra are marked with a humeral lunule and two bands of yellow pubescence. Ranges over the eastern half of U. S.

M. marginalis (Say), measuring 3-4 mm, is black with head and thorax reddish-yellow (frequently the head has a black spot and the thorax some black markings). It ranges widely and is often abundant.

3a (2) (a, b, c) First segment of hind tarsi with two oblique ridges and second segment with one ridge. Fig. 658. .... ***Mordellistena vapida*** **Lec.**

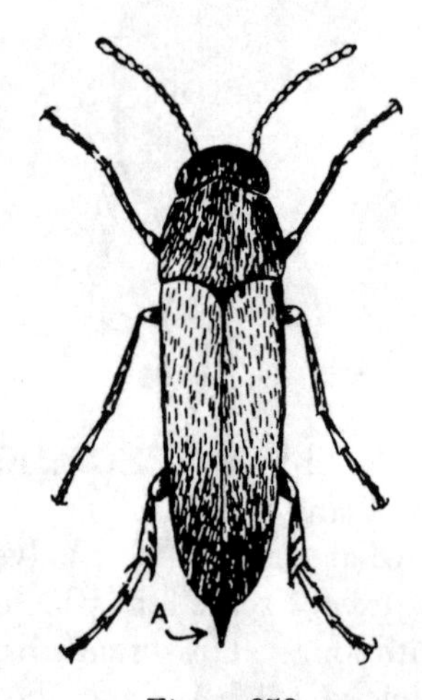

Figure 658

LENGTH: 2-3.5 mm. RANGE: eastern half of the U. S.

Clay yellow with scattered long yellow pubescence; elytra often darker at apical third.

M. lutea (Melsh.), (2.8-3.2 mm) and ranging throughout the eastern half of the United States is characterized by converging instead of parallel ridges on the hind tibiae. It is brownish yellow and thickly covered with yellow pubescence.

3b First segment of hind tarsi with three oblique ridges; second segments with two ridges; head black; elytra with a humeral spot. Fig. 659. .... ***Mordellistena scapularis*** **(Say)**

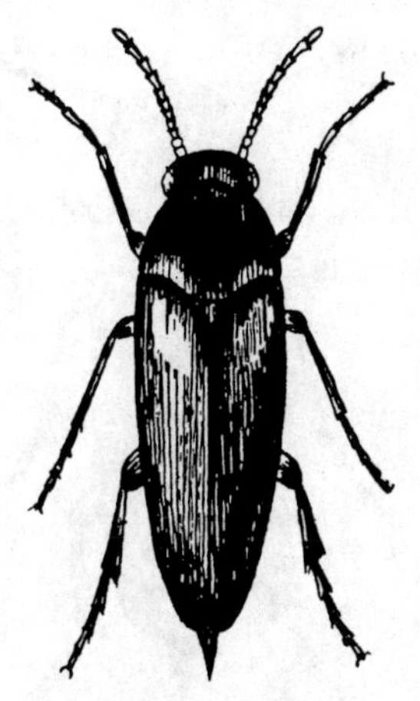

Figure 659

LENGTH: 4 mm. RANGE: No. U. S. and Southern Can.

Black; elytra marked with reddish humeral spot as pictured. In *M. splendens* Smith the ridges on the tibiae are unequal and the first segment of hind tarsus has 4 ridges. It is very slender, 6 mm. long, piceous, clothed with a silvery pubescence and ranges from Iowa to Florida.

3c **As in 3b except that head is red or reddish and elytra without a distinct humeral spot. Fig. 660. .... *Mordellistena comata* (Lec.)**

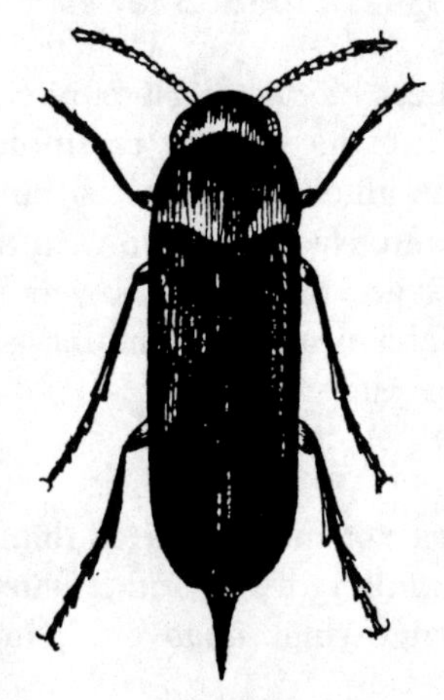

Figure 660

LENGTH: 2.8-3.3 mm. RANGE: So. U. S. from Fla. to Tx. and Az.

Body black; head mostly or wholly reddish; thorax brick-red, often with an oblong black spot near its base; legs in part dull yellow.

M. aspersa (Melsh.) a widely distributed species, differs from the above, having its head and thorax black. It is a bit smaller and is covered with a brownish-gray pubescence.

Mordellistena with over 150 species in our fauna is the largest member of this family.

4a **(1) Anal style long and slender; scutellum triangular. 5**

4b **Anal style short and blunt; scutellum notched behind. Fig. 661.**

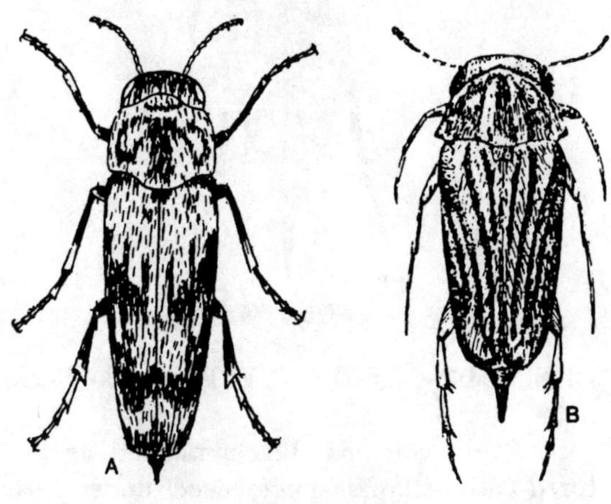

Figure 661

(a) *Glipa bidentata* (Say)

LENGTH: 10-13 mm. RANGE: N. Eastern U. S. from Iowa and Kan.

Brown, marked with pattern as pictured with ash-gray pubescence.

(b) *Tomoxia lineela* Lec.

LENGTH: 5-7 mm. RANGE: Eastern third of U. S. and Can.

Brown with markings of ash-gray as pictured.

Glipa hilaris (Say), (9-13 mm), ranges Ill. and Ind. south to Fla. and Tx. It is black with a pattern of ash-gray pubescence. Two brownish yellow pubescent bands, each margined with gray cross the elytra.

5a (4) Elytra with four yellowish pubescent spots. Fig. 662. .... *Glipa octopunctata* (Fab.)

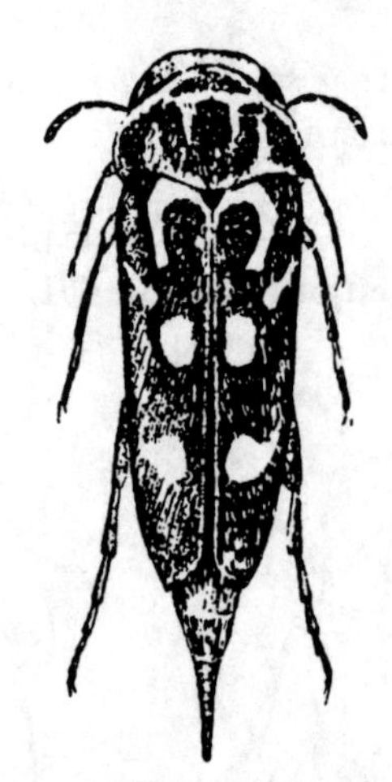

Figure 662

LENGTH: 6-7 mm. RANGE: East of the Rockies.

Fairly common. Black, marked as pictured with yellowish pubescence; under parts spotted with ash-gray pubescence.

Tomoxia discoidea (Melsh.) 2-3 mm long and ranging through the Central and Southern States is black with legs, antennae and top of head yellow. The elytra have a large humeral spot, a band back of the middle and the tips yellow; the yellow thorax has a center spot of black.

5b Elytra with numerous spots and patches of yellowish pubescence. Fig. 663. .... *Tomoxia serval* (Say)

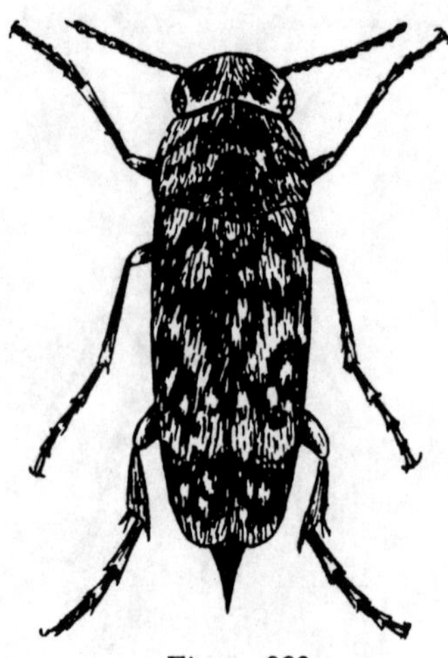

Figure 663

LENGTH: 4-4.5 mm. RANGE: Central and Eastern States.

Piceous brown; thorax and elytra mottled as pictured with yellowish pubescence; front legs and antennae dull yellow.

Mordella marginata Melsh., 3-4.5 mm long and common over a wide range, is black with variable markings of silver hairs on thorax and elytra. There are about 25 species of *Mordella* in the U. S. and Can.

FAMILY 100. RHIPIPHORIDAE
The Wedge-Shaped Beetles

The members of this small family are wedge shaped and rather closely resemble the Mordellids. The antennae are often fan shaped in the males and usually serrate with the females. The adults are found on flowers and under bark or under dead logs; the larvae are parasitic on bees and wasps.

1a Elytra not much shorter than abdomen, often falling away and leaving the membranous wings uncovered; lobe at base

of thorax covering scutellum; claws two-toothed. Fig. 664. ....
.... *Macrosiagon limbatum* (Fab.)

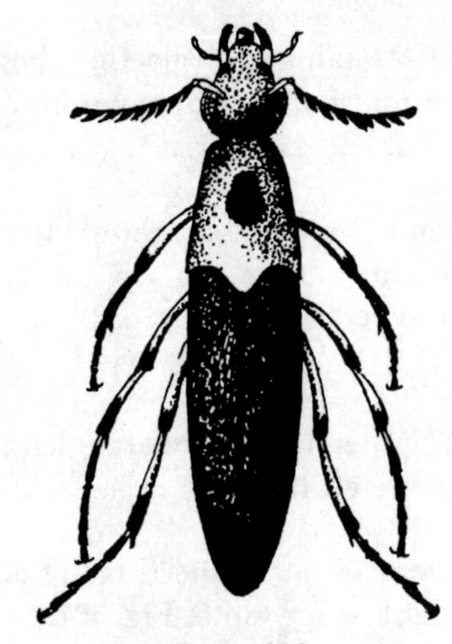

Figure 664

LENGTH: 6-10 mm. RANGE: Eastern half of U. S.

Head, thorax, under surface, femora and middle and hind tibiae reddish yellow; thorax with oval black spot on disk. Elytra either entirely black or with pale yellow at centers.

M. dimidiatum (Fab.) with about the same range and size, differs in that the front coxae are not separated by a narrow prosternum, and the insect is black with pale yellow elytra which are blackened at the tips. It may be often found on the flowers of Mountain Mint. Nine other species are in this genus in U. S.

1b **Elytra very small; thorax not covering the scutellum. Fig. 665.** ....
.... ***Rhipiphorus fasciatus*** **(Say)**

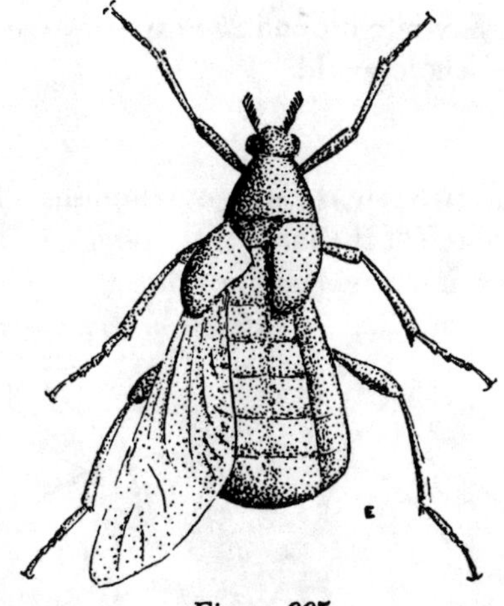

Figure 665

LENGTH: 4-6 mm. RANGE: Central to north eastern U. S.

Black. Legs and elytra yellowish. Vertex prominent with a median carina. Thorax conical with a median impressed line, densely punctate. Elytra widely separated and only ⅓ as long as abdomen. Adults are frequently taken from goldenrod blooms.

There are more than 30 species in this genus in our area.

Pelecotoma flavipes Melsh. at the other extreme has full length elytra. It is piceous-black with yellowish antennae and legs. It ranges throughout the eastern United States and measures 4.5 mm.

FAMILY 101. MELOIDAE
The Blister Beetles

Many of these medium to larger beetles produce *cantharidine* which blisters human skin and has been used medicinally. The antennae of the males often have some curiously enlarged segments. The adults are usually plant feeders but the larvae which pass through several stages frequently feed on the eggs of other insects. Black, gray and brown are the common colors. More than 300 species are known from the U. S. while around 2500 species are known for the whole world.

1a Elytra short and overlapping, leaving much of the large soft abdomen exposed; no inner wings. Fig. 666. .... *Meloe americanus* Leach

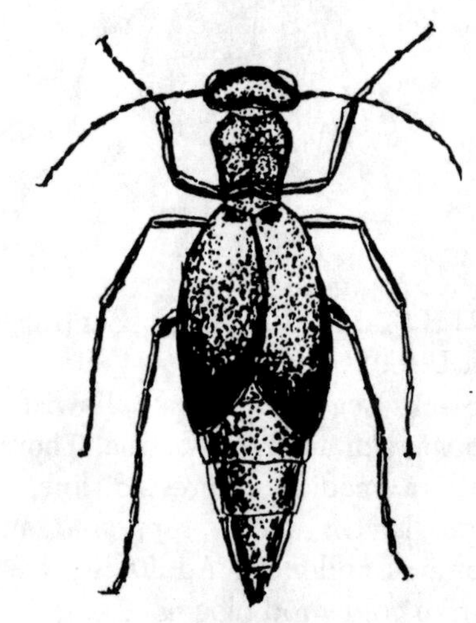

Figure 666

LENGTH: 15-24 mm. RANGE: eastern half of U. S.

Bluish black and shining throughout. Thorax about equal in length and width, coarsely and deeply punctured.

These "oil beetles" of which there are some 15 species known to our country are clumsy insects, and are usually found in early spring or late autumn. *M. impressus* Kby. is somewhat smaller and dull black. Range from Alaska across Can. and the northern half of U. S. Also Az. and N. M.

M. angusticollis Say, 12-15 mm long and widely distributed, has dark blue head and thorax with elytra and abdomen violet tinted. The thorax is about one-half longer than wide, but only two-thirds the width of the head. Range: So. Can. and northern U. S. There are about 20 species of Meloe in the U. S. and Can.

1b Elytra of normal length; wings usually present. 2

2a (1) Mandibles projecting beyond the labrum, acute at tip; upper blade of claw serrate. 9

2b Mandibles not reaching beyond the labrum, obtuse at tip; antennae not thickened towards tip. 3

3a (2) Antennae of moderate length; tarsal claws cleft to base. 4

3b Antennae rather short; tarsal claws with a short strong tooth. Fig. 667. .... *Tegrodera erosa* Lec. Soldier Beetle

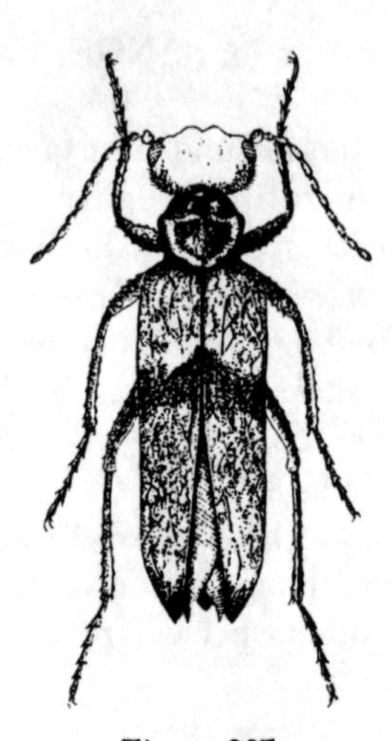

Figure 667

LENGTH: 16-30 mm. RANGE: the desert areas of our Southwest.

Head and prothorax red, the latter dusky; other parts shining black except the reticulations on the elytra which are golden.

T. latecincta Horn, having much the same range, is distinguished by the greater areas of black on the elytra. Both species are pests in alfalfa fields.

4a (3) **Antenna heavy at apex, segments spherical (moniliform).** **5**

4b **Antenna filiform.** **6**

5a (4) **Labrum with a deep notch. Fig. 668.** .... ***Lytta sayi* (Lec.)**

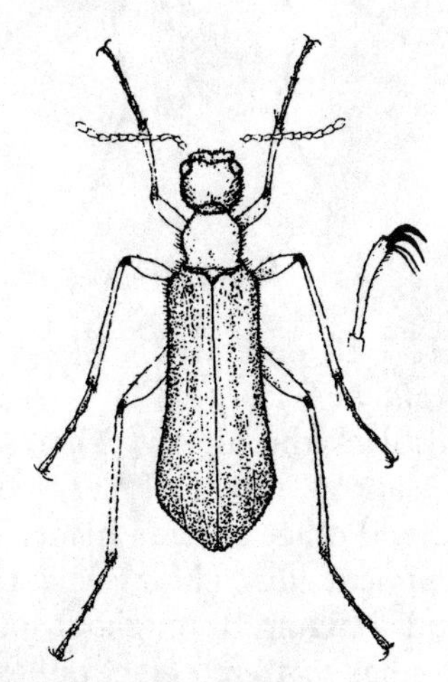

Figure 668

LENGTH: 15-18 mm. RANGE: Eastern U. S. and Canada.

Elytra glabrous, upper surface dull greenish, legs reddish-orange with black knees. On *Crataegus* and similar shrubs.

Pyrota engelmanni (Lec.), ranging from Indiana to the Southwest, is dull yellow with six black spots on the thorax and three black bars on the elytra. It attains a length of 20 mm.

Lytta aenea (Say) is a rather handsome beetle. It is 10-16 mm long. The head, thorax and under surface are greenish and covered with long gray hairs; elytra bronzed; antennae black; legs reddish-yellow with black knees. It is found on blossoms of apple and related plants and ranges over the Eastern half of the country.

5b **Labrum with only a shallow notch. Fig. 669.** .... ***Lytta nuttalli* (Say). Nuttall's Blister Beetle**

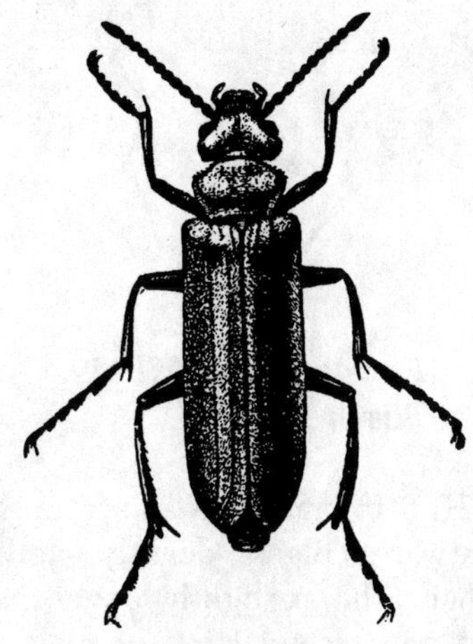

Figure 669

LENGTH: 16-28 mm. RANGE: Rocky Mt. region.

Metallic green often with purplish sheen; beautiful but very destructive.

Several other members of this genus are troublesome throughout the west. The "Infernal Blister Beetle" *L. stygica* (Lec.) 9-14 mm, ranges in spring along the West Coast while the "Green Blister Beetle" *L. cyanipennis* Lec. 13-18 mm is a western pest of leguminous crops. Both of these are green or bluish. *L. insperata* Horn 15-20 mm, which destroys sugar beets in California, is black. There are more than 50 species of *Lytta* in the U. S.

6a (4) **First antennal segment much enlarged. Fig. 670.**

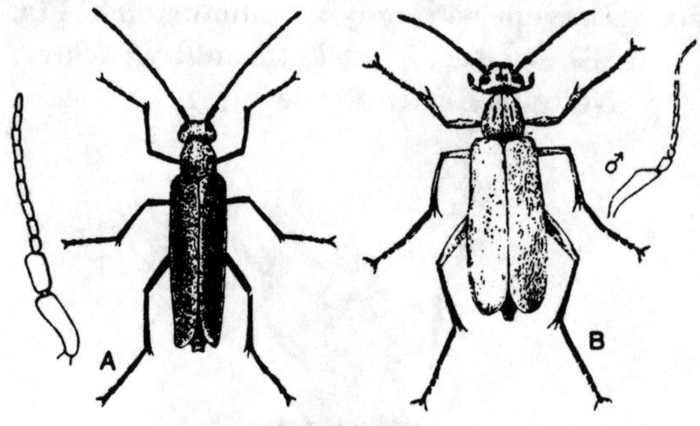

Figure 670

(a) ***Epicauta fabricii*** **Lec. Ash-gray Blister Beetle**

LENGTH: 8-15 mm. RANGE: from the Rocky Mts. eastward. Black, densely clothed with grayish hairs; thorax slightly longer than broad, both first and second joints of male antennae much thickened as pictured. A common garden pest.

(b) ***E. albida*** **(Say). Two-spotted Blister Beetle.**

LENGTH: 14-20 mm. RANGE: Kan., Tex., and N. M.

Gray or yellowish, the thorax marked with two nearly parallel lines; antennae of male as pictured. A serious pest of sugar beets, and garden crops.

E. torsa (Lec.), black, thinly clothed with grayish pubescence, 11-15 mm long, should be readily determined by the somewhat S-shaped first antennal segment of the males. In the females the second antennal segment is a bit longer than the third. It has been taken from Mass. to Fla. to Tex.

6b **First antennal segments of moderate length; mandibles short. 7**

7a **(6) Elytra wholly black. Fig. 671. .... *Epicauta pennsylvanica* (DeG.). Black Blister Beetle.**

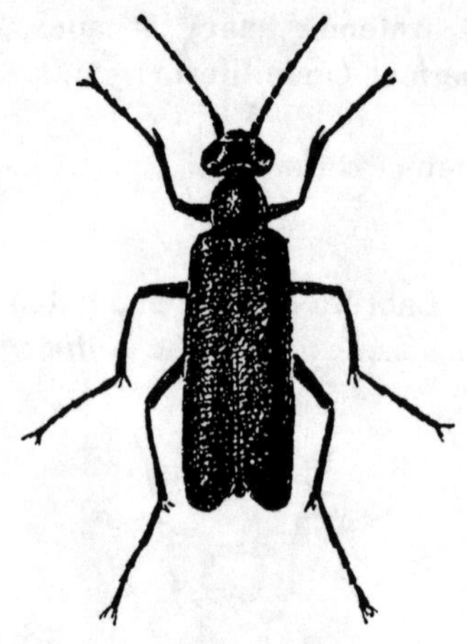
Figure 671

LENGTH: 7-13 mm. RANGE: Que. to Fla. west to Mont. and Tx.

Dull black throughout. Thorax quadrate with front angles rounded. Very common on goldenrod and other autumn plants.

E. puncticollis (Mann.) 7-11 mm in length and ranging throughout our Pacific coast states has scattered erect pubescence.

7b **Elytra clay-yellow with black stripes. Fig. 672. .... *Epicauta vittata* Fab. Striped Blister Beetle.**

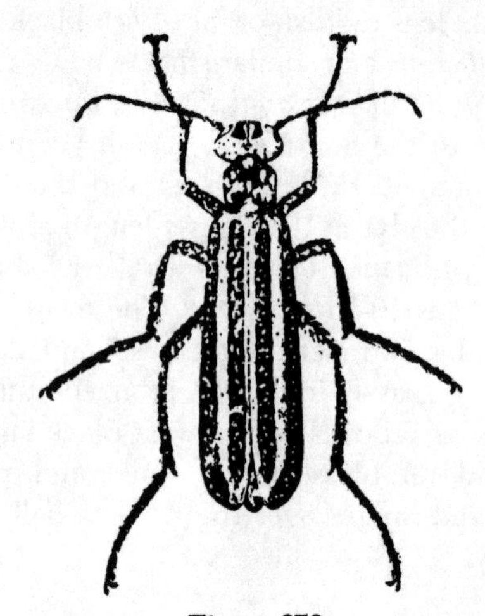
Figure 672

LENGTH: 12-18 mm. RANGE: Central and N.E. U. S. and Canada.

Under parts black, head, thorax and elytra clay-yellow; each elytron with 2 black stripes; head wider than thorax.

This "old fashioned potato bug" is a serious pest to gardens.

The Three-lined Blister Beetle *E. lemniscata* Fab. is distinguished from the former by a third black strip on the elytra. It ranges throughout our Southern States.

E. ferruginea (Say) is smaller, 6-9 mm. Clay-yellow hairs (sometimes gray) cover the black elytra. The other parts are black. Ranges from Idaho to Iowa south to Tex. and Az.

7c Elytra in part at least, gray. 8

8a (7) Elytra with numerous fine black spots. Fig. 673. .... *Epicauta maculata* (Say). Spotted Blister Beetle.

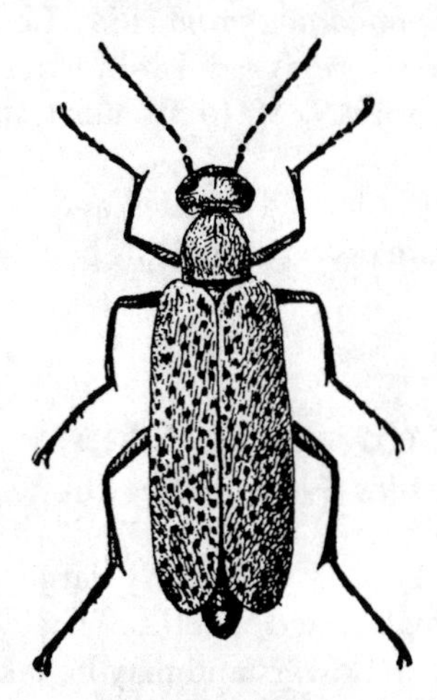

Figure 673

LENGTH: 10-14 mm. RANGE: west of Missouri River to the Rockies.

Entirely black but so covered with ash gray hairs that only numerous spots on the elytra show black. It is a pest of sugar beet, potato and clover.

E. pardalis Lec., 10 mm in length, is black with many white spots and short lines on its upper surface. It ranges throughout the southwest and into Mexico.

8b Elytra wholly gray or black, entirely margined with gray as pictured. Fig. 674.

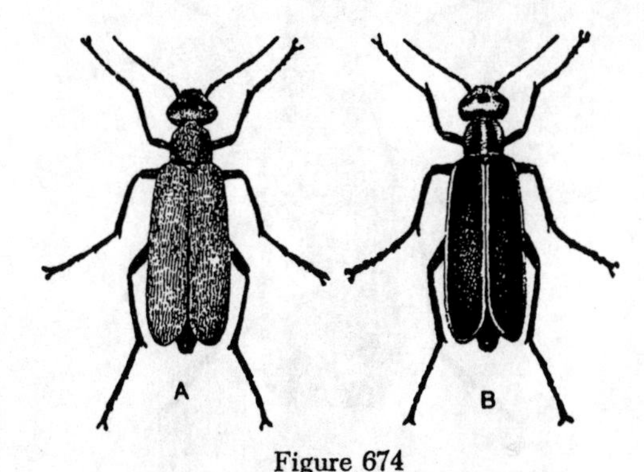

Figure 674

(a) *Epicauta cinerea* (Forst.). Gray Blister Beetle.

LENGTH: 10-18 mm. RANGE: Common in Coastal states, less so west to Iowa.

Robust. Black, clothed with gray hairs throughout. Two forms appear as in Fig. 674.

(b) *E. pestifera* Werner

LENGTH: 6-16 mm. RANGE: from the Atlantic to the Rockies, a pest of potatoes. The two species are separated by the shorter antennae of *cinerea* with a stout first segment. In this the hind tibial spurs are spiniform; in *pestifera,* the spurs are flattened, the outer one with a blunt tip.

E. atrata Fab. 8-11 mm, is black with fairly thick black pubescence with gray hairs making a marginal stripe and often one on the suture of the elytra. The head may be red back of the eyes, or with a small red spot in front, or it may be wholly black. It ranges from Texas to the Atlantic and into Canada. There are over 100 species of this genus in our territory.

9a (2) **Outer lobe of maxilla bristle-like and prolonged. Fig. 675. .... *Zonitis vittigera* Lec.**

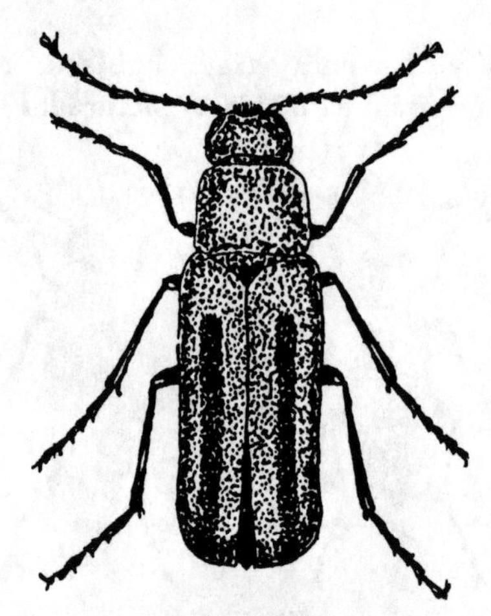

Figure 675

LENGTH: 9-11 mm. RANGE: Minn. and Mich. west to Neb. south to La. and Fl.

Reddish-yellow; elytra usually with broad discal black stripe; antennae, tibia and tarsi blackish.

Z. *sayi* Wick., similar in shape and size but wholly pale greenish-yellow, seems to confine itself to flowers of the tar weed, *Grindelia squarrosa*. It ranges largely west of the Missouri river into the Rocky Mt. area, but is found in prairie areas of Iowa and Minnesota.

Nemognatha punctulata LeC., measuring 8-11 mm, is orange-yellow, thinly covered with yellow hairs. Much of the underparts and of the antennae and tarsi is blackish. The front of the head is flattened and finely punctate. The sides of the thorax are straight or nearly so, its disk with scattered coarse punctations. The typical form is brownish-yellow with a black stripe on each elytron. Ranges from central Ill. and Ohio south to the Gulf and Fla.

9b Outer lobe of maxilla not prolonged. Fig. 676. .... *Zonitus bilineata* Say

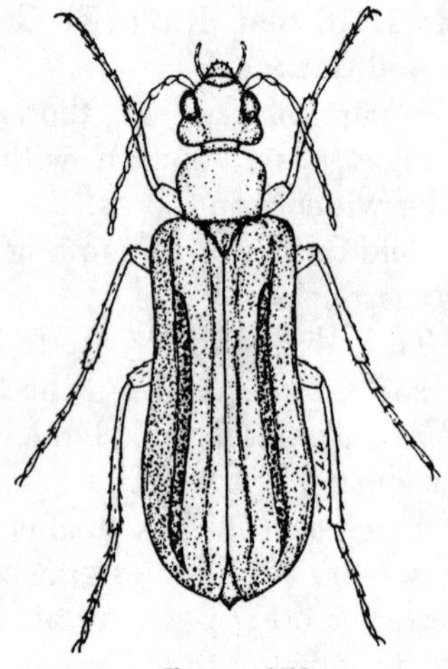

Figure 676

LENGTH: 7-9 mm. RANGE: from Conn. to Ont. to Id. to S. Cal. then to the upper South and the Midwest.

Robust, dull brownish-yellow; elytra yellowish-white each with a median blackish stripe not fully reaching base or apex.

Rhyphonemognatha rufa (LeC.), similar in size and shape, is reddish-brown throughout. It ranges from N. M. to Tx. north to Iowa and Ind.

Zonitis has 14 species and Nemognatha 26 in the area covered by this book.

FAMILY 102. ANTHICIDAE
The Ant-like Flower Beetles

The members of this fairly large family are mostly small sized beetles. They are often abundant on flowers and may be taken in large numbers with the sweeping net. Some species have a peculiar horn-like structure projecting forward on the thorax.

1a Thorax with horn-like part projecting forward over the head. See Fig. 677. 2

1b **Thorax normal, not as in 1a. 3**

2a **(1) Elytra pale with dark cross bar back of middle, horn broad with fine teeth on its sides. Fig. 677.**

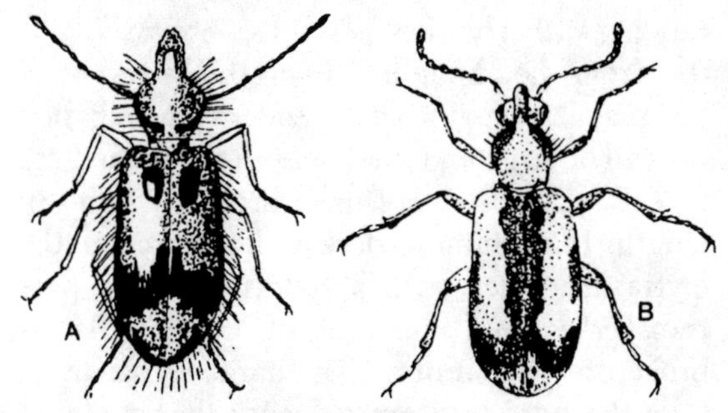

Figure 677

(a) ***Notoxus monodon*** **Fab.**

LENGTH: 2.5-4 mm. RANGE: coast to coast; abundant.

Dull brownish-yellow; thorax and elytra with piceous marks as pictured; elytra pubescent with both long and short hairs.

(b) ***N. anchora*** **Hentz**

LENGTH: 3-3.5 mm. RANGE: Atlantic coast west to Colorado.

Reddish-yellow with darker markings as pictured.

N. constrictus Csy., a similar species is known as the Fruit *Notoxus* in California and is common in orchards. Over 40 species in our area.

Mecynotarsus candidus Lec. (2 mm) differs from Notoxus in having the tarsi longer than the tibiae. It is uniform dull yellow and widely distributed. Eastern Canada and U. S.

2b **Elytra dark with oblique pale spots. Fig. 678.**

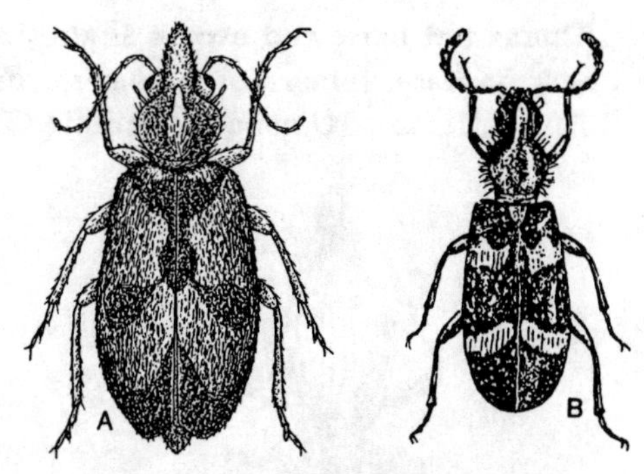

Figure 678

(a) ***Notoxus talpa*** **Laf.**

LENGTH: 3.5-4 mm. RANGE: the Rockies and eastward.

Thorax, antennae and legs dull reddish-brown; head, undersurface and elytra piceous; elytra with paler, somewhat pinkish spots as shown.

(b) ***N. bifasciatus*** **Lec.**

LENGTH: 3-4 mm. RANGE: Canada to Arizona.

Shining, piceous; legs and antennae reddish brown; elytra with pale bars as pictured.

N. murinipennis LeC. (3.5 mm) ranging in the eastern third of the United States has its head fuscous, thorax, legs and undersurface reddish-yellow and elytra purplish-black with prostrate gray pubescence.

3a **(1) Sides of mesosternum more or less abnormally dilated. 4**

3b **Sides of mesosternum straight, slightly oblique and coming to a point; not dilated. 6**

4a **(3) Thorax much constricted or narrowed behind the middle. 5**

4b **Thorax not narrowed except slightly so, near its base; femora club shaped. Fig. 679.** .... ***Omonadus floralis* (L.)**

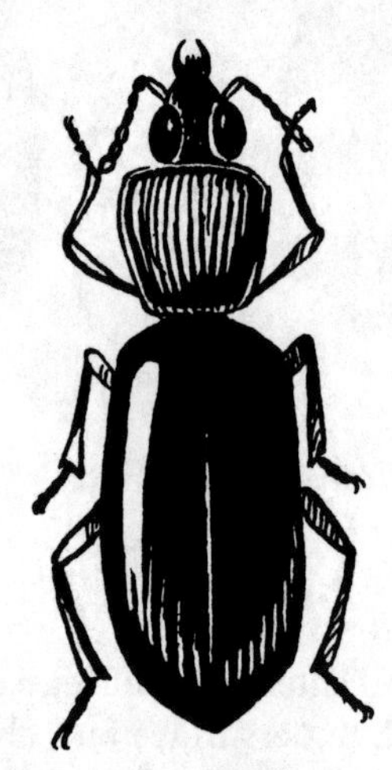

Figure 679

LENGTH: 3-3.5 mm. RANGE: A European species scattered world wide.

Reddish-brown, shining; head, abdomen and apical two-thirds of elytra piceous.

5a **(4) Antennae long and slender; constriction of thorax not extending across its upper side. Fig. 680.** ....
.... ***Ischyropalpus sturmii* (Laf.)**

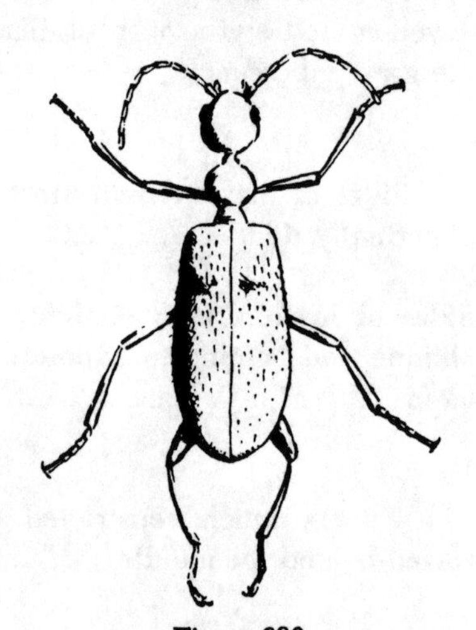

Figure 680

LENGTH: 2.5-3 mm. RANGE: Iowa to Conn.

Dark reddish-brown, covered with fine pubescence. Elytra piceous, the basal third reddish and marked with a crossband of longer gray pubescence. Sub-basal impressions of elytra distinct.

Malporus formicarius Laf. (3-4 mm) ranging rather widely is reddish-brown. It differs from the above in having the last joint of the maxillary palpi small and hatchet-shaped instead of large and dilated as in the preceding.

M. cinctus Say measures 3-4 mm in length. It is shining, dark reddish-brown; the elytra are black with a reddish base; a pale transverse band in front of the middle, is broken at the suture. The thorax is narrower than the head and much longer than wide. It belongs to the Central States.

A. obscurus Laf. measuring 3 mm or less, is shining black. Pale coarse hairs cover the basal third of elytra; other parts have fine pubescence.

The thorax is constricted near its base and is widest at its middle. Its known range is from Indiana eastward.

There are over 60 species of *Anthicus* in the U. S. and Can.

5b **Antennae thick, bead-like; constriction of thorax extending across the dorsal surface. Fig. 681.** ....
.... ***Tomoderus constrictus* (Say)**

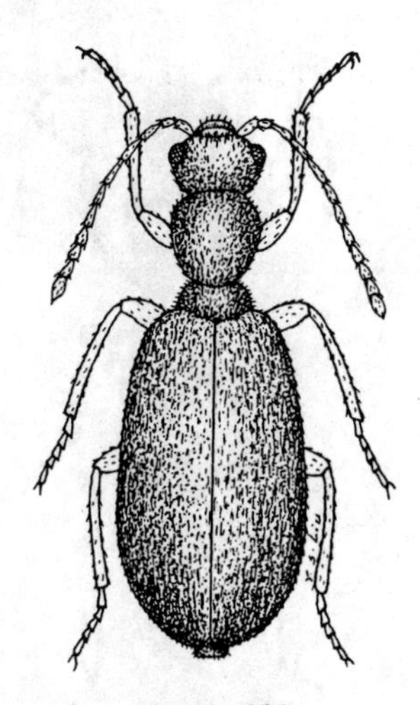
Figure 681

LENGTH: 2.5-3 mm. RANGE: Iowa to Florida and New York. Common.

Dark reddish-brown to piceous, shining; legs paler; elytra more piceous at apex.

Acanthinus myrmecops Csy., looks very much like an ant. It is almost 3 mm long, black with very short pubescence. The elytra have two light brown cross bars and an oblique band of silvery hairs. Ill. and Ind.

6a (3) Thorax evenly convex, moderately large. Fig. 682. .... *Anthicus cervinus* Laf.

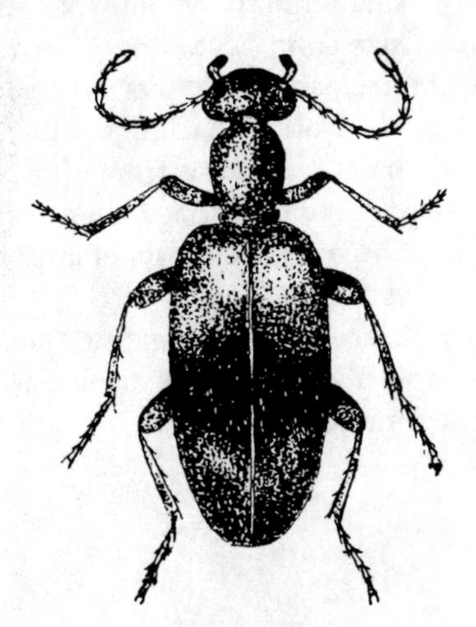
Figure 682

LENGTH: 2.4-2.7 mm. RANGE: U. S. and Canada. Common.

Reddish-brown, feebly shining with scant fine pubescence; antennae and legs dull yellow; elytra marked as pictured with piceous.

Sapintus pubescens Laf. (2.7-3 mm) and ranging through the eastern half of the U. S., has its thorax wider than long and elytra more than twice as wide as the thorax. It is black or dark brown with yellowish hairs.

6b Thorax abruptly sloping and flattened in front; head triangular. Fig. 683. .... *Amblyderus pallens* (Lec.)

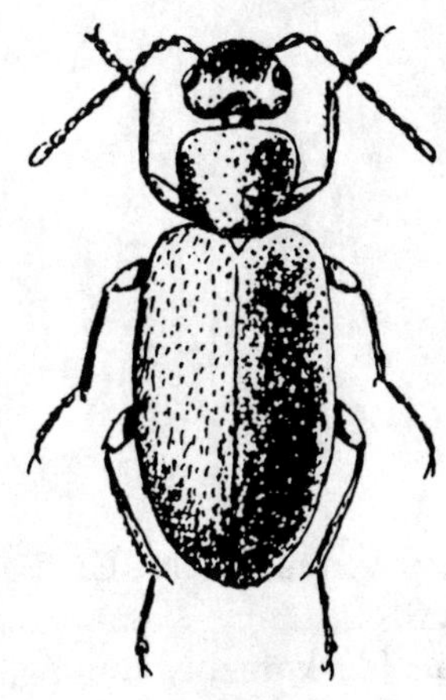
Figure 683

LENGTH: 2.7-3 mm. RANGE: Eastern half U. S.

Dull pale yellow throughout and covered densely with short coarse yellowish pubescence. Its color makes it practically invisible in its sandy habitat.

FAMILY 103. PEDILIDAE
The False Ant-like Flower Beetles

These rather small, usually black elongate cylindrical beetles number some 50 known species in North America. They were formerly included with the Anthicidae. They may be frequently found on the flowers of various plants, sometimes under leaves and, of course, are often taken in the sweeping net.

1a **Neck wide; eyes large. Fig. 684.** .... ***Pedilus labiatus*** **(Say)**

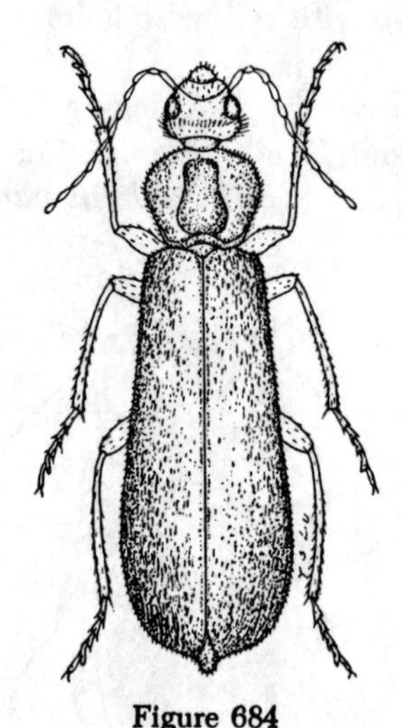

Figure 684

LENGTH: 6-8 mm. RANGE: Eastern two-thirds of U. S.

Piceous black, feebly shining; palpi, clypeus and labrum and first two segments of antennae pale yellow; thorax marked as pictured with reddish-yellow.

P. lugubris (Say) 6-8 mm, is wholly black except the first two antennal segments and the labrum which are reddish-brown. It ranges through Central Canada and the United States.

P. terminalis (Say) (5-7 mm) has a wholly red thorax. Range: Eastern U. S. and Canada.

The genus *Pedilus* has some 30 species in the U. S. and Canada.

1b **Neck narrow; less than 5 mm long. Fig. 685.** .... ***Macratria confusa*** **Lec.**

Figure 685

LENGTH: 4-4.5 mm. RANGE: Iowa to Conn.

Dark gray thickly covered with yellowish silky hairs; antennae and legs dull yellow.

Stereopalpus mellyi Laf. (7-8) ranges from Iowa to the Atlantic. The head is the same width as the thorax. The color is fuscous, modified with a dense grayish pubescence.

FAMILY 104. EUGLENIDAE
The Ant-like Leaf Beetles

This is a small family of ant-shaped beetles varying in size from 1.5 to 3 mm. In color they vary from testaceous to piceous. Some bear colored spots or other markings. They are covered with hairs that vary from short to long. The 11-segmented antennae have various modifications. The eyes are hairy, emarginate, and coarsely faceted.

Adults are taken by beating the leaves of trees and shrubs. About 40 species are known in North America.

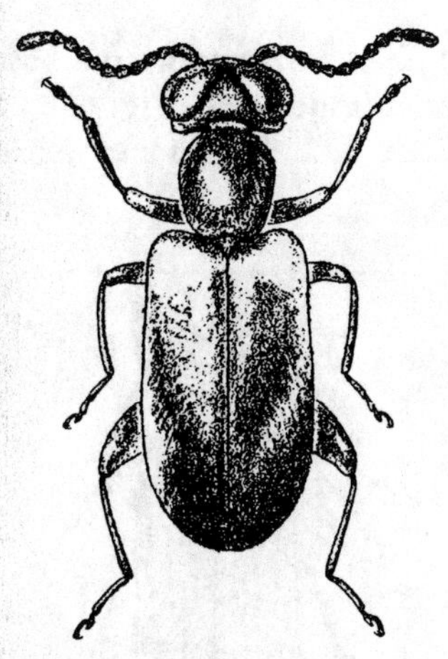

Figure 686

Figure 686 *Elonus basalis* (LeC.) LENGTH: 3-3.5 mm. RANGE: Ill. to Va. south to Fla. Brownish-black with base of elytra reddish-orange; legs pale to dark; elytra coarsely, densely punctate. In Ind. taken by sweeping herbage in woods.

Zonantes fasciatus (Melsh.) Length: 2-3 mm. Range: N. Y. to Fla. west to Ind. Shiny black with orange-red humeral and basal spots on the elytra; appendages yellow; pubescence rather long and dense.

Phomalis brunnipennis (Lec.) with length of 2 mm or less is dark brown. The head, thorax and middle and hind legs are piceous with short dense yellowish-gray pubescence. Range: Ind. eastward. Beaten from hickory in Ind.

FAMILY 105. STYLOPIDAE
The Parasitic Beetles

For many years these insects have been considered to make up an order of insects distinct from the beetles. However, today they are considered to be beetles that are related to the Meloidae and the Rhiphoridae.

The males are black, resemble flies, and range in size from .5 to 4 mm. They have flabellate antennae. The females are sac-like creatures capable only of digesting food and producing eggs. These females are parasitic on Hemiptera, Homoptera, Hymenoptera, and Orthoptera. The female while attached to its host has its eggs fertilized. These then develop within the body of the female. Later the larvae emerge, seek a host insect, and continue their development. The adult male flies away seeking another female to mate with.

About 60 species of this family are found in North America.

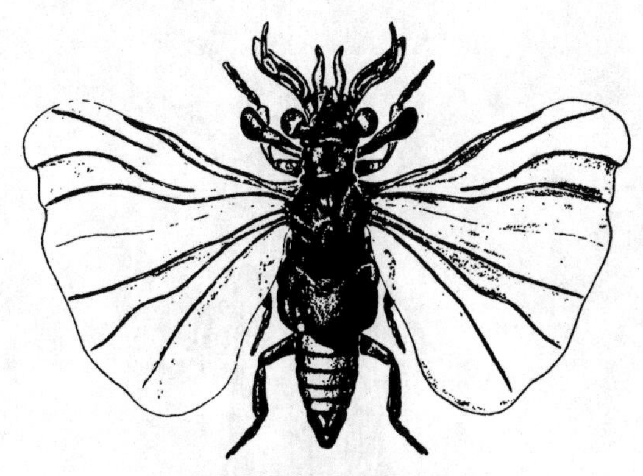

Figure 686a

FAMILY 106. CERAMBYCIDAE
The Long-horned Woodborers

Here is a great family of some 20,000 widely scattered species. The beetles usually possess long-to-very-long antennae. They vary much in size and color, some tropical species attaining a length of several inches. Brilliant metallic and pigmented colors are common.

1a **Pronotum with a margin; labrum attached at sides. 3**

1b **Pronotum without a margin; labrum free. 2**

2a (1) Front tibiae with a groove, palpi pointed at tip. 37

2b Front tibiae without a groove; palpi not pointed at tip. 6

3a (1) Antennae reaching beyond the thorax; segments not all alike. 4

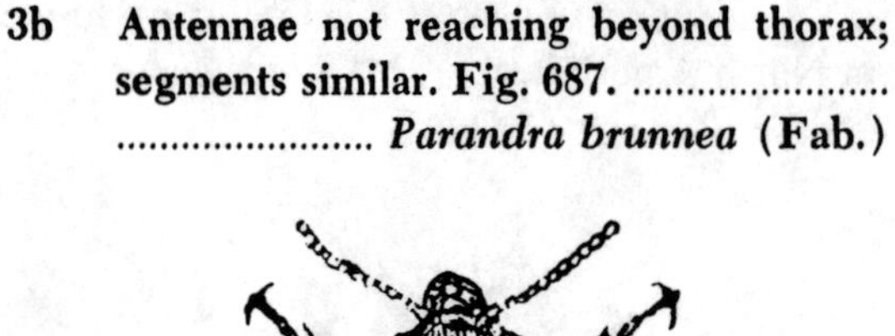

3b Antennae not reaching beyond thorax; segments similar. Fig. 687. ....
.... *Parandra brunnea* (Fab.)

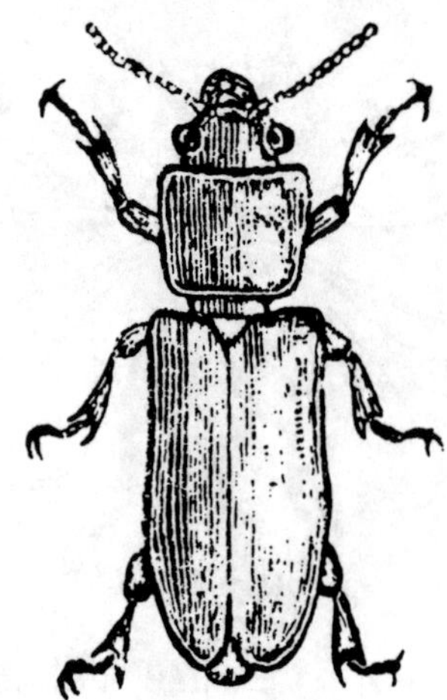

Figure 687

LENGTH: 9-18 mm. RANGE: Eastern and Central N. America.

Somewhat flattened, shining. Mahogany brown or reddish, the mandibles darker. Elytra margined, not striate.

This beetle which bores in soft maple and other trees does not look much like a long-horned woodborer.

P. marginicollis Schaef. is found in So. Cal. and Az. and *P. polita* Say ranges from Ind. to Central America. This last species has the eyes entire.

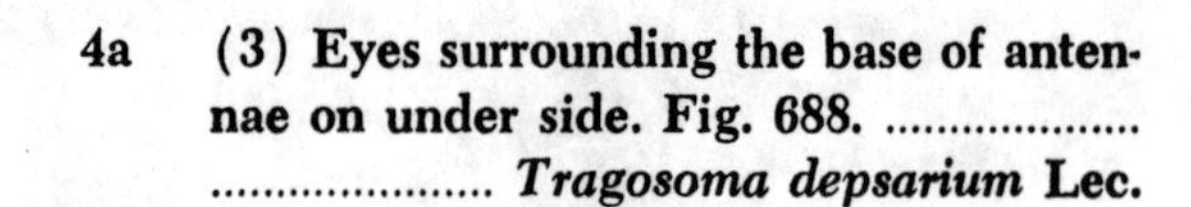

4a (3) Eyes surrounding the base of antennae on under side. Fig. 688. ....
.... *Tragosoma depsarium* Lec.

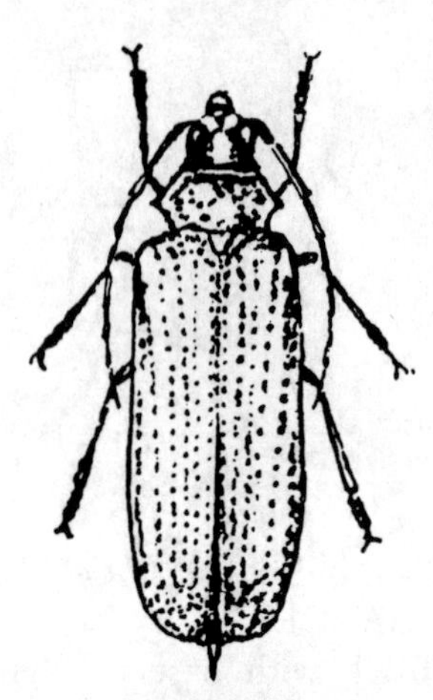

Figure 688

LENGTH: 30-34 mm. RANGE: Northeastern and Western United States and Canada.

Elongate, heavy, shining. Brown; antennae slender; elytra with raised lines and punctures. The eggs are said to be laid in pine stumps.

Two other species occur in the West.

4b Base of antennae not surrounded beneath by eyes; margin of pronotum with 3 teeth. 5

5a (4) Form rather slender; antennae with 11 segments. Fig. 689. ....
.... *Orthosoma brunneum* (Forst.)

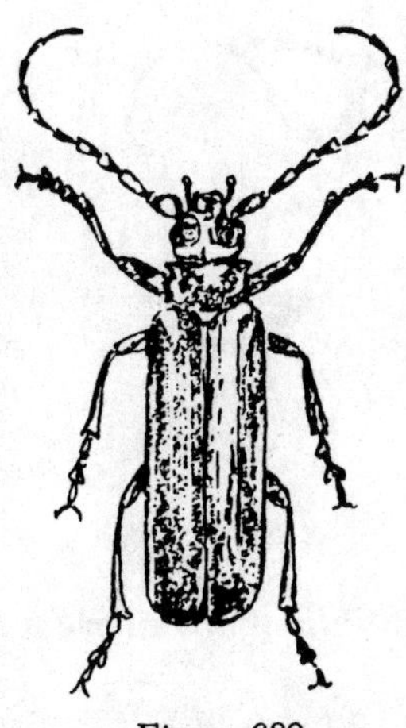

Figure 689

LENGTH: 22-40 mm. RANGE: Eastern U. S. and Canada.

Shining. Chestnut brown; head deeply indented between the antennae. Elytra each with three raised lines, surface with closely placed fine punctures.

Bores in pine and some of our broad leaves.

5b **Form broader; antennal segments mcre than 11. Fig. 690. .... *Prionus laticollis* (Drury)**

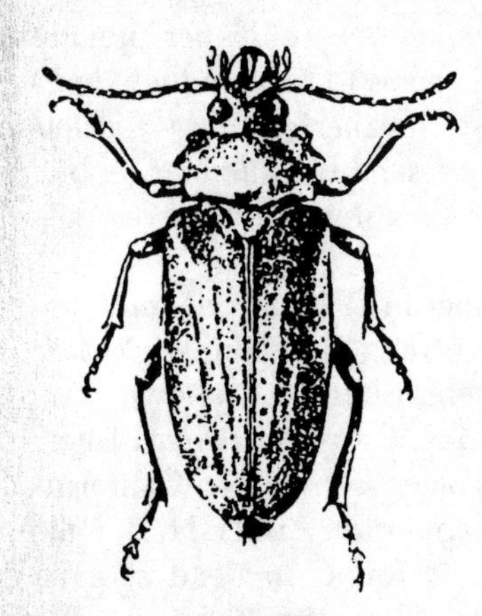

Figure 690

LENGTH: 22-48 mm. RANGE: Eastern half United States and Canada.

Shining. Blackish. Antennae of male longer than body. Its common name is "Broad-necked Root Borer."

"California Prionus," *P. californicus* Mots. 40-60 mm long, shining dark reddish brown is a native of our west coast. Its thorax has a much narrower margin than in the preceding.

"Tile-horned Prionus," *P. imbricornis* (L.), 20-47 mm and ranging from Kansas, eastward, is readily told by the 16-20 overlapping segments in its antennae. It is dark reddish-brown.

There are 12 species of *Prionus* in our area.

6a **(2) Base of antennae at least partly surrounded by the eyes; front coxae not conical. 24**

6b **Base of antennae not surrounded by the eyes. 7**

7a **(6) Front coxae conical. 13**

7b **Front coxae transverse, not conical. 8**

8a **(7) Ligula membranous, eyes finely granulate. 9**

8b **Ligula of chitin, base of pronotum not notched. 12**

9a **(8) Elytra with narrow raised white lines; femora club-shaped; deep median groove in thorax. Fig. 691. .... *Physocnemum brevilineum* (Say)**

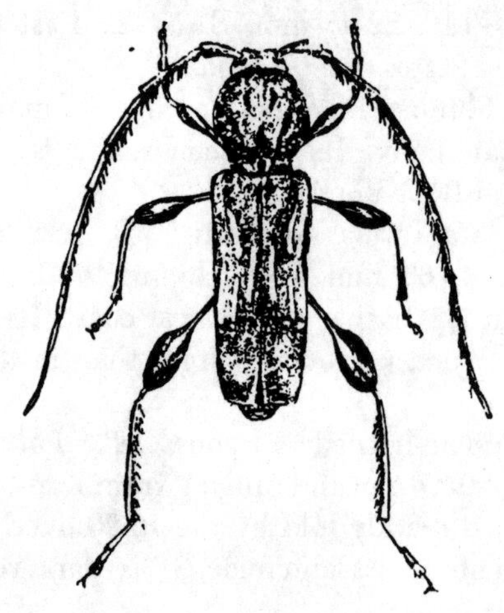

Figure 691

LENGTH: 12-16 mm. RANGE: Eastern United States and Canada.

Black. Elytra with bluish tinge, and raised whitish lines as shown. Bores in elm.

Ropalopus sanguinicollis (Horn), a related species, lacks the raised white lines on the elytra of the above. The thorax is very short and is colored red, while the remainder of the beetle is black. It measures 15-19 mm and ranges from Canada to Pennsylvania and Ohio.

9b Elytra without distinctly elevated whitish lines. 10

10a (9) Labial palpi much shorter than the maxillary palpi; thorax rounded. 11

10b Palpi about equal; mesosternum obtusely triangular. Fig. 692.

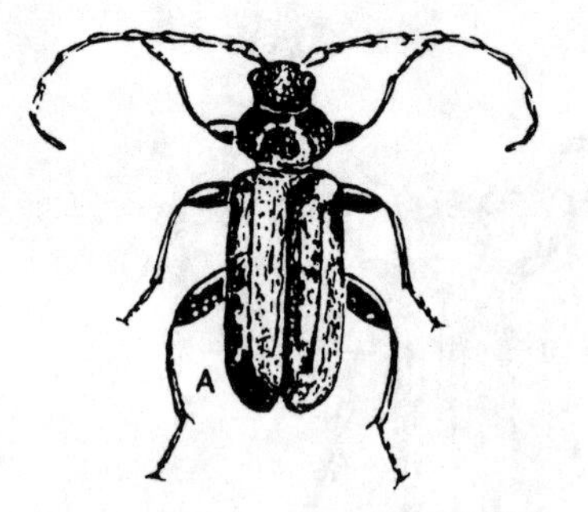

Figure 692

(a) *Callidium antennatum* Newn.

LENGTH: 13-14 mm. RANGE: United States.

Blackish-blue. Thorax and elytra thickly punctate, the punctures on the elytra being coarse. Feeds in pine.

(b) *Semanotus lignea* (Fab.)

LENGTH: 7-12 mm. RANGE: Coniferous belt of N. America.

Black. Elytra with yellow or reddish markings. The front coxae are separated in contrast to those of the former which touch.

Several other species of *Callidium* may be mentioned. *C. violaceum* L. 10-13 mm has blue elytra and thorax but is distinguished from *antennatum* by deeper punctures on the thorax. It seems to range widely in Europe and Northeastern America. *C. californicum* Csy. also closely resembles *antennatum* but is smaller with shorter antennae. It bores in junipers in California.

Fisher in 1920 named three other species of western range. *C. juniperi* Fisher 10 mm is dark greenish-blue and known from Ok., Ut., and N. Mex. *C. pseudotsuga* Fisher 10-13 mm and dull black is known in California and Oregon. *C. sequarium* Fisher 11-12 mm and wholly black, is found in dead sequoia limbs in California. Altogether there are 17 species of this genus in the U. S. and Can.

11a (10) Length 5-9 mm. Fig. 693.

(a) *Phymatodes varius* Fab.

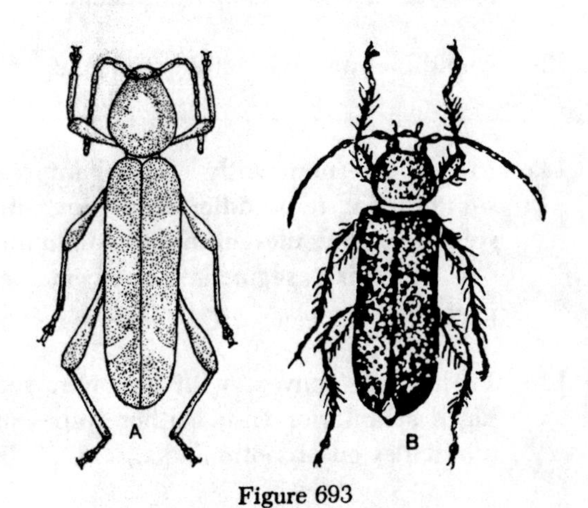

Figure 693

LENGTH: 6-9 mm. RANGE: Eastern N. America to Ut. and Az.

Black, with prostrate pubescence. Elytra marked with white as shown and often reddish-brown at base.

(b) *P. amoenus* (Say)

LENGTH: 5-8 mm. RANGE: from Kansas east and in Canada.

Reddish-yellow; elytra shining blue. The larvae bore in grape stems.

11b Length over 9 mm. Fig. 694.

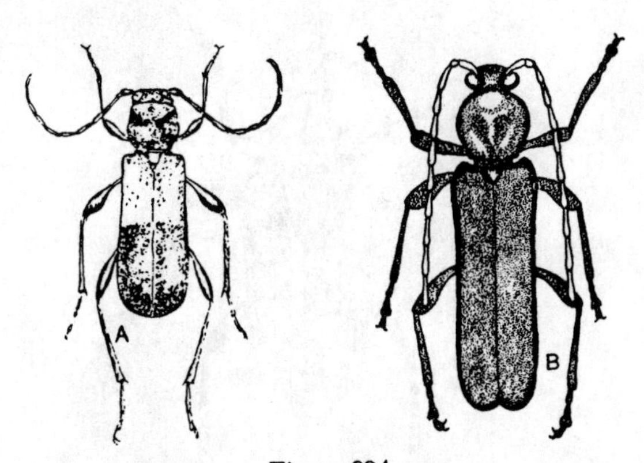

Figure 694

(a) *Phymatodes dimidiatus* (Kby.)

LENGTH: 9-13 mm. RANGE: Alas to N. Cal., east to Lake Superior.

Head, thorax and back part of wings black. Basal half of wings and the legs brownish. Feeds on spruce.

(b) *P. testaceus variabilis* (L.)

LENGTH: 12-13 mm. RANGE: Can. to Fla., west to Iowa. Also Europe and N. Africa.

Head blackish; thorax reddish-yellow, the center often darker. Elytra yellow, sometimes with blue at middle. Femora yellow or blue. Oak and other hardwoods are its food plants.

P. blandus Lec. is orange-yellow with metallic blue or purple elytra; 5-8 mm. Range: Wash. to Cal. to Nev. Feeds on willow and poplar.

P. lecontei Lins. Brown to black, the head and pronotum with erect pubescence; 8-15 mm. Pacific Coast from B. C. to Cal. Lives in oak. There are 25 known *Phymatodes* in the U. S. and Canada.

12a (8) Eyes divided; seemingly four. Fig. 695. ...... *Tetropium cinnamopterum* Kby.

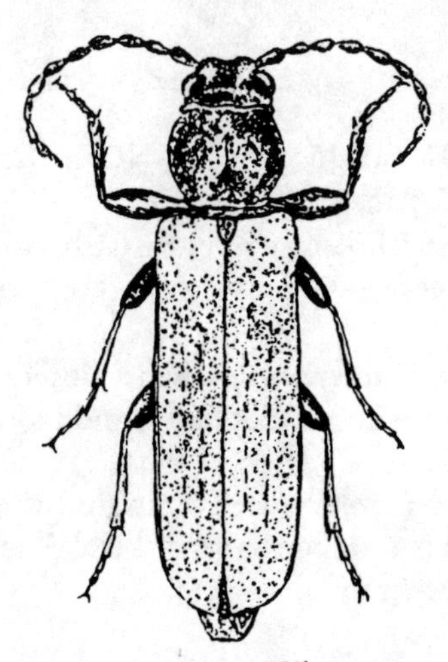

Figure 695

LENGTH: 12-14 mm. RANGE: United States, Canada and Alaska.

Slaty or sooty brown; under surface and often sides of elytra reddish-brown. Two slightly raised lines on each elytron. Feeds on conifers.

Two other members of the genus, *T. abietis* Fall 13-17 mm, brown, and *T. velutinum* Lec. 9-19 mm, blackish or brown, damage conifers. The latter ranges from Montana and British Columbia to Northern California, while *abietis* lives in the Pacific Coast States.

12b Eyes not wholly divided though frequently deeply notched. Fig. 696. .... ***Asemum striatum*** **(L.)**

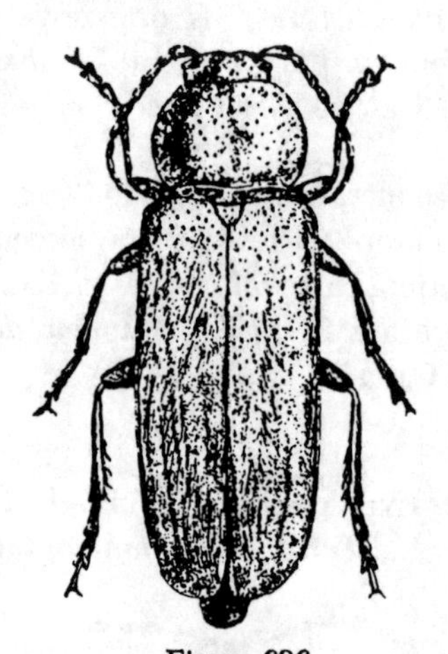

Figure 696

LENGTH: 12-15 mm. RANGE: Coniferous belt of N. Amer., Eur. and Asia.

Blackish-brown; elytra with two prominent ridges. It seems to feed on pine and spruce.

The genus *Arhopalus* is closely related. The eyes are larger, not hairy, and more coarsely granulate.

A. asperatus (Lec.) ranging throughout the West is a slender almost black beetle measuring 20-27 mm.

13a **(7) Mandibles with a fringe on inner margin. 14**

13b **Mandibles normal; without a fringe. .. 23**

14a **(13) Prosternum with a broad impression crossing its middle; pronotum with spines or tubercules on the side; hind tarsus with first segment pubescent beneath. 15**

14b **Prosternum convex, with an impressed place at anterior end; neither spines or tubercules on pronotum. 19**

15a **(14) Closed cell in anal region of wing. 16**

15b **Wing not as in 15a. 18**

16a **(15) Antenna slender and nearly uniform in diameter. 17**

16b **Antennal segments 5-11, much heavier than the preceding ones. Elytra with prominent ridges. Fig. 697. ....** ***Rhagium inquisitor*** **(L.)**

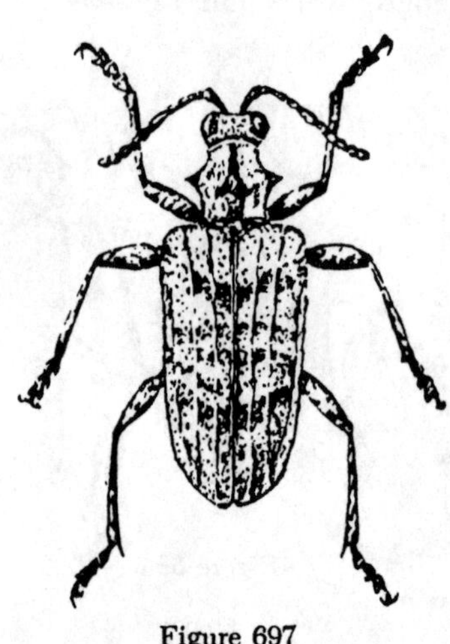

Figure 697

LENGTH: 13-18 mm. RANGE: Boreal N. America, Europe and Asia.

Robust. Black with markings of gray and reddish-brown pubescence as pictured. Each elytron with three elevated lines; intervals with scattered coarse punctures. The large white larvae burrow within the bark and sapwood of many conifers.

17a (16) Eyes prominent, large, coarsely granulate. Fig. 698. .... *Centrodera decolorata* (Har.)

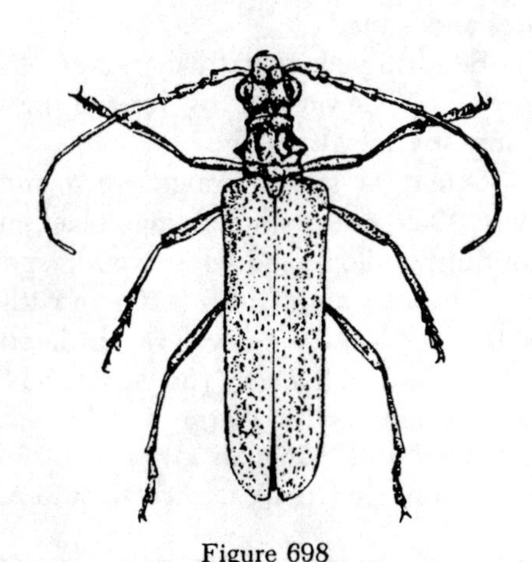

Figure 698

LENGTH: 25-28 mm. RANGE: Eastern half United States and Canada.

Reddish-brown with scattered pubescence. Elytra truncate at tip. Bores in beech and maple.

C. sublineata Lec. with a length of 14 mm and dark piceous, ranges from Pennsylvania to North Carolina, while *Evodinus monticola* Rand. is black with dull yellow elytra which bear black tips and four black spots near the suture. It measures 8-13 mm and ranges out of Canada to Wisconsin and North Carolina.

17b Eyes finely granulated, large, with notch in margin. Fig. 699. .... *Anthophylax attenuatus* (Hald.)

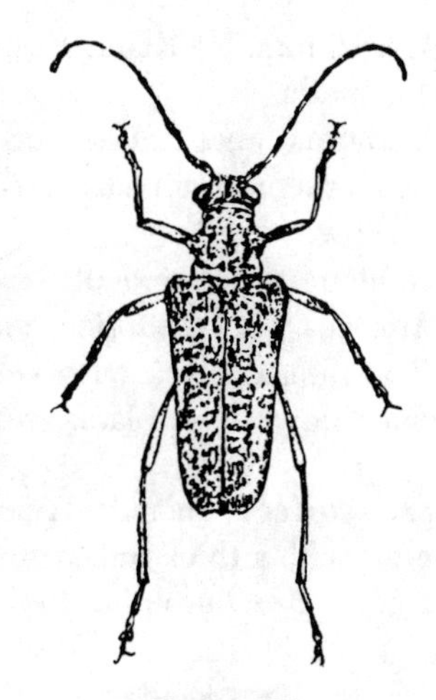

Figure 699

LENGTH: 15-17 mm. RANGE: Northeastern United States.

Elytra olive, mottled with grayish pubescence; head and thorax black with scattered golden hairs. Thorax with depressed line at middle. Lives in beech.

A. cyaneus Hald. measuring 12-16 mm is metallic blue to coppery and ranges from N. Amer. to Ga.

18a (15) Upper surface polished, shining; but slight if any pubescence. Fig. 700. .... *Gaurotes cyanipennis* (Say)

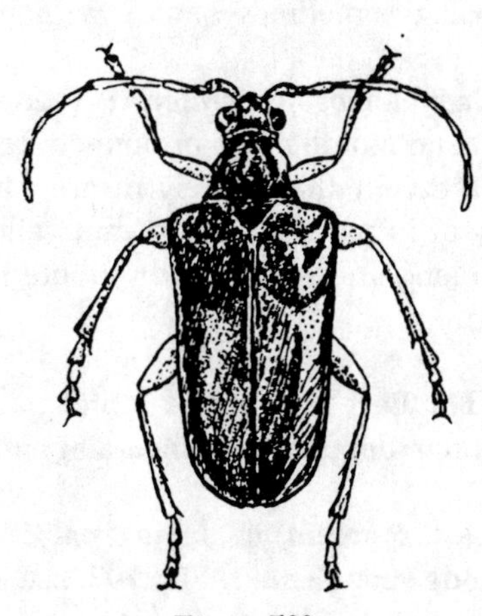

Figure 700

LENGTH: 9-10 mm. RANGE: Eastern United States and Canada.

Head, thorax and under surface black; elytra bluish-green; antennae and legs pale brownish-yellow.

Feeding on pines and other conifers in Boreal America is *Judolia montevagans* (Coup.) 8-12 mm, is black with yellow elytra, each of which have three black areas.

18b Upper surface heavily punctate and covered with a thick pubescent coat. Fig. 701. .... *Brachysomida bivittata* (Say)

Figure 701

LENGTH: 6-9 mm. RANGE: Eastern half United States.

Dull. Pale brownish-yellow; thorax and elytra marked in black as shown. Highly variable, being sometimes almost without markings, all dark.

Gnathacmaeops pratensis (Laich.) has its front and mouth much prolonged. It is wholly black except that the elytra are sometimes entirely or in part fuscous or dull yellow. It is 6-8 mm long and is widely distributed.

19a (14) Last segment of labial palpi short and wide; usually truncate at end. 20

19b Last segment of labial palpi slender; body very slender. Fig. 702. .... *Strangalia luteicornis* (Fab.)

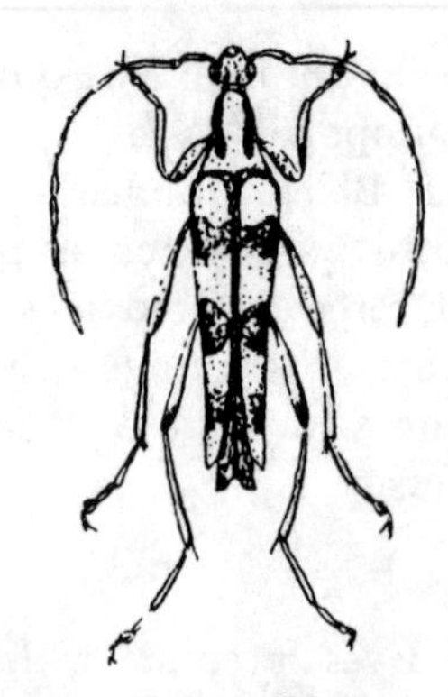

Figure 702

LENGTH: 9-13 mm. RANGE: Eastern United States and Canada.

Reddish-yellow with markings of black as pictured. The very narrow form of this genus attracts special attention.

Others of similar range are S. *famelica* Newn., 12-14 mm with antennae black instead of reddish-yellow and elytra with spots on sides, S. *acuminata* (Oliv.) 8-10 mm with head and thorax black; elytra yellow bordered with blackish, and S. *bicolor* (Swed.), 12-14 mm, reddish-yellow, with elytra wholly black; S. *sexnotata* Hald. 8-13 mm elytra with 6 black spots is common from Va. to Fl. west to Az.

20a (19) Apical antennal segments with definite pore-bearing spaces. Fig. 703.

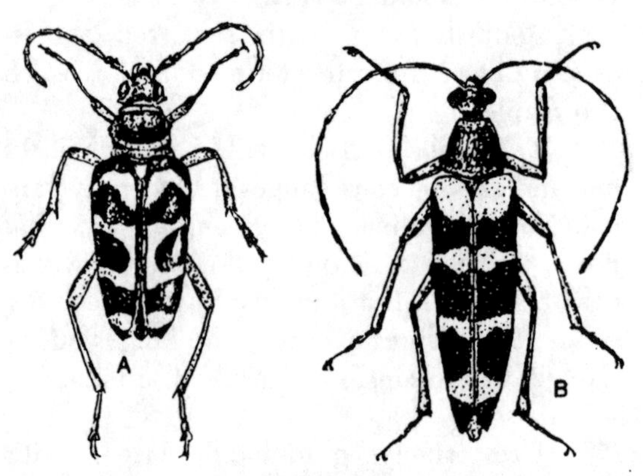

Figure 703

(a) *Typocerus zebra* (Oliv.)

LENGTH: 15-17 mm: RANGE: Eastern United States.

Upper surface black and golden yellow; under surface with dense golden pubescence; legs yellowish-red.

(b) *T. velutina* (Oliv.)

LENGTH: 10-14 mm. RANGE: Eastern United States.

Head, thorax and antennae black; elytra yellow, marked with reddish-brown as pictured; each puncture bearing a prostrate yellow hair.

There are 13 other *Typocerus* species in the U. S. and Can.

20b Antennae without pore-bearing areas. 21

21a (20) Hind angles of prothorax elongated. 22

21b Hind angles of prothorax not lengthened. Fig. 704. .... *Brachyleptura vagans* (Oliv.)

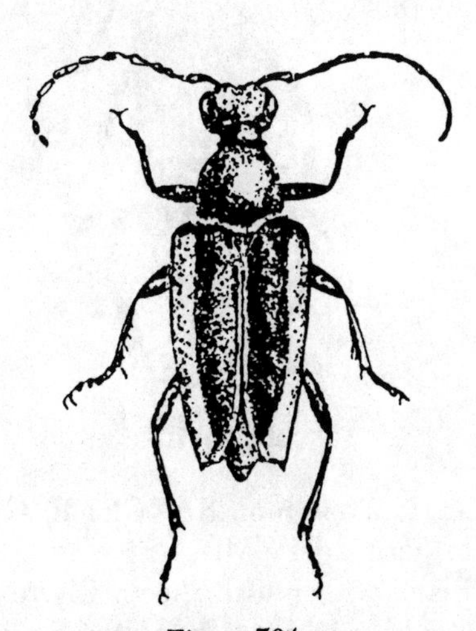

Figure 704

LENGTH: 9-12 mm. RANGE: Eastern United States and Canada.

Black, usually with yellowish or reddish margin on elytra; sometimes with elytra wholly black or entirely yellow.

B. rubrica (Say), 10-16 mm, has reddish elytra and abdomen in male. The abdomen of female is black as is also all the remainder of the beetle.

22a (21) Thorax with sides much rounded. Fig. 705. .... *Strophiona nitens* (Forst.)

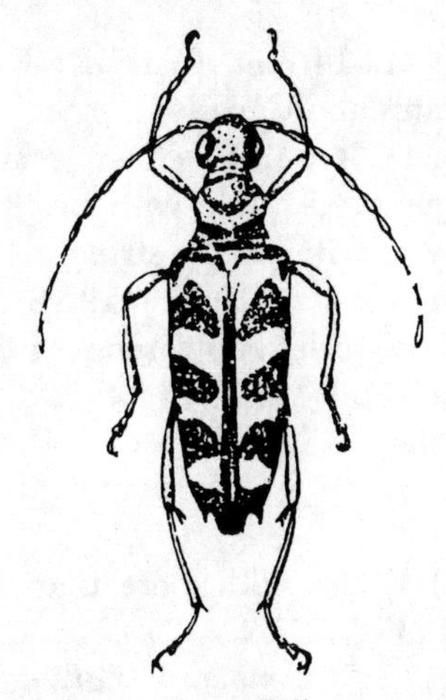

Figure 705

LENGTH: 10-13 mm. RANGE: Eastern half United States and Canada.

Black, marked with golden yellow pubescence; antennae dark reddish-brown and legs pale reddish-yellow.

In the West, *S. laeta* (Lec.), 10-13 mm golden yellow with three fairly wide black bands crossing the elytra, is abundant, breeding in oaks.

22b Thorax with sides not much rounded in front of middle. Fig. 706. .... *Leptura subhamata* Rand.

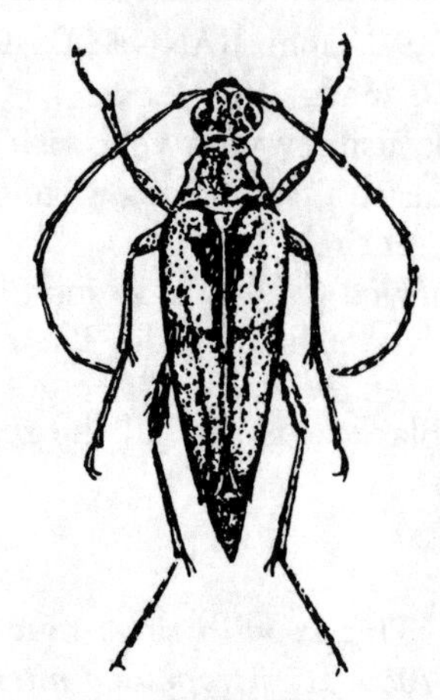

Figure 706

LENGTH: 12-14 mm. RANGE: Eastern half United States and Canada.

Female reddish-yellow, marked with black as pictured. Male with black head and thorax; elytra with yellow stripe.

L. obliterata (Hald.) 12-18 mm, black and pale brownish-yellow, ranges from B. C. to Cal., to Id. and Mont. 12 other *Leptura* are found in the U. S. and Can.

23a (13) Elytra with more than apical half blue. Fig. 707. ....
.... *Desmocerus palliatus* (Forst.)

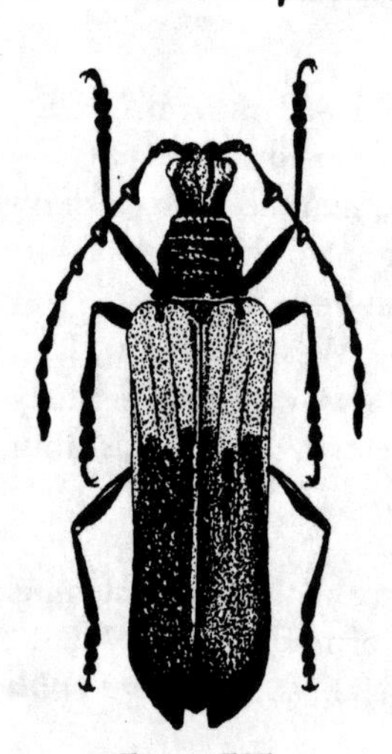

Figure 707

LENGTH: 17-23 mm. RANGE: Eastern United States and Canada.

Robust. Dark metallic blue. Elytra marked as pictured with golden yellow. This is a small genus of very attractive beetles. The other two species are western in range. All feed on elderberry, the larvae boring in the living stems. The adults congregate on the flowers.

Two species closely related to the above, yet appearing quite different, are of interest. In both, the elytra are very short and do not cover the wings. *Necydalis mellitus* Say 13-21 mm and found from Indiana eastward, is black and dull yellow with the tiny elytra reddish-brown or brownish-yellow.

Ulochaetes leoninus Lec. 20-26 mm known from British Columbia to California is black and yellow; the long, densely placed yellow hairs on the thorax give it the appearance of a bumblebee.

23b Elytra wholly golden orange or as pictured. Fig. 708. ....
.... *Desmocerus auripennis* Chev.

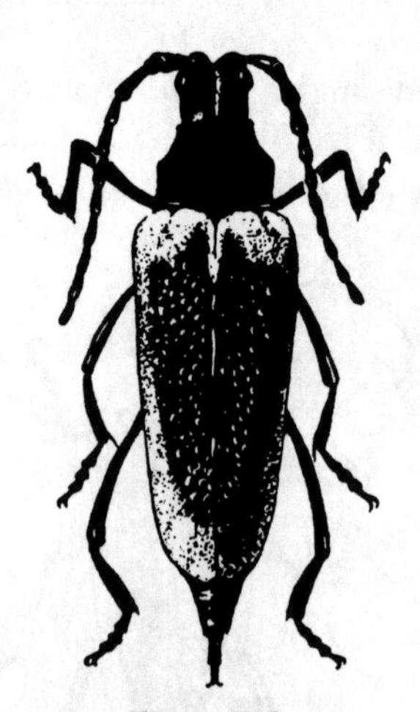

Figure 708

LENGTH: 23-28 mm. RANGE: B. C. to Cal. and northern Rocky Mts.

Black with bluish sheen. Elytra of male wholly golden orange; female with mid-dorsal variable greenish or bluish mark.

D. californicus Horn, has the elytra margined orange in both sexes, Coastal and Central Cal.

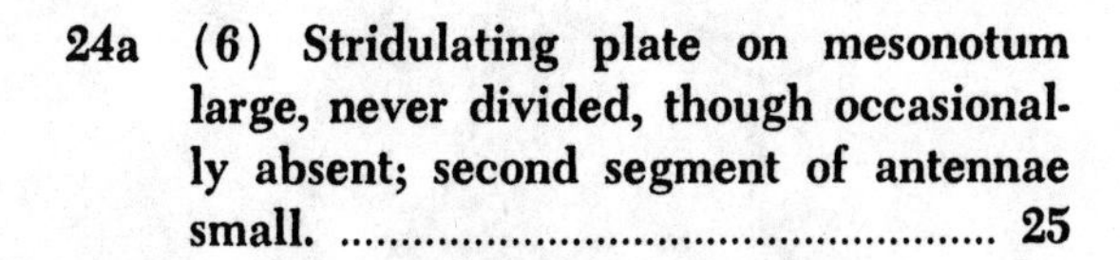

24a (6) Stridulating plate on mesonotum large, never divided, though occasionally absent; second segment of antennae small. 25

24b Stridulating plate on mesonotum divided by a smooth excavation; eyes nearly cut in two; second segment of antennae long. Fig. 709. .... *Atimia confusa* (Say)

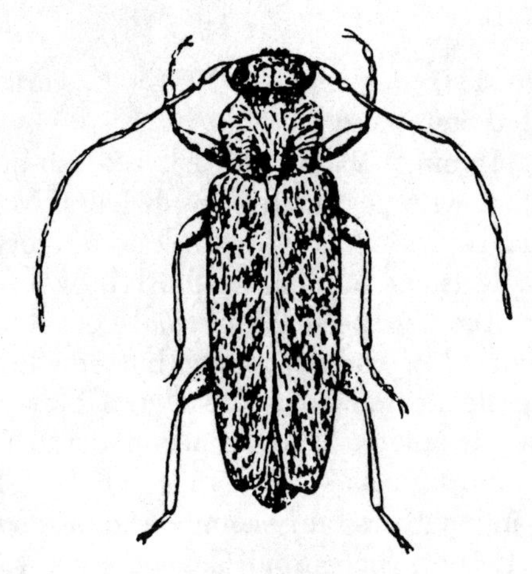

Figure 709

LENGTH: 7-9 mm. RANGE: Eastern half United States and Canada.

Subdepressed, dull. Black, marked as shown with fairly long yellowish pubescence. Elytra truncate at tips, with many fine punctures among scattered large ones. Bores in Red Cedar.

25a (24) Front coxal cavities open behind. 26

25b Front coxal cavities closed behind; eyes finely granulated. Fig. 710.

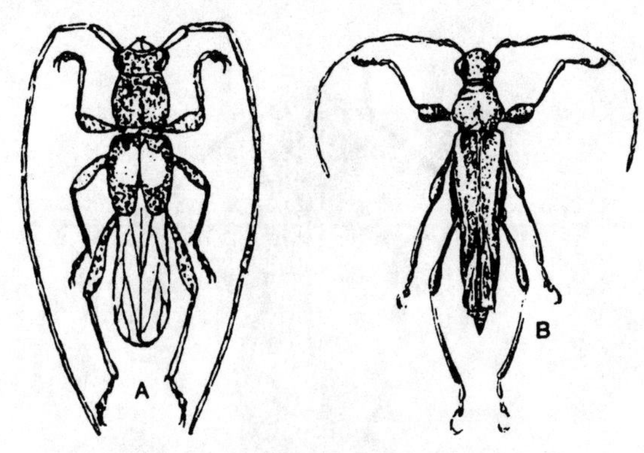

Figure 710

(a) *Molorchus bimaculatus* (Say)

LENGTH: 5-7 mm. RANGE: Eastern half United States.

Dull black, with scattered grayish hairs. Elytra short blackish with dull yellow spot at center; legs and antennae reddish-brown. On flowers of Hawthorn and early blooming shrubs.

(b) *Callimoxys sanguinicollis* (Oliv.)

LENGTH: 8-10 mm. RANGE: general.

Dull black. Thorax of female red with black margins. Hind legs yellow, with tips of segments black. Thorax of male black.

26a (25) Eyes finely granulated. 27

26b Eyes coarsely granulated. 33

27a (26) Scutellum broadly triangular or rounded behind. 28

27b Scutellum acutely triangular. Fig. 711. .. .. *Batyle suturale* (Say)

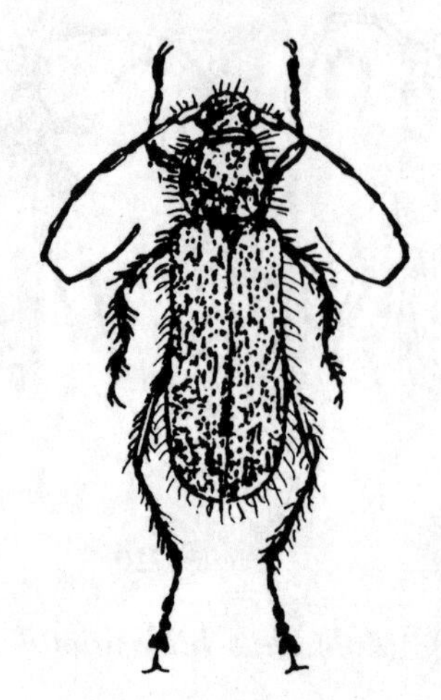

Figure 711

LENGTH: 6-8 mm. RANGE: Eastern N. America to N. Mex. and Sask.

Dark red; antennae, legs and sometimes elytral suture black or darkened; clothed with scattered long gray hairs. Elytra with tips rounded, punctures scattered and coarse.

Many of the long-horns gather on flowers to feed and meet their mates. This simplifies collecting. This one in common with several other species, has found the flowers of New Jersey Tea and of Dogwood much to its liking.

28a (27) Middle coxal cavities open behind. 29

28b Middle coxal cavities closed behind. Fig. 712. .... *Euderces picipes* (Fab.)

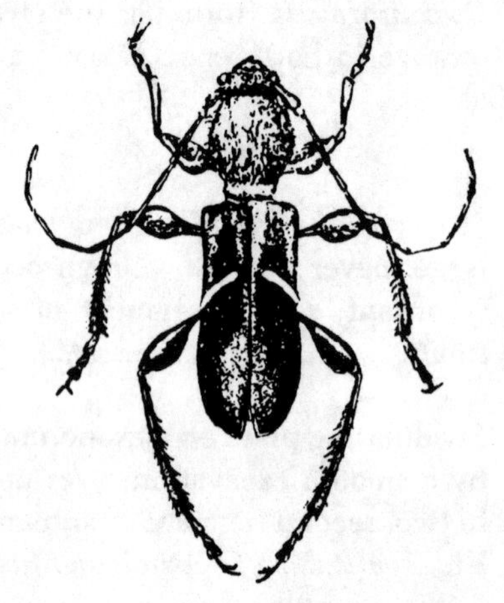

Figure 712

LENGTH: 5-8 mm. RANGE: Eastern half United States and Canada.

Shining. Black or dark reddish-brown; clothed with scattered grayish hairs. Marked on elytra as pictured with raised ivory-like bars. Legs and antennae reddish-brown.

Another species, *E. reichei* Lec., 4-5 mm, is much like the preceding but smaller and more slender, and has the elytral bars transverse. Its range extends from Indiana to Texas.

Purpuricenus humeralis (Fab.) 14-18 mm in length, is fairly common in our Central and Eastern States and Canada. It is robust, dull velvety black with a large triangular scarlet spot on the humerous of each elytron. The thorax has a small spine on each side. It is a beetle any collector will prize. The larvae develop in oak and hickory; the adults are found on mid-summer roadside flowers.

29a (28) Head small; front short; process of first ventral segment between the hind coxae rounded. 30

29b *Head* large; front long; process behind the hind coxae acute. Fig. 713.

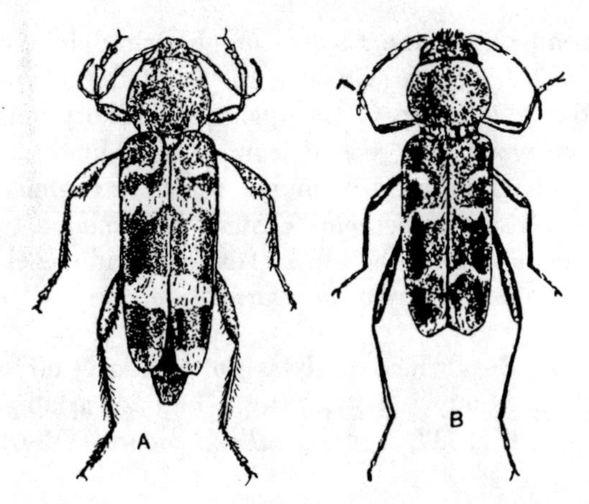

Figure 713

(a) ***Xylotrechus colonus* (Fab.) Rustic Borer.**

LENGTH: 8-16 mm. RANGE: Eastern half United States.

Thorax without apical and basal margins of paler hairs. Black or brown; pubescence yellowish or white marking the upper surface as pictured.

(b) X. *undulatus* (Say)

LENGTH: 11-21 mm. RANGE: East. Can. and Great Lakes to Yukon Terr. and Alas.

Thorax with apical and basal margins of light pubescence. Black or dark brown. Elytra marked with yellow or white pubescence as pictured.

A western species of general range, *X. nauticus* (Mann.) 8-15 mm is grayish-brown to blackish with three irregular transverse white lines crossing the elytra. English walnut and peach trees as well as other trees are victims of the larvae. Pacific Coast from Wash. to Cal. There are 22 species of *Xylotrechus* in our fauna.

30a (29) Antennal segments 3-6 without definite spines on outer side of apex. 31

30b Antennal segments 3-6 with definite spines on outer side at apex. Fig. 714. .... *Megacyllene antennata* (White)

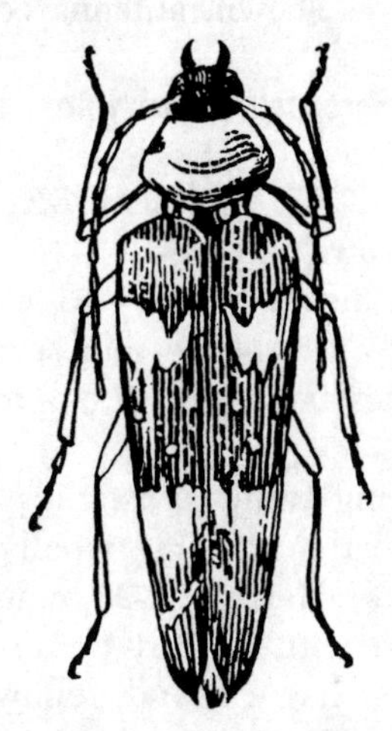

Figure 714

LENGTH: 12-30 mm. RANGE: Tex. to S. Cal.

Brown, with markings of gray scales as shown. Larvae a grub to 40 mm long bores in mesquite. It is known as the Round-headed Mesquite Borer.

31a (30) Thorax transversely excavated at sides near the base; prosternum perpendicular at tip. 32
thorax not excavated at sides. Fig. 715. ..

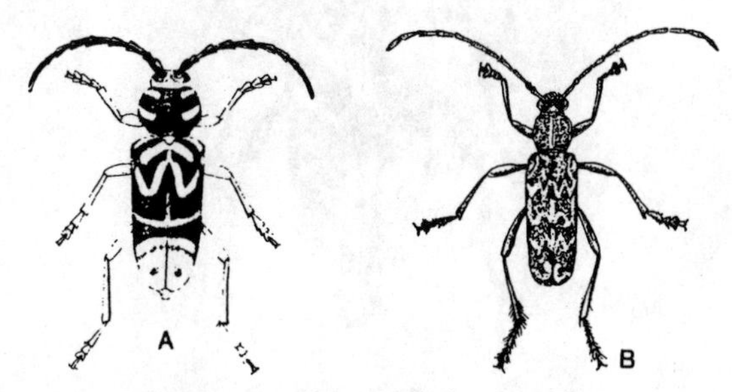

Figure 715

31b Prosternum sloping downward at tip; (a) *Glycobius speciosus* (Say) Sugar-maple Borer.

LENGTH: 23-25 mm. RANGE: Northeastern United States and Canada.

Black with markings of dense yellow pubescence as shown; antenna compressed.

(b) *Sarosesthes fulminans* (Fab.)

LENGTH: 12-18 mm. RANGE: Eastern N. America to Iowa and Kan.

Black; head, thorax and elytra marked as pictured with grayish pubescence; antenna filiform. The larvae live in oak, chestnut and butternut.

Differing from *Sarosesthes* in its larger size and the thorax being wholly black, *Calloides nobilis* (Harris) 20-23 mm is dull black and covered with a short dense pubescence. Each elytron has a round yellow spot at its base.

32a (31) Basal third of elytra with three narrow yellow bands, the last W shaped. Fig. 716. .... *Megacyllene robiniae* (Forst.) Locust Borer

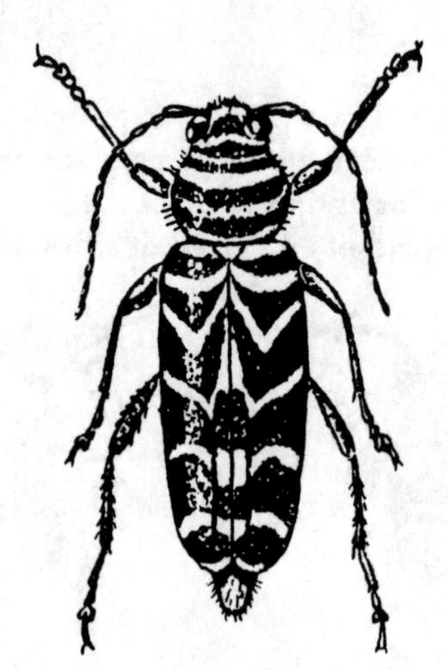

Figure 716

LENGTH: 14-20 mm. RANGE: East. N. Amer. to the Great Plains.

Velvety black with markings of golden yellow. It is a handsome beetle. The adults are common on the flowers of goldenrod in the fall. *M. caryae* (Gahan), the Painted Hickory Borer, is found in the spring on hickory logs and wood. The second segment of its hind tarsus is glabrous at its middle where as *robiniae* is densely pubescent. Eastern N. America to Tex. Found in the fall in Tex. No band of yellow along ant. margin of pronotum.

32b Basal third of elytra not as above; often wholly yellow, though highly variable. Fig. 717. .... *Megacyllene decora* (Oliv.)

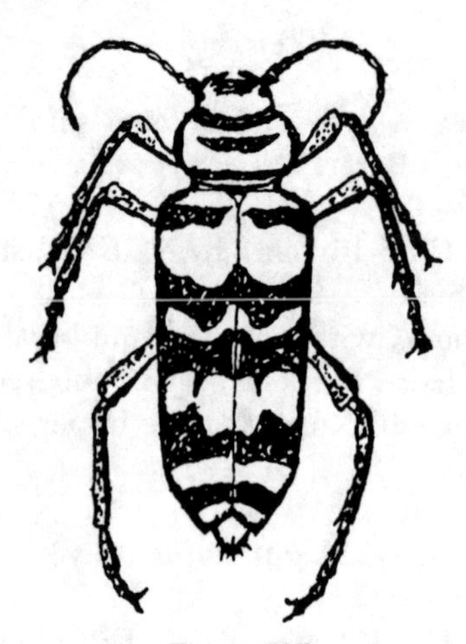

Figure 717

LENGTH: 14-20 mm. RANGE: Great Plains from Man. and Alta. to Ind. south to Tex. and Ala.

Golden yellow with variable markings of velvety black. Under parts yellow. Found on goldenrod flowers in Fall. Any sizable collection of this species is likely to show several patterns varying in greater or less degree from the one here pictured.

33a (26) Not more than one yellow cross-band on body. 34

33b Body with six yellow cross-bands. Fig. 718. .... *Dryobius sexfasciatus* (Say)

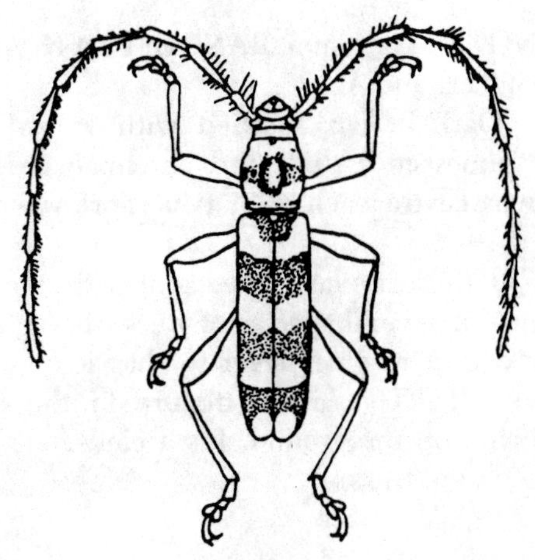

Figure 718

LENGTH: 19-27 mm. RANGE: From N. Y. west to Mich. and Okla.

Shining, black. Marked as shown with cross bands of bright yellow pubescence. Black spaces on elytra smooth but with scattered fine punctures.

Paranoplium gracilis (Lec.) is deep brown and measures 12 mm. The thorax is oval with its narrowest part at the base. It is found in California; the grubs feed in dead coast live oak.

34a (33) Scutellum rounded behind. 35

34b Thorax with spine on each side; scutellum triangular, acute. Fig. 719. .... *Knulliana cinctus* (Drury) Banded Hickory Borer.

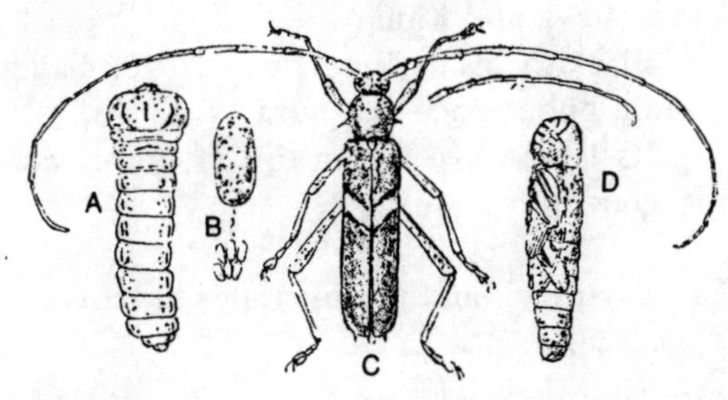

Figure 719

LENGTH: 16-32 mm. RANGE: Eastern N. Amer.

Blackish-brown, with scattered grayish pubescence. Elytra usually marked with yellow as shown. It seems to prefer hickory for the larvae. (a, larva; b, eggs; c, adult; d, pupa.)

Hesperophanes pubescens (Hald.) measuring 15 mm is known to occur in Pennsylvania. It is pale yellowish brown.

35a (34) Antenna not bearing spines. Fig. 720. .... *Eburia quadrigeminata* (Say)

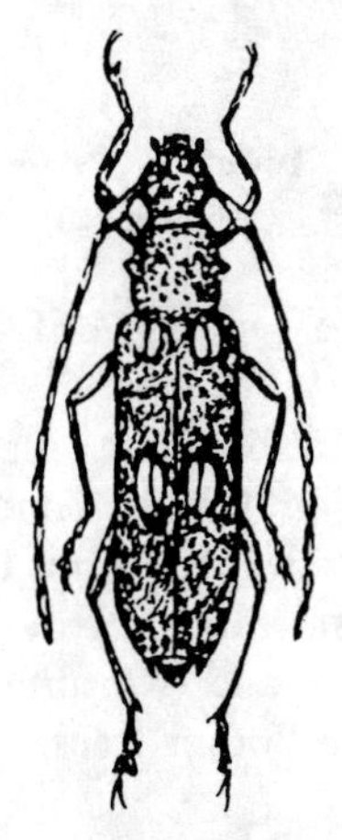

Figure 720

LENGTH: 14-24 mm. RANGE: Eastern half North America.

Pale brownish-yellow marked as shown with ivory-like white spots on elytra. The larvae bore in honey locust, oak, etc. and seem to live unusually long,—forty years or more. We have had several of these "patriarchal" grubs brought in and have reared them to adult stage. Ten other members of this genus are found in the U. S.

35b Antennae with some segments bearing sharp spines at outer apical end. 36

36a (35) **Spines long. Fig. 721.** .... ***Elaphidion mucronatum*** **(Say)**

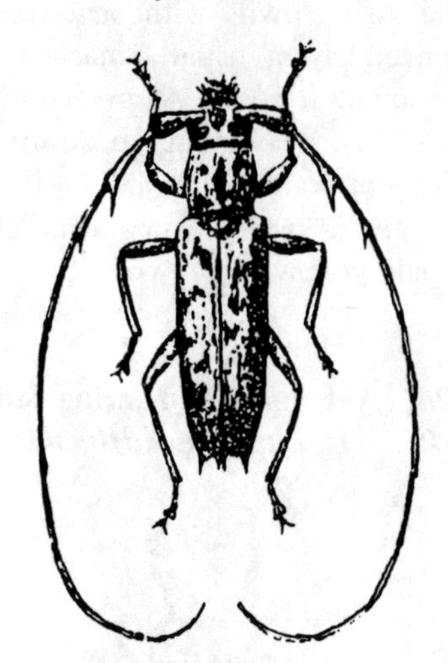

Figure 721

LENGTH: 15-20 mm. RANGE: Eastern half United States.

Dull reddish-brown, marked as shown with yellowish-gray pubescence. Thorax with median elevated ridge and two lateral elevated spots smooth. Elytra with scattered coarse punctures and two spines at tip of each. Bores in oak and other trees.

36b **Spines short. Fig. 722.** .. ***Elaphidionoides villosus*** **(Fab.) Twig Pruner**

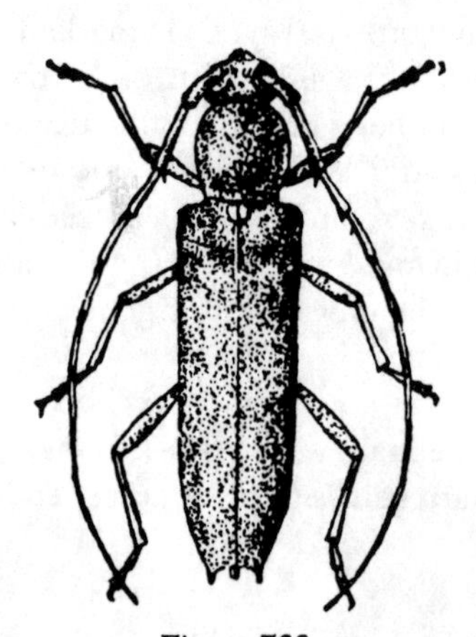

Figure 722

LENGTH: 11-18 mm. RANGE: East N. America to Tex. and Az.

Dark brown, mottled with grayish-yellow pubescence; legs and antennae reddish-brown. Elytra each with two short spines at tip.

This beetle lays its eggs near the tip of a branch of several species of trees, then girdles the twig at a lower level so that it dies and breaks off. The larvae mature in the dead branches on the ground. It's a clever act, but hard on the trees.

37a (2) **Elytra with an oval protuberance near the scutellum. Fig. 723.** .... ***Psenocerus supernotatus*** **(Say) Currant Tip Borer**

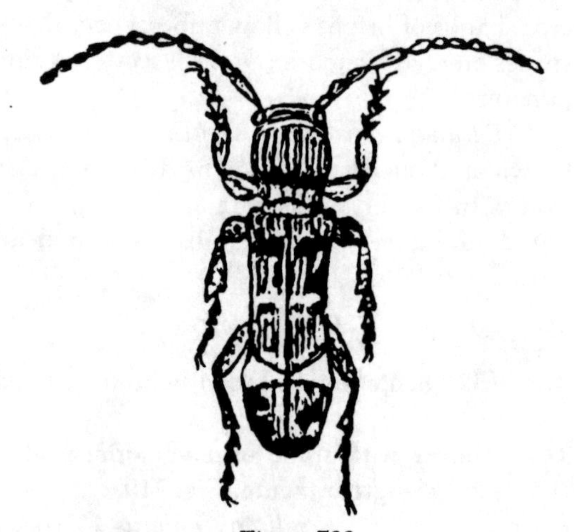

Figure 723

LENGTH: 3-6 mm. RANGE: East. U. S. and Can. to Iowa and Kan.

Blackish or dark reddish-brown; bands of white pubescence on elytra as shown.

Its larvae live in the tips of grape, currant, gooseberry, etc.

37b **Elytra without protuberance at base.** **38**

38a (37) **Scape of antennae with elevated spot (cicatrix) at apex.** **39**

38b **Scape of antennae without cicatrix at apex.** **41**

39a (38) **Legs long; front pair of legs and the antennae longer in male. Fig. 724.** .... ***Monochamus titillator*** **(Fab.)**
Southern Pine Sawyer

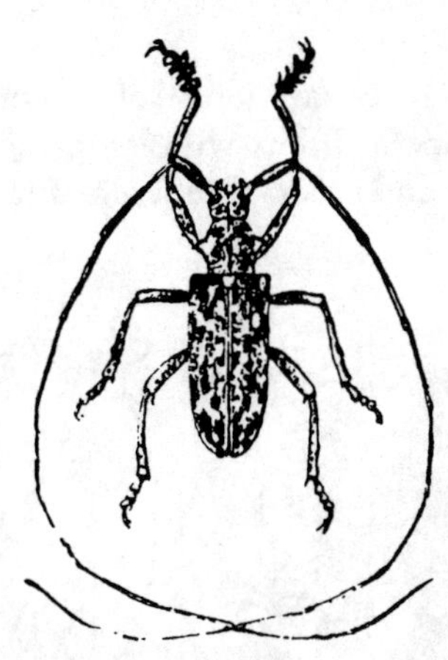

Figure 724

LENGTH: 20-32 mm. RANGE: East. Can. and U. S. east of Miss. River; Tex. and Kan.

Brown. Elytra mottled irregularly with gray and white pubescence. Lives in pine and frequently emerges from pine lumber. There are 10 other species in our area.

The antennae of the male are often two or more times as long as the beetles body. Such specialization is surely a hazard to its owner. They are sometimes found in new buildings where they have emerged from the lumber.

39b **Legs shorter and about equal.** **40**

40a (39) **Color black and white. Fig. 725.** .... ***Plectrodera scalator*** **(Fab.)**
Cottonwood Borer

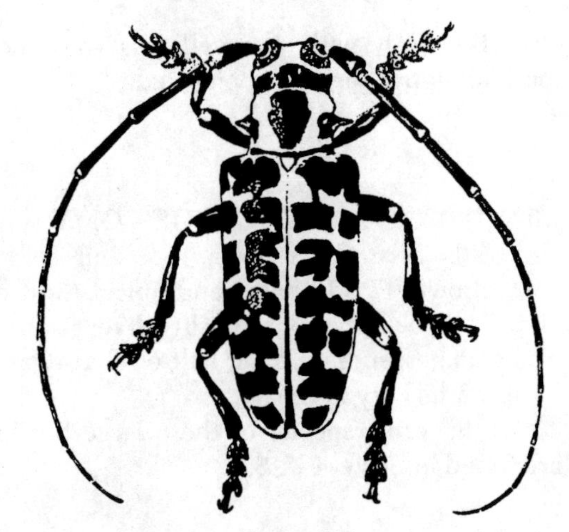

Figure 725

LENGTH: 25-35 mm. RANGE: N. Y. to Mich., south to Ga. and Tx., Colo.

Robust, shining. Black and white as pictured. A very attractive beetle which breeds in quaking aspen and other poplars.

They are not very abundant but their large size and unusual pattern make them prize specimens in a collection.

40b **Color brown and gray; thorax with lateral spines. Fig. 726.**

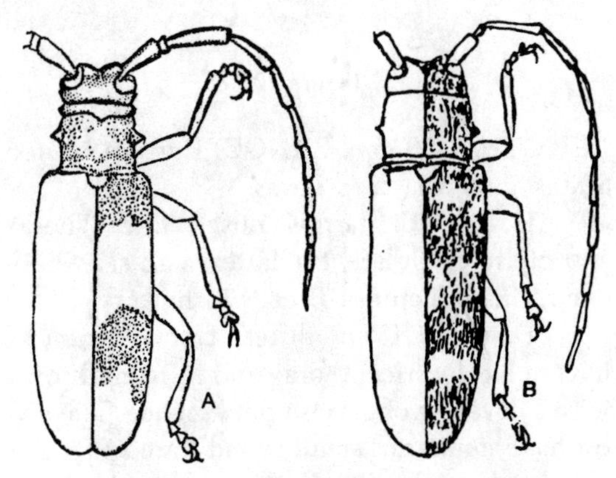

Figure 726

(a) ***Goes pulchra*** **(Hald.)**

LENGTH: 22-25 mm. RANGE: Mass. and Ont., west to Wisc., south to Va. and La.

Brownish with clay-yellow pubescence marking upper surface as pictured.

(b) *G. debilis* Lec.

LENGTH: 11-15 mm. RANGE: Ont.; state east of the Miss. River; Iowa, Mo.; and Az.

Brown. Head, thorax and apical third of elytra clothed with reddish-yellow pubescence. Pubescence on basal half of elytra gray. Lives on hickory and oak.

Five other species of these large beetles are found in Eastern U. S.

41a (38) Color black. Fig. 727. .... *Dorcaschema nigrum* (Say)

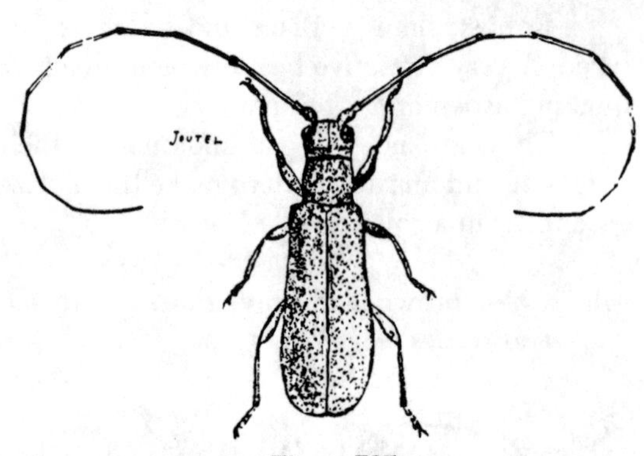

Figure 727

LENGTH: 8-10 mm. RANGE: Eastern United States and Canada to Texas.

Head with narrow raised line. Thorax and elytra granulate, the latter with scattered, deep, fine punctures. Breeds in hickory.

D. wildii Uhler differs from *nigrum* in having a cylindrical thorax and in being brown with a covering of grayish pubescence. The elytra have numerous small round bare spots and a large irregular one behind the middle. It measures 15 to 22 mm, breeds in mulberry and osage-orange, and is found from Kentucky to Massachusetts. *D. alternatum* (Say) differs from *wildii* by being smaller, 8-12 mm; pronotum with 4 stripes of clay yellow pubescence; elytra with four rows of similar spots; common on mulberry; eastern half of U. S. to Tex.

41b Not as in 41a. 42

42a (41) Front coxal cavities rounded, middle cavities closed (or nearly so). 43

42b Front coxal cavities angulate; middle cavities open behind. 47

43a (42) Basal joint of antennae club-shaped; thorax with lateral spine at middle and a dorsal tubercle. Fig. 728.

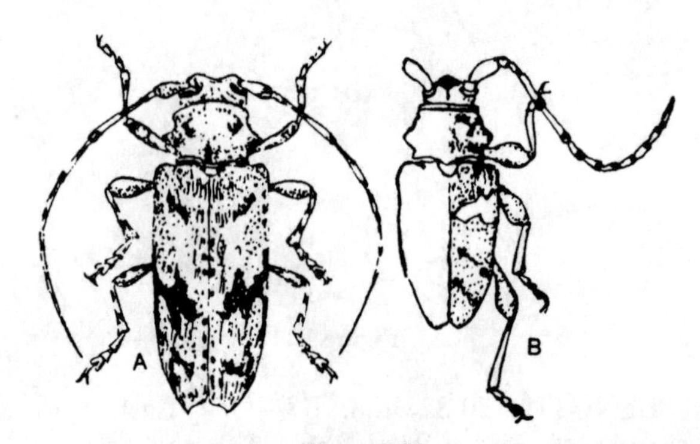

Figure 728

(a) *Aegoschema modesta* Gyll.

LENGTH: 10-13 mm. RANGE: Eastern United States and Canada.

Dark brown, marked with prostrate yellowish-gray pubescence.

(b) *Psapharochrus quadrigibbus* (Say)

LENGTH: 10-13 mm. RANGE: Eastern half United States.

Dark brown, marked with yellow-brown prostrate pubescence. White markings on elytra as shown. Two elevations on each elytron near base. Antennae with white rings.

43b **Lateral spines on thorax behind middle; basal joint of antennae cylindrical. 44**

44a **(43) Females with elongated ovipositor. Fig. 729.**

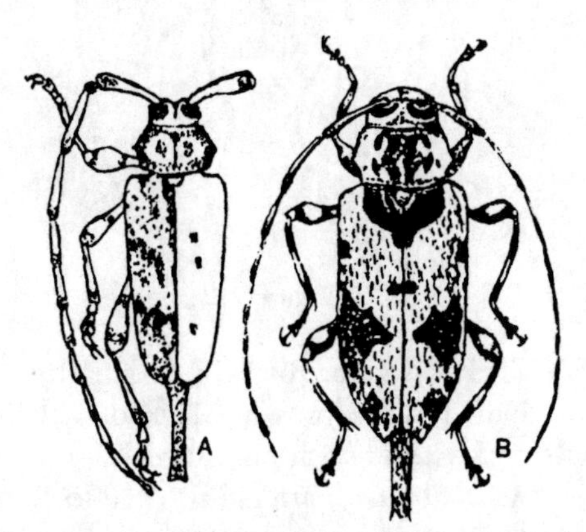

Figure 729

(a) ***Graphisurus fasciatus*** **(DeG.)**

LENGTH: 8-14 mm. RANGE: Eastern half United States and Canada.

Thorax and elytra gray with black dots and blotches. Antennae and legs with rings.

(b) ***G. triangulifera*** **(Hald.)**

LENGTH: 13-15 mm. RANGE: Southern States.

Brown marked with pubescence of clay-yellow.

An outstanding species ranging through the west, *Acanthocinus spectabilis* LeC., 19-23 mm, has antennae several times as long as the body. Gray, with back elytral band and many small black dots.

44b **Females without elongated ovipositor. 45**

45a **(44) Mesosternum broad, thorax with small lateral tubercles a bit behind the middle. Fig. 730. .... *Leptostylus transversus* Gyll.**

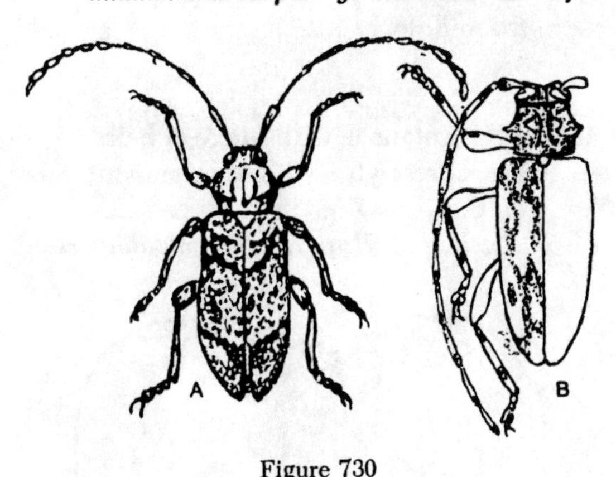

Figure 730

LENGTH: 7-10 mm. RANGE: Eastern half United States and Canada.

Robust. Blackish-brown, marked as shown with short prostrate grayish hairs. Five blunt tubercles on disk of thorax. Common.

(b) ***Amniscus sexguttatus*** **(Say)**

LENGTH: 7-10 mm. RANGE: Eastern half United States and Canada. On pine.

Robust. Brownish with pattern as shown of grayish pubescence. Tibiae and antennae annulate.

A. macula (Say) 4-9 mm brownish, thorax with a broad whitish stripe on sides banded by a narrow blackish stripe. Each elytron with a wide irregular white spot behind the middle and six lines of small black dots. Known East. N. America south to Va., west to Kansas. On hardwoods.

Urgleptes querci (Fitch) 3.5-6 mm differs from the above in lacking the ridge on the side. The head and thorax are dark brown; elytra pale purplish-brown marked with a wide irregular cross-band covering a good part of the apical half but leaving the tip mostly light, and with 3 or 4 dark spots on the humeral half. Antennae more than twice as long as the body. It ranges from Iowa eastward and northward.

45b **Mesosternum triangular or narrow; thorax angulate or with sharp spine behind the middle.** **46**

46a **(45) Antennae with fringe of hairs on under side; elytra with a prominent ridge on the side. Fig. 731.** .... ***Hyperplatys maculata* Hald.**

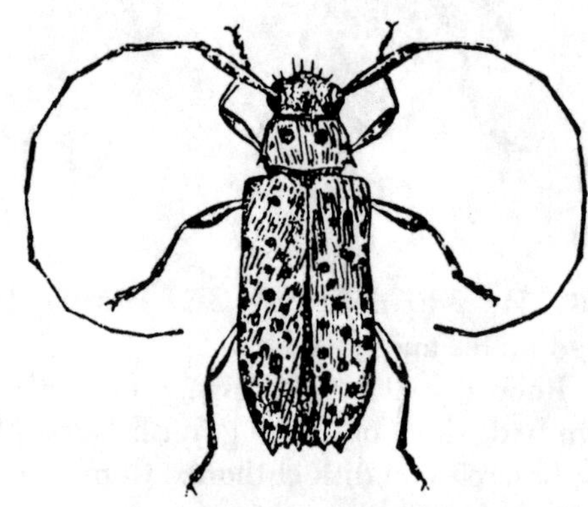

Figure 731

LENGTH: 4-6 mm. RANGE: general.

Reddish-brown, a dense coat of grayish pubescence and black spots form the pattern, usually a large spot on apical two-fifths; elytra with black vitta along sides.

H. aspersa (Say) is similar in size, but has longer antennae and heavier pubescence. Elytra without post median large spots.

46b **Antennae without fringe of hairs. Fig. 732.** .... ***Sternidius alpha* (Say)**

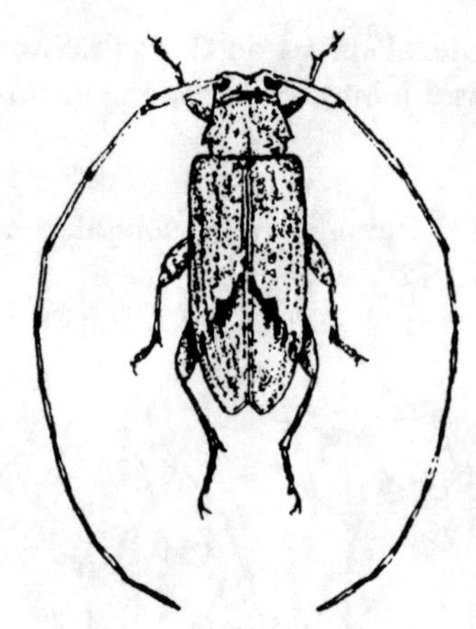

Figure 732

LENGTH: 5-7 mm. RANGE: general.

Dull reddish-brown. Marked with black spots and with grayish pubescence.

Astyleiopus variegatus (Hald.), 8-12 mm, differs from the preceding in having the front flat instead of convex. It is purplish-brown mottled with black and clothed with a short pubescence. The antennae and tibiae are ringed with gray. It ranges from Eastern N. America west to N. Dakota, south to Al. and Tx.

47a **(42) Tarsal claws divided or bearing an appendage.** **52**

47b **Tarsal claws not as in 47a.** **48**

48a **(47) Thorax with a spine or tubercle; small flattened species. Fig. 733.** .... ***Pogonocherus mixtus* Hald.**

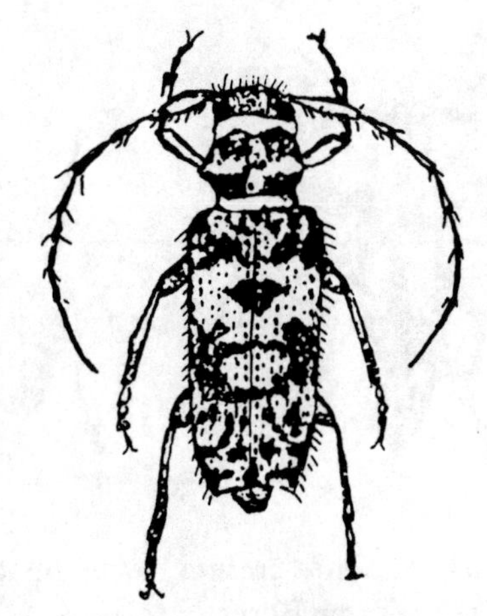

Figure 733

LENGTH: 4-7 mm. RANGE: United States and Canada.

Head and thorax piceous; elytra blackish marked as shown with dull brownish-yellow. Elytra with scattered erect black hairs and band of white pubescence. Lives on pine, pear, willow, etc.

P. penicellatus Lec., 6 mm, is piceous with the elytra back of the base covered with gray pubescence. It ranges widely through Eastern Canada and the United States.

48b Thorax without spines or tubercules; larger. 49

49a (48) Antennae with colored rings (annulate). Fig. 734. .... *Saperda obliqua* Say Alder Borer

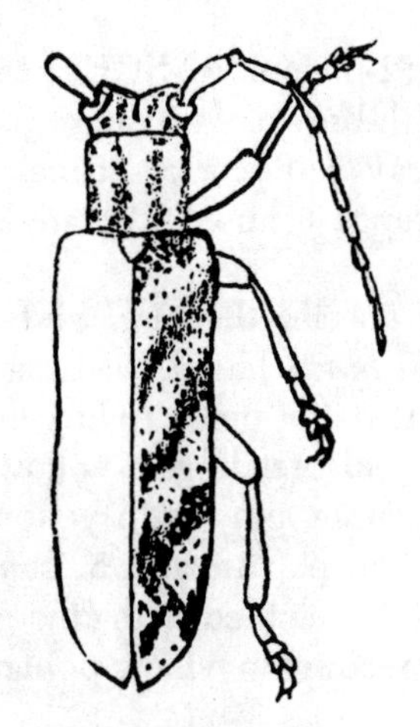

Figure 734

LENGTH: 21-26 mm. RANGE: Eastern half United States and Canada.

Yellowish-brown, marked with oblique darker bands. Elytra with spine at tip.

We picture and key several members of this genus as it is of high economic importance and widely distributed, as well as their being very attractive beetles.

49b Antennae of one color, not ringed. 50

50a (49) Elytra with spine at suture on rounded tip. Fig. 735. .... *Saperda calcarata* Say Poplar Borer

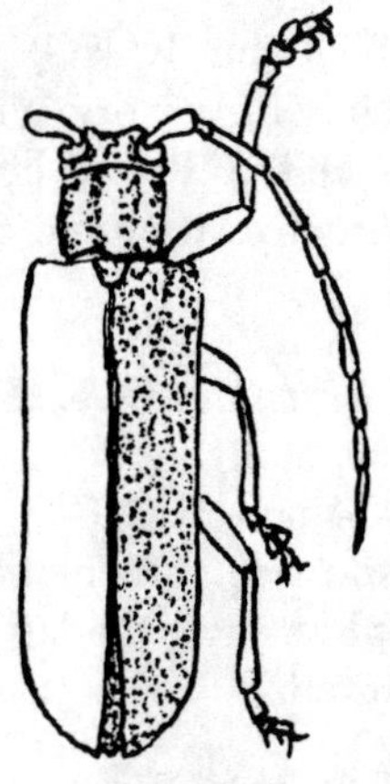

Figure 735

LENGTH: 21-30 mm. RANGE: general.

Reddish-brown, with dense prostrate yellow and gray pubescence. Lines and other markings orange-yellow and many small black dots.

Three species that are western in their range are S. *horni* Joutel, 16-20 mm, shining black with a coat of dense light yellowish gray pubescence and breeding in willow; S. *populnea* (L.), 12 mm, gray with yellow markings and feeding on poplar; and S. *concolor* Lec., 6-20 mm, black, clothed with fine gray pubescence and breeding in willow poplar.

50b Elytra without spine at suture. 51

51a (50) Elytra with stripes. Fig. 736.

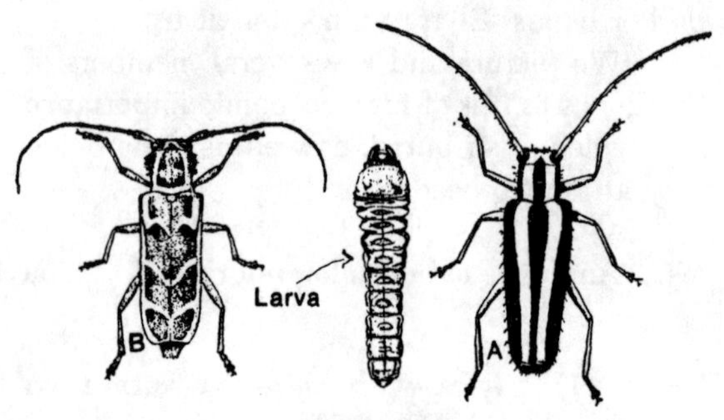

Figure 736

(a) *Saperda candida* Fab.

LENGTH: 15-20 mm. RANGE: general.

Brownish with silvery white pubescent stripes. Its larvae the "Round-head Apple Tree Borer" is destructive to apple trees and their near relatives.

(b) *S. tridentata Oliv.* Elm Borer

LENGTH: 9-14 mm. RANGE: general.

Brownish-black, ornamented with gray and orange pubescence, the latter forming the stripes as pictured.

51b Elytra with spots or solid color. Fig. 737.

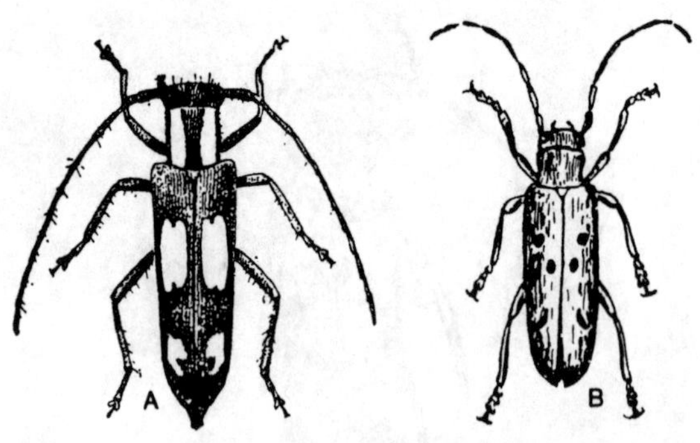

Figure 737

(a) *Saperda cretata* Newn. Spotted Appletree Borer

LENGTH: 12-20 mm. RANGE: general.

Cinnamon brown with white spots as pictured. Breeds in *Craetegus*.

(b) *S. vestita* Say Linden Borer

LENGTH: 12-21 mm. RANGE: Eastern half United States.

Dark reddish-brown, densely covered with prostrate olive-yellow pubescence. Marked with black as shown.

There are a half dozen more of these colorful beetles in our fauna.

52a (47) Eyes completely divided into two widely separated parts; red with black spots. Fig. 738. .... *Tetraopes tetrophthalmus* (Forst.) Red Milkweed Beetle

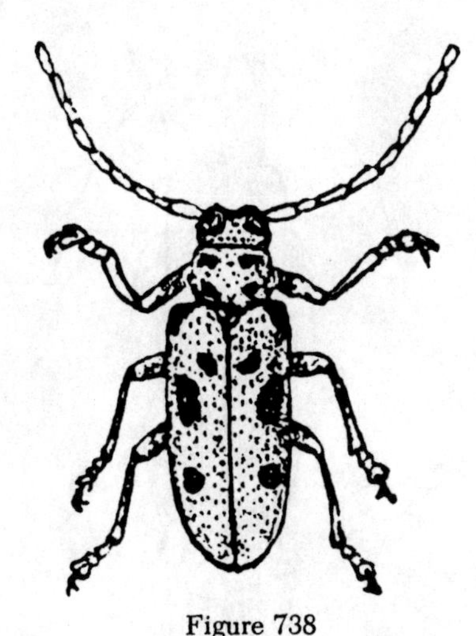

Figure 738

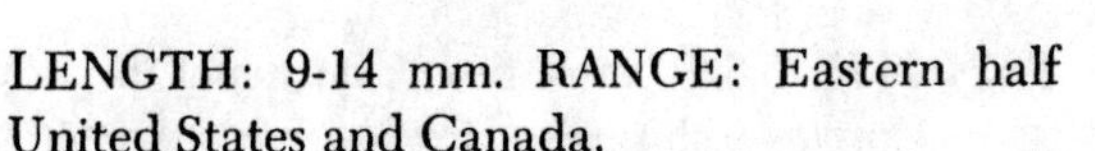

LENGTH: 9-14 mm. RANGE: Eastern half United States and Canada.

Black. Head, thorax and elytra red, marked with black dots as shown. Antennae not ringed, it and scutellum black.

The members of this genus of which there are about 25 species feed on milkweed.

T. femoratus Lec. 10-15 mm, ranges throughout our country. Its antennae have a white ring on each segment.

This genus includes 12 species in the U. S. and Can.

52b Eyes not wholly divided, though deeply notched. 53

53a (52) Body above uniform gray; claws with small tooth or slightly cleft. Fig. 739. .... *Mecas pergrata* (Say)

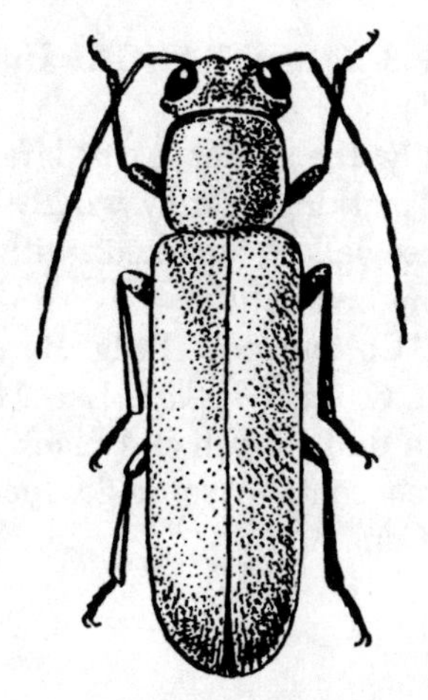

Figure 739

LENGTH: 8-15 mm. RANGE: Great Plains to southeastern U. S.; N. Mex. Length: 6-11 mm. It is black with a heavy coating of gray and whitish hairs. The thorax has five black spots, and the antennae are ringed; reddish femora.

This genus contains 10 species in our area.

53b Not uniformly gray. Fig. 740. .... *Oberea tripunctata* (Swed.) Dogwood Twig Borer

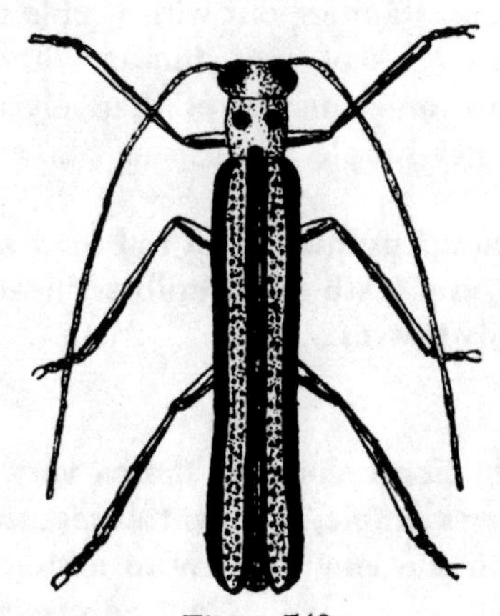

Figure 740

LENGTH: 8-16 mm. RANGE: United States and Canada.

Usually the under parts largely yellow with black markings; rarely wholly black. Upper surface yellow, marked with black as shown, though variable.

The "Cottonwood Twig Borer," of the West Coast, *O. quadricallosa* Lec. 11-14 mm, is orange with dull grayish and black markings.

Oberea contains some 25 species in the U. S. and Can.

FAMILY 107. CHRYSOMELIDAE
The Leaf Beetles

Almost one-tenth of the known beetles belong to this great family. Any general collection of insects is practically certain to contain leaf beetles. Their size is never very large; some are so tiny as to be almost microscopic. What they lack in size many species more than make up for in their beauty of pattern and brilliant colors. No wonder they are favorites with many collectors.

1a **Head standing out with visible neck-like part back of eyes; thorax without margins, much narrower than elytra; mandibles simple. 2**

1b **Thorax usually about as broad as elytra; several teeth on mandibles; head visible only to eyes. 5**

2a **(1) Form elongate; thorax very narrow; claws simple; first ventral segment about half the length of entire abdomen. Fig. 741. .... *Donacia piscatrix* Lac.**

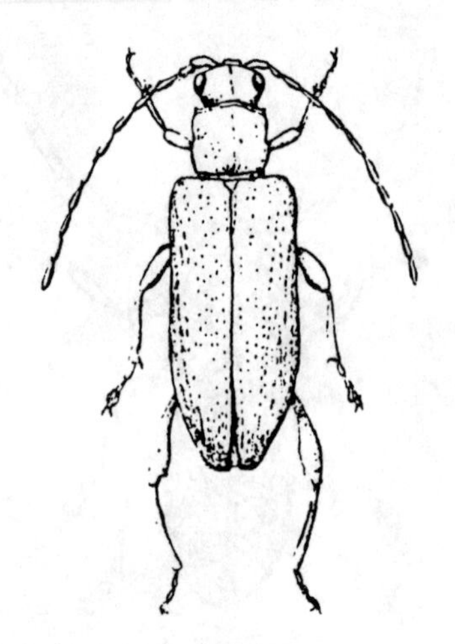

Figure 741

LENGTH: 6.5-10 mm. RANGE: general.

Convex, shining. Bronze with varying sheen of green, coppery or brownish-yellow. Legs and antennae reddish-yellow.

A large member of the genus is *D. pubicollis* Suffr., measuring 10-12 mm. It is distinguished by the heavy pubescence on both head and thorax. Ont. to the Dakotas, south to Ill. and Ind. Taken on water lillies. The members of this genus range widely but feed quite largely on marsh plants and water lillies.

Plateumaris emarginata (Kby.) Length: 6-8 mm. Ranges across South. Can. and N. U. S. from coast to coast. Color cupreous black, green, blue, or purple; legs dark.

Both of the above genera are very similar in appearances and habits. Members of the genus *Donacia* have the sutural margin of the elytra straight to the apex, whereas in *Plateumaris* the margins are sinuate near the apex.

2b **First ventral abdominal segment not longer than the next two segments. 3**

3a **(2) Claws simple; thorax narrow; punctures on elytra in rows. 4**

3b **Claws cleft or toothed; junctures on elytra scattered. Fig. 742.**

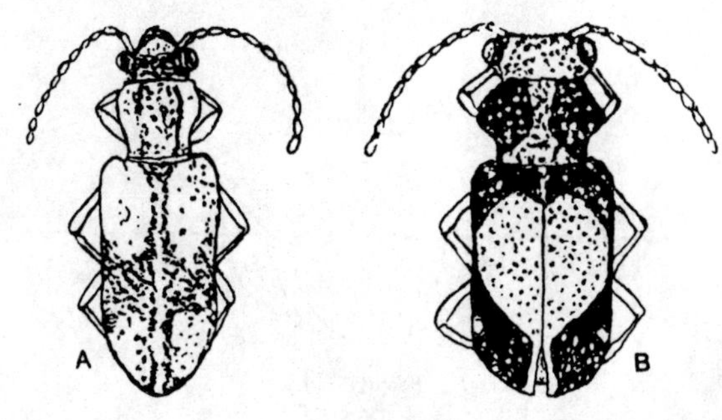

Figure 742

(a) ***Orsodacne atra childreni* Kby.**

LENGTH: 4-7 mm. RANGE: United States and Canada.

Elytra dark brown marked with yellow. Six or more varieties are recognized varying much in color and markings.

(b) ***Zeugophora varians* Cr.**

LENGTH: 3-4 mm. RANGE: general.

Black marked with yellowish. Legs and antennae dull reddish-yellow. Found on leaves of poplar trees. The front coxae touch in this genus but are separated in the preceding.

Z. scutellaris Suffr. 3.5-4 mm with wide distribution in both hemispheres, has head, thorax and legs yellow, with the abdomen and elytra shining black. The short, rather heavy antennae segments 1-3 yellow, the others piceous.

4a (3) Thorax constricted near its middle. Fig. 743. .... *Lema trilineata* (Oliv.) Three-lined Potato Beetle

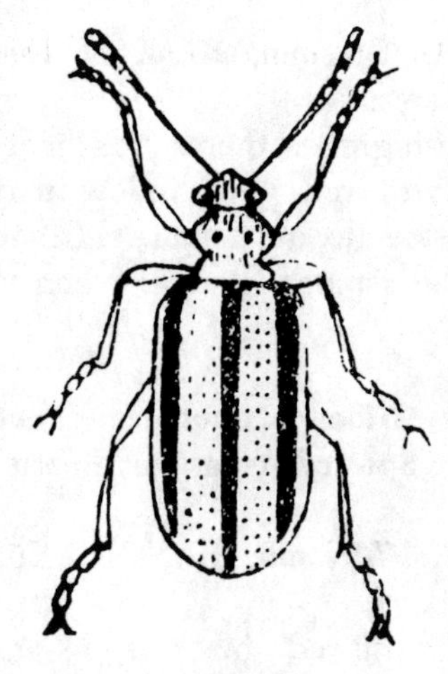

Figure 743

LENGTH: 6-7 mm. RANGE: general.

Reddish-yellow marked with black. Years ago this was regarded as a serious pest of potatoes. Ordinarily they are not now sufficiently abundant to even attract attention. Many of our worst insect pests, as with the weeds, are introduced species.

This genus contains over 20 colorful species in our fauna.

4b Thorax not constricted. Fig. 744.

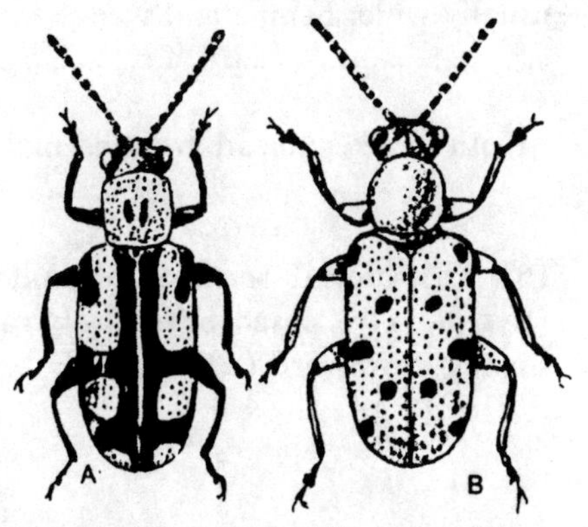

Figure 744

(a) ***Crioceris asparagi* (L.) Asparagus Beetle**

LENGTH: 7-8 mm. RANGE: Europe and United States.

Bluish-green; thorax red marked with green; elytra red with yellow markings. A handsome beetle; quite squirrel like in dodging around the asparagus stems when you try to catch it.

(b) *Crioceris duodecim-punctata* (L.)
Spotted Asparagus Beetle

LENGTH: 7-9 mm. RANGE: Europe and United States.

Brownish-red black spots as shown. These two beetles do heavy damage to asparagus.

5a **(1) More or less broadly expanded margins on thorax and elytra; head often not visible from above. 63**

5b **Head usually visible from above; thorax and elytra without widely expanded margins. 6**

6a **(5) Mouth turned down; front inflexed; usually wider behind and wedge-shaped. 61**

6b **Mouth directed ahead; front normal. 7**

7a **(6) Last dorsal segment of abdomen (pygidium) exposed beyond elytra and sloping downward (a). Fig. 745. .... 10**

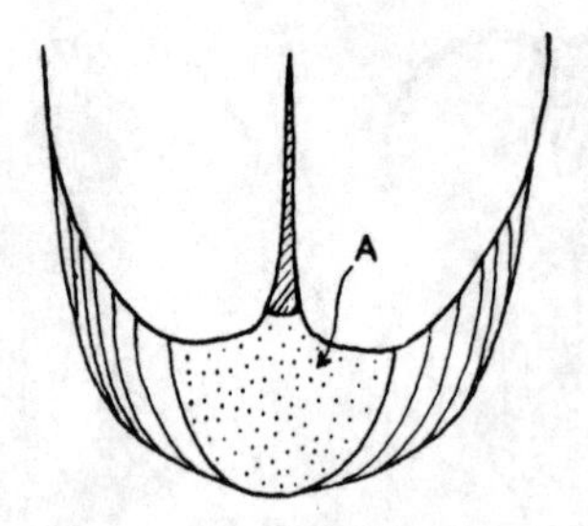

Figure 745

7b **Last dorsal segment not exposed, middle ventral segment not narrow. 8**

8a **(7) Antennae placed close together on the front; front coxae conical. 34**

8b **Antennae widely separated at base (usually greater distance than length of first joint). 9**

9a **(8) Front coxae transverse; third tarsal segment usually not divided. 24**

9b **Third tarsal segment with two lobes; front coxae rounded. 15**

10a **(7) Surface of body much roughened above. Fig. 746. .... *Neochlamisus gibbosa* (Fab.)**

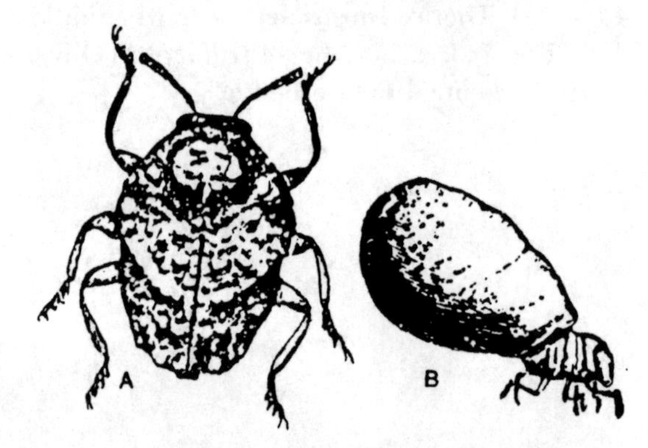

Figure 746

LENGTH: 4 mm. RANGE: Eastern half United States, Tex. and N. Mex.

Brownish, bronzed often with purplish sheen, 12-15 tubercles on each elytron. The larvae carry a case for protection (b).

This and related general include 27 very similar appearing species.

10b Upper surface without tubercles. 11

11a (10) Antenna serrate, short; front coxae touching. Fig. 747. .... *Coscinoptera dominicana* (Fab.)

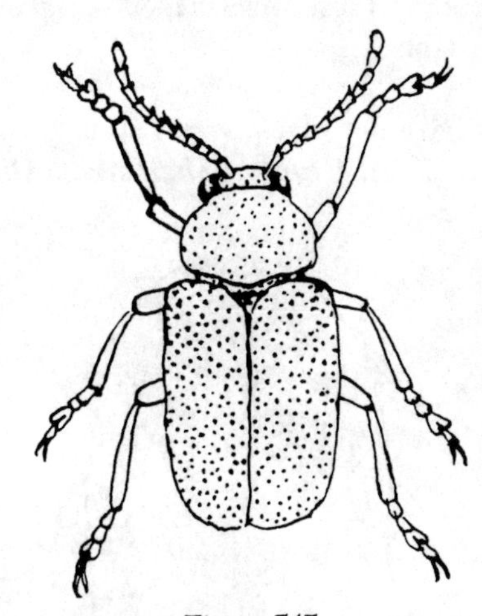

Figure 747

LENGTH: 4-7 mm. RANGE: general.

Black, with ash gray pubescence on both upper and lower surface, very dense below.

Babia quadriguttata (Oliv.) measuring 3.5-4 mm and ranging throughout the Eastern and Central United States, is convex, shining, and black with two reddish-yellow spots on each elytron. Other species are found in Tx. and the Southwest.

11b Front coxae separated by prosternum antennae usually long and slender. 12

12a (11) Thorax margined at base; front femora heavier than hind ones.

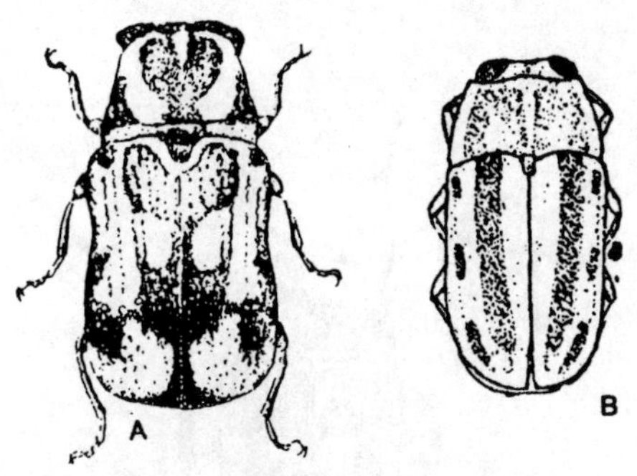

Figure 748

(a) *Pachybrachis tridens* (Melsh.)

LENGTH: 2-3 mm. RANGE: Eastern half United States.

Robust. Yellow with black markings; the pattern is highly variable. Under surface black, legs and antennae yellow. Feeds on hickory and elm as well as some roadside weeds.

(b) *P. bivittatus* (Say)

LENGTH: 3.5-6 mm. RANGE: general.

Robust. Yellow, marked with black. Legs and under parts mostly reddish. It is found on grasses in moist places.

The genus is a large one with some 200 named species. *P. melanostictus* Suffr. 3 mm mottled black and yellow, feeds on willow in the states of the Rockies and westward. *P. varicolor* Suffr. 4.25 mm yellow with mixed black markings feeds on white fir in Colorado and southward. *P. donneri* Cr. 3-4.5 mm black with yellow markings, feeds on willows in the Pacific Coast states.

12b Thorax without a basal margin; front femora not heavier. 13

13a (12) **With tooth on front edge of thoracic flank. Fig. 749.** .... ***Bassareus clathratus*** **(Melsh.)**

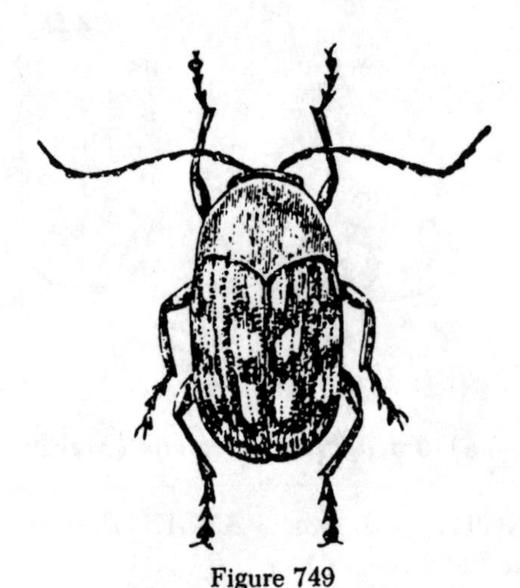

Figure 749

LENGTH: 4-5 mm. RANGE: Eastern half United States.

Robust. Dull red marked as shown with pale yellow. Antenna and legs reddish at base becoming blackish at apex.

B. lituratus (Fab.) 3-4.5 mm has a black head, thorax reddish yellow with three black spots and elytra each with three yellow stripes; the one on the fifth interval being short and at the base only.

Bassareus contains 8 species and many varieties.

13b Without tooth as in 13a. **14**

14a (13) **Elytra black. Fig. 750.**

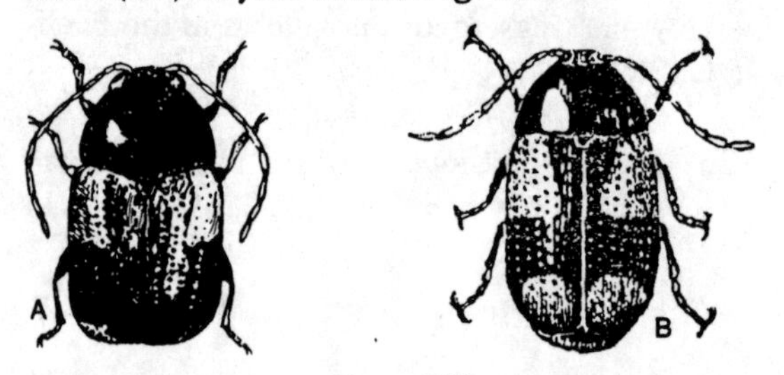

Figure 750

(a) ***Cryptocephalus notatus quadrimaculatus*** **Say**

LENGTH: 4-5 mm. RANGE: N. Eng. to the Carolinas west to Colo., Utah, and Idaho.

Robust, shining. Black with red markings. Base of antennae pale.

(b) ***C. quadruplex*** **Newn.**

LENGTH: 2.5-3 mm. RANGE: Eastern United States.

Shining. Black marked with red. Base of antennae yellow.

Both of these species feed on several species of plants.

14b Elytra not black. Fig. 751. .... ***Cryptocephalus venustus*** **Fab.**

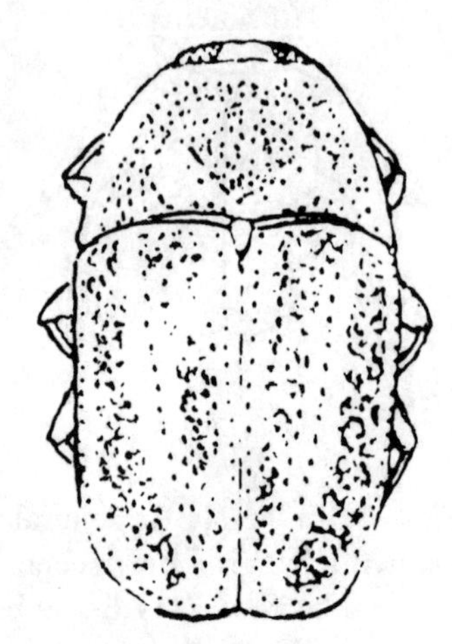

Figure 751

LENGTH: 4.5-5.5 mm. RANGE: Eastern half United States.

Robust. Head and thorax reddish-brown. Elytra yellow with black or brown stripes as pictured; the pattern and coloring are quite variable.

This beetle may often be found in abundance on the flowers of Fleabane in meadows.

This is another large genus with some 65 species in our territory.

15a (9) Under part of front margin of thorax curved to form a lobe behind the eye. 20

15b Under part of front margin of thorax straight. 16

16a (15) Thorax margined on the sides. 17

16b Thorax without distinct side margins. 19

17a (16) Middle and hind tibiae with notch near apex; distinct grooves above the eyes. Fig. 752. .... *Metachroma marginale* Cr.

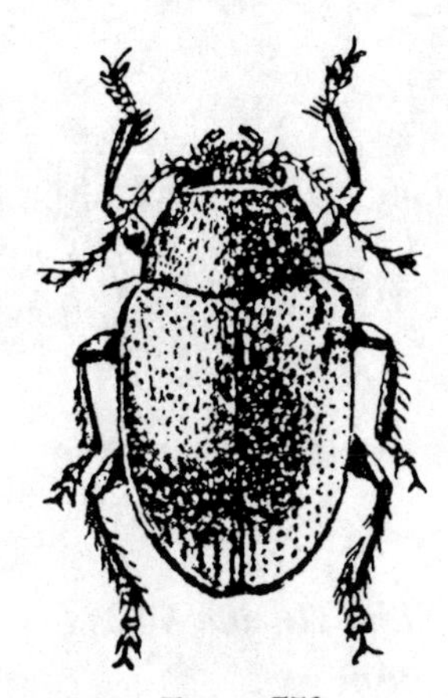

Figure 752

LENGTH: 6-8 mm. RANGE: N. C. to Fla. to Tex.

Light brown; head and thorax darker than elytra. Found on hard pine.

M. angustulum Cr., 4-7 mm pronotum and elytra varying in color from pale to piceous, the later often with a wide dark lateral stripe. Ranges from Ont. and Ohio to Mont. and Nebr.

Metachroma is a large genus of 34 species in the U. S.

17b No grooves on head above the eyes. .. 18

18a (17) Thorax with regular and entire side margins; third joint of antennae longer than second. Fig. 753. .... *Nodonota puncticollis* (Say)

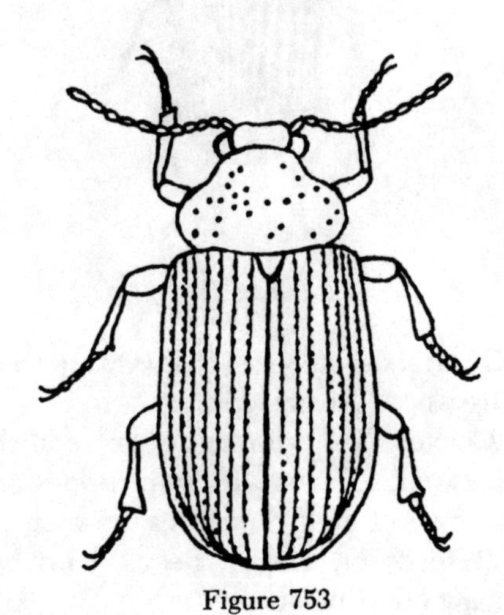

Figure 753

LENGTH: 3.5-4.5 mm. RANGE: East of the Rockies.

Convex, moderately shining. Bronzed, greenish or bluish; antenna and sometimes the leg reddish-yellow.

A similar species, *N. tristis* (Oliv.), 3-4 mm, is bronze, greenish blue or metallic blue, is widely distributed and is more oval in form than *puncticollis*. East of the Rockies.

18b Side margins of thorax in waves or irregular. Fig. 754. .... *Colaspis brunnea* (Fab.)

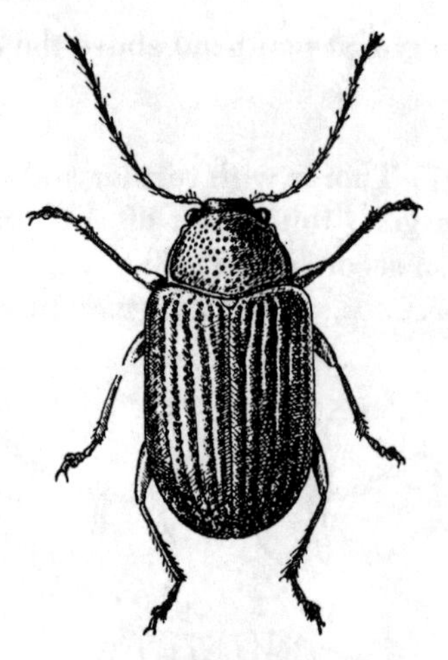

Figure 754

LENGTH: 4.5-6 mm. RANGE: Mostly east of the Mississippi River, Ariz.

Convex. Dull brownish or reddish-yellow; legs pale. Elytra with faint ridges and irregular rows of deep punctures between.

It feeds on grapes, beans, strawberries and many other plants.

19a (16) Body bronzed or metallic above; groove above eye. Fig. 755. .... *Graphops pubescens* (Melsh.)

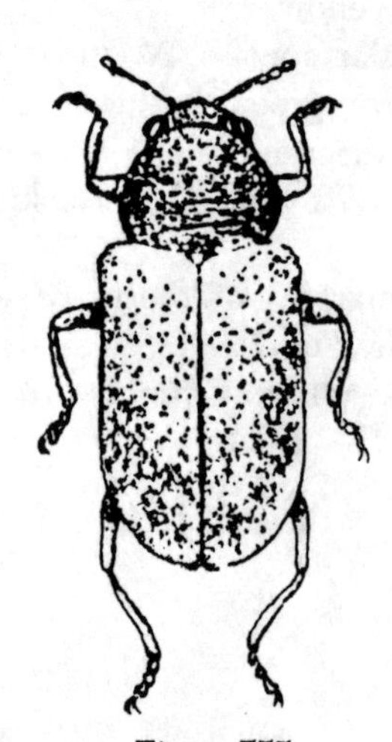

Figure 755

LENGTH: 3.5-4 mm. RANGE: Eastern half United States and Canada.

Cuperous bronzed with scant grayish pubescence. Fine punctures on elytra.

During the summer it seems to feed on evening-primrose; it hibernates as an adult during the winter.

G. curtipennis (Melsh.) 2-2.5 mm in length, is heavier set than the above. It is coppery-brown with scattered short gray hairs and is known from Iowa, New York, Florida and Texas.

This is a fairly large genus with 18 species in our fauna.

19b Body not metallic above. Fig. 756.

Figure 756

(a) *Fidia viticida* Walsh Grape Rootworm

LENGTH: 6-7 mm. RANGE: Eastern half United States.

Dull reddish-brown with coat of yellowish-gray pubescence. Legs and antennae paler. Feeds on leaves of grapes; the larvae on the roots.

(b) ***F. longipes* (Melsh.)**

LENGTH: 5-6.5 mm. RANGE: Eastern United States.

Brownish or smoky with prostrate coat of gray hairs. Also feeds on grape.

Xanthonia decemnotata (Say), 3 mm dull yellow to brownish-red, with 8 to 10 blackish spots on elytra is common Texas to Canada.

20a (15) Upper surface without pubescence or scales. 23

20b Upper surface covered with scales or pubescence. 21

21a (20) Plain margins at sides of thorax. 22

21b Thorax without distinct side margins; elytra brown or black. Fig. 757. .... *Bromius obscurus* (L.)
Western Grape Rootworm

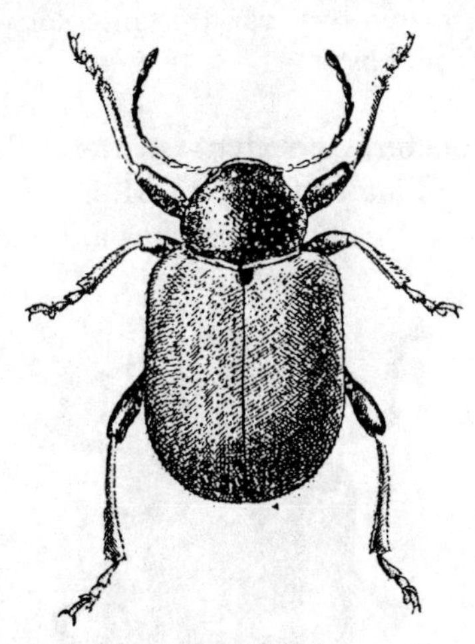

Figure 757

LENGTH: 5-6 mm. RANGE: Eurasia, North America.

Elytra brown or brownish-yellow with scattered prostrate yellowish hairs. Head, thorax and under surface brown to black. Elytra with irregular rows of fine punctures.

This species has several named varieties. The beetles eat characteristic holes in the leaves of grapes while the larvae feed on the roots of these plants.

22a (21) Margin of thorax toothed; front tibia with teeth near apex on inner side. Fig. 758. .... *Myochrous denticollis* (Say)
Southern Corn-leaf Beetle

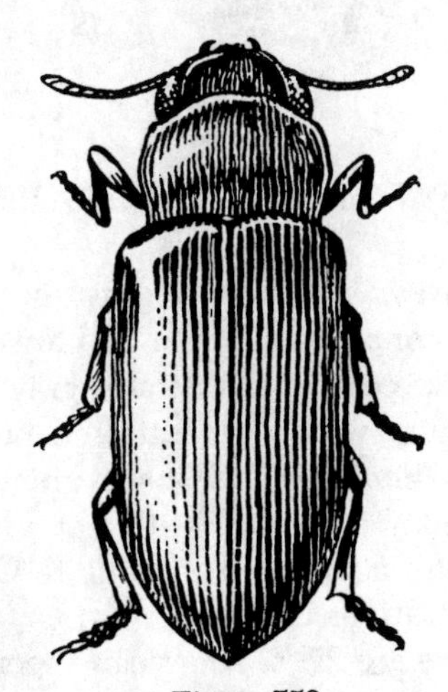

Figure 758

LENGTH: 5-7 mm. RANGE: Eastern half of the U. S.

Convex. Brown, bronzed; antennae dull red at base. Thorax with three teeth at side and coated with grayish yellow scales.

The Long Leaf Beetle, *M. longulus* Lec. similar in size and color ranges through the Southwest and does damage to cotton, muskmellon and grape buds. There are 13 species in this genus, Eastern and Southern U. S.

22b Tibiae and thoracic margins without teeth. Fig. 759. .... *Glyptoscelis albicans* Baly.

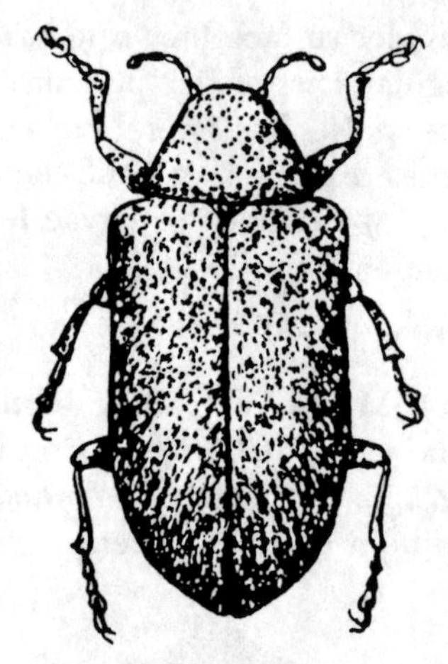

Figure 759

LENGTH: 7-9 mm. RANGE: So. Ind. and Ill. to Tex.

Convex, shining. Reddish-brown with sheen of bronze. Both upper and lower surface with thick coat of prostrate gray hair-like scales. Elytra with fine, scattered punctures.

G. septentrionalis Blake, 6-8 mm, is metallic bronzed and covered with fine gray scales. It ranges from British Columbia to California. Taken on various conifers.

There are 22 N. American species in this widely distributed genus.

23a (20) Elytral punctures in distinct rows; outer edge of middle and hind tibiae notched near apex. Fig. 760. .... *Paria quadrinotata* Say

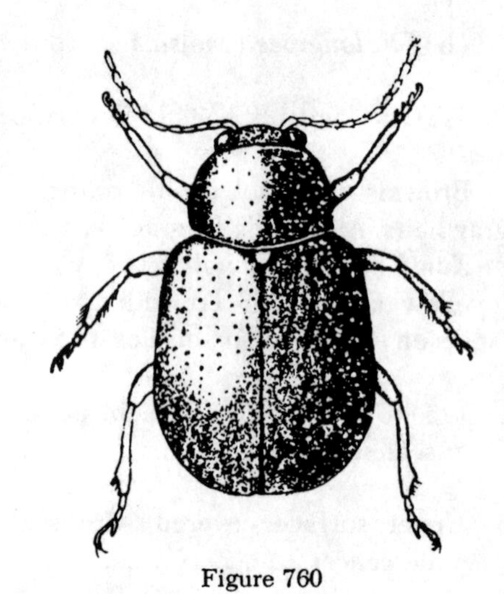

Figure 760

LENGTH: 3-4 mm. RANGE: Eastern United States and Canada.

Convex, usually shining and smooth. Color varies from dark head and pronotum with pale elytra with four dark spots to all dark or with the elytra dark with the pronotum pale. Primary host is black walnut. Taken on hickory and other hardwoods. Wilcox described 16 species, many very variable in color and having specific hosts.

23b Punctures on elytra scattered irregularly; tibiae entire. Fig. 761. .... *Chrysochus auratus* (Fab.)

Figure 761

LENGTH: 8-11 mm. RANGE: Eastern half United States and Canada.

Convex, shining. Brilliant green with brassy or coppery sheen. Under parts and antennae bluish-black. A very attractive beetle sometimes exceedingly common on dogbane and milkweed.

In the West it is replaced with *C. cobaltinus* Lec. somewhat shorter and stouter, and darker colored, a metallic dark blue or greenish.

24a (9) Tibia without teeth, slim claws with two prongs or a tooth. Fig. 762. .... *Gonioctena americana* (Schffr.)

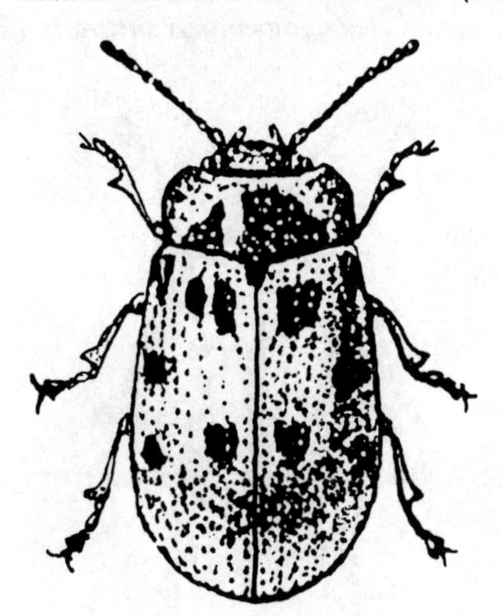

Figure 762

LENGTH: 6-8 mm. RANGE: Eurasia and United States.

Convex. Shining. Pale yellowish-brown marked as pictured with black. This "Poplar leaf beetle" feeds on both willow and poplar.

The willow family has many insect pests; another willow-feeding species, *Phratora americana* (Schffr.), 4-5 mm, is shining purple, blue, or green.

24b Claws simple. 25

25a (24) Third joint of tarsi notched or bilobed. 32

25b Third joint of tarsi uncut or but feebly notched (bilobed in *Prasocuris*). 26

26a (25) No margin at base of thorax; body elongate, striped. Fig. 763. .... *Prasocuris phellandrii* (L.)

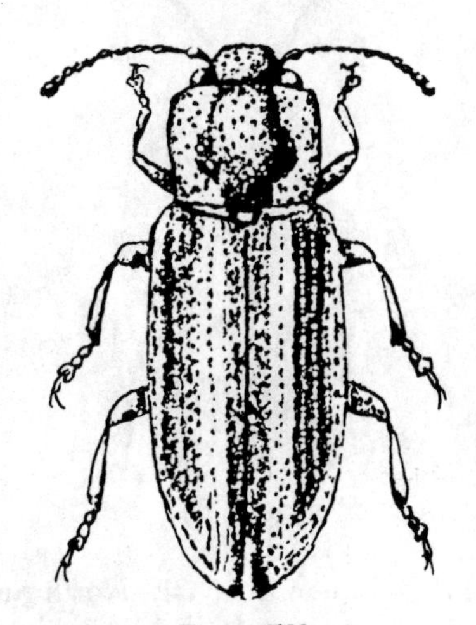

Figure 763

LENGTH: 5-6 mm. RANGE: Europe and United States.

Blackish with bronze sheen marked with dull yellow as shown. Fine punctures on elytra set closely in rows.

A similar but considerably smaller species *Hydrothassa vittata* Oliv. 3.5-4.5 mm, is greenish-black with markings of reddish-yellow. The two elytral stripes unite at the base and the thorax is but little wider than long. Common on buttercup.

26b Margin at base of thorax; stout; convex. 27

27a (26) Last joint of palpi cut square, shorter than preceding. 28

27b **Last joint of palpi no shorter than the preceding one. 29**

28a **(27) Mesosternum enlarged into a tubercle between the middle coxae. Fig. 764. .. .. *Labidomera clivicollis* (Kby.)**

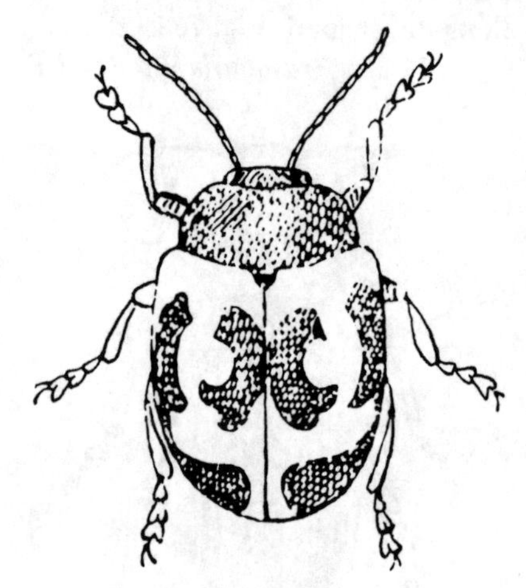

Figure 764

LENGTH: 8-12 mm. RANGE: North America.

Robust, convex. Dark bluish-black; elytra reddish-yellow with markings of black as shown. Often common on milkweeds.

28b **Mesosternum on same level with the prosternum. Fig. 765. .... (a) *Leptinotarsa decimlineata* (Say) Colorado Potato Beetle**

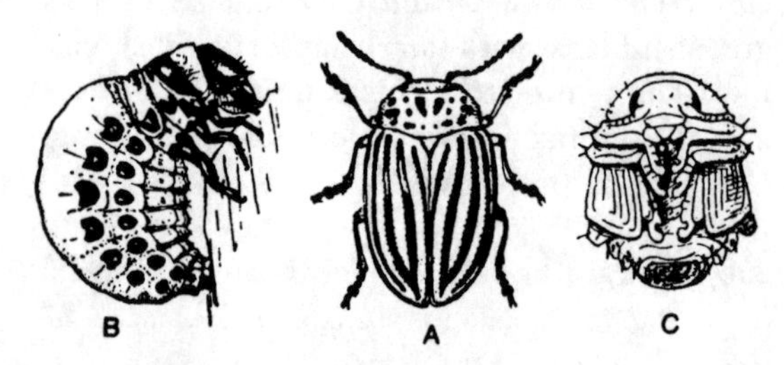

Figure 765

LENGTH: 5-11 mm. RANGE: United States and Canada.

Robust convex. Dull yellow, marked as shown with black. This beetle was likely the first insect to be controlled with an arsenical spray.

L. juncta (Germ.) False Potato Beetle with a length of 10-12 mm and ranging through the Southern States, closely resembles the above. The elytra have single instead of double rows of punctures and the outer face of the femora has a black dot.

29a **(27) Claws diverging; joint simple. .. 30**

29b **Claws united at base, parallel; joint with tooth beneath. Fig. 766. .... *Zygogramma suturalis* (Fab.)**

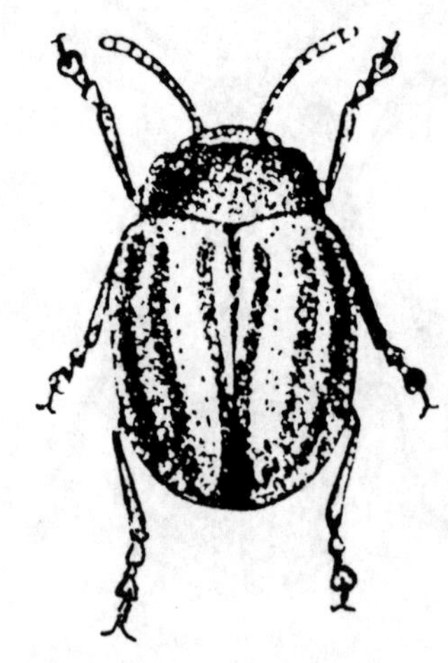

Figure 766

LENGTH:5.5-6.5 mm. RANGE: Colorado to East.

Convex, shining. Brown, with metallic sheen; elytra yellow marked with brown as pictured.

There are several other similar species of the genus most of them occuring in the West.

30a **(29) Elytra with brown and yellow stripes. Fig. 767. .... *Calligrapha praecelsis* Rog.**

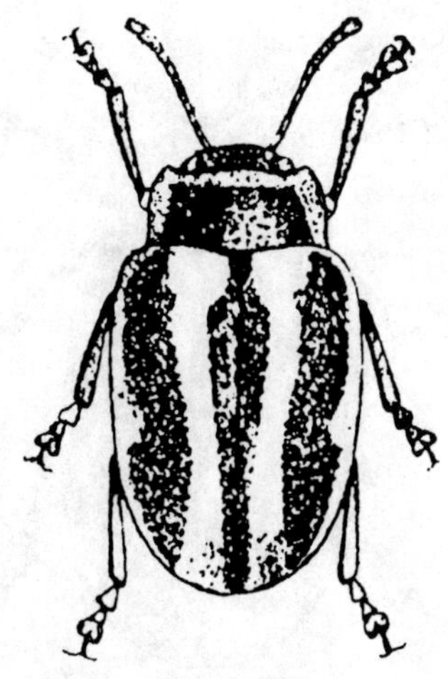

Figure 767

LENGTH: 7-8 mm. RANGE: Central United States.

Convex. Dark reddish-brown, bronzed, marked as shown with yellowish.

C. similis Rogers, 5-7 mm, a rather common species with much the same range, has its thorax wholly brown.

30b Elytra with various spots. 31

31a (30) Thorax wholly dark green. Fig. 768.

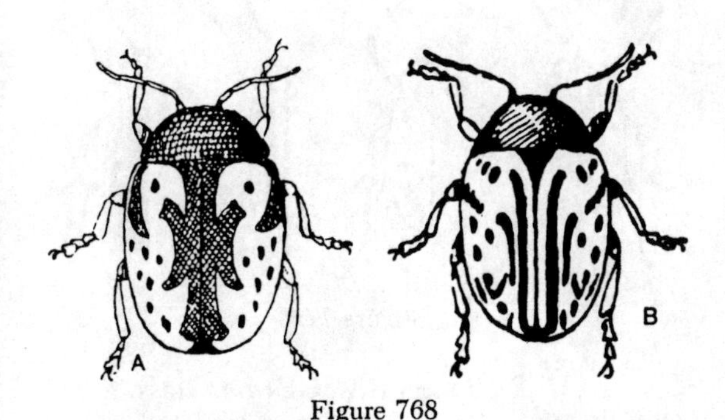

Figure 768

(a) ***Calligrapha scalaris* (Lec.) Elm Calligrapha**

LENGTH: 8-10 mm. RANGE: Eastern half United States.

Robust, convex, shining. Dark metallic green; elytra creamy-white marked with green as pictured. Live in basswood.

(b) ***C. philadelphica* (L.)**

LENGTH: 8-9 mm. RANGE: Eastern half United States.

Robust, convex, shining. Dark olive metallic green. Elytra creamy-white marked with dark green as shown. There are several other species closely similar to this one.

**31b Thorax partly pale. Fig. 769. ....
.... *Calligrapha multipunctata bigsbyana* (Kby.)**

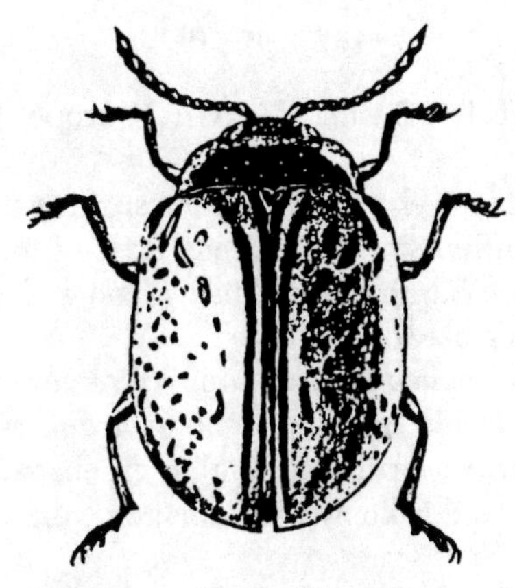

Figure 769

LENGTH: 6.5-8.5 mm. RANGE: United States and Canada.

Robust, convex. Brown to greenish, somewhat bronzed. Upper surface pale, marked as shown with brown or dark green.

It differs from *C. multipunctata* (Say) in having the pronotum pale with reddish-brown spots at center; the many elytral spots are black instead of greenish.

Calligrapha includes 36 species in the U.S. and Can.

32a (25) **Sides of thorax unthickened. Fig. 770.** .... ***Gastrophysa polygoni* L.**

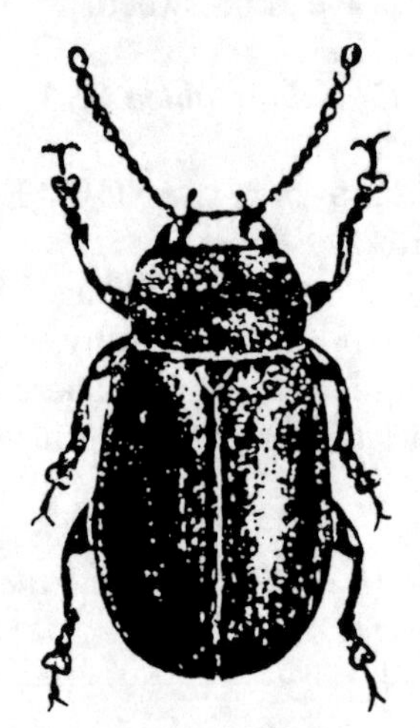

Figure 770

LENGTH: 4-5 mm. RANGE: Europe, United States.

Head, elytra and under surface metallic green or bluish. Thorax, legs, base of antennae and tip of abdomen reddish. Tarsi and part of antennae black.

G. cyanea Melsh., popularly known as the Green Dock Beetle, is found from coast to coast. It is a brilliant metallic green, measures 4-5 mm and is known to damage rhubarb.

32b Sides of thorax thickened. **33**

33a **(32) (a, b, c) Elytra reddish with black spots. Fig. 771.** .... ***Chrysomela knabi* Brown**

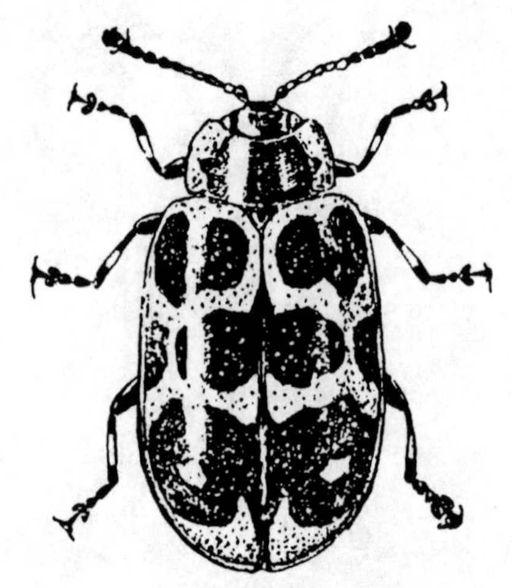

Figure 771

LENGTH: 6-9 mm. RANGE: So. Ont. to So. N. Eng. to N. Car. west to Alberta and N. Mex.

Black. Upper surface reddish with black markings. Adults and larvae both feed on poplar. Often mistaken by beginners for a lady beetle.

33b Elytra greenish-yellow or dull reddish with elongate spots. Fig. 772. .... ***Chrysomela scripta* (Fab.)**

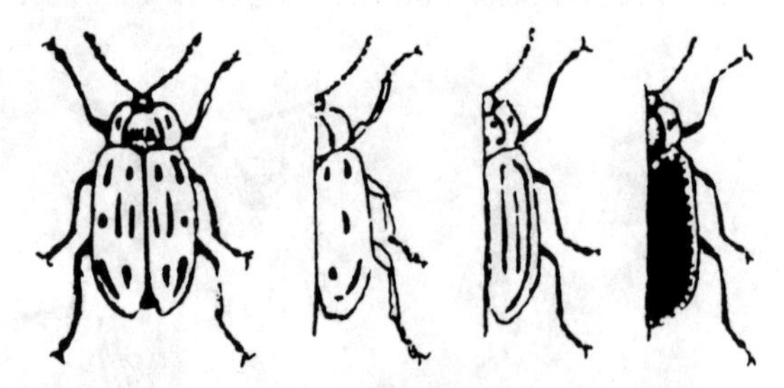

Figure 772

LENGTH: 7-10 mm. RANGE: So. Mass. to So. Ont., south to Fla., west to N. Mex., Utah, and So. B. C.

Colors as in key. Markings as pictured. Highly variable. Both adults and larvae feed on poplar and willow and are often very abundant. The figure shows four variations in the pattern.

33c Elytra dull yellow, without spots. Fig. 773. .... *Chrysomela crotchi* **Brown.**

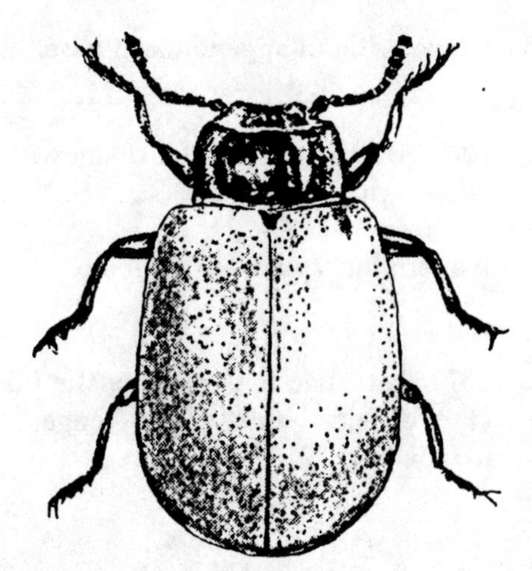

Figure 773

LENGTH: 7-9 mm. RANGE: Nova Scotia to Va., to Idaho and central Alaska.

Head and thorax green; elytra yellowish-brown.

Defoliates poplars.

This, like some other leaf beetles, could be mistaken for a lady beetle. The number of tarsi clears up any doubt. This genus contains 19 species, mostly in the northern U. S. and Canada.

34a (8) Hind legs for leaping, with much thickened femora (a). (Flea Beetles) Fig. 774. 42

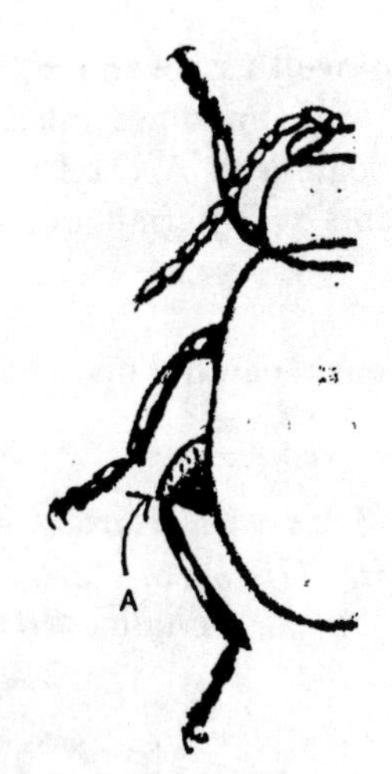

Figure 774

34b Hind legs for walking; femora slender. 35

35a (34) Front coxal cavities closed behind. Fig. 775. .... *Cerotoma trifurcata* (Forst.) Bean Leaf Beetle

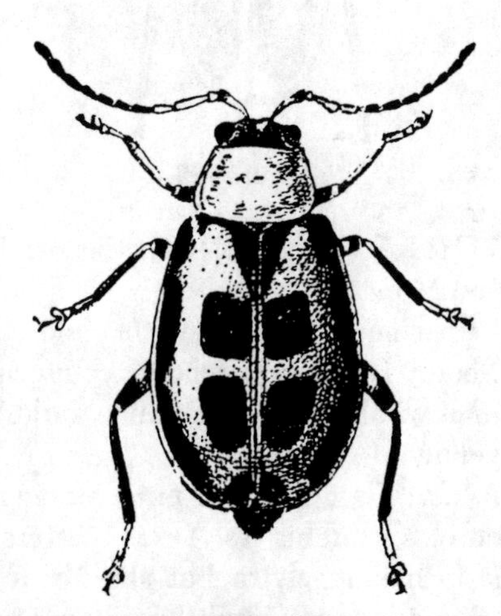

Figure 775

LENGTH: 3.5-5 mm. RANGE: Eastern half United States and Canada.

Convex. Under parts black. Upper surface dull yellow, occasionally reddish, marked as pictured with black.

This as well as several other species of leaf beetles are sometimes mistaken for lady beetles by beginners. A careful check of the tarsal segments will permit one to be certain about that.

35b **Front coxal cavities open behind.** **36**

36a **(35) Claws with a broad appendage at base. Fig. 776.** .... ***Phyllecthris gentilis* Lec.**

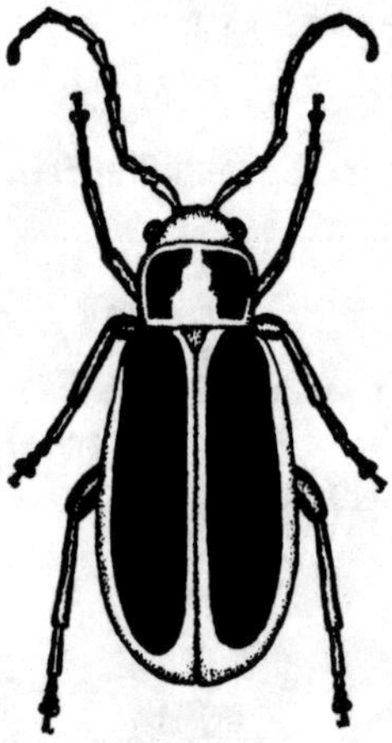

Figure 776

LENGTH: 2.5-4 mm. RANGE: Eastern half of U. S. and Mont.

Upper surface black with yellow markings. Thorax frequently wholly yellow; elytra sometimes wholly black. Antennae dark. Under parts yellow.

P. dorsalis (Oliv.), 6 mm, ranging from District of Columbia to Texas, differs from *gentilis* in having elytra but slightly if at all wider than the thorax and the antennae of the male with 11 segments instead of 10. The under parts are black.

Scelolyperus cyanellus Lec., 3.5-4.5 mm differs from the above in having spurs on the tibiae and in the elytra being a solid dark blue instead of being marked in two colors; the antennae and legs are yellow. It is known in N. C., Ohio, Pennsylvania. It feeds on wild roses.

36b **Claws without appendage at base.** **37**

37a **(36) At least part of the tibiae with terminal spurs.** **41**

37b **No terminal spurs on the tibiae.** **38**

38a **(37) Antennae not reaching the middle of the body, third joint longer than fourth. Fig. 777.**

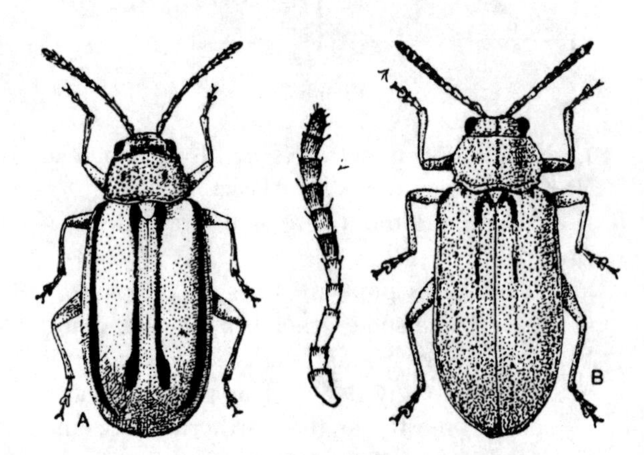

Figure 777

(a) *Erynephala puncticollis* (Say)

LENGTH: 7-9 mm. RANGE: Manit., Idaho, Mex.

Wholly pale yellow to dark brown; the elytral margins and suture sometimes marked with black.

(b) *Monoxia consputa* (Lec.)

LENGTH: 3.5-4.5 mm. RANGE: Tex, Wyo., Cal.

Wholly pale yellowish-brown often darker on dorsum. The adults injure the leaves of the sugar beet. It is known as the "Western Beet Leaf Beetle." This genus includes 15 species, the majority western.

38b Antennae extending beyond middle of the body, claws with two prongs. 39

39a (38) Fourth segment of antennae longer than third. Fig. 778. .... *Trirhabda canadensis* Kby.

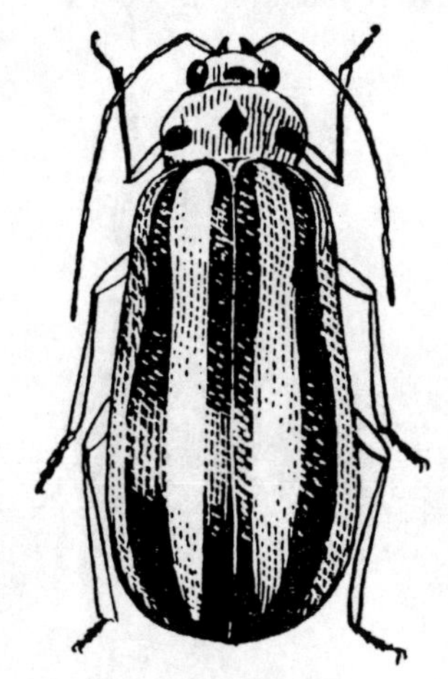

Figure 778

LENGTH: 7-10 mm. RANGE: So. Can. and No. U. S. in west, south to Cal. and Ariz.

Upper parts yellowish, marked as shown with black. Common name "Goldenrod Beetle."

T. luteocincta (Lec.) of similar size and living in Cal. is iridescent green to blue to deep purple. The larvae feed on wormwood.

Included here are 25 species spread over the continent.

39b Fourth segment of antennae shorter than third. 40

40a (39) Elytra with long black markings or narrow stripes. Fig. 779. .... *Ophrailla notata* (Fab.)

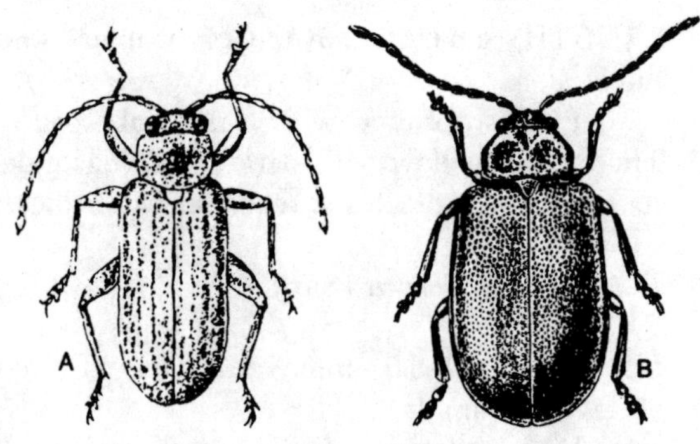

Figure 779

LENGTH: 3.5-5 mm. RANGE: United States and Canada.

Pale yellow with scattered prostrate pubescence. Elytra and thorax marked as shown in black.

Eight rather common beetles are included in this genus.

Xanthogaleruca luteola (Mull.). The Elm Leaf Beetle, 5 mm yellow or orange with two black stripes and a spot on each elytron, reached this country from Europe around 1834 and has spread through many of our states. Both larvae and adults are serious pests in defoliating elm trees.

40b No elongated black marks on elytra. Fig. 780 and Fig. 779b.

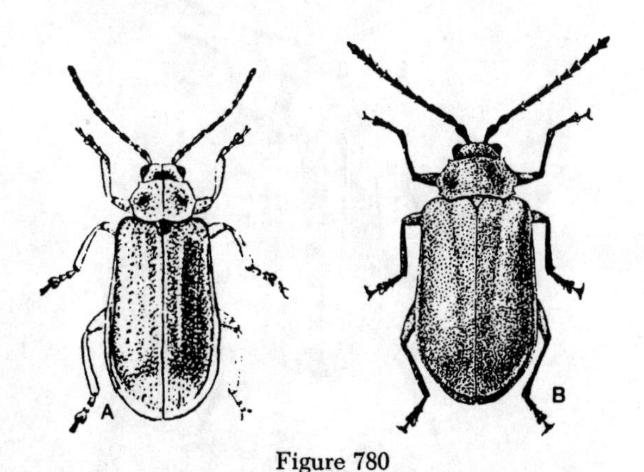

Figure 780

(a) *Galerucella nymphaeae* (L.) Water Lily Leaf Beetle

LENGTH: 4.5-6 mm. RANGE: Canada and the U. S.

Blackish-brown with fine pubescence. Thorax dull yellow with dark spots; legs pale. As the name indicates it feeds on waterlilies.

(b) *G. decora* (Say)

LENGTH: 4.5-5.5 mm. RANGE: United States and Canada.

Dull yellow to brown with yellowish pubescence. Antennae dark; legs pale. Lives on willows. Fig. 779b.

(c) ***Tricholochmaea cavicollis* (Lec.)**
Cherry Leaf Beetle

LENGTH: 4.5-5.5 mm. RANGE: Eastern United States and Canada.

Bright red with eyes, antennae and tips of legs black. It ranges throughout our country and eats the leaves of peach and cherry. Some 14 rather common species are included in this genus. Fig. 780b.

41a (37) Elytra with black stripes. Fig. 781.
.... *Acalymma vittatum* (Fab.)

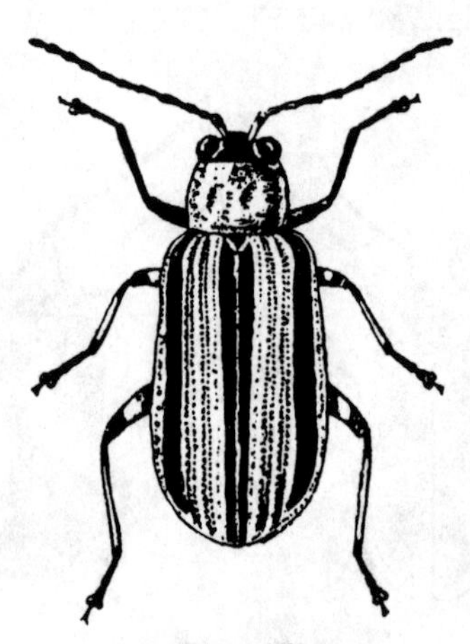

Figure 781

LENGTH: 4.5-6 mm. RANGE: Colorado east and Canada.

Upper surface yellow, marked as shown with black. This "Striped Cucumber Beetle" is a general pest. In the West it is replaced by *A. trivittatum* (Mann.) about the same size, but having the thorax darker and only the basal half of the first antennal joint, and the base of the femora pale yellow. Altogether there are 5 members of this genus in our fauna.

41b Elytra with spots. Fig. 782. ....
.... *Diabrotica undecimpunctata* Mann.
Spotted Cucumber Beetle

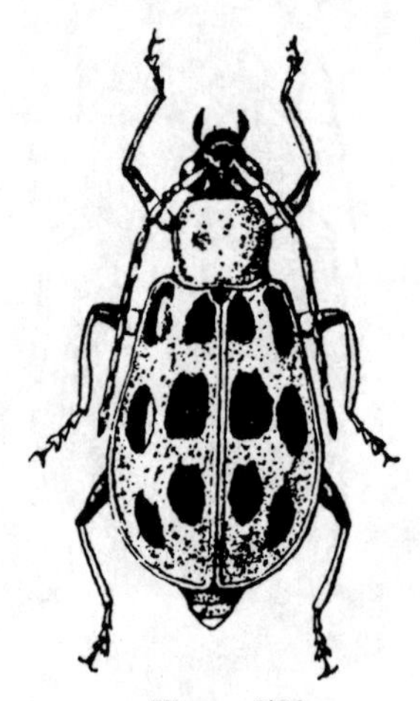

Figure 782

LENGTH: 6-7.5 mm. RANGE: United States.

Pale greenish-yellow marked with black. Three basal joints of antennae and base of femora pale yellow. It is also known as the Southern Corn Rootworm.

41c Elytra with neither spots or stripes. Fig. 783. .... *Diabrotica longicornis* (Say)
Corn Rootworm

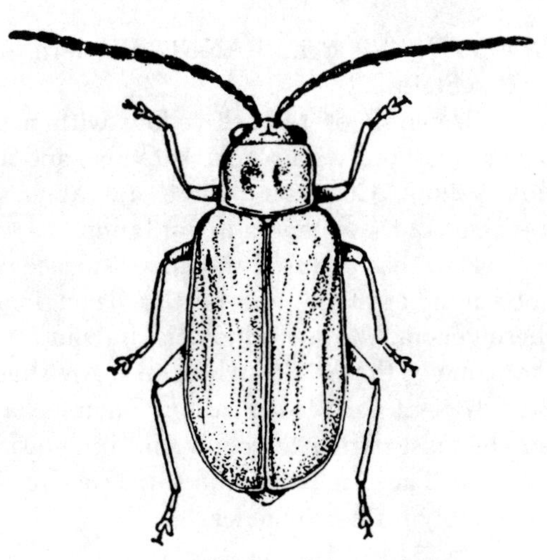

Figure 783

LENGTH: 5-6 mm. RANGE: Eastern and Central States.

Pale green or greenish-yellow. Antennae and sometimes the head and thorax with reddish-brown tinge. This "Northern Corn Rootworm" is often abundant feeding on corn silk, and on the flowers of golden-rod and sunflowers. Since the eggs are laid in corn fields rotation effectively controls the larvae.

Ten species of *Diabrotica* are found in the U. S. and Canada.

42a (34) Front coxal cavities open behind. 43

42b Front coxal cavities closed behind. 52

43a (42) Last joint of hind tarsus hemisphere shaped; elytral punctures scattered. 44

Beetles under 44 were until recently all included in the genus *Oedionychis*. This genus has been broken into two genera using the characteristics in 44a and 44b.

43b Last joint of hind tarsi not as in 43a; usually slender. 45

44a (43) Antennae stouter, short; elytral margins not flattened. Fig. 784. .... *Kuschelina vians* (Ill.)

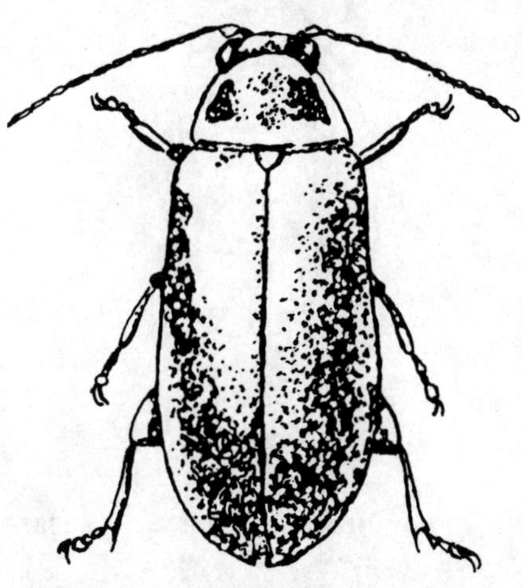

Figure 784

LENGTH: 4.5-7 mm. RANGE: West of Colorado.

Black, often with greenish or violet sheen. Thorax reddish-yellow with black spot which sometimes is an inverted W. Abdomen yellowish at margins and apex.

There are slightly less than 20 members of this genus in the U. S. and Can.

44b Antennae slender; reaching at least to middle of body. Elytral margins flattened. Fig. 785. .... *Capraita sexmaculata* (Ill.)

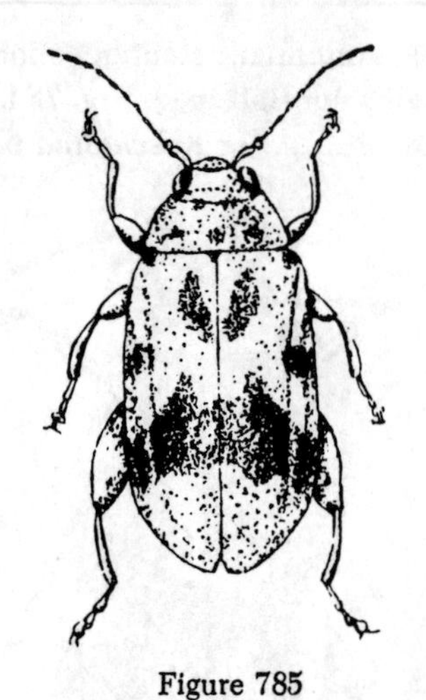

Figure 785

LENGTH: 3-4 mm. RANGE: Eastern half United States.

Brownish or reddish-yellow with markings as pictured with black. Antennae and legs dull yellow. Common. There are about 15 members of these genera in our fauna.

Horn has pictured the upper surface of a good number of the flea beetles belonging to these genera. We reproduce his drawings with the name of the beetles which in many should be sufficient for identification. Their color is for the most part pale yellow and brownish or blackish. The catalog numbers used in Fig. 786 are merely for convenience.

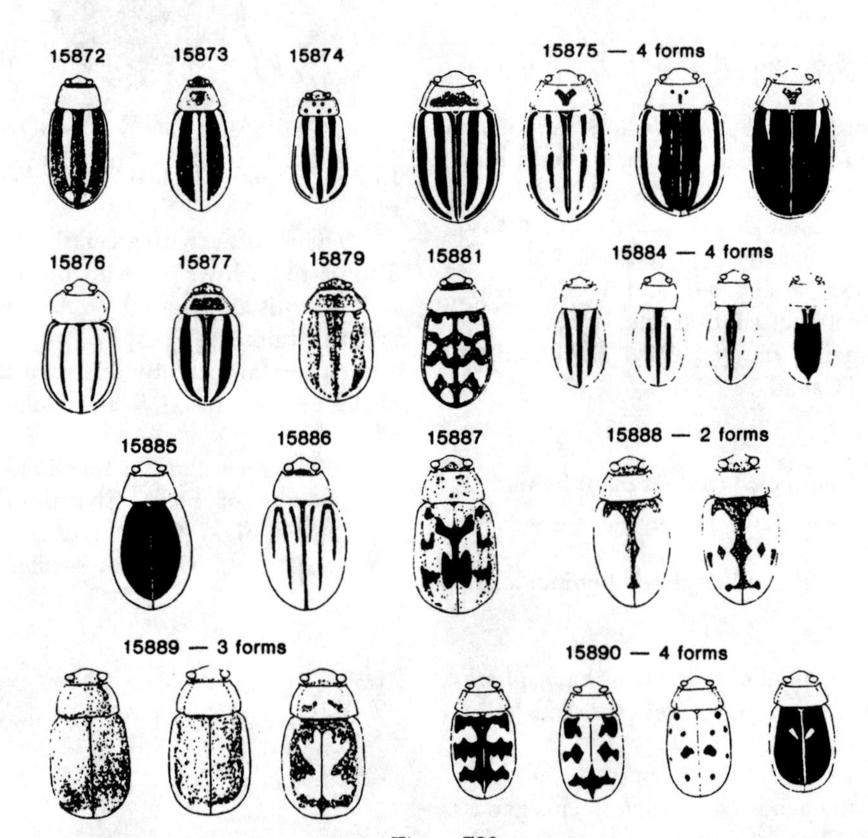

Figure 786

Cat. No.	Name	Length	Range
15872	*K. interjectionis* Cr.	6-6.5 mm	Texas, Mexico
15873	*K. fimbriata* (Forst.)	5-7.5 mm.	From Texas east

15874....*K. aemula* Horn ....4.5-5.5 mm .... Arizona
15875....*K. peturista* (Fab.) 4 forms ....5.5-8.5 mm .... From Texas east
15876....*K. tenuilineata* Horn ....6.5-7 mm .... So. Arizona
15877....*K. miniata* (Fab.) ....5-6.5 mm .... From Texas east
15879....*K. ulkei* Horn ....4.5-5 mm .... Florida
15881....*K. jacobiana* Horn ....6.5 mm .... So. Arizona
15884....*C. texana* Cr. 4 forms ....4-4.5 mm .... Texas
15885....*C. thyamoides* Cr. ....4-4.5 mm .... Eastern half U. S.
15886....*C. circumdata* Rand. ....3.5-5 mm .... Texas and eastward
15887....*C. sexmaculata* (Ill.) ....3.5-4 mm .... Eastern half U. S.
15888....*C. suturalis* (Fab.) 2 forms ....3.5-4 mm .... Georgia, Florida
15889....*C. quercata* (Fab.) 3 forms ....3.5-4 mm .... From Colo. eastward
15890....*C. scalaris* Melsh. 4 forms ....4.5-5 mm .... Texas and eastward

K. Kuschelina
C. Capraita

45a (43) Thorax with a transverse impressed line across its base. 48

45b Thorax without a transverse impression as in 45a. 46

46a (45) First segment of hind tarsi comparatively short and broad; claws with appendages. 47

46b First segment of hind tarsi long and slender; claws simple. 49

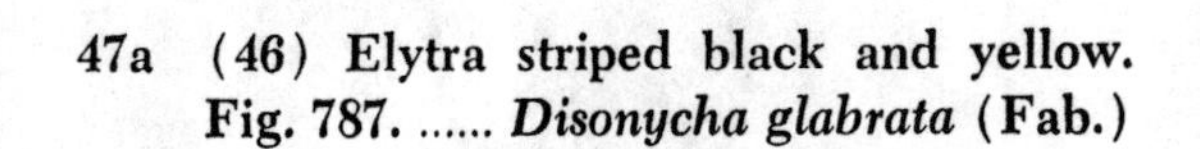
47a (46) Elytra striped black and yellow. Fig. 787. ...... *Disonycha glabrata* (Fab.)

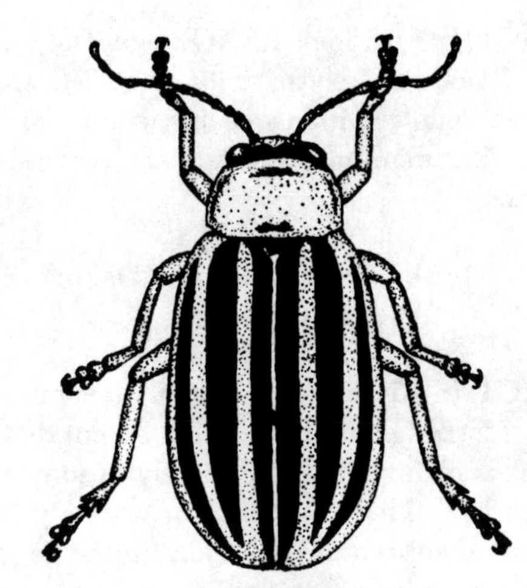

Figure 787

LENGTH: 5-5.5 mm. RANGE: N. Y. to Fla., west thru Ind. to Tex. and Ariz.

Highly shining. Upper surface yellow marked with varying amounts of black. Common.

D. alternata Ill. ranging throughout much of the United States is larger 5-9 mm and usually has four black dots on thorax. It feeds on willow. *D. pensylvanica* (Ill.) Eastern third of U. S. and Tex. Striped black and white or pale yellow; 5-6 mm.

47b Elytra not striped. Fig. 788.

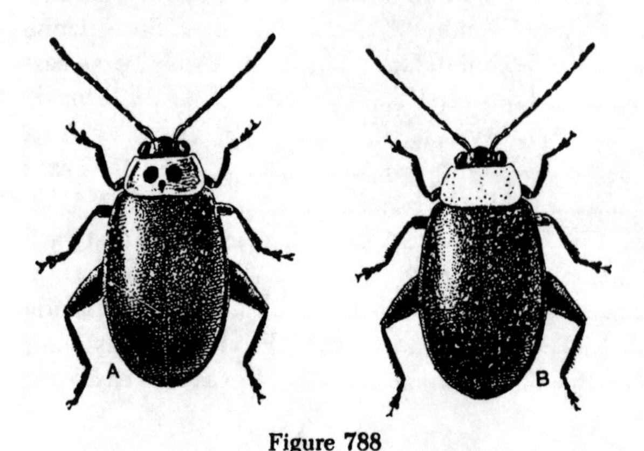

Figure 788

(a) *Disonycha triangularis* (Say)

LENGTH: 5-6.5 mm. RANGE: general.

Black, the elytra with a bluish sheen. Thorax yellow with three black dots as pictured. Common on beets, spinach and related plants.

(b) *D. xanthomelaena* (Dalm.) Spinach Flea Beetle

LENGTH: 4.5-5.5 mm. RANGE: general.

Marked like the preceding except that the thorax is plain yellow and the elytra somewhat greenish. The legs are partly or wholly yellow, instead of entirely black. It has much the same food tastes as its near relative.

Disonycha includes about 30 species in our fauna.

48a (45) Usually not over 4 mm long. Fig. 789. .... *Altica ignitia* Ill.

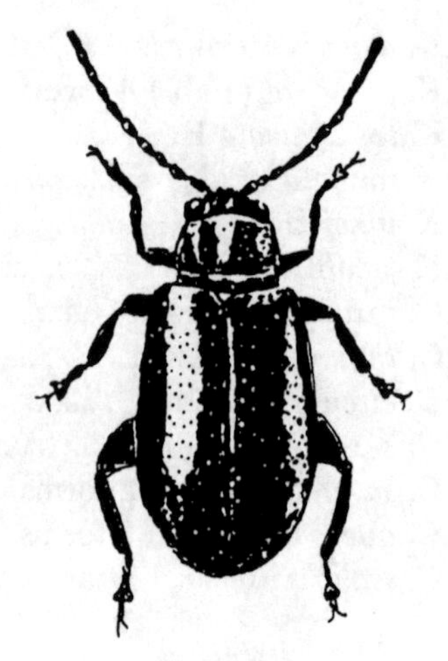

Figure 789

LENGTH: 3-4 mm. RANGE: United States.

Metallic green, brassy, bronzed or sometimes purplish. It is sometimes known as the Strawberry Flea Beetle.

A pest of grapes in the Southwest, *A. torquata* Lec. 3-4 mm is metallic blue or purple while the "Apple Flea Beetle" another western species, *A. foliaceae* Lec. is similar in color and size.

48b Length over 4 mm. Fig. 790. .... *Altica chalybea* Ill.

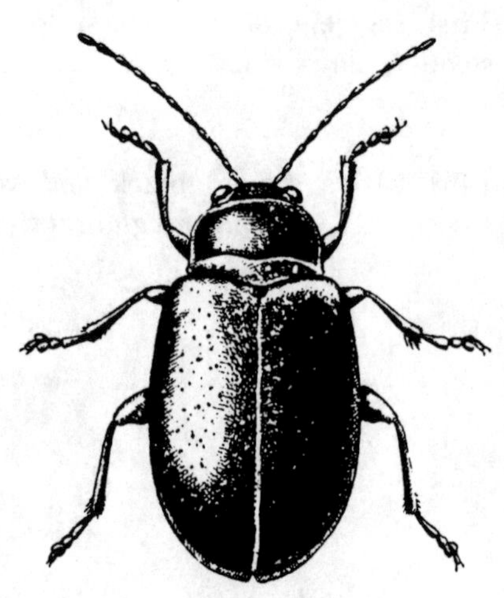

Figure 790

LENGTH: 4-5 mm. RANGE: general.

Metallic blue. Feeds on wild grape and poison ivy.

The "Alder Flea Beetle" A. *bimarginata* Say ranging throughout North America and measuring 5-6 mm, dark steel blue is readily distinguished by a longitudinal fold on the sides of the elytra.

Altica contains about 70 very similar species in our area. Probably there are also numerous unnamed ones.

49a (46) Elytra usually with yellow spots or stripe; hind tibiae not grooved on outer edge. 50

49b Elytra without yellow spots or stripe; hind tibiae with a groove on the tip of outer edge. 51

50a (49) Elytra black and yellow. Fig. 791. ..

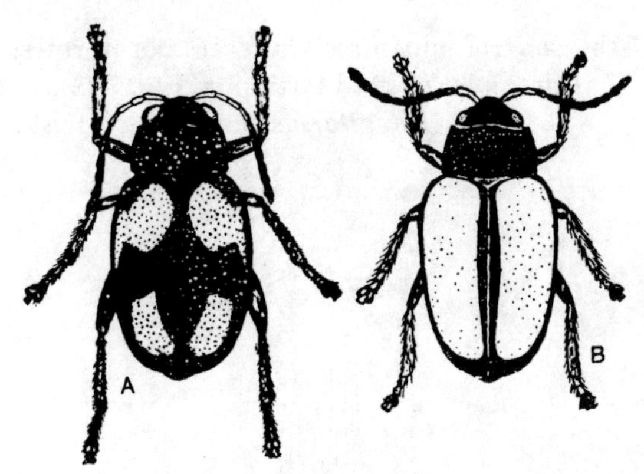

Figure 791

(a) *Phyllotreta bipustulata* (Fab.)

LENGTH: 2-2.5 mm. RANGE: Europe and eastern half United States.

Robust. Non-metallic black and marked as pictured with yellow.

(b) *P. armoraciae* (Koch.) Horseradish Flea Beetle

LENGTH: 3-3.5 mm. RANGE: Europe and United States.

Convex, shining. Black marked as pictured with dull yellow.

50b Elytra of one color. Fig. 792. .... *Phyllotreta pusilla* Horn Western Black Flea Beetle

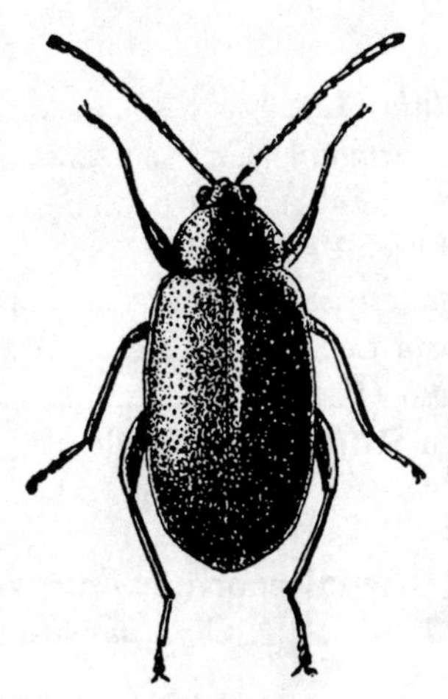

Figure 792

LENGTH: 1.5-2 mm. RANGE: Western half United States.

Robust, shining. Black or dark olive-green. Travels in swarms and causes heavy losses to mustard crops.

The Colorado Cabbage Flea Beetle *P. albionica* (Lec.) is a western pest of the mustard family as well as other plants. It is wholly black with a faint brassy sheen and measures 1.5-2 mm.

There are about 40 members of this genus in the U. S. and Canada. Many are blackish and unmarked.

Others of the Genus Phyllotreta

The left elytron and the characteristic antenna of six other species of the Genus *Phyllotreta* as pictured by Horn are reproduced in (Fig. 793). The name, size and distribution is associated with the catalog number. The ground color is black and the markings pale yellow in each case.

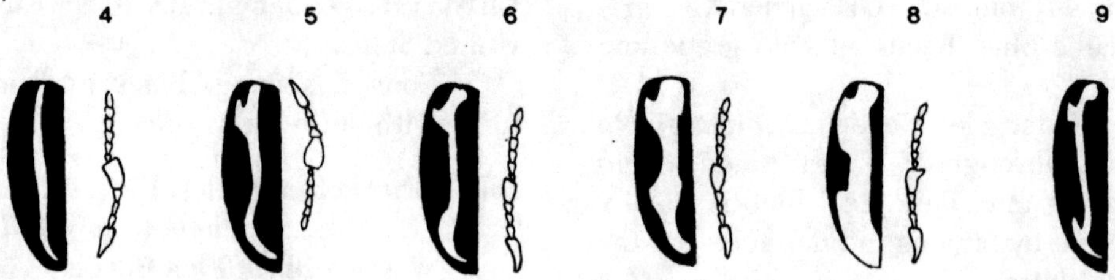

4 5 6 7 8 9

Figure 793

4....*P. lepidula* (Lec.) ....2-2.5 mm .... Southwest
5....*P. zimmermanni* (Cr.) ....2.5 mm .... N. Eng.-Man.
6....*P. vittata* (Fab.) ....2 mm .... Eastern half U. S.
Striped Flea Beetle
7....*P. oregonensis* (Crotch) ....2.56-3 mm .... Oregon, Texas, Cal.
8....*P. robusta* Lec. ....2 mm. .... Mich., Man.
9....*P. ramosa* (Cr.) ....2 mm .... Calif., Nev., N. Y.
Western Striped Flea Beetle

51a (49) Elytral punctures in regular rows. Fig. 794. .... *Glyptina brunnea* Horn

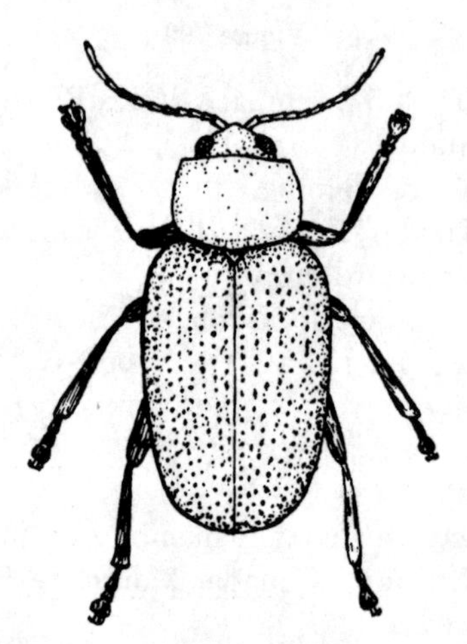

Figure 794

LENGTH: 1.8-2 mm. RANGE: Eastern half United States.

Convex; shining. Uniformly reddish-brown above, under parts brown; antennae and legs paler.

G. cerina (Lec.) a southwestern species is pale reddish-yellow with darker head and measures 2-2.5 mm. It feeds on potatoes.

51b Elytral punctures scattered, not in rows; first joint of hind tarsi long. Fig. 795. .... *Longitarsus melanurus* Melsh.

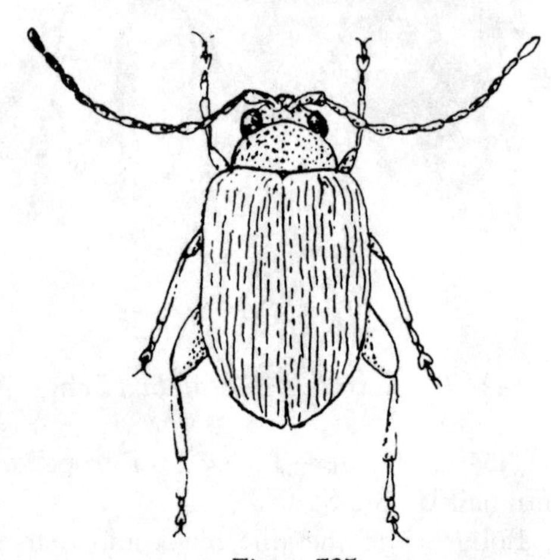

Figure 795

LENGTH: 2-2.5 mm. RANGE: Eastern half United States and Canada.

Convex, shining. Blackish or dark reddish-brown. First and second legs dull yellow. Elytral punctures deep, coarse and abundant.

This is another difficult genus of some 40 similar appearing species.

52a (42) (a, b, c) **Antennae with 12 segments; claws divided. Fig. 796.** .... ***Blepharida rhois* (Forst.)**

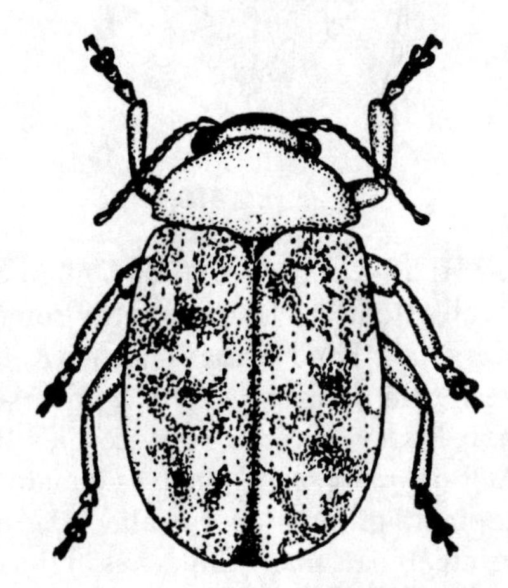

Figure 796

LENGTH: 5-6.5 mm. RANGE: general.

Robust, convex. Upper surface dull yellow with irregular blotches of reddish-brown, head and thorax darker; under parts reddish-brown. Feeds on sumach and known as "sumach flea beetle or the "jumping sumach beetle." It is the only member of its genus.

52b **Antennae with only 10 segments; hind tarsi attached above the apex. Fig. 797.** .... ***Psylliodes convexior* Lec.**

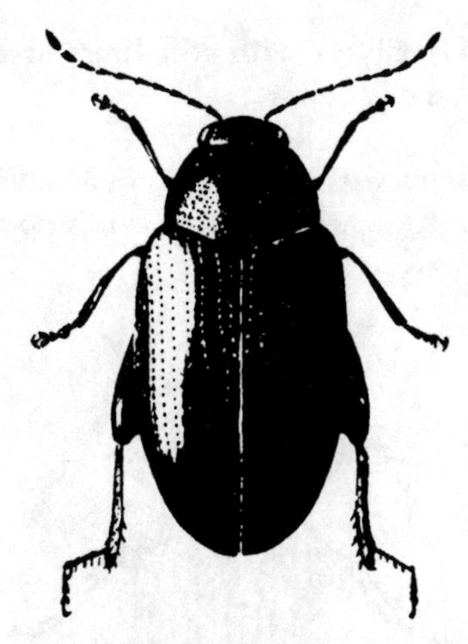

Figure 797

LENGTH: 2-2.5 mm. RANGE: general.

Convex, shining. Piceous with sheen of bronze, antennae piceous, the first three joints paler.

The Hop Flea Beetle *P. punctulata* Melsh. is similar in size, color and range. The larvae feed on hop roots, while the adults eat the leaves of many plants. Its impressed striae and more abundant elytral punctures distinguish it from the preceding.

52c **Antennae with 11 segments.** **53**

53a (52) **Hind tibiae sinuate near apex; their outer margin with a tooth. Thorax not impressed at base.** **59**

53b **Neither tooth or sinuation on hind tibiae.** **54**

54a (53) **Distinct impression across thorax near its base; punctures in rows on elytra.** **55**

54b **No impression across thorax in front of base; punctures scattered on elytra.** **58**

55a (54) Elytra with stiff hairs arranged in rows on the intervals. 56

55b Elytra without hairs as in 55a. Fig. 798. .. .. *Derocrepis erythropus* Melsh.

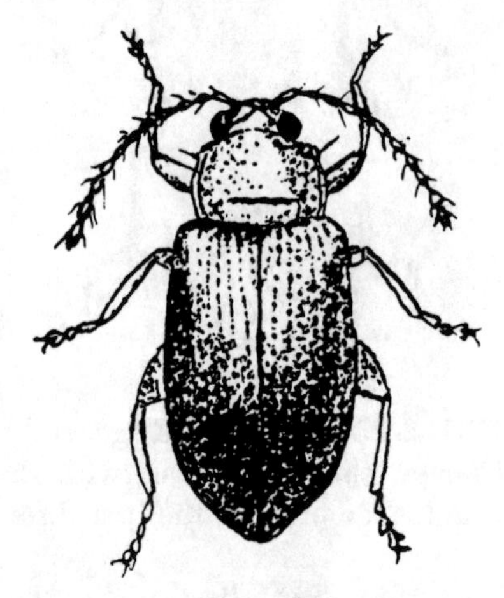

Figure 798

LENGTH: 2.5-3 mm. RANGE: general.

Elytra dark blue shining; head, thorax and legs reddish-yellow; under surface blackish. Feeds on black locust and other trees. Introduced from Europe.

Crepidodera nana Say, is a widely scattered species running into several varieties. It measures 2.5-3.5 mm, is colored with metallic bronzes, blues or greens (including the thorax) and feeds on the leaves of various trees, especially willow.

56a (55) Upper surface piceous. 57

56b Upper surface not piceous; lighter. Fig. 799. .... *Epitrix hirtipennis* Melsh.

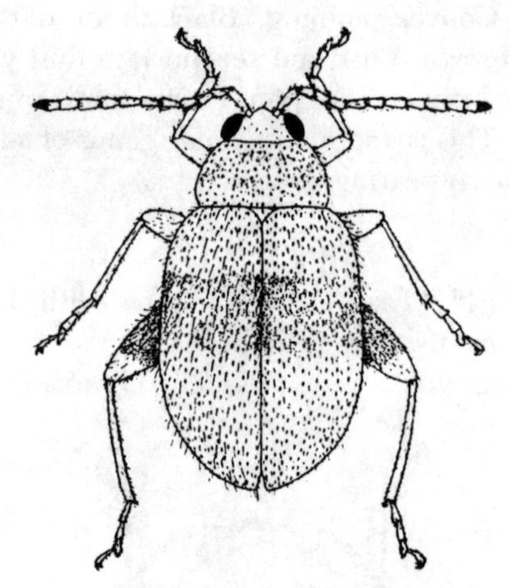

Figure 799

LENGTH: 1.5-2 mm. RANGE: United States.

Dull. Reddish-yellow; elytra sometimes with darkened band across middle. Antennae and legs pale reddish-yellow. Under surface brown. It is known as the Tobacco Flea Beetle, but feeds on many plants, such as tomato, potato, eggplant, ground-cherry, almond, squash, orange, etc. It eats many tiny holes in the leaves and is often very destructive.

57a (56) Thorax with light punctures, scattered. Fig. 800.

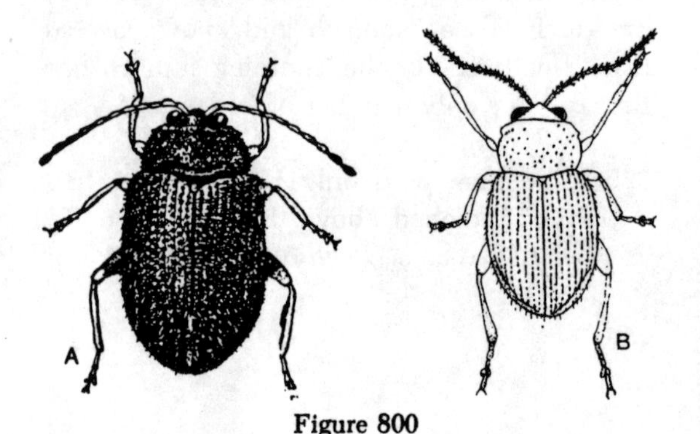

Figure 800

(a) *Epitrix cucumeris* Har.

LENGTH: 1.5-2 mm. RANGE: United States.

Shining. Black, legs and antennae reddish, elytral striae rather shallow. This Potato Flea Beetle feeds on practically all members of the nightshade family as well as some other plants. It is a serious pest.

(b) ***E. brevis*** **Sz.**

LENGTH: 1.5 mm. RANGE: Eastern half of United States.

Shining. Piceous. Antennae and legs pale reddish-brown; hind femora piceous.

57b Punctures on thorax heavy and crowded. Fig. 801. .... *Epitrix fuscula* Cr.

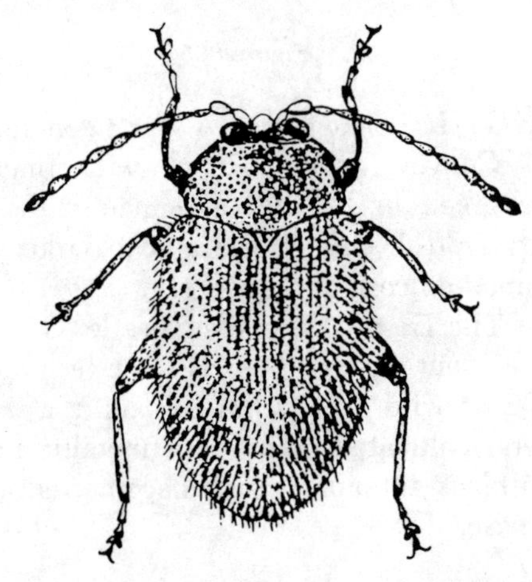

Figure 801

LENGTH: 2 mm. RANGE: Eastern half of United States.

Robust, convex, feebly shining. Piceous. Antennae and legs except femora reddish yellow. Elytral intervals with seta bearing punctures. This is the flea beetle that works havoc with egg plants.

The genus *Epitrix* is not large but makes a serious menace to garden crops. They are so small as to often pass unnoticed. They eat small holes in the leaves and are difficult to control.

This genus includes 12 species in our area.

58a (54) Elytra wholly dark. Fig. 802.

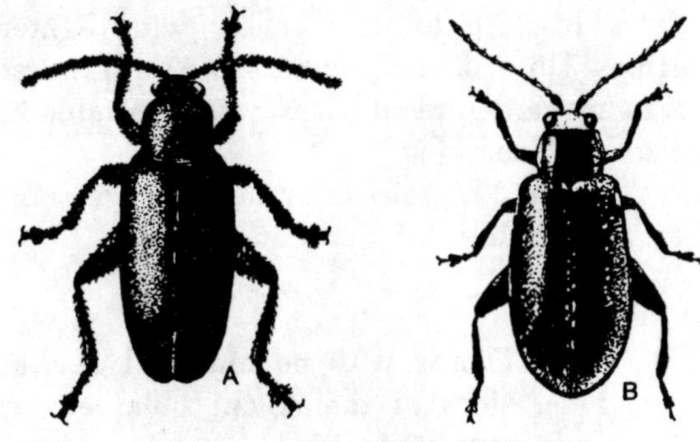

Figure 802

(a) ***Systena hudsonias*** **(Forst.)**

LENGTH: 4-4.5 mm. RANGE: general.

Shining. Black; segments 3-4-5 of antennae dull yellow, the others dark; legs black. Elytra with crowded coarse punctures.

(b) ***S. frontalis*** **(Fab.)**

LENGTH: 3.5-4.5 mm. RANGE: general.

Head dull reddish-yellow, otherwise much like the preceding but less shining.

58b Elytra striped. Fig. 803. .... *Systena taeniata* (Say)

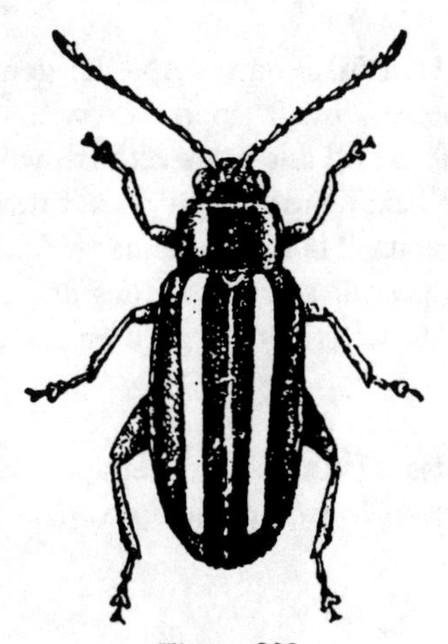

Figure 803

LENGTH: 3-4.5 mm. RANGE: general.

Head and thorax usually reddish brown; elytra blackish to pale yellow with lighter stripes. This is a very common and serious pest of many garden plants. It is highly variable in shades and marking.

About 20 species of *Systena* are presently known from the U. S. and Canada.

59a (53) Thorax with no marginal line at base; sides of thorax cut obliquely at front angles. Fig. 804. .... *Chaetocnema confinus* Cr.

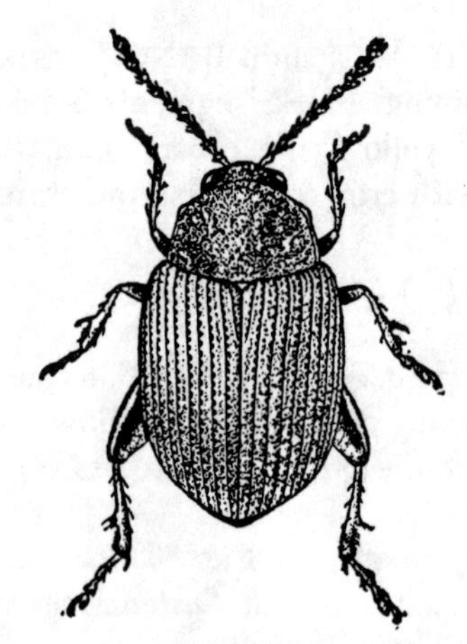

Figure 804

LENGTH: 1.5-1.8 mm. RANGE: general.

Robust, oval. Piceous, with tinge of bronze; legs and antennae reddish yellow; hind femora black; head without punctures but finely alutaceous. Thorax twice as wide as long. A common pest of sweet potatoes and other garden plants where it eats channels along the veins.

59b Sides of thorax not cut as in 59a but curved from base to apex. 60

60a (59) Head without punctures. Fig. 805. .... *Chaetocnema pulicaria* Melsh.

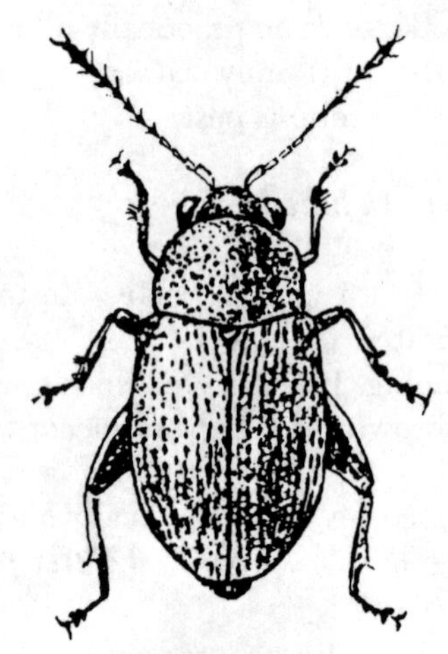

Figure 805

LENGTH: 1.5-1.8 mm. RANGE: general.

Convex, shining. Black with tinges of blue green or bronze; antennae with basal joints reddish yellow, the apical darker. It is common on corn and grasses.

The Desert Corn Flea Beetle, *C. ectypa* Horn about 1.5 mm long is a western species living on wild grasses but becoming a pest of several cultivated crops. It is metallic bronze with black antennae which become reddish at the base.

60b Head with punctures. Fig. 806. .... *Chaetocnema denticulata* (Ill.)

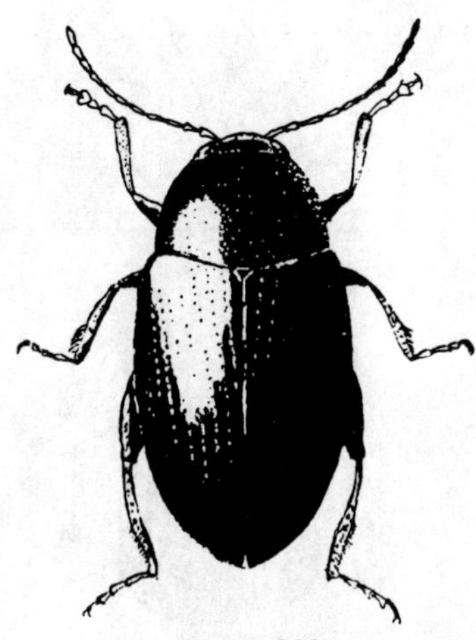
Figure 806

LENGTH: 2-2.5 mm. RANGE: United States.

Robust, sub-opaque. Piceous with bronze or brassy sheen. Antennae reddish-yellow at base, darker at apex. Tibiae and tarsi reddish-yellow. This Toothed Flea Beetle often damages corn, beets, sorghum and other grass crops.

C. cribrifrons Lec., 2.5 mm. and widely distributed, is quite similar to the above but differs in being shining and in having the punctures on the head coarse and well separated.

This genus contains 30 or more species in our fauna.

61a (6) Antennae with only eight segments. Fig. 807. ...... *Octotoma plicatula* (Fab.)

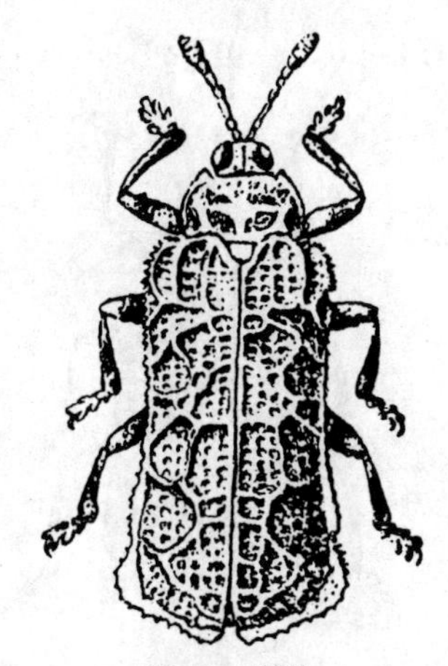
Figure 807

LENGTH: 4.5-5 mm. RANGE: Eastern half United States.

Depressed, dull. Piceous, with tinge of bronze. Antennae reddish brown, club darker. Thorax in front, scutellum and base of femora dull reddish-yellow. Elytra and thorax with a design of ridges as pictured. On trumphet vine.

This and the two genera which follow are made up of a few peculiar wedge-shaped beetles, very interesting in form and coloring but of little economic importance.

Another species, *O. marginicollis* Horn, is found in Ariz.

61b Antennae with 11 segments. 62

62a (61) Elytra with but three entire ridges. Fig. 808. .... *Odontota dorsalis* (Thunb.)

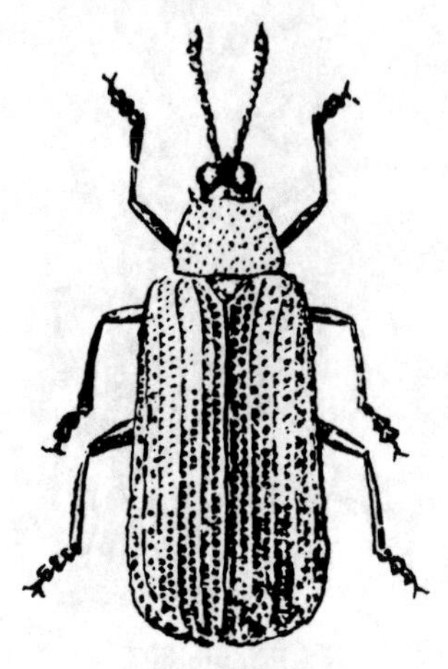

Figure 808

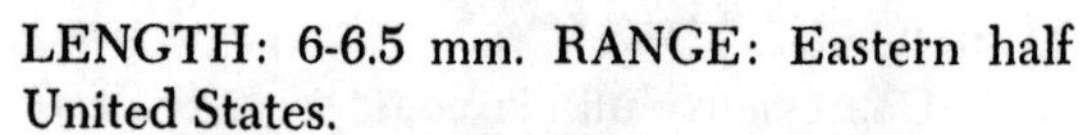

LENGTH: 6-6.5 mm. RANGE: Eastern half United States.

Shining bluish black with thorax and part of elytra, scarlet. A mid dorsal stripe on elytra black widening from base to apex. It seems to find its interest in leguminous plants.

Chalepus scapularis (Oliv.) 5-7 mm, differs in having black elytra with red shoulders and the underparts wholly black. The thorax is red, but darkened at center. It ranges from Arizona eastward.

C. bicolor (Oliv.), 6 mm, is red beneath with thorax wholly red and elytra entirely black.

62b Elytra with a partial fourth ridge inserted between the second and third at both apex and base. Fig. 809. .... *Baliosus ruber* (Web.)

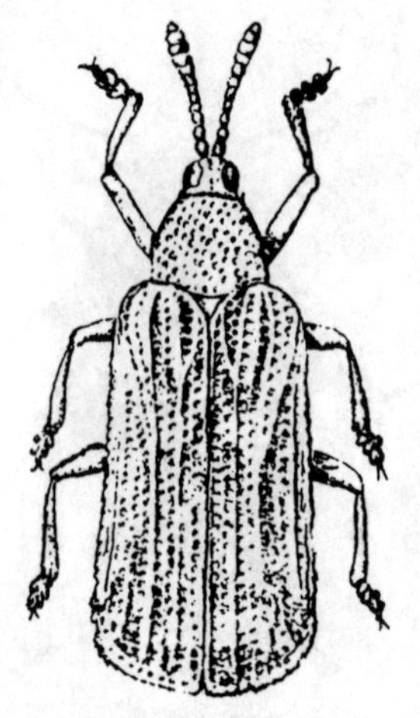

Figure 809

LENGTH: 6-6.5 mm. RANGE: general.

Much flattened. Reddish-yellow to rosy. Under parts reddish, legs yellow. Reported to live on maple, oak, basswood and locust trees.

B. californicus Horn belongs to the Southwest.

Still another closely related genus, *Stenispa,* has two fairly common species. The members of this genus possess eleven segments in the antennae, but the elytra lack the ridges displayed by these two other genera. *S. metallica* (Fab.) is wholly black with a slightly bronze tinge. It ranges throughout the eastern half of the United States and measures 5-6 mm.

S. collaris Baly is a bit larger (6-7 mm) and differs in having its thorax red, and the other parts with a bluish tinge to the black. It seems to belong to our Southern States but has been reported also from Michigan.

63a (5) Head partly exposed; thorax not so much rounded in front. Fig. 810. .... *Chelymorpha cassidea* (Fab.)

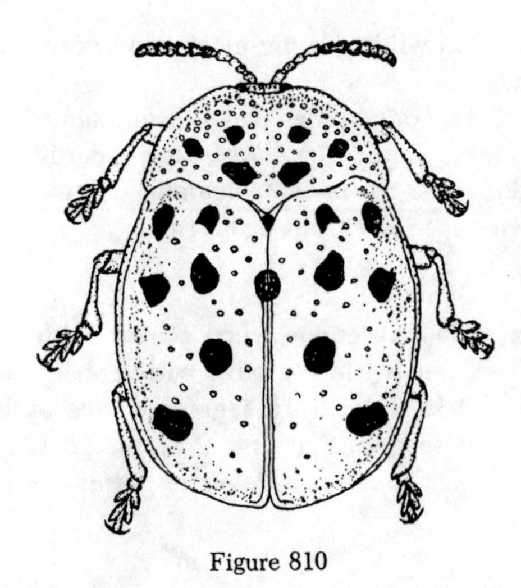

Figure 810

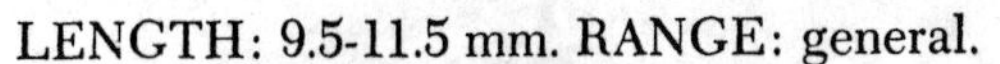

LENGTH: 9.5-11.5 mm. RANGE: general.

Convex. Brick-red marked with black. Underparts black. Beginners sometimes mistake it for a lady beetle. Four varieties have been named. It feeds on plants of the morning-glory family and is sometimes found on milkweed.

It is known popularly as the Argus Tortoise Beetle and is one of the largest of our leaf beetles and often very abundant.

63b **Head concealed beneath the margin of the rounded thorax. 64**

64a **(63) Length 9 mm or more; margin thickened. Fig. 811. .... *Physonota unipunctata* (Say)**

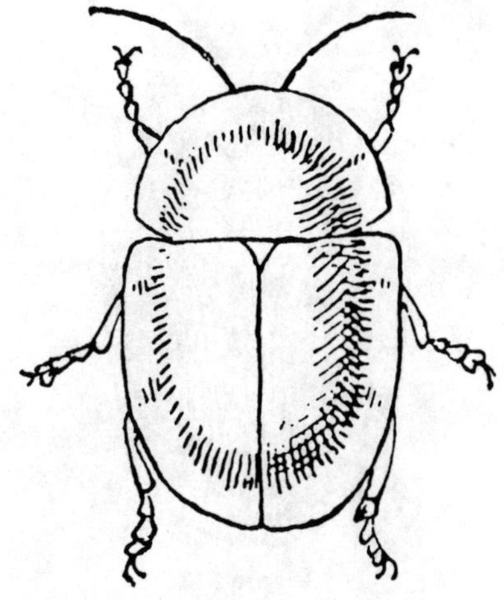

Figure 811

LENGTH: 9-11.5 mm. RANGE: general.

Convex. Pale greenish yellow; thorax with three black dots, the middle one oblong; underparts largely black. Sometimes very abundant on rosin weed and other plants. The larva are oddly shaped slimy creatures.

In collections the specimens often fade to an unmarked straw color quite different from the living ones.

64b **Smaller; margins thin and flattened. .. 65**

65a **(64) Claws simple. 66**

65b **Claws with angular dilations at base. 67**

66a **(65) Depressed. Fig. 812. .... *Cassida nebulosa* L.**

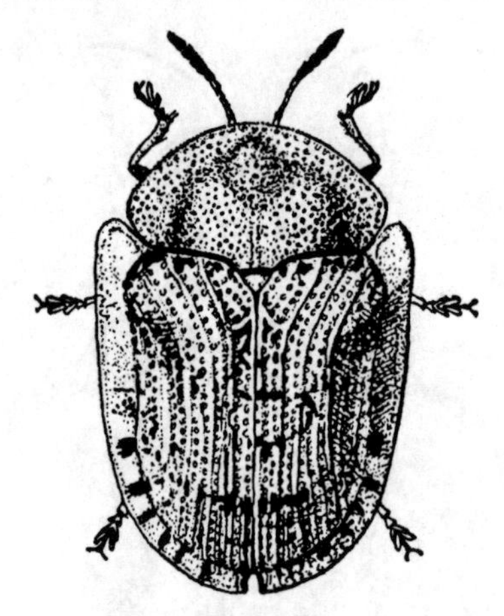

Figure 812

LENGTH: 6-7 mm. RANGE: Europe and introduced.

Yellowish-gray to pale green with blackish markings. This "Beet Tortoise Beetle" is a serious pest in European beet fields but while, supposedly, having been introduced here it has not caused serious damage.

66b Convex, red. Fig. 813. ....
.... *Jonthonota nigripes* (Oliv.)

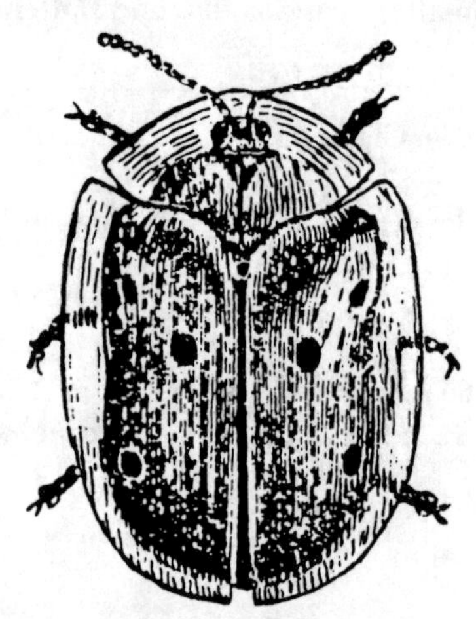

Figure 813

LENGTH: 6-7.5 mm. RANGE: general.

Dull red; three indistinct black spots on each elytron. Legs and under surface black. Feeds on wild morning-glory and sweet potato vines.

Three varieties have been named (a) *atripes* Lec. and (b) *ellipsis* Lec. occur in Colorado and (c) *novemmaculata* (Mann.) has been described from California.

67a (65) Antennae when at rest, with basal segments in a groove with a sharp ridge at its side; third segment twice as long as second. Fig. 814. ....
.... *Deloyala guttata* (Oliv.)

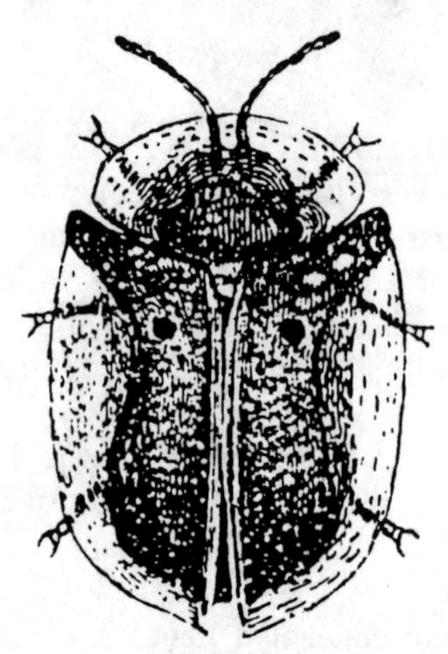

Figure 814

LENGTH: 5.5-6 mm. RANGE: general.

Dull yellow; thorax usually with sizeable black spot enclosing two lighter ones; elytra black with yellow spots, and pale margins. Common on members of the morning-glory family.

67b Third segment of antennae but little longer than second; no antennal groove. 68

68a (67) Elytra roughened. Fig. 815. ....
.... *Plagiometriona clavata* (Fab.)

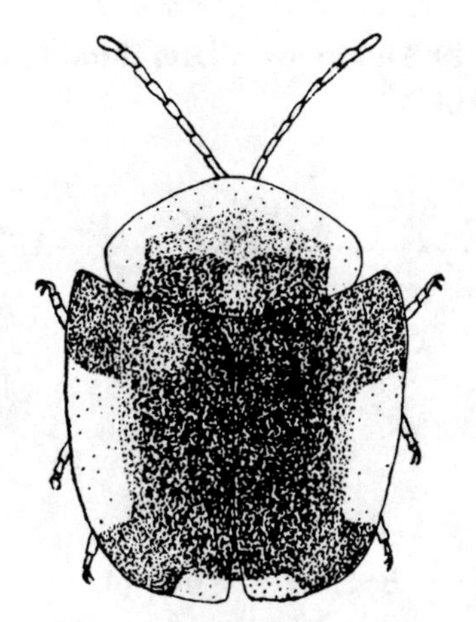

Figure 815

LENGTH: 7-7.5 mm. RANGE: general.

Broadly oval. Margins of thorax and elytra translucent with brown markings as pictured. Under surface pale yellow. A large and very interesting species reminding one of a turtle even more than the other species of the family. It is said to damage chili peppers in New Mexico and Arizona.

68b Elytra smooth. Fig. 816.

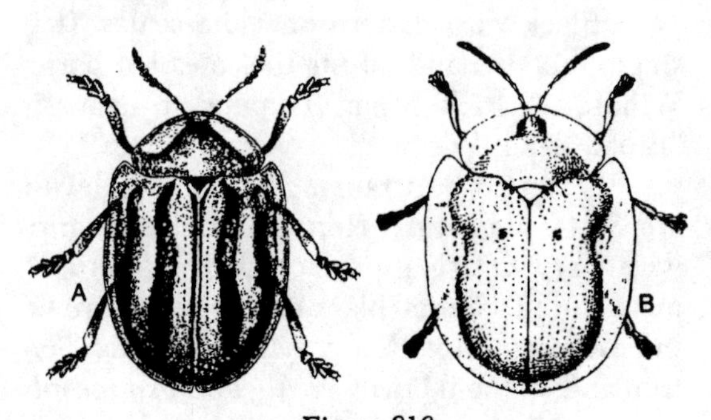

Figure 816

(a) *Agroiconota bivittata* (Say)

LENGTH: 4.5-6 mm. RANGE: general.

Thorax yellowish with center area brown; elytra dull, yellow with black stripes as pictured. Common on sweet potato and wild morning-glory plants.

(b) *Metriona bicolor* (Fab.)

LENGTH: 5.5-6 mm. RANGE: Eastern half of United States.

This very brilliant brassy-gold-while-living beetle attracts much attention but proves a disappointment to collectors for it becomes a dull reddish-yellow when dead. Under parts black. It is often called "Gold Bug."

This genus contains 5 other similar species.

FAMILY 108. BRUCHIDAE
The Pea and Bean Weevils

Nearly 1000 species are known. Practically all (with the exception of a few feeding on the seeds of palm trees) develop within the seeds of legumes. Their size is necessarily usually small. The colors are often dull though a few species are ornamented with red.

1a Hind tibiae with two jointed spurs. 2

1b Hind tibiae without jointed spurs; front coxae prominent. 3

2a (1) Claws simple; not over 3 mm. Fig. 817. ...... *Zabrotes subfasciatus* Boheman

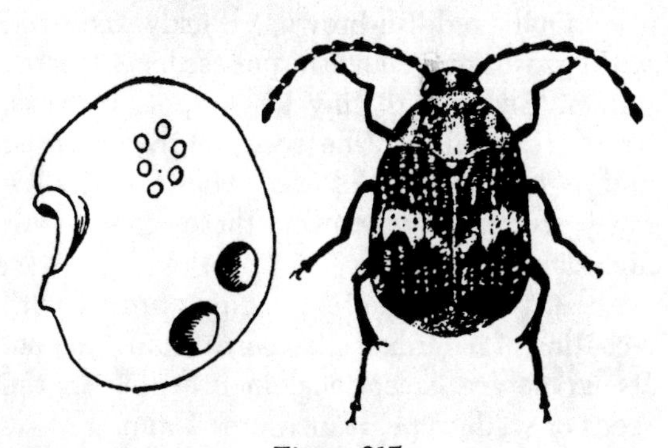

Figure 817

LENGTH: 1.5-2 mm. RANGE: Southwestern United States.

Pale to dark brown or black, marked as shown with white. It is known as the "Mexican Bean Weevil" and while originally a pest of the black and red Mexican beans it seems to thrive on beans of any variety and breeds continuously. The figure shows a bean with six eggs and two emergence holes.

Several other species are known for this genus, most of which are western in their range.

2b Claws not simple; over 6 mm. Fig. 818. .. .. *Amblycerus hoffmannseggi* (Gyll.)

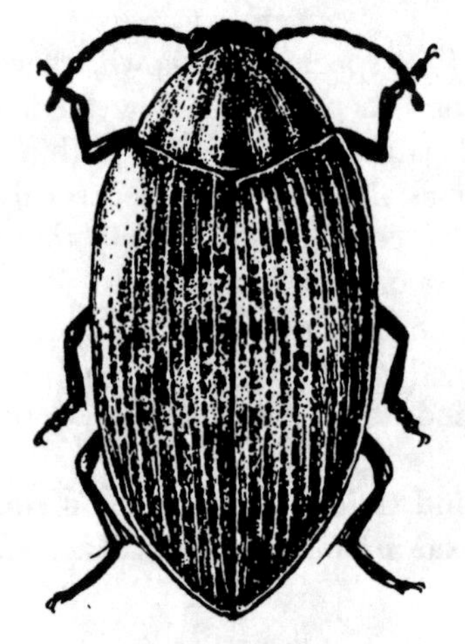

Figure 818

LENGTH: 7-8 mm. RANGE: Eastern half United States.

Dull reddish-brown, thickly covered with prostrate yellowish pubescence; thorax darker. Elytra with tiny black spots in rows. The larvae inhabit the seeds of honeylocust and perhaps the black locust. These beetles are fairly common wherever these trees fruit abundantly.

A. vitis (Schaeffer) reddish brown with a coating of paler hairs, is known from Arizona. Its larvae are exceptional in breeding in the seeds of wild grape. It measures 4 mm.

3a (1) Thorax with lateral tooth at middle. Fig. 819.

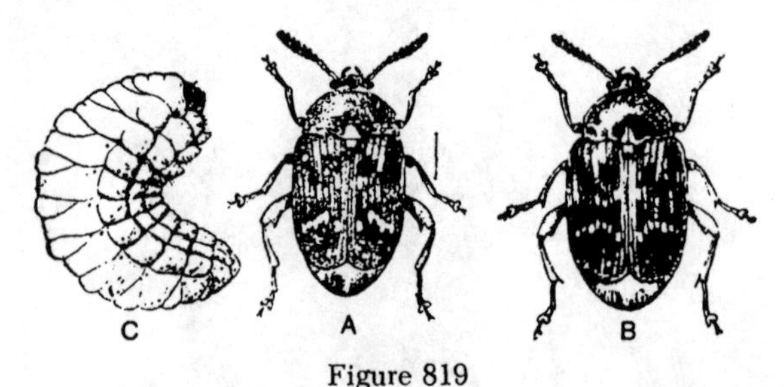

Figure 819

(a) *Bruchus pisorum* (L.)

LENGTH: 4.5-5 mm. RANGE: cosmopolitan.

Black, thickly covered with reddish-brown, yellowish and whitish hairs to form the pattern pictured. Pygidium with two black spots on gray background. This is the Pea Weevil. The eggs are laid on young pods and the larvae (c) develops only in growing peas. The adults spend the winter within the pea seed but do not lay any eggs in stored peas.

(b) *Bruchus rufimanus* Boheman

LENGTH: 3-4 mm. RANGE: cosmopolitan.

Black with pattern of white scales. It is single brooded and infests the so-called horse, Windsor or tick bean. Its popular name is "Broad Bean Weevil."

Other species ranging in our Southwest are *Stator limbatus* (Horn) 3 mm, dull gray with 8 brown stripes on each elytron S. *pruininus* (Horn) 2-3 mm, black covered with brown pubescence and with a gray spot on each elytron and on the thorax and *Algarobius prosopis* (Lec.) 3-5 mm, in varying shades of reddish with whitish pubescence and sometimes brown markings breeds in the seeds of mesquite and related plants.

3b Thorax without lateral tooth. 4

4a (3) **Hind femur with a tooth on both the inner and outer margins. Fig. 820.**

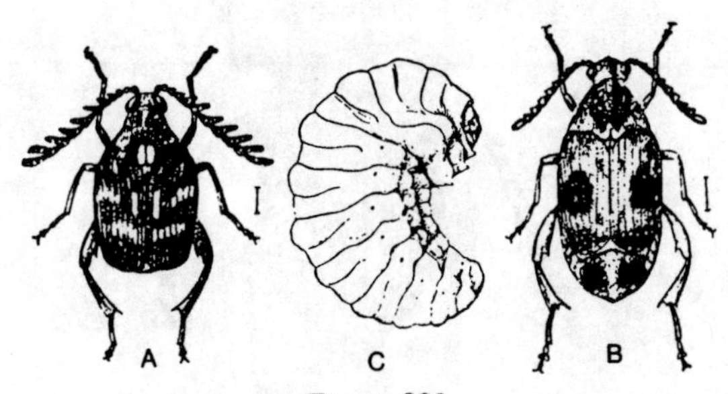

Figure 820

(a) ***Callosobruchus chinensis* (L.)**

LENGTH: 3-4 mm. RANGE: cosmopolitan.

Grayish-black with white markings and two raised ivory spots. This weevil is commonly imported alive in stored products, but is not definitely known to be established in our country.

(b) ***C. maculatus* (Fab.)**

LENGTH: 3-4.5 mm. RANGE: cosmopolitan.

Reddish-brown with thorax black and gray. This "Cowpea Weevil" breeds continuously in beans and cowpeas, and has been known as the Southern Cowpea Weevil and the "Four-spotted Bean Weevil." Figure c shows its larvae.

4b **Hind femur with but one tooth and two small denticles. Fig. 821. .... *Acanthoscelides obtectus* (Say) Bean Weevil**

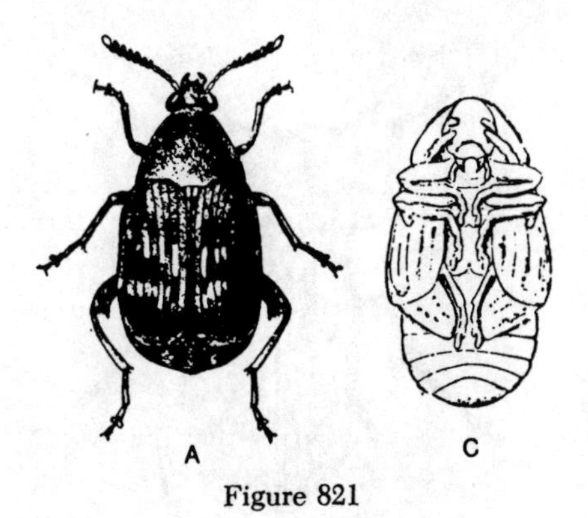

Figure 821

LENGTH: 2.5-3.5 mm. RANGE: cosmopolitan.

Robust, flattened above. Gray on brown, marked as shown; the legs reddish. It attacks practically all varieties of beans and breeding continuously in stored seeds quickly reduces them to worthlessness. The pupa (c) is much the same as in all the members of this family.

FAMILY 109. ANTHRIBIDAE
The Fungus Weevils

Less than 100 species of this family occur in our North American region. The beetles are gray, brown or blackish. The larvae feed in woody fungi and dead wood.

The beak is flat and usually very short. Two types of antennae prevail, long filiform structures or short antennae with a 3-segmented club. The tarsi are 4-segmented but since the triangular second segment obscures the third, there appears to be but three segments in all.

1a **Antennae arising from the sides of the beak. 2**

1b **Antennae arising from the front of the head or from the base of the beak, eyes not notched. Fig. 822. .... *Araecerus fasciculatus* (DeG.)**

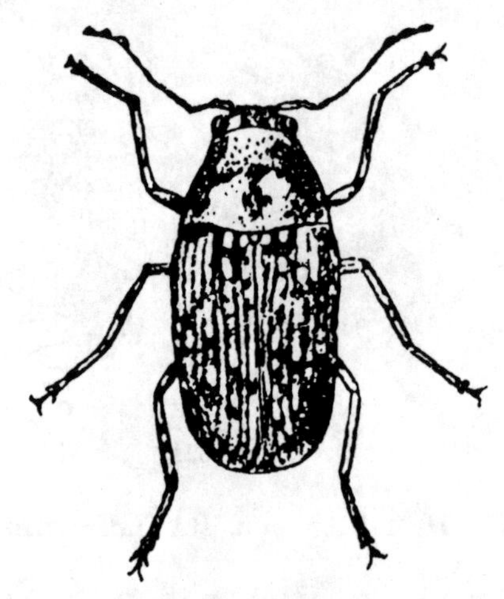
Figure 822

LENGTH: 2.5-4.5 mm. RANGE: cosmopolitan.

Dark brown or blackish with pubescence of yellow or brown. Antennae and legs in part reddish-brown.

This "coffee-bean weevil" has been scattered widely with coffee and cacao seeds. In our country it has found the seeds of cotton, pokeweed and other plant products to its liking.

2a (1) Thorax having the transverse ridge straight at middle and separated a short distance from the base; sides of beak spreading over the antennal cavities. Fig. 823. .... *Tropideres fasciatus* (Oliv.)

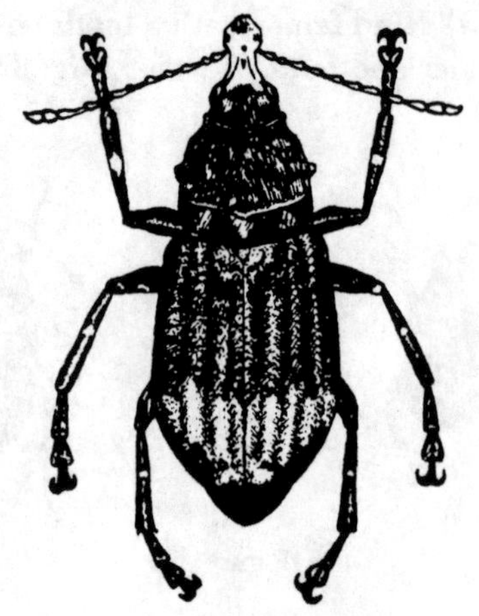
Figure 823

LENGTH: 6.5-10 mm. RANGE: Much of the United States and southern Canada.

Dark brown or blackish; marked with patches of white pubescence on beak and elytra; thorax roughly sculptured and with broad impression in front of middle. On dead twigs and fungi.

Ormiscus saltator Lec. is another widely scattered species. Its brownish-black surface is marked with numerous spots and bands of ash-gray pubescence. The antennal cavities are not covered by its sides of the beak as in the above species. It is a little fellow measuring only 1.2-1.6 mm.

Another related species known from Iowa eastward and from Canada south to Florida is *O. walshii* (Lec.) 2.3-3 mm in length. It is elongate-oval, brownish-black with a coat of ash-gray or brownish pubescence and with scattered patches of rather coarse yellowish-white hairs.

2b Transverse ridge arising at the base of thorax, and its posterior side being perpendicular. 3

3a (2) The sides of the beak parallel or nearly so. Fig. 824.

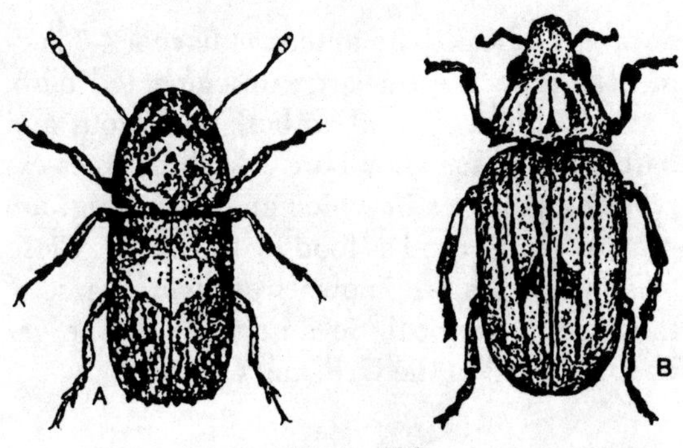

Figure 824

(a) *Toxonotus cornutus* (Say)

LENGTH: 3.5-6 mm. RANGE: Eastern half of United States and Canada.

Brown with conspicuous markings of darker hairs and large white spot on elytra, as pictured. Beak with feeble ridge above.

(b) *Euparius marmoreus* (Oliv.)

LENGTH: 4-9 mm. RANGE: Eastern half United States and Canada.

Dusky brown, marked with irregular pattern of pale brown and gray scales on upper surface; legs ringed with gray and black. Common in woody fungi and under bark.

E. lugubris (Oliv.) with a length of 4-7 mm, is reported from Florida, Texas, Arkansas, Indiana and District of Columbia. It is oblong, robust and velvety black with white markings on beak, front third of thorax and an irregular patch on each elytron. The apex of the elytra is white and the legs have alternating white and black rings.

3b The sides of the beak narrowing from the eyes forward; thoracic ridge turning forward at hind angles. Fig. 825. .... *Trigonorhinus tomentosus* (Say)

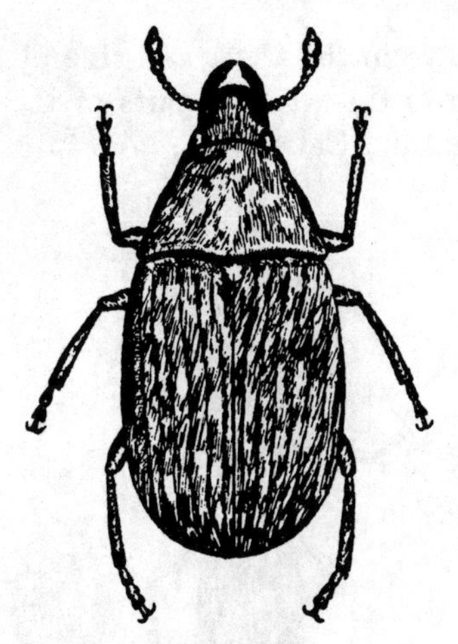
Figure 825

LENGTH: 2-2.5 mm. RANGE: Eastern half United States.

Dark reddish-brown; tibiae and antennae lighter, upper surface with dense grayish yellow or brownish pubescence. Often found on common ragweed.

T. rotundatus (Lec.), with a length of 1.5-2.5 mm and ranging widely, differs from the above genus in being more slender and in the thoracic ridge not turning forward. It is dark brown with bright gray and brown pubescence. The commonest of these is *T. sticticus* Boh. It is dark, brown each elytron with 2 black spots; 2.5-3 mm; on corn and wheat smut. This is a very variable and widely distributed species.

FAMILY 110. BRENTIDAE
The Straight-Snouted Beetles

Members of this family have elongated bodies with a long, straight, projecting beak. In color they are reddish with or without markings and the body size varies from 7 to 50 mm. Antennae are 10 or 11-segmented and filiform or moniliform in shape. Adults are found under bark where they feed on wood-eating insects, fungi, or on sap flowing from wounds in trees. Six species live in the U. S. and five of these are confined to the warmer parts of the country-Fla., Tex., and Cal.

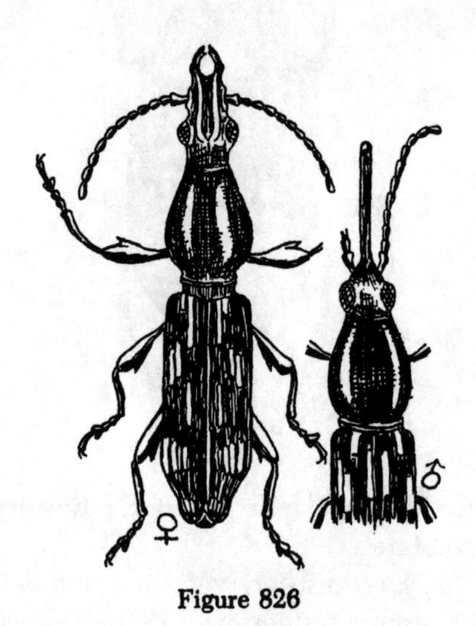

Figure 826

Figure 826 *Arrhenodes minuta* (Drury)

LENGTH: 5-22 mm. RANGE: Eastern N. Amer. south to Fla. and Tex.

Subcylindrical. Dark reddish-brown with yellowish, somewhat variable markings on elytra. The larvae bore through the heart wood of oak and other trees; the adults are found under the bark.

Brentus anchorago (L.), 10-52 mm, known from Florida and Lower California, is shining black or reddish black. The head is longer than broad in this genus.

There are nearly 1300 species worldwide, most of them tropical.

FAMILY 111. PLATYPODIDAE
The Ambrosia Beetles

The Ambrosia Beetles are small, 2-8 mm, pale to dark brown, elongate, cylindrical beetles with the first tarsal segment elongate. The head is wider than the pronotum, deflexed, and does not form a beak. The antennae have a 4-7 segmented scape and a large unsegmented club.

Adults are found on both coniferous and hardwood trees. They bore into these trees excavating galleries in which ambrosia fungi are grown. This is used as food by the larvae. Close to 1000 species are known worldwide, most of these being tropical. Seven species in one genus are found in the U. S. and Canada.

Figure 827

Figure 827 *Platypus wilsoni* Swaine.

LENGTH: 5-5.7 mm. RANGE: B. C. to Cal. Reddish-brown, moderately shiny; elytra finely striate, the declivity in the male modified and spined.

P. flavicornis (Fab.), 5.2-5.7 mm long, blackish with third interval of elytra raised and roughened at its base and elevated in the male; ranges from New Jersey to Florida, Texas and Mexico.

FAMILY 112. SCOLYTIDAE
The Bark Beetles

About 1200 species of this economically important family are known worldwide. They are plant feeders and attack all parts of plants but so many species live within the cambium of trees that the group is known as "bark-beetles," or as "engraver beetles," because of the designs they cut in tree trunks. They are especially destructive to conifers. The larger species measure only a few millimeters in length, while the majority are very small.

1a **Front tibiae with long curved hook on the outer apical angle; elytra nearly glabrous, but with many rows of punctures. Fig. 828. .. *Scolytus rugulosus* (Mueller) Shot-hole Borer**

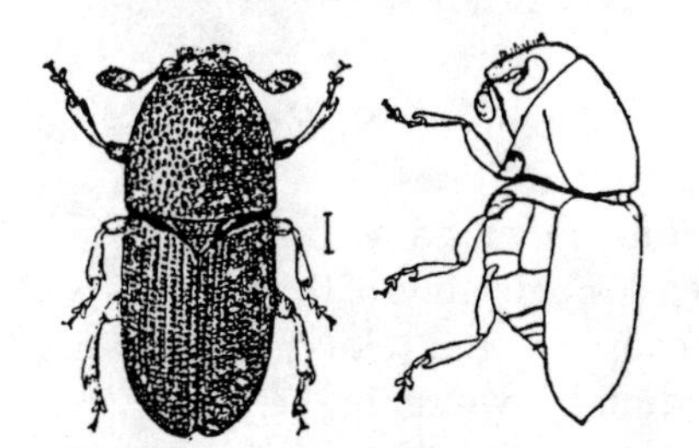

Figure 828

LENGTH: 2-2.5 mm. RANGE: Europe and much of United States.

This introduced beetle is dull blackish with the apex of the elytra reddish-brown. The head of the male is broader and more flattened than that of the female. Several species of fruit trees belonging to the rose family frequently suffer from their attacks on the less healthy trunks and limbs. The side view and pupa are also shown.

S. *quadrispinosus* Say Hickory Bark Beetle measuring 4-5 mm and ranging widely; feeds under the bark of several species of hickories. The abdomen of the male is armed with several spines.

S. *multistriatus* Marsh., the European Elm Bark Beetle, 2.2-3 mm, dark reddish-brown, glabrous, sternite 2 with a spine arising from anterior third. This is the vector of the Dutch Elm Disease that has destroyed most American elms.

1b **Front tibia not as in 1a. 2**

2a **(1) The head covered from above by the thorax; the pronotum usually more roughened at its front. 7**

2b **The head visible from above; rounded. 3**

3a **(2) Antennal club of but one elongated segment which is attached by its side to the five segmented funiculus. Fig. 829. .... *Chramesus hicoriae* Lec.**

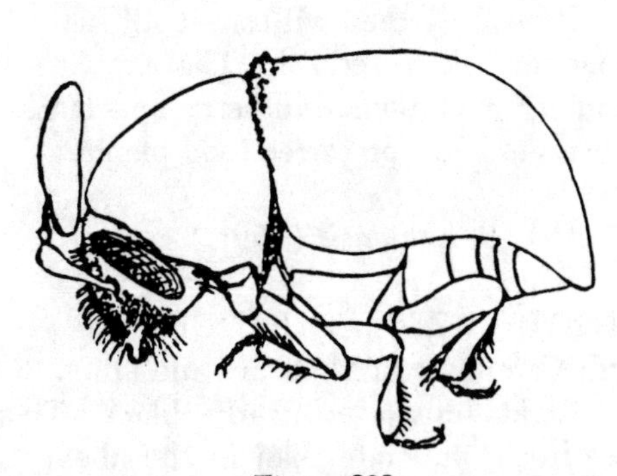

Figure 829

LENGTH: 1.5 mm. RANGE: Eastern United States and Canada.

Dull black; clothed above with stiff ash-gray hairs and erect bristles on the elytra. Thorax broader than long; elytra about twice as long as thorax; antennae yellow. Lives on hickory.

Polygraphus rufipennis (Kby.) 2-2.5 mm, black with dull-red elytra and rufus legs and antennae, has its front coxae touching instead of separated as in *C. hicoriae*.

It is abundant in Canada and ranges widely in the United States.

3b **Antennae not as in 3a. 4**

4a (3) **The segments of the antennal club loosely joined and lamellated. Fig. 830.**
(a) *Phloeotribus frontalis* (Oliv.)

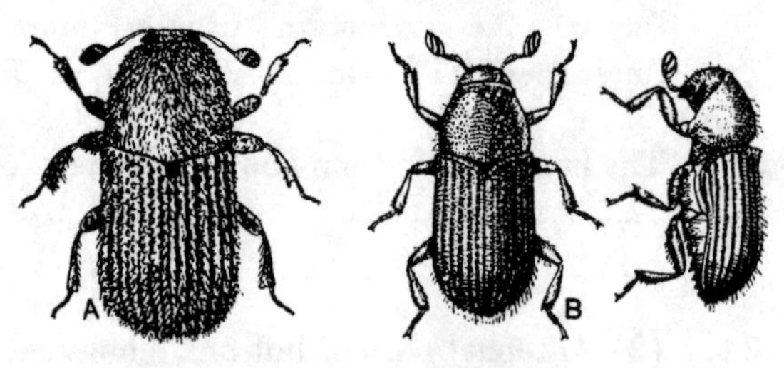

Figure 830

LENGTH: 1.5-2 mm. RANGE: Eastern half of United States.

Brown. Clothed with short stiff hairs; antennae and legs reddish. Thorax with fine granular punctations. Mulberry and Hackberry seems to be its preferred food plants.

(b) *P. luminaris* (Harr.)

LENGTH: 2-2.2 mm. RANGE: Canada to North Carolina and Michigan and Ohio.

Light brown to nearly black. Thorax more finely punctate than in the above, and with much longer lamelate joints in the club of the antennae. It is the Peach Tree Bark Beetle.

4b **The segments compacted and not at all lammelate. 5**

5a (4) **Funicle of antennae with seven segments; club with four segments; front coxae widely separated. Fig. 831. .... *Hylastinus obscurus* (Marsh.) Clover Root Borer**

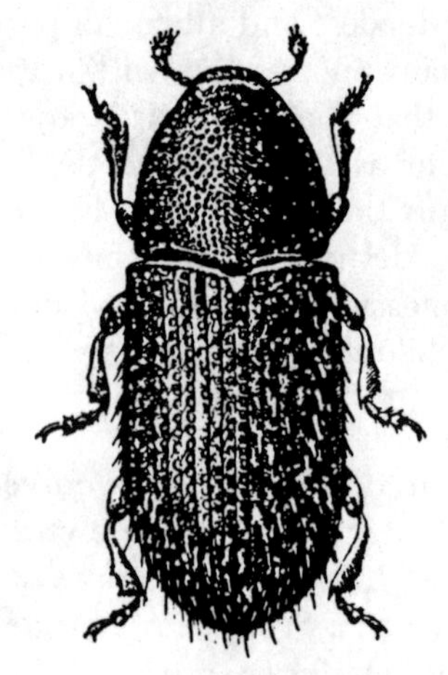

Figure 831

LENGTH: 2-2.5 mm. RANGE: Europe and much of the United States.

Dark brown; elytra with deeply punctured striae; a single row of bristles on each interval. Feeds on red clover and related plants, spending the winter in the roots.

Hylurgops rugipennis (Mann.) 4 mm is a western species. It differs from the above in having the first and fifth ventral segments longer than the others.

Hylurgopinus rufipes (Eich.), known rather commonly from Iowa to Virginia and Canada, measures 2-2.5 mm and is dull brownish-black and thinly covered with short, yellowish hairs. The beak is short, the funicle has 7 segments and the club is about twice as long as wide. This is the native Elm Bark Beetle.

5b **Funicle of antennae with but five segments; front coxae very narrowly separated. 6**

6a (5) **Pronotum somewhat elongate, not much narrower in front; thorax as broad as elytra. Fig. 832.**

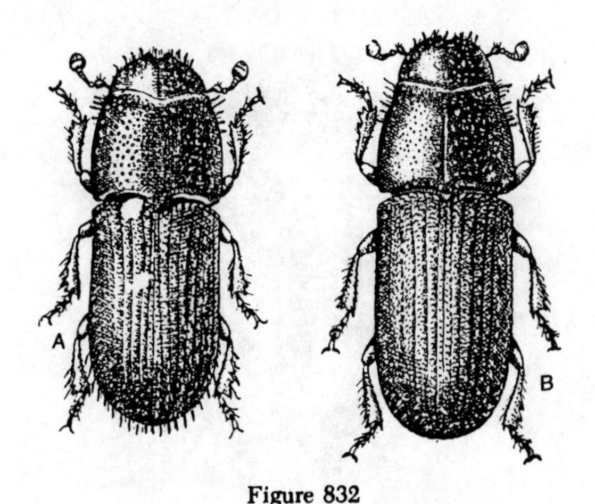

Figure 832

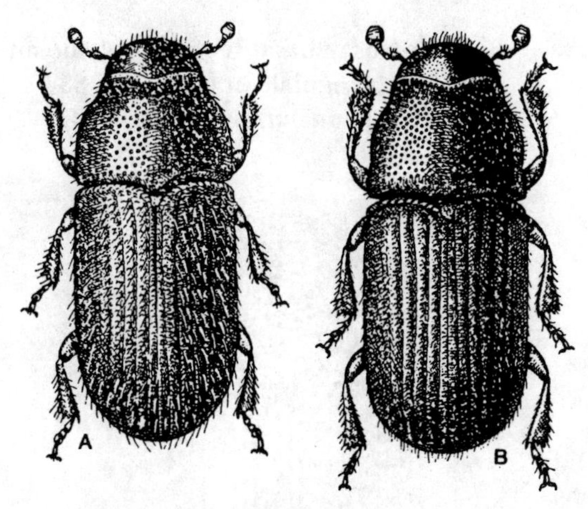

Figure 833

(a) ***Dendroctonus frontalis*** **Zimm. Southern Pine Beetle**

LENGTH: 2.5-4 mm. RANGE: Eastern half of United States.

Light to reddish-brown or blackish. Sides and apical ends of elytra with long hairs. An enemy of pines and spruce.

(b) ***D. brevicomis*** **Hopk. Southwestern Pine Beetle**

LENGTH: 4.5 mm. RANGE: Western States.

Dark brown or black, without long hairs on sides of elytra. It attacks yellow pine and Douglas fir and is highly destructive.

6b Pronotum definitely narrowing in front; dorsal face of elytra with long hairs on basal half. Fig. 833.

(a) ***Dendroctonus simplex*** **Lec.**

LENGTH: 3.5-5 mm. RANGE: Maine and West Virginia to Alaska.

Reddish to reddish-brown, in part nearly black.

Thoracic punctures large and small mixed. Lives on the tamarack and is known as the Eastern Larch Beetle.

(b) ***D. ponderosae*** **Hopk.**

LENGTH: 4-6 mm. RANGE: Montana, Wyoming, Idaho, Nevada and California, and Pacific Northwest.

Pure brown to black; anterior part of pronotum transversely impressed. It is known as the Mountain Pine Beetle.

There are 22 or more species of this genus in our fauna.

7a (2) Eyes divided; no distinct sutures on the antennal club. 8

7b Eyes not divided; antennal club with at least some visible sutures. 9

8a (7) **The horny basal segments of the antennal club angulate in front. Fig. 834. .... *Trypodendron lineatum* (Oliv.)**

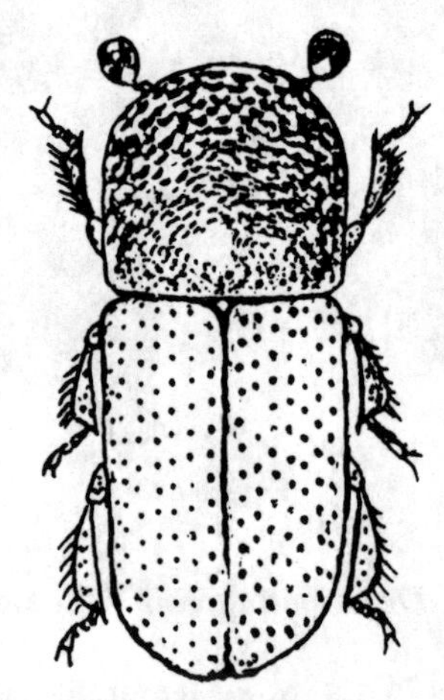

Figure 834

LENGTH: 3-3.4 mm. RANGE: United States and Canada, also Alaska and Europe.

This Spruce Timber Beetle is known to infest several species of conifers. The galleries of the larvae are rather wider apart and are parallel to the wood fibers, the walls being blackened. The adults are brownish or black and sculptured as pictured.

T. retusum (Lec.), 4.5 mm, is shining blackish-brown; the elytra have well defined striae. Eastern U. S. and Canada.

8b **The horny basal segments of the antennal club rounded in front. Fig. 835. .... *Xyloterinus politus* (Say)**

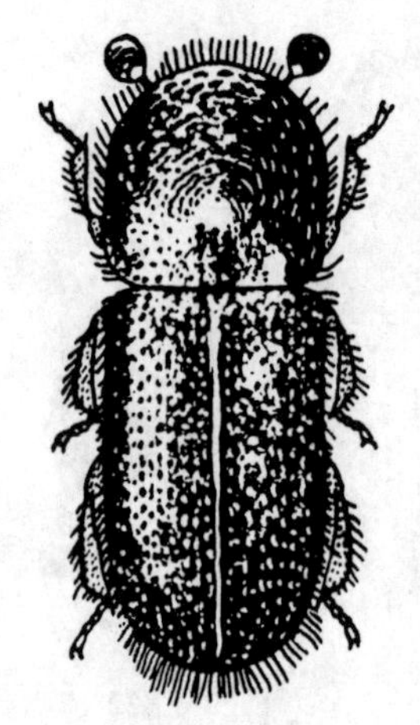

Figure 835

LENGTH: 2.8-3 mm. RANGE: Eastern United States and Canada.

This species, while usually infesting a number of broad-leaf trees, including beech, oaks, maples, elms, ash and hickory, is also known to occur in spruce and hemlocks. The galleries often extend deep into the wood and are blackened. The adults vary from light brown to nearly black.

9a (7) **Funicle of antennae of more than three segments. 11**

9b **Funicle of less than three segments. 10**

10a (9) **Funicle with two segments, the second, small and closely joined to the club; front tibiae with transverse ridges on its face. Fig. 836.**

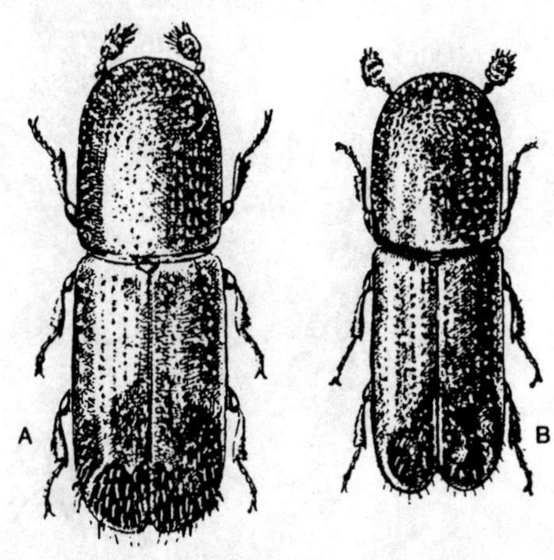

Figure 836

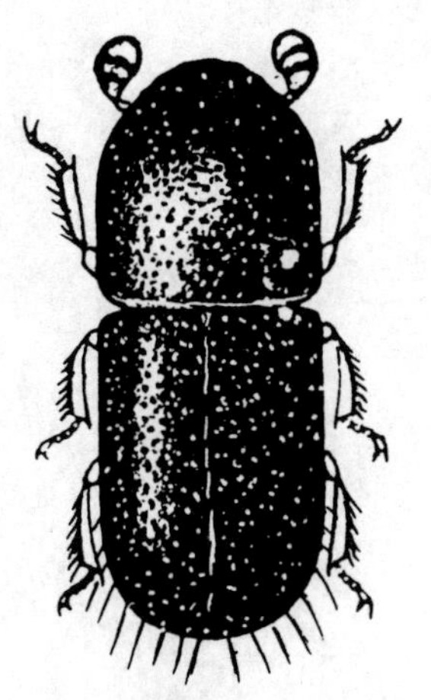

Figure 837

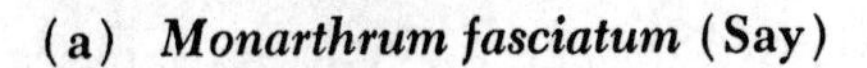

(a) *Monarthrum fasciatum* (Say)

LENGTH: 2.5-3 mm. RANGE: Eastern Canada to Florida.

Piceous; antennae and legs rusty-yellow, front of elytra yellowish; shining and smooth; elytra at tips clothed with yellow hairs. Infests both broad-leaves and conifers of several species.

(b) *Monarthrum mali* (Fitch)

LENGTH: 2-2.2 mm. RANGE: Eastern Canada to Florida.

Differs mostly from the above in its smaller size and in two small teeth at the apex of the elytra as pictured. It too, has a wide range of recorded host trees. It is known as the Apple Wood-stainer.

10b Funicle with but one segment; front tibiae without ridges as above. Fig. 837. .... *Corthylus punctatissimus* (Zimm.)

LENGTH: 3.7-4 mm. RANGE: Much of the United States.

Dark brown to black, with reddish legs and antennae. Thorax smooth and shining behind; elytra with prominent scattered punctures. It has been called the Sugar Maple Timber Beetle but is known also to infest huckleberry, the American hornbeam and the eastern hop hornbeam. Like several other species in this family the larvae are fed by the adults on a fungus food, ambrosia.

11a (9) Club of antennae with pubescence and rings on both sides. Fig. 838. .... *Gnathotrichus materiarius* (Fitch)

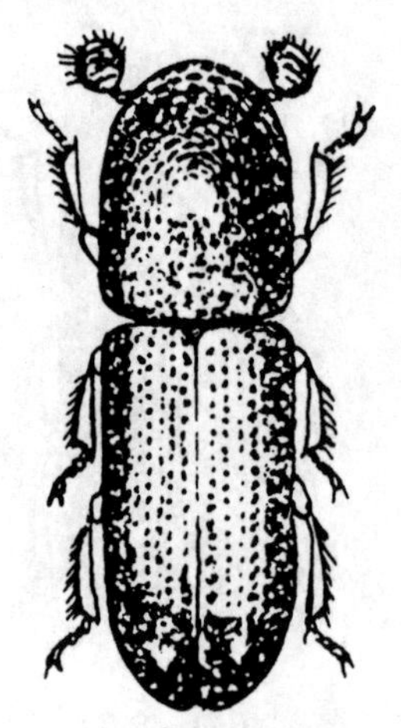
Figure 838

LENGTH: 3 mm. RANGE: Eastern half United States and Canada.

Shining, brown; elytra paler at base; undersurface and head black; feet and antennae yellow; elytra thinly clothed with short yellow hairs. It feeds on pine.

Ips calligraphus (Germ.), a widely scattered species feeding on pine, is considerably larger (4.5-6.5 mm) than most members of the family. It is piceous or brownish with yellow antennae. *Ips* is a widely distributed genus of over 30 species.

11b Club of antennae with pubescence and rings at most on but one side. Fig. 839. .... *Xyleborus celsus* Eich.

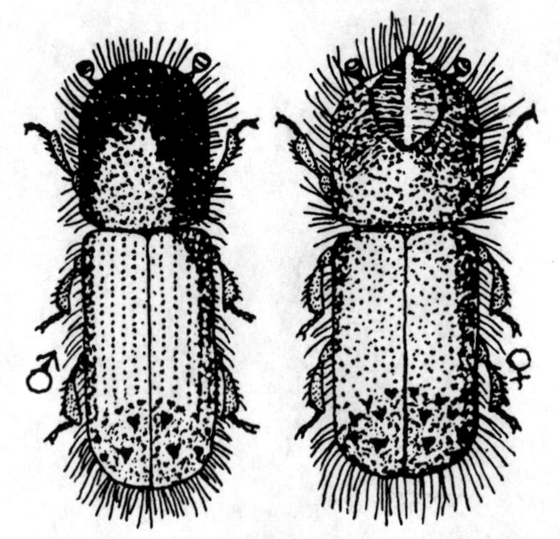

Figure 839

LENGTH: 2.7 mm. RANGE: New Jersey and South Carolina to Indiana.

Shining, brownish-black. Elytral striae feeble, the fourth interval broader than the others. It seems to feed only on hickory.

X. xylographus (Say), widely scattered throughout much of North America and Europe, is shining, reddish-brown, with thin covering of short hairs. The elytrae are striate with deep punctures. It lives on a rather wide range of broadleaf and coniferous trees, and measures 2.5 mm or a little more. This genus has about 20 species in our fauna..

FAMILY 113. CURCULIONIDAE
The Snout Beetles and True Weevils

This is one of the most important families of beetles, both in numbers and as economic pests. Over 40,000 species are known, ranging in size from near microscopic to large handsome species. The larvae (weevils) for the most part spend their life within plant fruits or stems although some are external plant feeders. They commonly "play possum" when disturbed.

1a **Antennae straight, without a protecting groove on the beak. 2**

1b **Antennae with compact club, elbowed; beak with a protecting groove (a) to receive the scape. Fig. 840. .... 8**

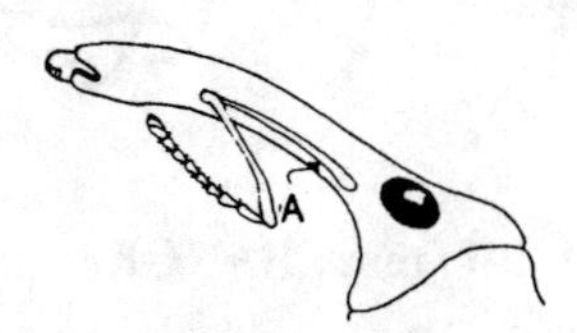

Figure 840

2a **(1) Antennae definitely clubbed, the segments being closely united. Fig. 841. .... 6**

Figure 841

2b **Antennal club absent or if present, with segments distinctly separated. Fig. 842. .... 3**

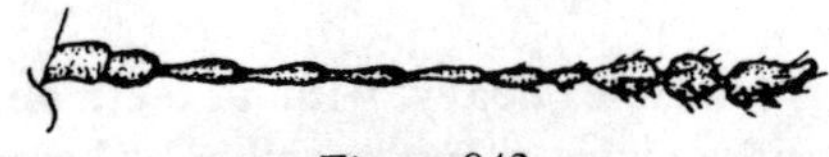

Figure 842

3a **(2) Prothorax with a definite acute margin: three dorsal abdominal segments exposed by shortened elytra. Fig. 843. .... *Pterocolus ovatus* (Fab.)**

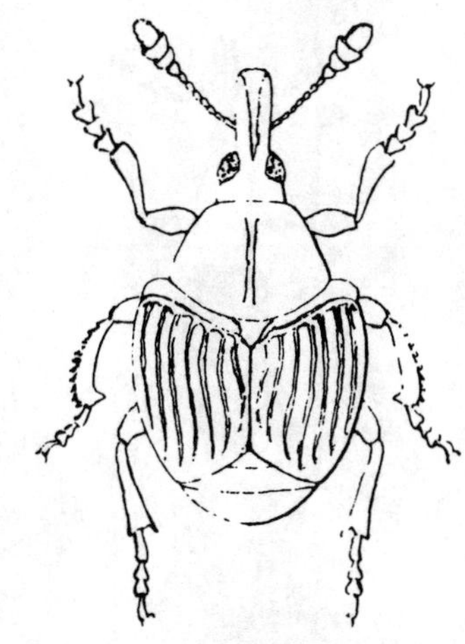

Figure 843

LENGTH: 2.5-3.5 mm. RANGE: Eastern half United States and Canada.

Indigo blue; legs, antennae and snout darker. Elytra flattened, the striae with irregular punctures. Middle coxae small, rounded and widely separated. It has been taken on oak, peach, plum and grape.

Except for the snout this sole member of its subfamily somewhat resembles a clown beetle. The three abdominal segments exposed by the shortened and diagonally cut elytra are densely punctured.

3b **Thorax without a marginal ridge on its side, labrum absent; palpi short and stiff. 4**

4a **(3) Antennae not having a club, the last segment cylindrical and elongated, especially in the male; ant-shaped beetles. Fig. 844. .... *Cylas formicarius elegantulus* (Sum.)**

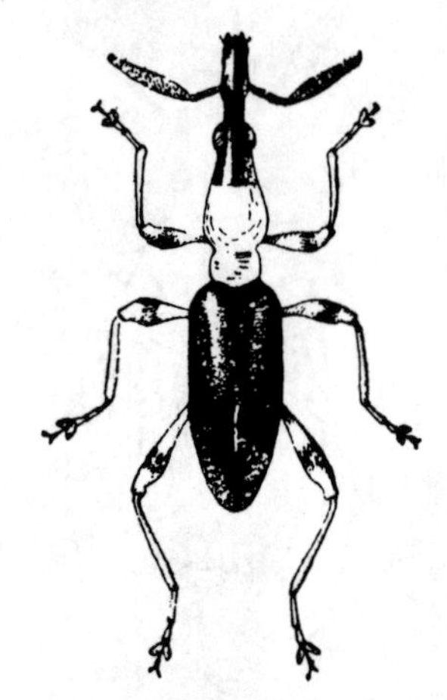

Figure 844

LENGTH: 5-6 mm. RANGE: Florida to Texas.

Smooth and highly polished; salmon or reddish-brown; elytra bluish-black; head and snout blackish. Elytra with weakly punctured fine striae. Last segment of antennae oval in female; in male long and slender.

This Sweet Potato Weevil came up from the tropics and has long been a pest of yams and sweetpotatoes in the Gulf States. Its distinct form and coloring makes it readily recognized. Taken in Tex. on wild morning glory. Apparently various wild *Convolvulaceae* serve as host.

4b Antennae distinctly clubbed. 5

5a (4) Mandibles flattened with one or more teeth on both the inner and outer sides (a). Claws two pointed or toothed (b). Fig. 845. .... *Rhynchites bicolor* Fab.

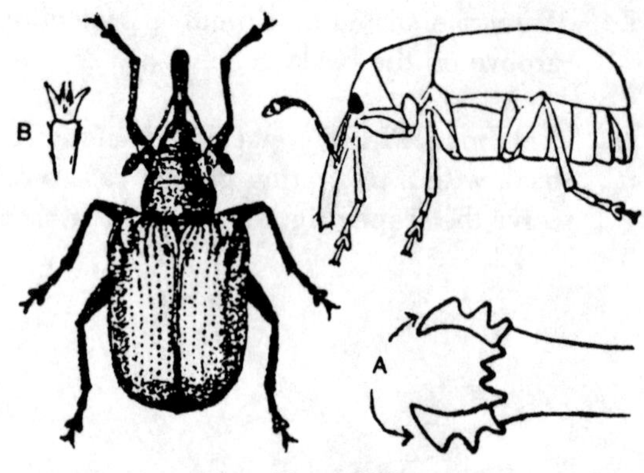

Figure 845

LENGTH: 5-7 mm. RANGE: much of the United States.

Base of head and thorax and elytra bright red; other parts black. This "Rose Curculio" is often very common within the flowers of wild roses and occasionally becomes a pest of cultivated roses and some other members of the rose family.

Eugnamptus collaris Fab. measuring 3.5-5 mm has its abdomen black and elytra blue-black, but its other parts variable in color (a number of varieties have been named). It is common on several species of trees and other plants. It ranges widely from Texas east to the Atlantic. The genus *Rhynchites* includes some 30 species and *Eugnamptus* 8 in the U. S. and Canada.

5b Mandibles heavy, without teeth on outer side; claws grown together at base; tibia with two sturdy hooks at tip. Fig. 846.

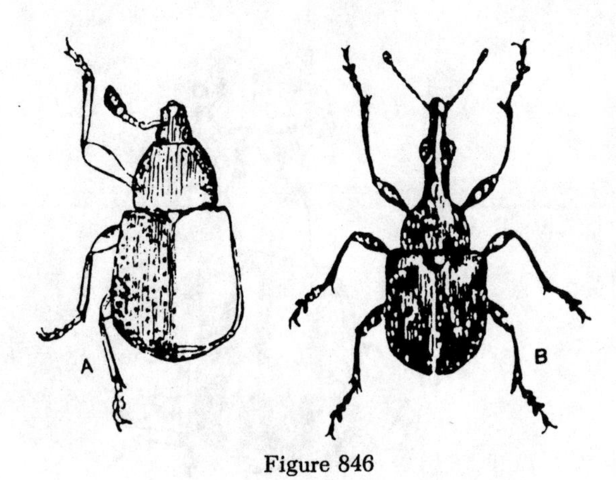

Figure 846

(a) ***Attelabus bipustulatus*** **Fab.**

LENGTH: 3-4 mm. RANGE: Eastern half United States and Canada.

Black with bluish tinges. Each elytron with spot of red as pictured. Found chiefly on oak trees.

(b) ***A. analis*** **Ill.**

LENGTH: 5-6 mm. RANGE: Texas, Florida, Canada, and Eastern U. S.

Base of head, thorax, abdomen, and elytra bright red; other parts bluish-black. Surface not at all pubescent. Found on hickory, oak, and walnut. Very similar is *A. nigripes* LeC. Front femora are stout and the host plant is sumac.

5c **Mandibles without teeth on outer side; Tarsal claws free with an acute tooth at base. Stout gray and black checkered beetle, 12-18 mm long. Antennae not elbowed but with small oval club (a). (New York Weevil) Fig. 847. ....**
.... *Ithycerus noveboracensis* (Forst.)

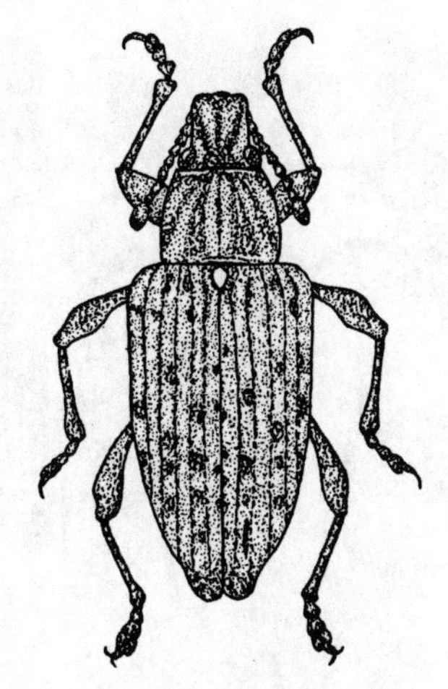
Figure 847

LENGTH: 12-18 mm. RANGE: Eastern Can. and the U. S.

Black, clothed with gray and brown prostate hairs arranged in interrupted stripes on thorax and elytra. Beak with ridges, which are darkened; scutellum whitish.

Breeds in oak and is destructive to apple, peach, plum, pear and cherry trees. The larvae appear in early spring and eat both young buds and foliage. They may be found hiding in the crevices in the bark of trees. It is one of our largest snout beetles.

6a **(2) Elytra short exposing pygidium; trochanters small. 7**

6b **Pygidium covered by elytra. Femora attached to the apex of the large trochanters. Small pear-shaped beetles. Fig. 848.**

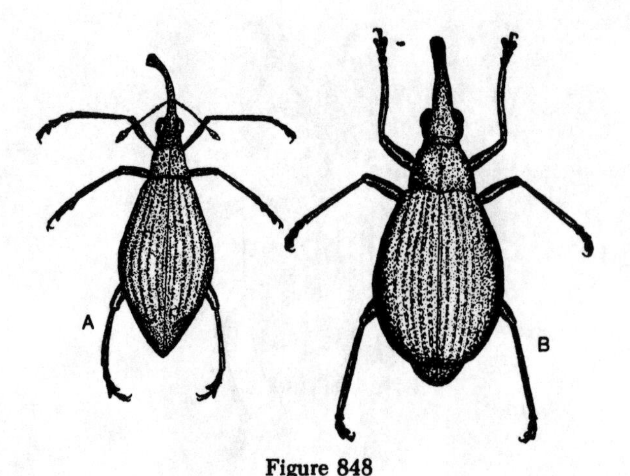

Figure 848

(a) ***Apion impunctistriatum*** **Sm.**

LENGTH: 2 mm. RANGE: much of eastern half of United States.

Shining black; antennae piceous, very slender; front femora of male with large tubercule near tip.

(b) ***A. porcatum*** **Boh.**

LENGTH: 2.4-2.7 mm. RANGE: Eastern half United States.

Robust, shining, black. Front femur of male without tubercule and tibiae without teeth at tip.

Over 100 species of quite similar little beetles belong to this genus.

7a (6) First segment of antennae longer than second (a); hind femora broad (b). Fig. 849. .... *Allocorphynus slossoni* Schffr.

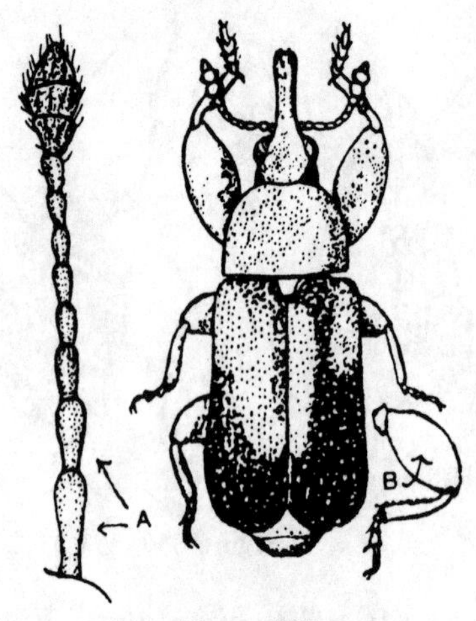

Figure 849

LENGTH: 3.5-4.2 mm. RANGE: Florida.

Reddish-yellow; elytra black, marked with reddish yellow as pictured. Beak punctured in male, smooth, shining and longer in female; head transverse, with scattered coarse punctures; thorax about twice as wide as long, broadly rounded in front. Both adults and larvae feed in the flowers of Florida arrow-root.

7b First segment of antennae no longer than second segment; hind coxae widely separated. Fig. 850. .... *Tachygonus lecontei* Gyll.

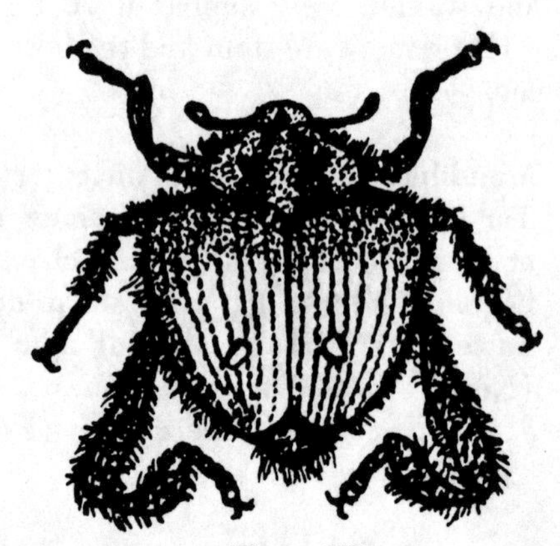

Figure 850

LENGTH: 2-2.5 mm. RANGE: New Jersey to Texas.

Reddish-brown marked with black as pictured. Surface with coarse white and black hairs and fine buff ones. Lives on the leaves of young oak trees.

T. gracilipes Csy. 2.2-2.5 mm, and ranging in the eastern part of our region, is shining black with pale reddish-brown sutural stripe and median cross-bar on the elytra.

8a (1) Abdomen of male with one more anal segment than female; club of antennae usually with rings, not shining; tarsi usually dilated with the third segment bilobed and brush-like beneath (a). Fig. 851. 9

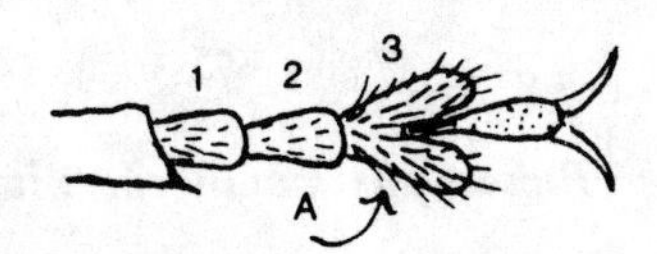

Figure 851

8b Tarsi usually narrow and not brush-like beneath; anal segments the same number in both sexes; basal segment of club of antennae usually either shining or enlarged or both. 61

9a (8) Prosternum in front of the coxae, triangular; beak when at rest lying within the breast; tarsi not widened. Fig. 852. .... *Thecesternus humeralis* (Say)

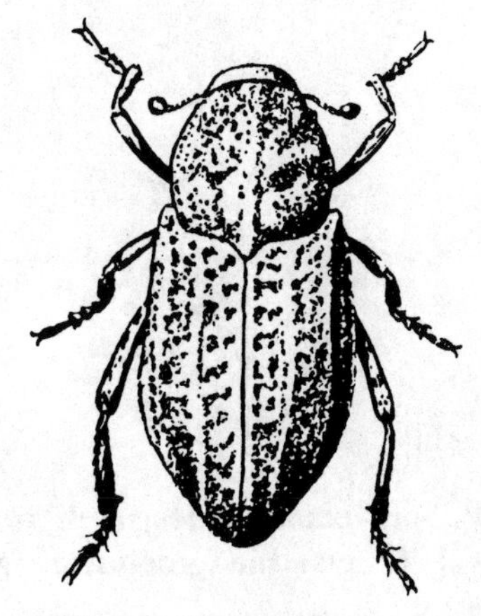

Figure 852

LENGTH: 6.5-9 mm. RANGE: Indiana to Colorado and south.

Robust; dull black, often densely covered with fine brownish-yellow scales.

The small subfamily here represented has but the one genus and less than 10 species. They are sometimes known as the "Bison Snout Beetles."

9b Prosternum not triangular in front, sometimes with a groove to receive the beak. 10

10a (9) Beak usually long and slender; but if short and slender then held within the breast when in repose, mandibles not marked with a scar. 19

10b Beak always short and rather thick; mandibles marked with a scar. (A round or oval depression on the face of the mandible; sometimes on a projection.) 11

11a (10) Front of thorax with ocular lobes sometimes partially covering the eyes (a); eyes often elongate with the greater length longitudinal. Fig. 853. 12

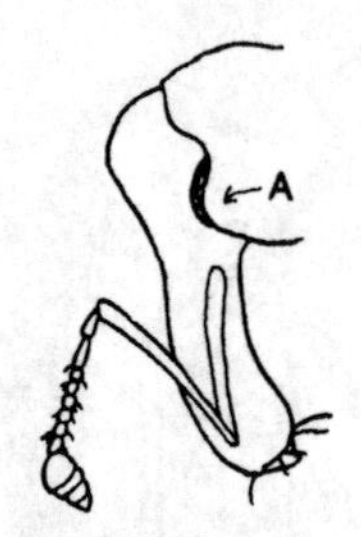

Figure 853

11b Without ocular lobes; eyes rounded or oval, if oval the greatest length transverse. 14

12a (11) Mentum sufficiently large to cover the maxilla. 13

12b Mentum small and retracted, antennal grooves deep, sometimes uniting beneath. Fig. 854. .... *Eudiagogus pulcher* Fahr.

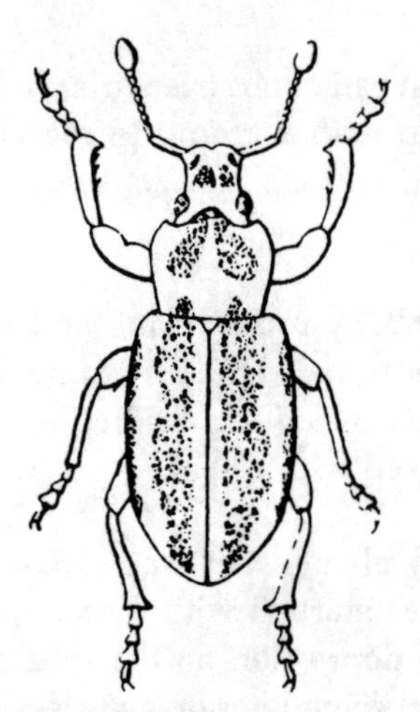

Figure 854

LENGTH: 5-8 mm. RANGE: Florida to Texas.

Black, heavily clothed with black and bright green or copper colored scales on suture and sides as pictured. Feeds on legumes.

Aracanthus pallidus (Say), ranging widely through the eastern half of our country, is dusky brown with gray and reddish scales. It is around 3 mm in length and is characterized by a very small scutellum.

13a (12) Antennal grooves turning downward under beak or towards the eyes; beak rounded; eyes oval. Fig. 855.

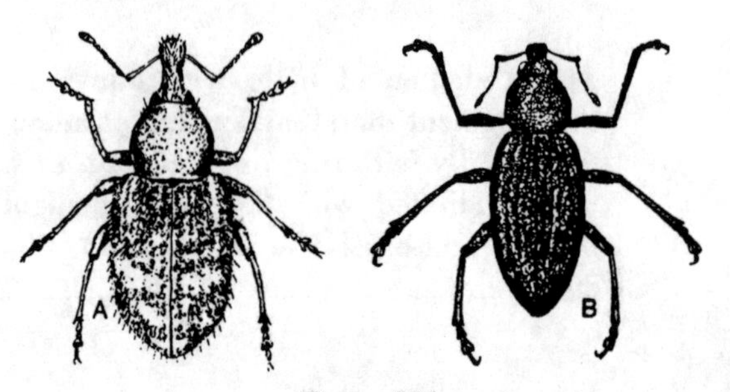

Figure 855

(a) *Panscopus maculosus* Blatch.

LENGTH: 5.5-7.5 mm. RANGE: Indiana to New York.

Dark brown with pattern of silvery and brown scales as pictured. This genus contains 27 spp. in our fauna.

(b) *Anametis granulata* (Say)

LENGTH: 5-7 mm. RANGE: Eastern half United States.

Dark brown, thickly clothed with grayish scales. Antennae and tarsi reddish-brown. Common on trees and shrubs.

13b Grooves short and visible from above; strongly curved. Fig. 856. .... *Phyxelis rigidus* (Say)

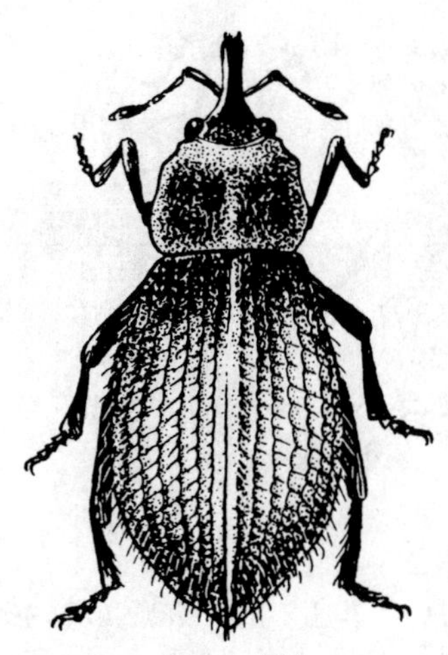

Figure 856

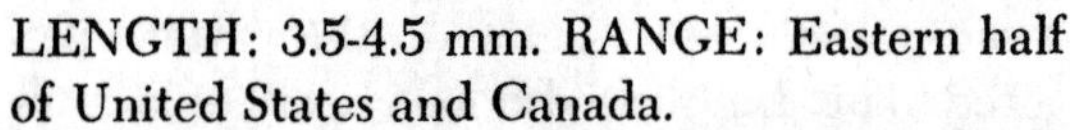

LENGTH: 3.5-4.5 mm. RANGE: Eastern half of United States and Canada.

Dusky brown with coating of grayish scales and often a further waxy coat, yellowish or brownish.

Eupagoderes geminatus Horn, lives in the desert areas of the Southwest. It measures 7-12 mm and is wholly covered with white scales and is known as the White Bud Weevil.

14a (11) Grooves linear and directed beneath the beak. 16

14b Grooves not both linear and directed beneath. 15

15a (14) Claws united at the base; second joint of funicle equalling the first. Fig. 857. .... *Aphrastus taeniatus* Say

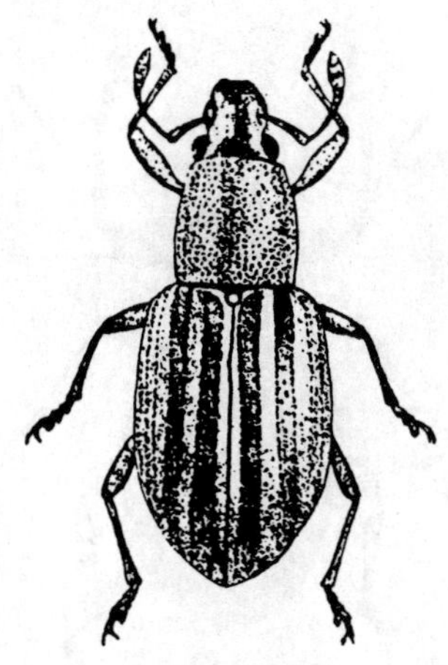

Figure 857

LENGTH: 5.5-6.5 mm. RANGE: Eastern half of United States.

Convex. Densely clothed with tiny pale gray scales, and marked on head and elytra as pictured with brown scales. Common on ragweed.

Thricolepis inornata Horn, black and gray above and measuring 3-5 mm, ranges widely in the West and is fairly common.

Paraptochus sellatus (Boh.) is an abundant pest of apricots, peaches, apples, prunes, etc. from British Columbia to California. It is dark brown and 3.5-5 mm long.

15b Claws not fused, antennae long and slender; hind tibiae with two short spurs. Fig. 858. .... *Brachyrhinus rugostriatus* Goeze

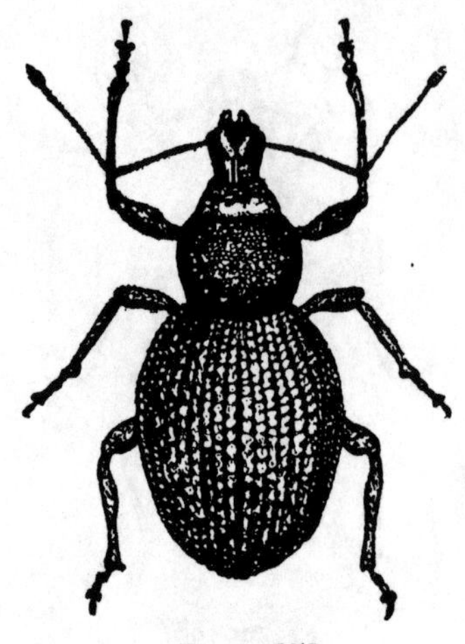

Figure 858

LENGTH: 7.5-8 mm. RANGE: United States and Canada.

Brownish-black, rather dull, the surface covered with rounded tubercules, each bearing a hair. Femora club-shaped. There are 8 similar appearing species in our territory.

Sciopithes obscurus Horn ranges from California to British Columbia where it is known as the Obscure Root Weevil. It is 5-7 mm in length, and is brown with lines of whitish.

16a (14) Thorax with a fringe of short bristly hairs, or if without fringe, then the front coxae separated by a narrow sternum. 18

16b Front coxae touching; thorax without fringe as in 16a. Claws not united. 17

17a (16) Mesoplural pieces very unequal, the episternum larger and reaching the margin of elytra; antennae slender. Fig. 859. .... *Epicaerus imbricatus* Say

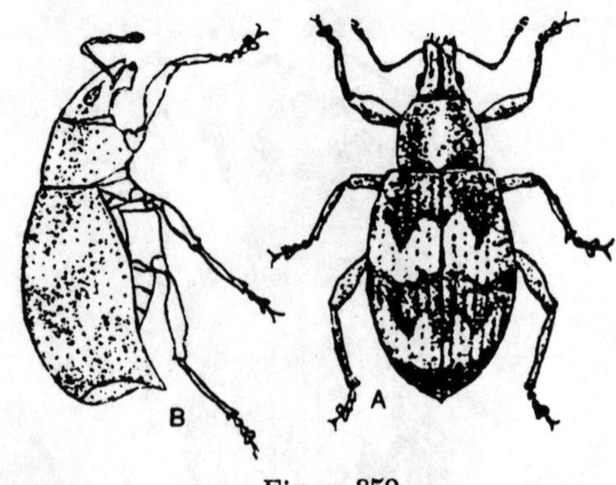

Figure 859

LENGTH: 7.5-11.5 mm. RANGE: East of the Rockies.

Upper surface densely clothed with small fuscous and grayish scales in the pattern pictured. This Imbricated Snout Beetle is a rather well known pest of economic plants (a, dorsal view; b, side view), such as apple, cherry, gooseberry, raspberry, melons, cabbage, cucumbers, sugar beet, etc.

17b Mesopleural pieces diagonally and about equally divided; 10th elytral striae entire; segments of the flagellum elongated. Fig. 860. .... *Pantomorus cervinus* Boh.

Figure 860

LENGTH: 6.5-8.5 mm. RANGE: much of the United States.

Dark brown, rather lightly clothed with gray or pale brown scales; elytra marked with whitish as pictured. This "Fullers Rose Beetle" has been a pest in greenhouses. It feeds on many species of plants and in California is thought to spread citrus blast. Since the adults cannot fly, banding of trees and shrubs offers a means of control.

18a (16) Front coxae touching, or nearly so, thorax with fringe of ocular hairs. Fig. 861. .... *Tanymecus confertus* (Gyll.)

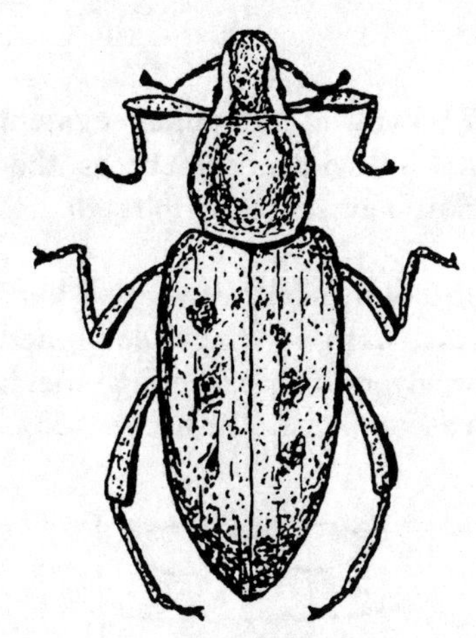

Figure 861

LENGTH: 5-7 mm. RANGE: East of the Rockies.

Dusky brown with scattered grayish and cuperous scales; elytra with spots of brown as pictured.

T. lacaena (Hbst.), similar in color and shape, is considerably larger (8-9 mm). It occurs from Florida to Texas.

18b Front coxae distant, front femora larger than others; front tibiae with tooth on inner edge. Fig. 862. .... *Pandeleteius hilaris* (Hbst.)

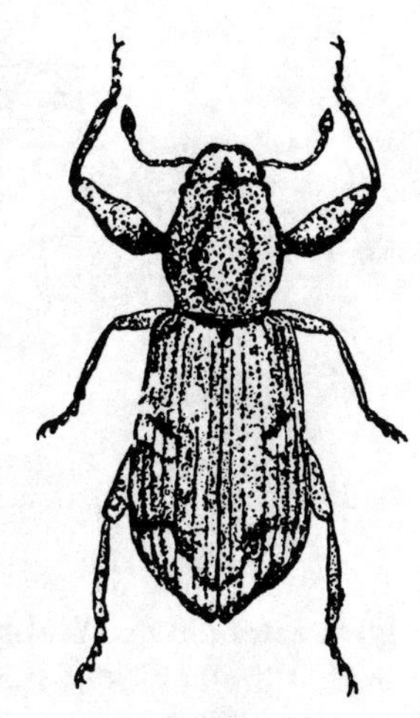

Figure 862

LENGTH: 4-5 mm. RANGE: Florida to Texas and north to Canada.

Brown, densely clothed with fine fuscous and grayish scales in pattern as pictured. A common species.

It breeds in oak, making small burrows. The adults hibernate under logs and bark and may be taken by beating or sweeping New Jersey tea, smartweed and oak leaves. It has been called the Gray-sided Oak Weevil. There are a dozen more members of this genus in North America.

19a (10) Beak free, not received by the prosternum. 20

19b Beak when at rest lying within the prosternum; upper surface often much roughened. 53

20a (19) Mesopleura protruding between the thorax and elytra and appearing from above as a wedge or as rounded knobs (a), and obliquely truncating the shoulder of the elytra. Fig. 863a. 51

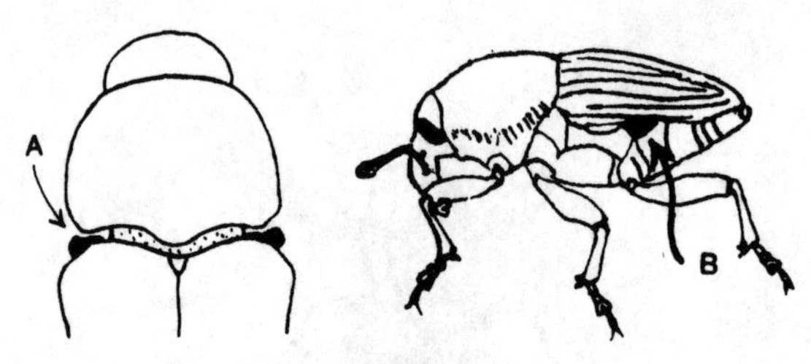

Figure 863

20b Not as in 20a. 21

21a (20) Elytra extending so feebly over the abdomen that the angles of the first ventral abdominal segments are exposed to view (b). Fig. 863b. 49

21b Angles of the first ventral abdominal segment covered by the elytra. 22

22a (21) Beak usually at least as long as the prothorax; gular peduncle usually long. 23

22b Beak broad and short, claws simple, ventral abdominal segments about equal; guler peduncle short and broad; mentum large and quadrate. Fig. 864. .... *Sitona hispidulus* (Fab.)

Figure 864

LENGTH: 3-5 mm. RANGE: Europe, Canada and United States.

Shining. Black with small oval scales, gray and cuperous forming the pattern pictured.

This is the "Clover Root Curculio."

Sitona flavescens Marsh. with a length of 4.5-5.5 mm and ranging in Europe and much of the United States and Canada is black and densely clothed above with reddish-brown hair-like scales forming a pattern in fresh specimens. The under surface is covered with prostate grayish hairs. Both the adults and the larvae are injurious to white and alsike clover. About 15 more *Sitona* are found in our territory.

23a (22) Ventral abdominal segments about equal in width; sometimes the first is wider; claws often with teeth. 34

23b Third and fourth ventral abdominal segments narrow, their combined width scarcely or not exceeding the width of the second or the fifth. Fig. 865. 24

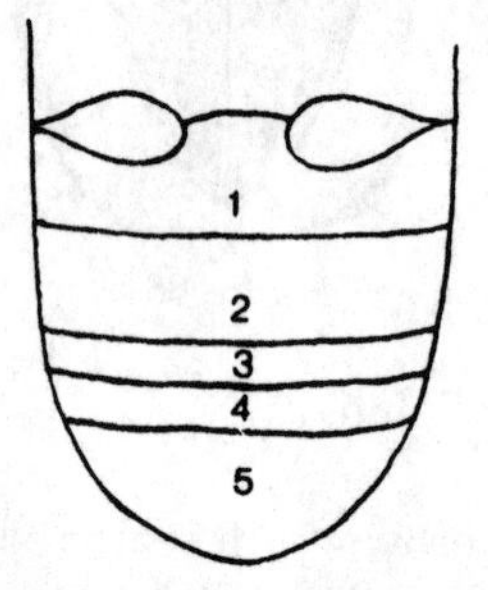

Figure 865

24a (23) Gular margin prominent, its peduncle and mentum retracted; beak rather heavy no longer than the prothorax; mandibles flat and punctured on the outer side. Fig. 866. .... *Lepidophorus setiger* Ham.

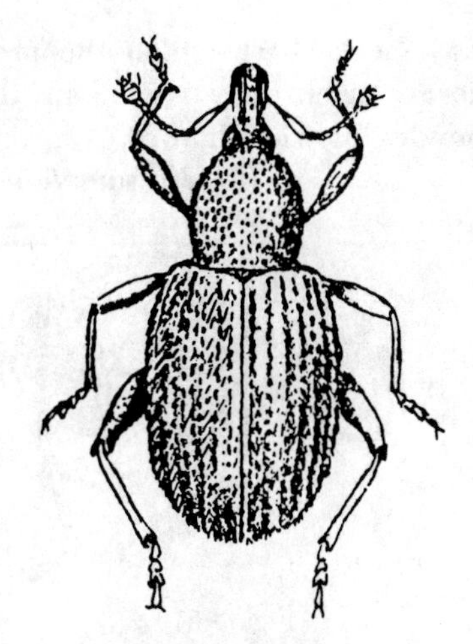

Figure 866

LENGTH: 3-3.5 mm. RANGE: Eastern United States.

Black, antennae and legs reddish-brown; densely clothed above with small round brownish scales interspersed with scattered white ones to form three pale stripes on thorax and mottling on elytra.

24b **Gular margin not prominent, its peduncle usually long; front coxae touching or almost so; claws simple. 25**

25a **(24) Mandibles usually with a notch at tip forming two teeth. Fig. 867a. 26**

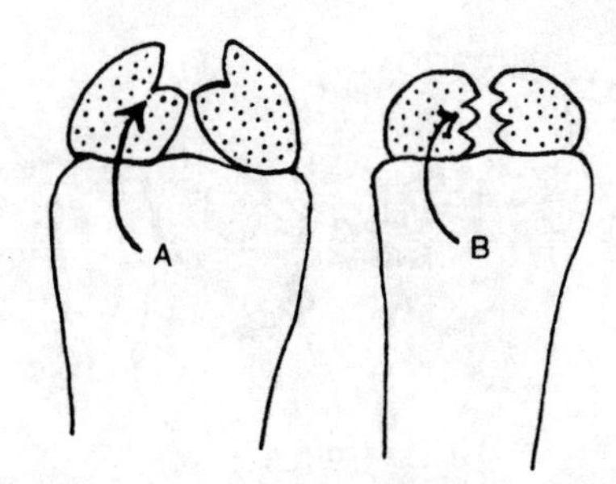

Figure 867

25b **Mandibles with two notches and three teeth at tip. Fig. 867b. 29**

26a **(25) Third and fourth ventral segments narrow, the two together but equalling one of the others. 27**

26b **Ventral abdominal segments not very unequal; space between eyes not more than their diameter. 28**

27a **(26) Second segment of funicle (a) much longer than the first; tibia sharp pointed at apex. Fig. 868.**

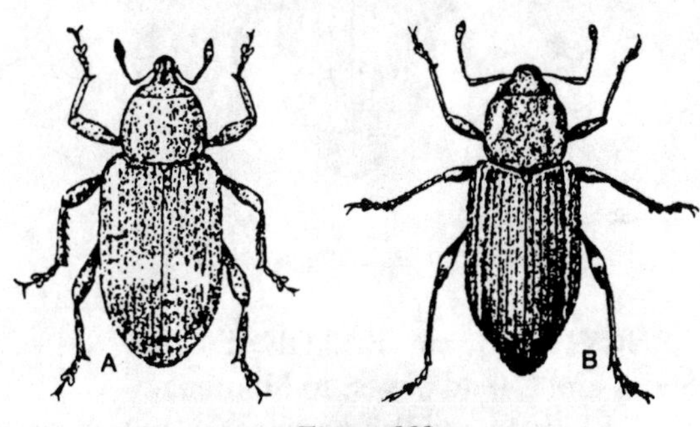

Figure 868

(a) ***Listronotus latiusculus*** **(Boh.)**

LENGTH: 5-6 mm. RANGE: Eastern United States and Canada.

Black, usually covered with grayish, brownish, or reddish scales, forming a rather obscure pattern, as pictured; antennae and tarsi reddish-brown. Sometimes destructive to parsley.

(b) ***L. palustris*** **Blatch.**

LENGTH: 6-8.5 mm. RANGE: Florida.

Black; upper surface with pattern as pictured set with brown and silver-gray scales; antennae, tibiae and tarsi dark reddish-brown; beak fairly heavy, as long as thorax.

Listronotus includes about 30 species in the U. S. and Can.

27b **Second segment of funicle scarcely longer if at all and slimmer than the first. Fig. 869. .... *Hyperodes solutus* (Boh.)**

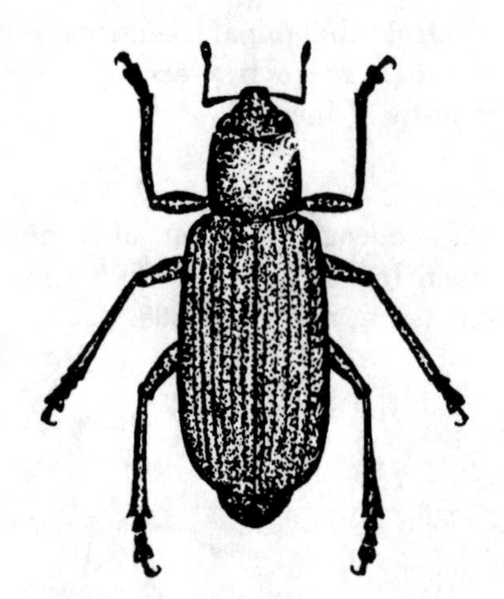

Figure 869

LENGTH: 3-5 mm. RANGE: Eastern United States and Canada, west to Montana.

Black to reddish-brown, the upper surface with rounded clay-yellow scales in pattern as pictured; antennae and legs reddish brown.

H. hyperodes (Dietz), measuring 4.5-6 mm, black with covering of dark brown and grayish-brown scales, has been reported as boring within the stems of tomatoes in California.

Hyperodes is a relatively large genus with about 40 species. *H. montanus* (Dietz) is known from Indiana and Iowa to the Dakotas. It measures 3.5 mm and is blackish, with a covering of grayish-brown and white scales. The antennae and legs are reddish-brown. *H. sparsus* (Say) ranges widely. It measures 4-5 mm. and is reddish-brown or piceous with a covering of dull gray scales.

H. humilis (Gyll), measuring 4-5 mm, apparently ranges over much of our territory. It is dull black; the thorax has a stripe on each side, while the antennae, tibiae, tarsi and front of thorax are dark reddish-brown.

28a **(26) Beak shorter than the prothorax; thorax much narrower than the well rounded elytra. Fig. 870. .... *Hypera punctata* (Fab.)**

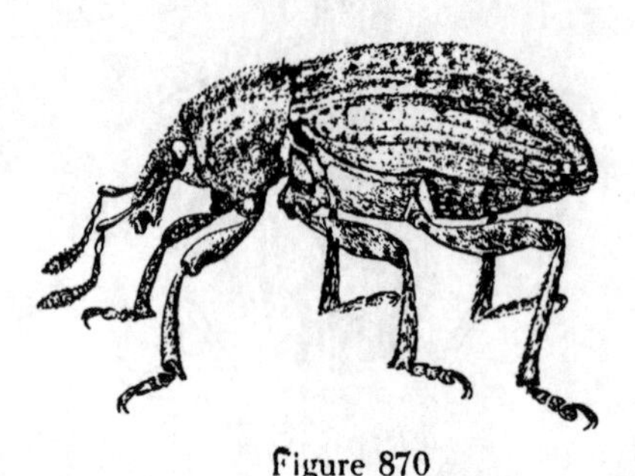

Figure 870

LENGTH: 5-8.5 mm. RANGE: introduced from Europe; now widely scattered through the United States.

Black, thickly clothed with scales of varying shades of gray, brown and blackish. Beak two-thirds as long as thorax.

This "Clover Leaf Beetle" may be found in almost any clover field; the green and pinkish larvae climb up the plants at night and feed on the leaves. External feeding by the larvae is rather unusual in this family.

28b **Beak longer than the prothorax; thorax not more than one-third narrower than the elytra. Fig. 871.**

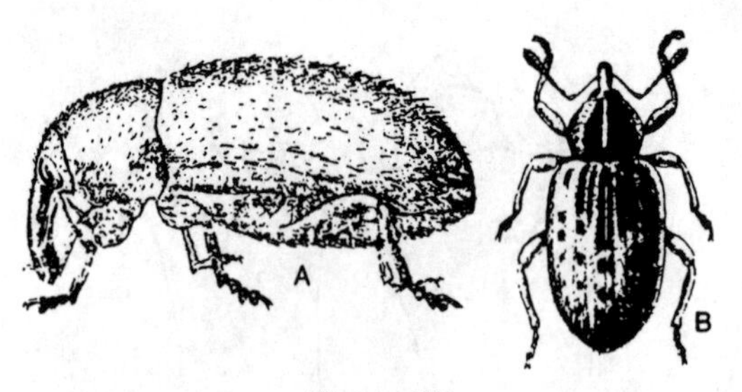

Figure 871

(a) ***Hypera nigrirostris* (Fab.)**

LENGTH: 3.5-4.5 mm. RANGE: Europe and most of United States.

Reddish brown to black, appearing greenish or bluish-black or yellow from the heavy coat of scales and hairs; beak, head and legs black. It is often abundant in clover fields and is known as the "Lesser Clover Leaf Weevil."

(b) ***H. postica*** **(Gyll.). Alfalfa Weevil**

LENGTH: 3-5 mm. RANGE: introduced from Europe, abundant over much of U. S.

Pale brownish to black with design of scales and hairs as pictured; antennae and legs paler. A serious pest of alfalfa and clover.

29a **(25) Tibia with corbels (an ovate area near apical end which is bounded by minute bristles) and with a long curved claw at apex, which is oblique, dilated and ciliate; length at least 4 mm. 33**

29b **Tibiae without corbels, truncate at tip; usually less than 4 mm long, never over 7 mm. 30**

30a **(29) Body with usual coating of scales and hair; without a waterproof coat. .. 31**

30b **Body covered with a waterproof, varnish-like coat of scales; beak cylindrical and much longer than head; legs long and slender. Fig. 872. ....**
.... *Lissorhoptrus oryzaephilus* Kusch.

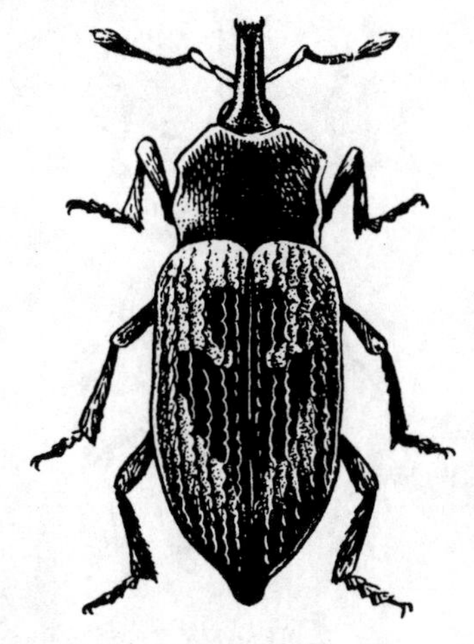

Figure 872

LENGTH: 2.8-3.2 mm. RANGE: Canada to the Gulf, west to Texas.

Black with coating of large olive-gray scales. Beak as long as thorax, heavy. Basal half of antennal club smooth and shining.

This "Rice-Water Weevil" feeds on various water plants and is often very destructive to rice. The larvae are known as the "rice-root maggot." The adults can live under water for a long period of time without drowning.

31a **(30) Beak constricted at base, a transverse line separating it from the head; claws fused; club of antennae small. .. 32**

31b **Claws diverging; beak not constricted at base. Fig. 873. ....**
.... *Dorytomus mucidus* (Say)

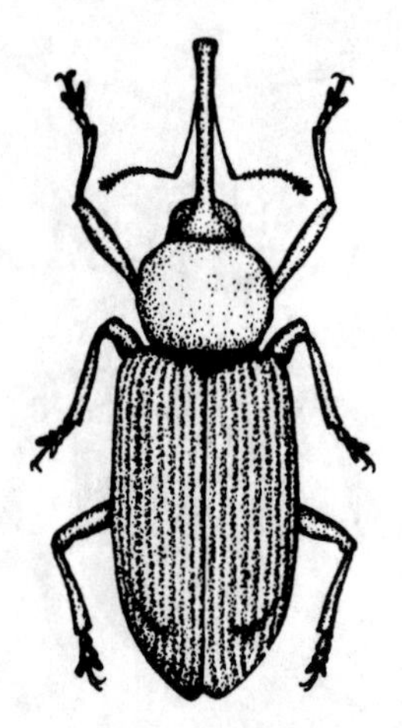

Figure 873

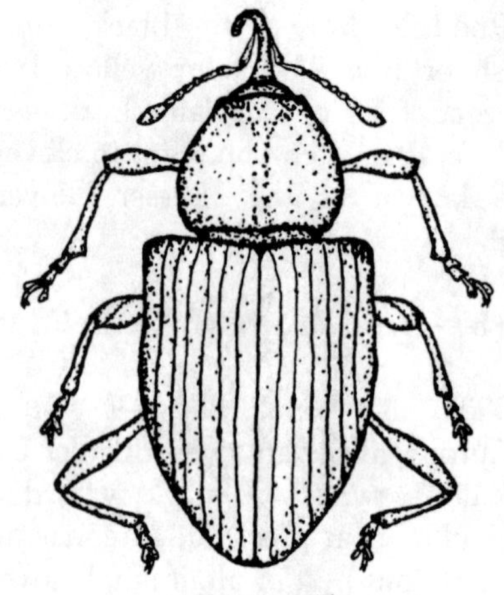

Figure 875

LENGTH: 5-6.5 mm. RANGE: Canada to Mississippi, New England to Kansas.

Dark reddish-brown, thinly clothed with prostrate yellowish hairs; antennae reddish-brown; femora toothed. Beak slender and curved only half as long as thorax. Breeds in the seed pods of cottonwood. There are about 30 *Dorytomus* in the U. S. and Can.

Notaris puncticollis (Lec.) is substantially the same in size and color but differs in having no teeth on the femora. It ranges from Newfoundland to Iowa.

32a (31) Eyes close together on underside of head (Fig. 874a); tibiae with hooks or claws at apex. Fig. 875. .... *Smicronyx tesselatus* Dietz

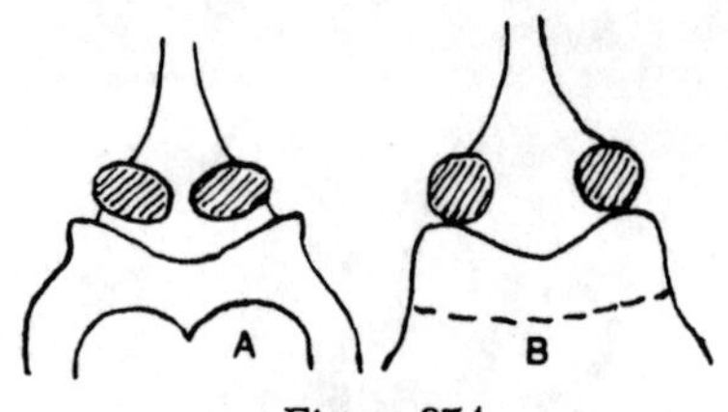

Figure 874

LENGTH: 2-2.4 mm. RANGE: Texas and eastward.

Very convex. Black to reddish-brown; with covering of brown and white scales, forming pattern as pictured. First segment of funicle thick.

S. *apionides* Csy., ranging widely from Texas to the East, may be readily told by its small size 1.7-2 mm. It is black with some white scales on sides of thorax and base of elytra.

S. *perpusillus* Csy. known from Florida, may be distinguished by its small size, 1.6 mm, and by the widely scattered large white scales on the elytra. The color is blackish with the legs reddish.

S. *quadrifer* Csy. seems to be distributed well across the southern part of the United States. It measures about 2.5 mm and is black with a coat of large, pale brown or whitish scales. A large squarish, velvety-black spot near the middle of the elytral region makes it readily determined.

Smicronyx includes some 75 species in our area.

32b Eyes more widely separated on underside of head (Fig. 874b). Hind tibiae ending in a sharp point. Fig. 876. .... *Smicronyx scapalis* (Lec.)

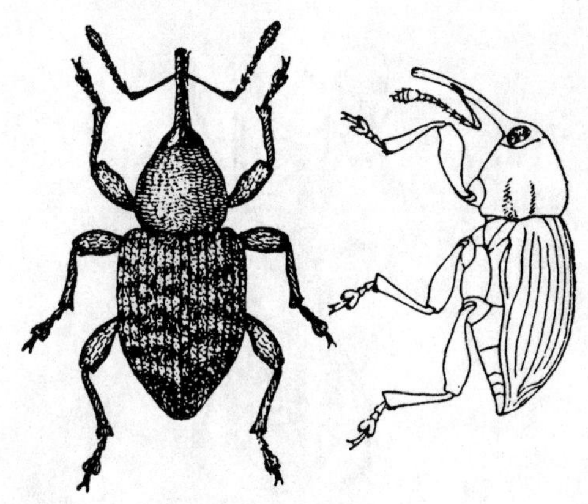

Figure 876

LENGTH: 4-5 mm. RANGE: Texas to New Hampshire.

Dark brown to black, with a dense coat of grayish-yellow scales. Legs dull red; male with first two segments of funicle about equal; female with the first segment considerably longer.

S. sordidus Lec. known from Illinois to Arizona, is colored much as the above but it is noticeably smaller. (2.5-3 mm).

33a (29) Antennae attached wholly on the sides of the more slender cylindrical beak; their point of attachment not visible from above; front coxae not quite touching. Fig. 877. .... *Pissodes strobi* (Peck)

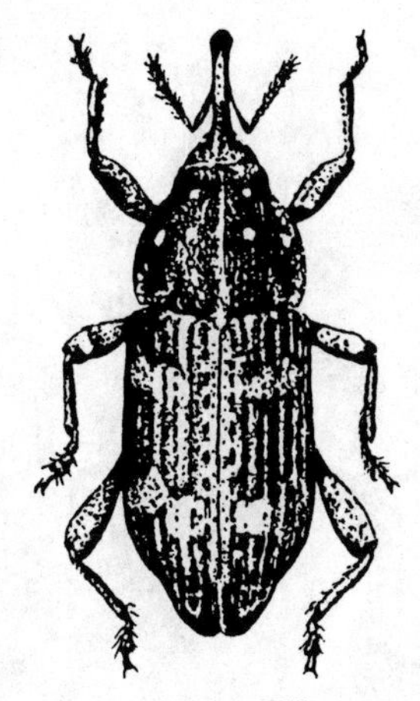

Figure 877

LENGTH: 4.5-6 mm. RANGE: Wisconsin and eastward.

Light to dark brown, marked as pictured with yellowish. Elytra wider than thorax; striae punctured; third and fifth intervals somewhat higher than others. This "White Pine Weevil" infests several species of pines but is particularly destructive to the White Pine.

P. costatus Mann., measuring 5.5-7 mm and light to dark brown, is known as the Ribbed Pine Weevil and ranges from Alaska to California. It is one of several species of the genus that attack pines, spruces and other conifers. There are over 30 *Pissodes* in the U. S. and all very similar.

33b Antennae attached toward the sides of the thicker and stronger beak, their point of attachment visible from above; front coxae touching. Fig. 878. .... *Hylobius congener* Dalla T.

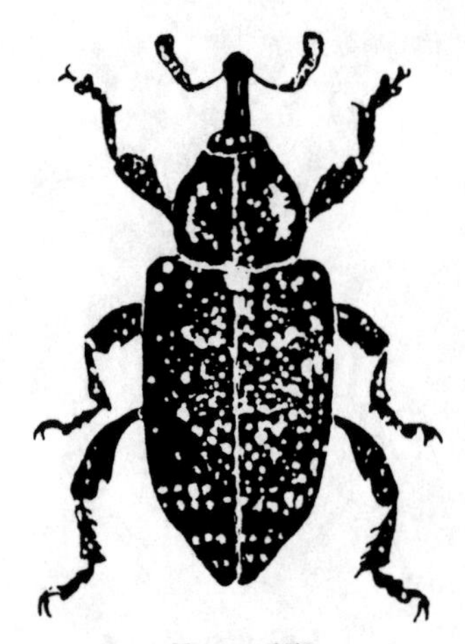

Figure 878

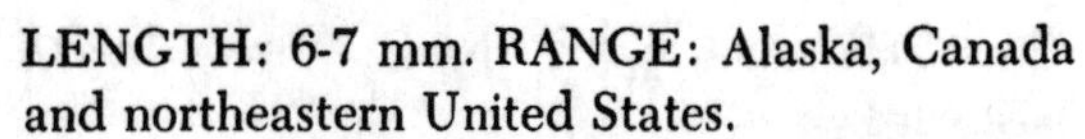

LENGTH: 6-7 mm. RANGE: Alaska, Canada and northeastern United States.

Oblong, robust. Dark reddish-brown with scant pattern of whitish hairs, as pictured.

Eudociminus mannerheimi Boh., 10.5-15 mm long, is another member of this tribe. It is black with side stripes on thorax and four small spots of brown scales on the fourth interval of the elytra. It ranges from New York to Louisiana and is known as the Cypress Weevil.

34a (23) Thorax narrowed at base into a neck-like stem. Fig. 879. .... *Myrmex chevrolati* (Horn)

Figure 879

LENGTH: 3.5-4.5 mm. RANGE: New England to Iowa and Texas south to the Gulf.

Convex. Highly shining, black. Densely covered with white hairs on sides but nearly hairless on top. Elytra less than twice as long as wide; disk with rows of closely-placed fine punctures. Front of head as pictured (a).

M. floridanis (Csy.) and *M. myrmex* Hbst. are similar in size and color but both have a large triangular tooth on the femora. The first has rows of white hairs mingled with black ones on the elytra while in *Myrmex* the hairs are wholly black. *Myrmex* includes some 25 other species in the U. S. and Can.

34b Thorax not as in 34a. 35

35a (34) Hind angles of thorax obtuse. 36

35b Hind angles of thorax acute or produced; outline wedge-shaped or broader behind. Fig. 880.

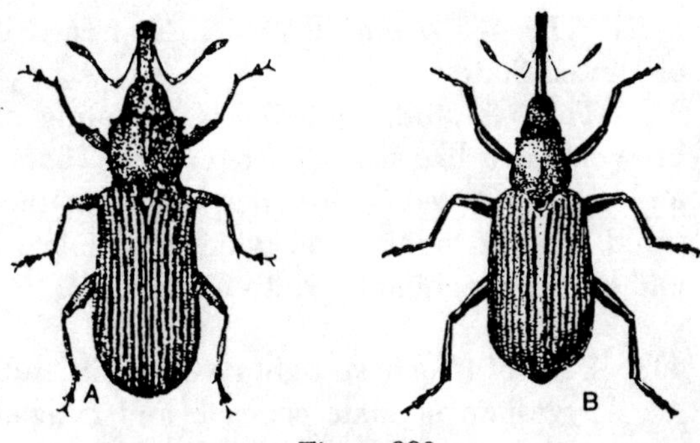

Figure 880

(a) ***Magdalis aenescens*** **Lec.**

LENGTH: 4-6 mm. RANGE: Alaska and California.

Black with bronzy-metallic sheen; elytra distinctly striate. This "Bronze Apple Tree Weevil" is a pest of apple and some other fruit trees.

(b) ***Magdalis barbita*** Say

LENGTH: 4-6 mm. RANGE: Canada and New England to Texas.

Black, feebly shining; antennae and tarsi piceous. The larvae live under the bark of oak, hickory and elm. *Magdalis* is another fairly large genus with 25 species in our fauna.

36a (35) The mandibles moving vertically; snout often very long and slender, (nut and acorn weevils). 37

36b Mandibles moving in a horizontal plane; claws divided, toothed or with an appendage. 42

37a (36) Beak never longer than the body (usually shorter), curved. 41

37b Beak of female definitely longer than the body. 38

38a (37) First segment of the funicle (in antennae) (d, Fig. 882) usually much longer than the second, never shorter. 39

38b Second segment of funicle longer than the first. Feeds on chestnuts. Fig. 881. .... *Curculio proboscideus* Fab.

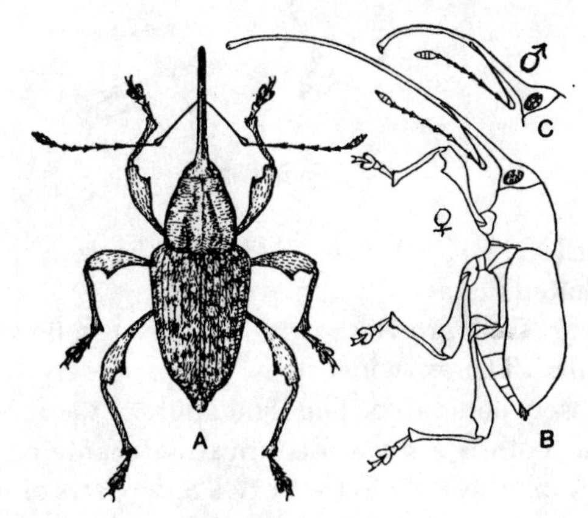

Figure 881

LENGTH: 8-11 mm. RANGE: Eastern half United States.

Robust. Dark brown, clothed with dense coat of yellowish or golden hairs; thorax and elytra with fuscous spots; fifth ventral segment square cut, with a smooth area at apex and a tuft of hairs on each side. It is known as the Large Chestnut Weevil. The female (b) has a much longer snout than the male (c).

This fairly sizable genus is a most important one in the damage caused by its larvae. The female drills a hole deep into developing nuts or acorns. An egg is then placed at the bottom of the hole deep within the fruit where the larvae spends most of its life, safely out of reach of poisons.

39a (38) Surface covered with hairs; femoral tooth rectangular; large; feeds on hickories and pecan. Fig. 882. .... *Curculio caryae* (Horn)

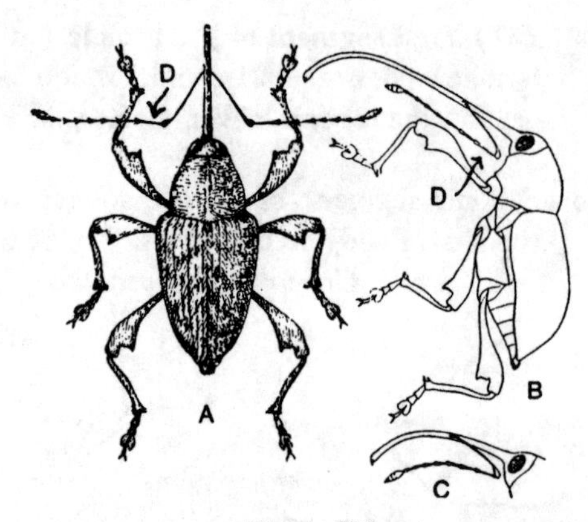

Figure 882

LENGTH: 7-9 mm. RANGE: Eastern half United States.

Dull brownish with scattered yellowish hairs. Thorax wider than long, densely and closely punctured. The right angle of the femoral tooth is a good identification character. It is known as the Pecan Weevil. Side views of the female (b) and male (c) are shown.

39b Surface covered with scales. 40

40a (39) Beak of female (b) curved from end to end, pygidium of male with tuft of hair; feeds on chestnut. Fig. 883. .... *Curculio sayi* (Gyll.)

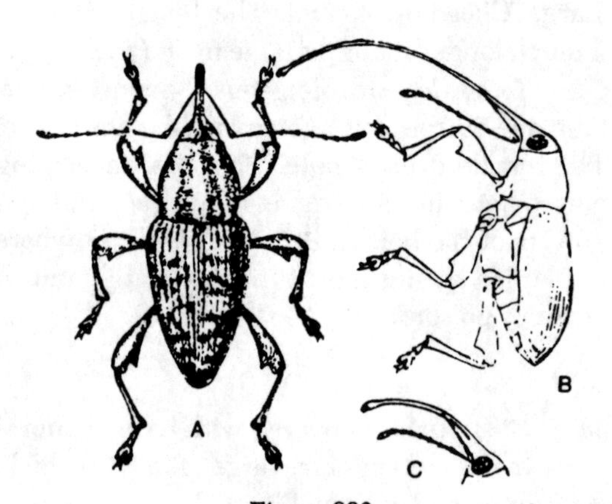

Figure 883

LENGTH: 4-.5 -7 mm. RANGE: Eastern half of United States.

This Chestnut Weevil has a coating of brownish hair-like scales over piceous. Thorax and elytra marked with yellow scales as pictured. Thorax longer than wide in female (a and b); length equal to width in male (c).

40b Beak of female straight nearly to its end, pygidium of male concave and fringed with hairs; feeds on oaks. Fig. 884. .... *Curculio proboscideus* (Fab.)

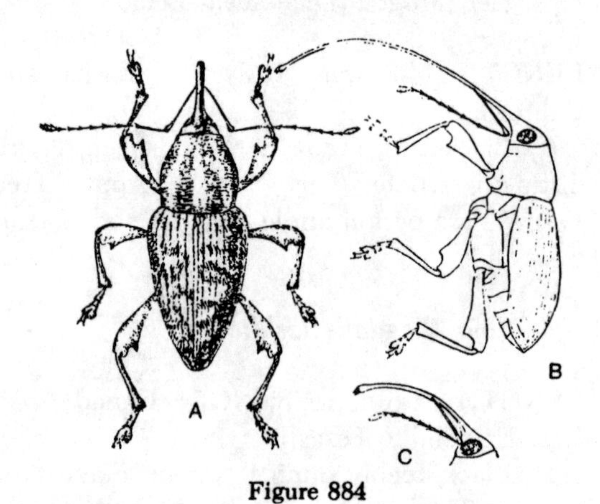

Figure 884

LENGTH: 8.5-9 mm. RANGE: Eastern half United States.

This pest of the acorns of the black oak group is brownish with a covering of pale brown hair-like scales. The beak of the male is somewhat shorter than the body while that of the female is nearly twice as long. The tooth on the femur is small.

41a (37) Feeds on hazel-nuts. Outer edge of femoral tooth oblique; beak about equal in both sexes. Fig. 885. .... *Curculio neocorylus* Gibson

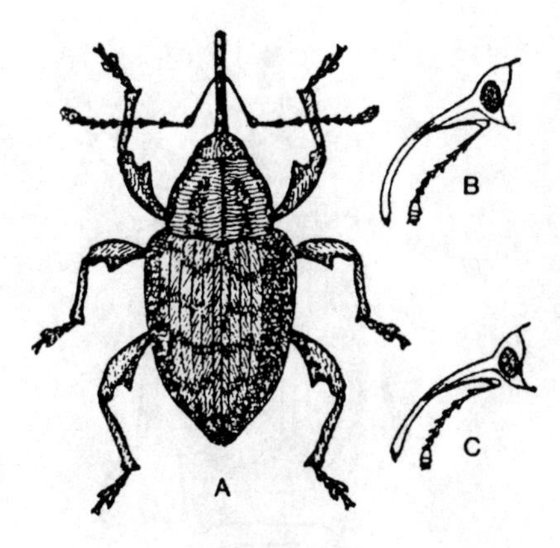

Figure 885

LENGTH: 6-8 mm. RANGE: Canada to Texas.

Blackish brown, clothed above with ash-gray to yellowish scale-like hairs; thorax and elytra marked with paler scales as pictured. Beak of female (b) only two-thirds as long as body; that of male (c) still shorter; pygidium of male with long yellow hairs. Appendices of claws broad, nearly or quite rectangular.

The eggs are laid in hazel-nuts where the larvae develop. The infested nuts fall early.

41b Breeds in acorns. Outer edges of femoral tooth standing at a right angle to the femur; beak of male slightly less than half the length of the body, that of female three-fifths as long as the body. Fig. 886. .... *Curculio sulcatulus* (Csy.)

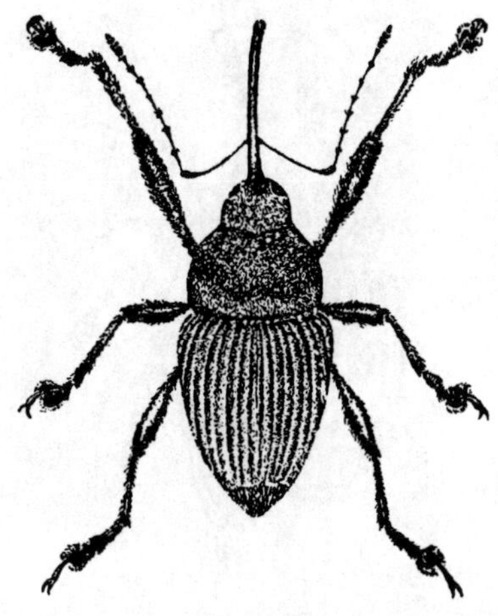
Figure 886

LENGTH: 5.5-7.5 mm. RANGE: Eastern half of U. S. and Utah and Col. south to Mexico.

Sooty to reddish brown, the body scantily covered with short pale gray scales; elytra with gray hairs which rub off to leave bare areas. First joint of funicle longer than second; pygidium of male with tuft of long yellowish hairs.

It seems to infest several species of oaks of both the white oak and black oak groups, but apparently prefers the latter group which takes two years to mature its acorns. Currently 27 nut-weevils are known from our area.

42a (36) Ventral sutures angling at sides. 43

42b Ventral sutures straight and parallel. .. 44

43a (42) Pygidium exposed; flagellum with but five segments. Fig. 887. .... *Gymnetron tetrum* (Fab.)

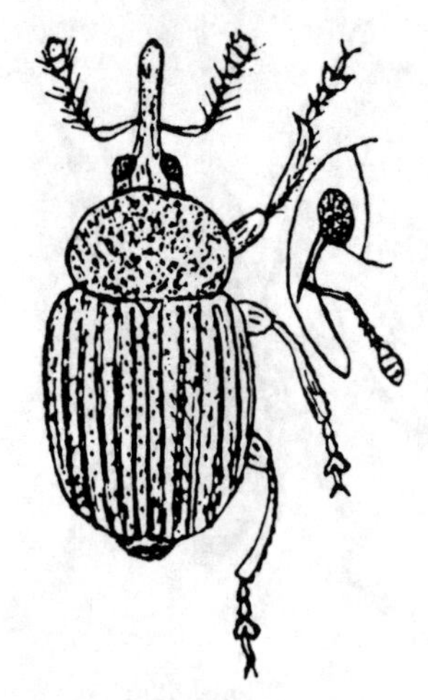

Figure 887

LENGTH: 2.5-4 mm. RANGE: Europe and much of North America.

Black with a clothing of yellowish-gray pubescence; thorax with many fine punctures; femora toothed below. This species is exceedingly common and can almost always be found around its food plant, the common mullein, especially within the capsules. There are 3 other similar imports in our area.

Miarus hispidulus Lec. is a bit smaller. It breeds in the seed capsules of some of the lobelias.

43b Pygidium covered; flagellum with 6-7 segments. Fig. 888. .... *Odontopus calceatus* Say

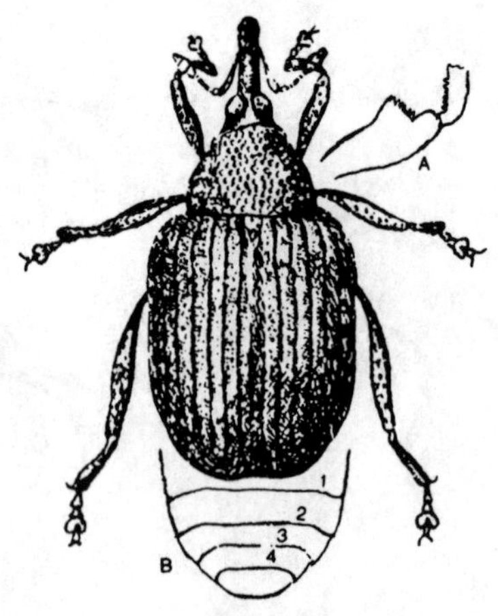

Figure 888

LENGTH: 3-4 mm. RANGE: Eastern half United States.

Black, slightly shining; tarsi and antennae red-brown. Thorax heavily set with coarse punctures. Striae of elytra deep and coarsely punctured; intervals flat. Front femora with serrations for holding (a); ventral segments angulate at sides (b). Feeds on sassafras and tuliptree.

Piazorhinus scutellaris (Say), measuring 2-2.5 mm and ranging through the eastern half of the United States, is closely related. It is black except for a white scutellum and yellow hairs on the thorax with white hairs on the elytra.

44a (42) Claws with a sharp tooth. 45

44b Claws with a short broad tooth at base; hind femora thickened for leaping. Fig. 889. ...... *Rhynchaenus pallicornis* (Say)

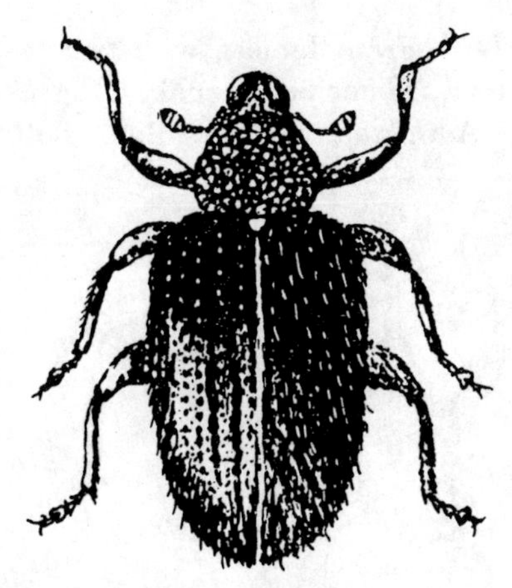

Figure 889

LENGTH: 2.5-3 mm. RANGE: Newfoundland to Oregon and Texas.

Shining, black, thinly clothed with grayish-yellow hairs; tarsi and antennae reddish-brown; beak short and heavy; elytral striae shallow, with coarse punctures, intervals flat.

Its larvae mine in the leaves of several species of trees including apple and cherry.

R. ephippiatus (Say), similar in size has a seven-jointed funicle instead of six as in the above. Black; elytra with a sub-basal band of white and brownish pubescence and a small post-median band, these connected on interval 2; the appendages, except the hind femora, reddish-brown. From coast to coast on willow.

45a (44) Hind tibiae ending in a sharp point (mucronate); scrobes long. 46

45b Hind tibiae with hook at tip; pygidium covered; claws with a long tooth; beak long and slender. Fig. 890. .... *Tachypterellus quadrigibbus* (Say)

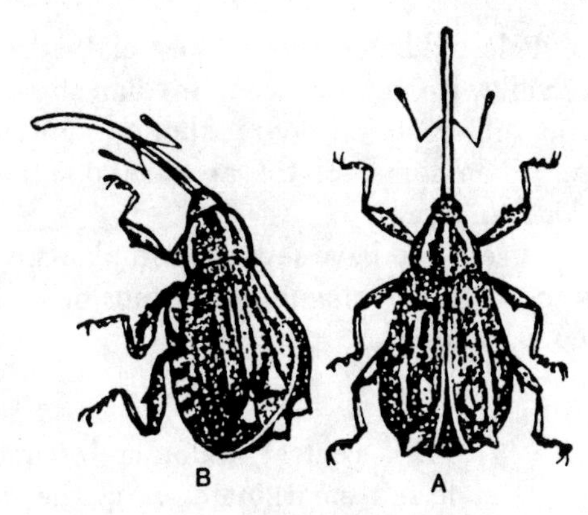

Figure 890

LENGTH: 3-4.5 mm. RANGE: general throughout the United States and Canada.

Dark red; beak, antennae and legs, paler; thin coat of grayish pubescence on elytra and thorax forming three lines on the thorax; elytra with two large tubercles on each, as pictured.

This Apple Curculio feeds not only on apples and pears but on other closely related plants also.

46a (45) Antennal grooves directed against the eye; antennal club elongated. 47

46b Antennal grooves directed below the eye; pygidium not elongated. Fig. 891. .... *Pseudanthonomus crataegi* Walsh

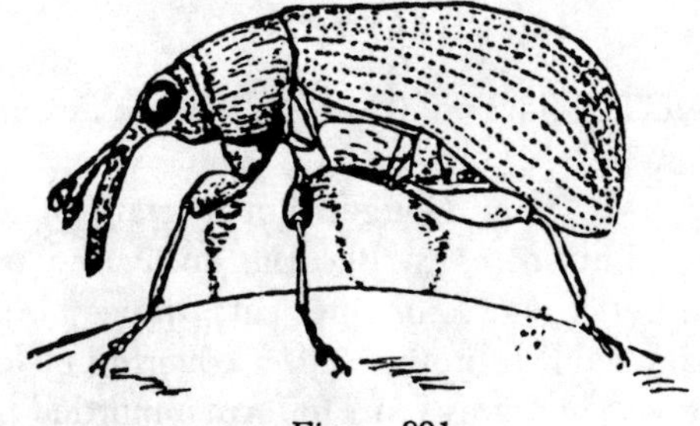

Figure 891

LENGTH: 2.5-2.8 mm. RANGE: Canada to Wisconsin and Florida.

Pale reddish-brown, thinly clothed with long yellowish pubescence in median streak on thorax and spots on elytra; thorax one-fourth wider than long. Elytra at base one-fourth wider than thorax.

It seems to have several food plants, various species of *Craetegus* and *Prunus* being frequently attacked.

47a (46) Fifth ventral abdominal segment of male not emarginate, or at the most but slightly so. 48

47b Fifth ventral of male short, with deep emargination behind. (a) Fig. 892. .... *Anthonomus scutellaris* Lec.

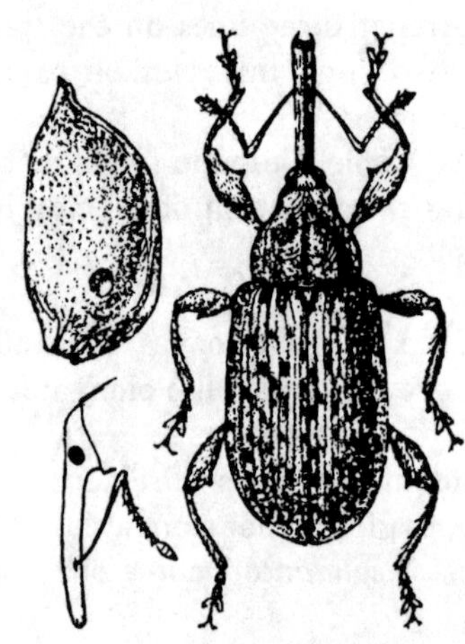

Figure 892

LENGTH: 5-6 mm. RANGE: Massachusetts to Texas.

This "Plum Gouger" got its start on wild plums but seems to like the cultivated ones even better and sometimes eats apples also. It is dark reddish-brown with a covering of long yellowish and gray hairs to form a mottled pattern on thorax and elytra. The striae on the elytra are narrow with fine punctures.

This genus is a large one with more than 100 named species, several of which are highly economic.

48a (47) Front femora with two teeth, the forward one being smaller. Fig. 893. .... *Anthonomus grandis* Boh. Boll Weevil

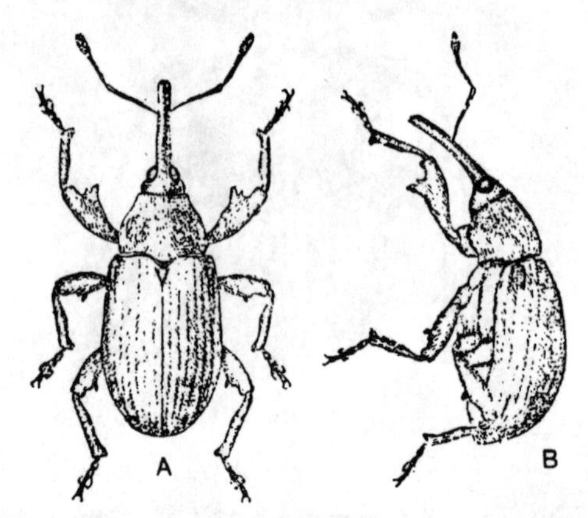

Figure 893

LENGTH: 4-7.5 mm. RANGE: Mexico and much of our American Cotton belt.

This little pest literally removes the shirt from many backs. It came up from Mexico in the 1890's and has worried numerous cotton growers into an early grave. It is reddish-brown to blackish with a covering of pale yellow scales forming many small dots on the elytra. The eggs are laid within the cotton bolls and the larvae destroy the seeds and fiber.

48b Front femora with but one tooth; elytra with darker mark back of middle. Fig. 894.

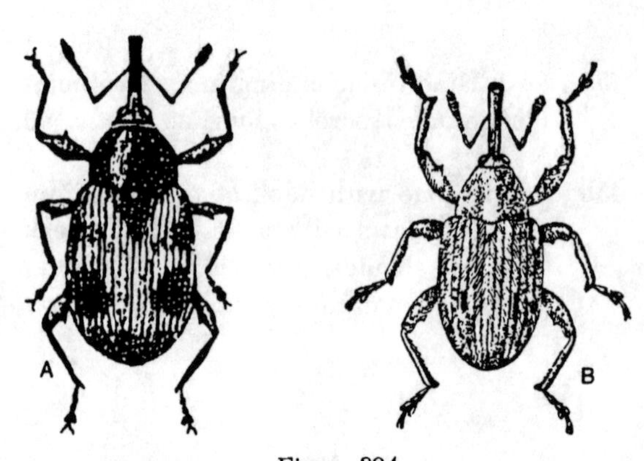

Figure 894

(a) ***Anthonomus signatus*** **Say**

LENGTH: 2-3 mm. RANGE: Canada to Florida and Texas.

Piceous with thinly placed whitish hairs; the dark red elytra are marked with two bare spots as pictured. The antennae are dull yellow with darker club. It occurs on various plants but has been named the Strawberry Weevil because of its damage to this plant.

(b) ***A. eugenii*** **Cano Pepper Weevil**

LENGTH: 2.5-3 mm. RANGE: Mexico and our Southwest.

Reddish-brown or blackish with a brassy tinge. Pale yellowish brown scales make the design on thorax and elytra. It is highly destructive to peppers.

A. musculus Say about 2 mm long and known from Colorado to New England and Ontario to Florida is dark reddish-brown or piceous and covered with a thin white pubescence. The beak is slightly curved and is as long as the head and thorax taken together. There are more than 110 *Anthonomus* in the U. S. and Canada.

49a (21) Elongate slender beetles with a slender curved beak. 50

49b Elongate but more robust beetles with thicker beak which becomes dilated near apex. Fig. 895. .... *Cleonus quadrilineatus* (Chev.)

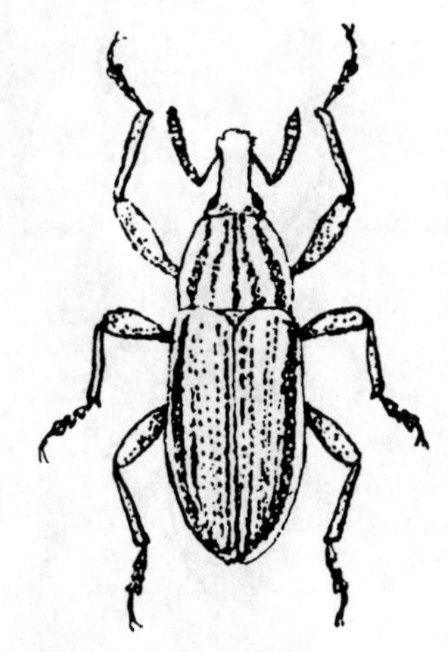

Figure 895

LENGTH: 9-11 mm. RANGE: Southwestern United States.

Shining black, striped with yellowish pubescence as pictured. Feeds on the roots of some legumes.

The Radish Weevil, *C. sparsus* Lec., another western species, is smaller (6.5 mm). It has four bluish-gray stripes on each elytron.

A larger, more robust species, *C. carnicollis* Lec., known from Colorado and Florida, has four broad gray stripes on each elytron. It measures 12-14 mm. Some 35 *Cleonus* are known from the U. S., and Can.

50a (49) Thorax broadly flattened with an impression extending most of its length. Fig. 896. .... *Lixus concavus* Say Rhubarb Curculio

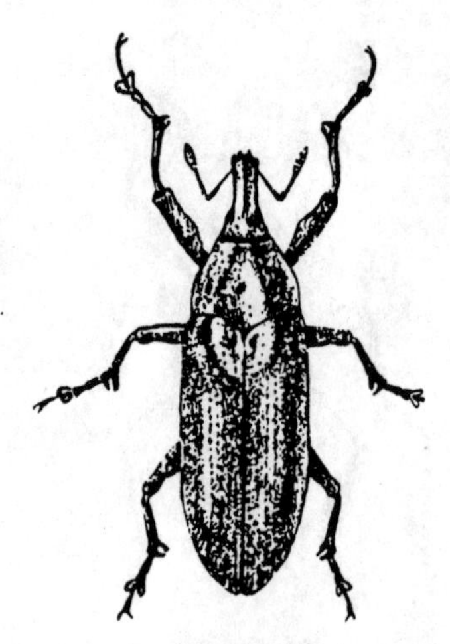

Figure 896

LENGTH: 10-13 mm. RANGE: much of the United States.

Black, sparingly clothed with fine ash-gray pubescence; often covered with rust-red pollen. Lives on curled dock and becomes a pest of rhubarb, puncturing the leaf stems.

This genus contains about 40 readily recognizable beetles.

50b Thorax with impression only near basal end; beak shorter than thorax. Fig. 897. .. .. *Lixus scrobicollis* Boh.

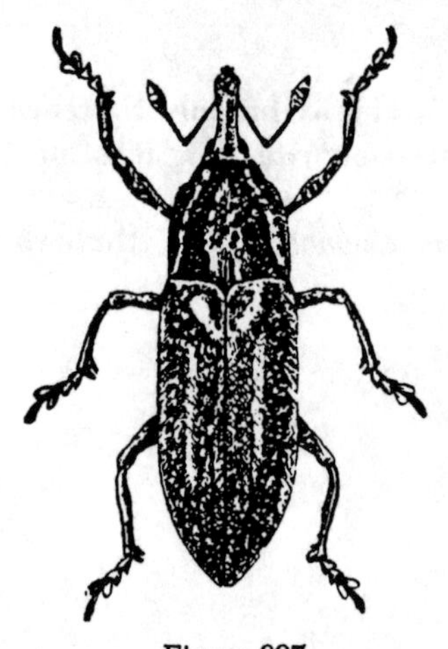

Figure 897

LENGTH: 6.5-9 mm. RANGE: Texas to Atlantic Coast.

Slender, black, with scanty gray pubescence forming pale stripes on thorax and elytra; frequently (freshly emerged specimens) coated with a rust-red dust. Breeds in ragweed.

L. terminalis Lec., known from Iowa to New England and Florida, measures 8-12 mm is shining reddish-brown and has the second joint of the funicle longer than the third and fourth combined. *Lixus* is a large genus with about 70 members in our fauna.

51a (20) Claws united at base; parallel or nearly so. 52

51b Claws diverging, not united; club of antennae nearly round and polished at its base. Fig. 898. .... *Baris interstitialis* (Say)

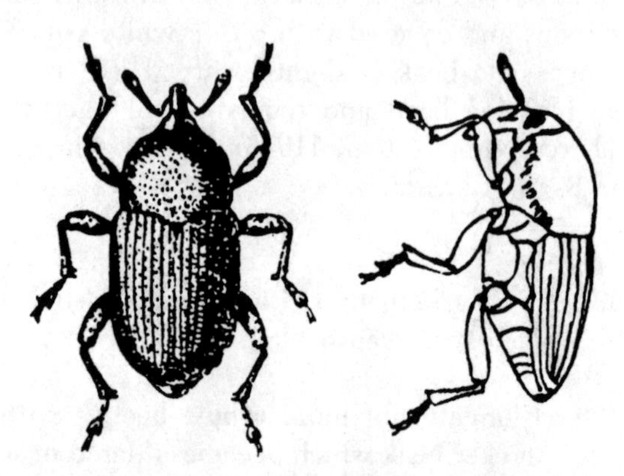

Figure 898

LENGTH: 3.5-5 mm. RANGE: New England to Florida, Texas and Colorado.

Black, antennae and legs reddish-brown; beak half as long as thorax, much curved. Thorax and elytra of equal width.

More than 140 species of *Baris* are known. A species introduced from Europe, *B. scolopacea* Germ., 3.5 mm, blackish, with white and pale brown scales, lives on ragweed and is becoming widely distributed.

B. umbilicata (Lec.) 3-5 mm in length and ranging from Texas and Colorado to Wisconsin and Michigan is shining black with dark reddish-brown tarsi and antennae.

52a (51) Front coxae widely separated; prosternum with a deep, wide excavation in front. Fig. 899.

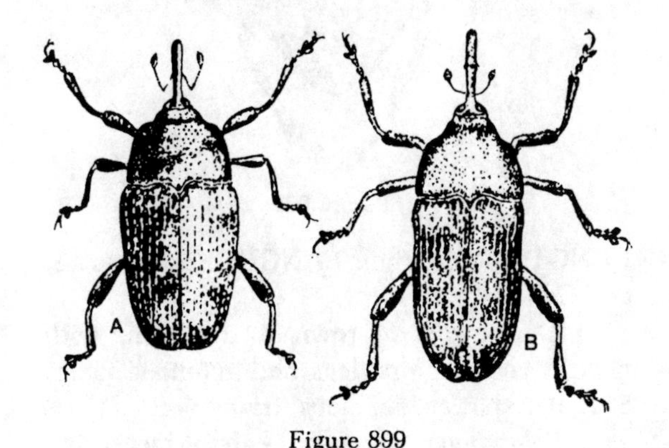

Figure 899

(a) ***Ampeloglypter ater* Lec.**

LENGTH: 2.5-3 mm. RANGE: New England to Indiana.

Shining, black; tarsi and antennae reddish-brown. First joint of the funicle as long as the next three combined. Taken on Virginia creeper.

(b) ***A. sesostris* (Lec.) Grape Cane Gall Maker**

LENGTH: 2.5-3 mm. RANGE: Eastern half United States.

Shining, pale reddish-brown. Beak one-half longer than thorax, much curved at base. Elytral striae fine, without punctures. It feeds on grapes.

52b Front coxae narrowly separated, beak short and robust; antennal club small. Fig. 900. ...... *Trichobaris trinotata* (Say)

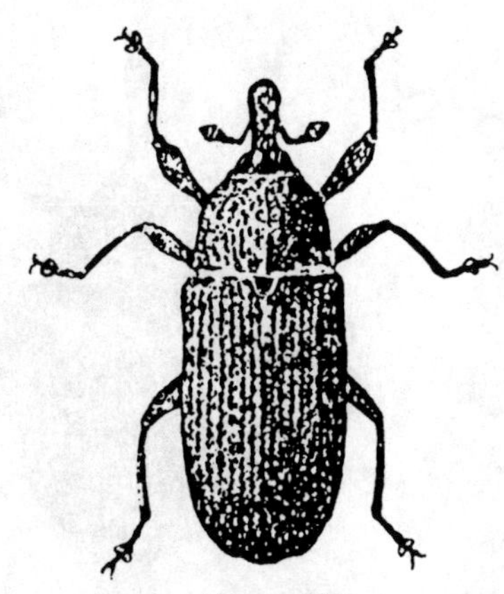

Figure 900

LENGTH: 3-4 mm. RANGE: Canada to Colorado and Florida.

Black, thickly covered with white prostrate scale-like hairs; scutellum black; elytral striae very fine. This is the Potato Stalk Borer. It feeds on several species of the night-shade family as well as on the cocklebur and some other plants.

53a (19) Elytra short, exposing pygidium; body broad. 54

53b Elytra covering pygidium; body oval. 56

54a (53) Pectoral groove (in sternum) extending behind the front coxae into the mesosternum; thorax with four large tubercles. Fig. 901. .... *Craponius inaequalis* (Say)

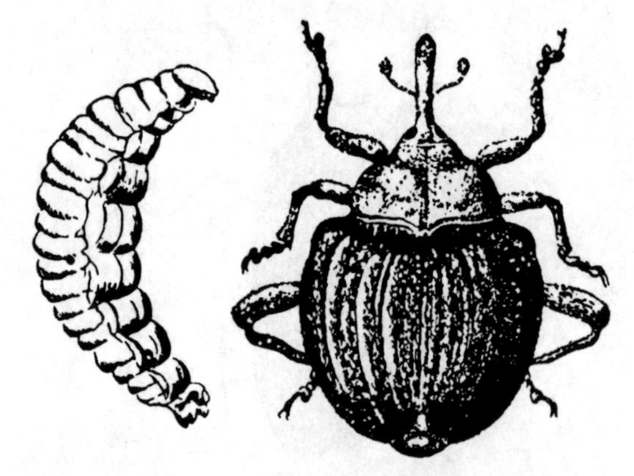

Figure 901

LENGTH: 2.5-3 mm. RANGE: New England to Florida to Midwest.

Dark brown or piceous, with whitish and brown hairs forming spots and lines on elytra. Antennae and legs dark reddish-brown. A pest of grapes, known as the "Grape Curculio."

Acanthoscelis acephalus (Say), 3-4 mm, is blackish with gray-yellow scales. The front tibia has a toothed process on outer apical angle but has no teeth near the base; the tarsi is broad. It has no tubercules on the elytra.

54b Not as in 54a. 55

55a (54) Beak short and heavy; no postocular lobes; funiculus with seven segments. Fig. 902. ...... *Rhinoncus triangularis* Say

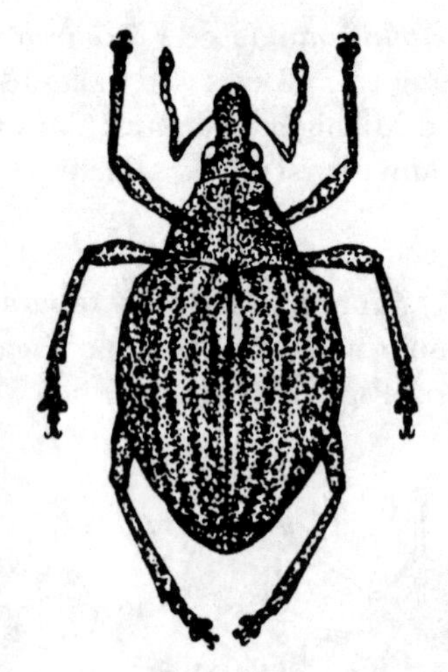

Figure 902

LENGTH: 3.5-4 mm. RANGE: Eastern U. S. and Canada.

Dark reddish-brown to blackish, with grayish-yellow hairs, legs and antennae paler. Scutellar spot conspicuous, triangular.

Pelenomus sulcicollis (Fahr.), measuring 2.5 mm, is black with brassy-brown scales. The prosternum lacks the antecoxal ridges of the above.

55b Beak slender and longer, eyes hidden at least in part by postocular lobes, hind femora toothed beneath. Fig. 903. .... *Ceutorhynchus rapae* Gyll.

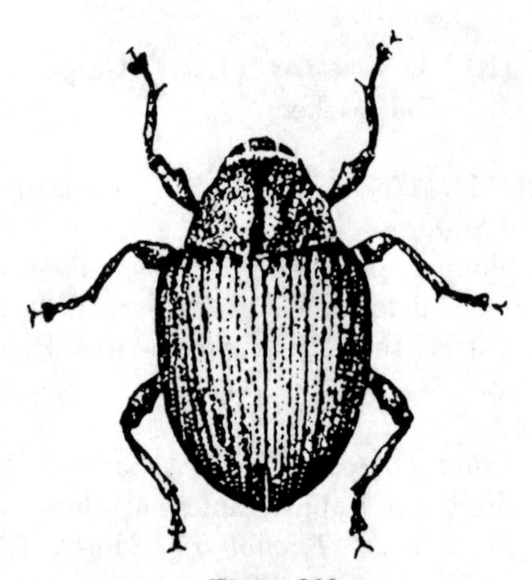

Figure 903

LENGTH: 2.7-3.2 mm. RANGE: Europe and America.

This "Cabbage Curculio" is black with scales first yellowish but becoming white above and larger white scales beneath; funicle 7-segmented, first and second segments each about as long as third and fourth measured together.

This genus has more than 75 species, small, but widely distributed.

C. septentrionalis Gyll., colored much as above, is smaller (2-2.2 mm) and has but six segments in the funicle. Its range is general and it too feeds on various mustards.

56a (53) Elytra with prominent humeri, abruptly wider than thorax; pectoral groove in prosternum only, open behind. 57

56b Elytra without prominent humeri; pectoral groove passing the front coxae and entering the mesosternum. 60

57a (56) Upper surface with scales, claws toothed or cleft. 58

57b Upper surface without pattern of scales; claws simple; middle coxae but slightly separated. Fig. 904.

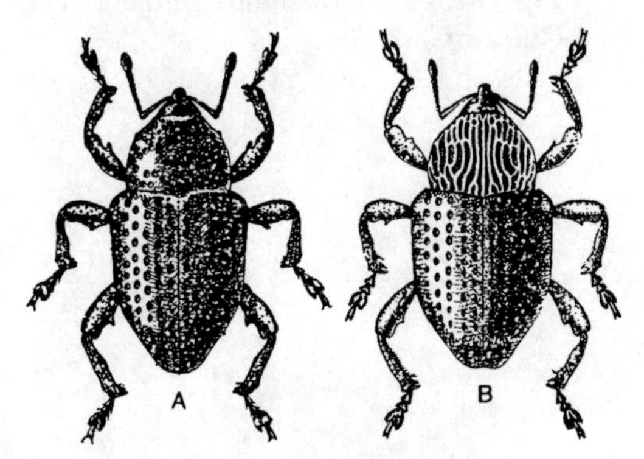

Figure 904

(a) *Chalcodermus aeneus* Boh.

LENGTH: 4.9-5.5 mm. RANGE: Middlewest to Texas.

Black, often with bronze sheen. Beak longer than the thorax and nearly straight. Eyes close together; elytra with large punctures and intervals alutaceous. This is the Cowpea Curculio.

(b) *C. collaris* Horn

LENGTH: 4.7-5.5 mm. RANGE: Massachusetts, Iowa, Florida, and Texas.

Much like *aeneus* except that the thorax is deeply marked with wavy lines. Elytra brown marked with a lighter reddish-brown; scutellum, smooth and bronzed.

Rhyssematus is a closely akin genus, distinguished from the above by claws that are toothed or cleft. *R. lineaticollis* Say, about 6 mm long and widely distributed, breeds in the seed pods of milkweeds. The elytra are black with brownish spots and the antennae and tarsi reddish-brown. Ranges over eastern half of the U. S.

58a (57) Thorax with a broad shallow groove, with two low crests in front; elytra marked as pictured. Fig. 905. .... *Conotrachelus anaglypticus* (Say)

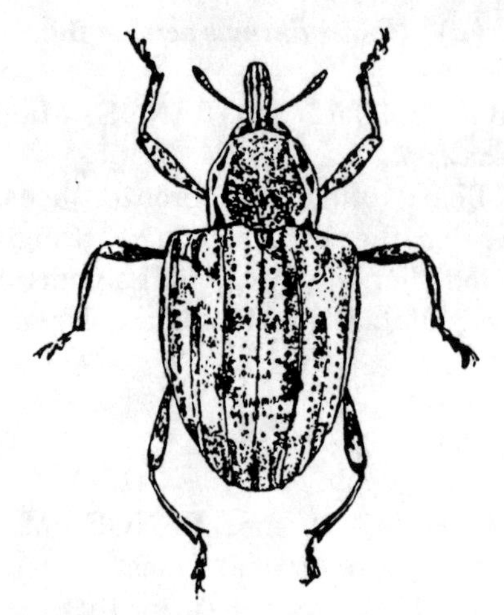

Figure 905

LENGTH: 3.5-4.7 mm. RANGE: Iowa to New England.

Dark reddish-brown to black with scant covering of yellowish and whitish hairs, forming the pictured pattern on thorax and elytra.

C. erinaceous Lec., about 3 mm in length, dark reddish-brown with pattern of grayish-yellow and white scales, has cleft claws. Ranges from Ohio and Ind. to Tex. and Fla.

About fifty American species of this genus have been described.

58b Thorax without groove as in 58a. 59

59a (58) Femora with two teeth; elytra with an interrupted longitudinal ridge. Fig. 906.

(a) *Conotrachelus nenuphar* (Hbst.)

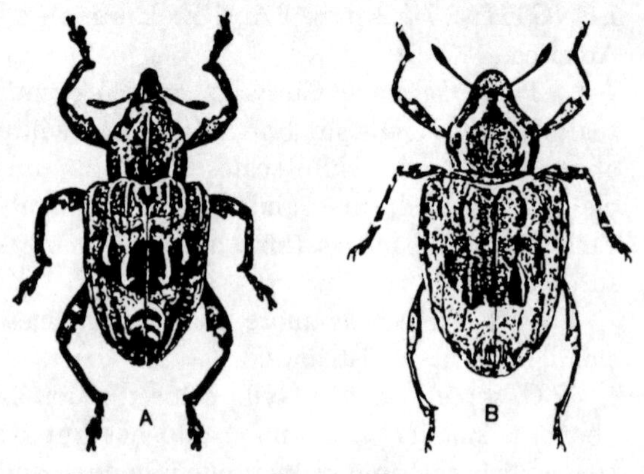

Figure 906

LENGTH: 4.5-6.5 mm. RANGE: general east of the Rockies.

Dark brown and black with markings as pictured of yellow and white. This "Plum Curculio" is a pest also of peach, cherry and apple.

(b) *C. juglandis* Lec.

LENGTH: 5-7 mm. RANGE: Eastern half United States and Canada.

This "Butternut Curculio" breeds in walnuts, butternuts and hickory nuts, causing the nuts to fall when about half grown. It is dark brown with grayish-yellow markings on legs, thorax and elytra.

59b Femora with but one tooth; elytral humeri with small blunt tooth on outer side. Fig. 907. .. *Conotrachelus crataegi* Walsh Quince Curculio

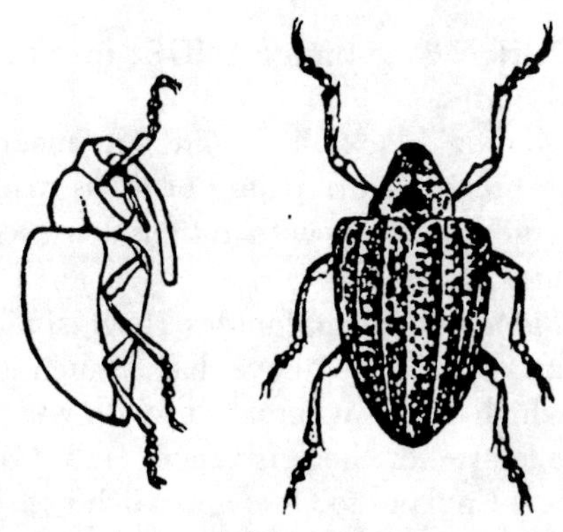

Figure 907

LENGTH: 4-6 mm. RANGE: Iowa to New England and southward.

Brownish-black with clay-yellow and ash-gray scales. Each elytron with four prominent carinae. It lives on *Crataegus* and related plants.

C. posticatus Boh., similar in size, color, habits and range to the above, has its beak heavy, short and curved instead of long and slender as in *crataegi*.

There are over 60 species of *Conotrachelus* in the U. S. and Canada.

A closely related genus, *Phyrdenus*, differs from *Conotrachelus* in having the head sulcate and with a prominent tubercle on each side; the tibiae are but feebly spined at tip. *P. divergens* (Germ.) 4.5-4.8 mm and known from New Jersey to Florida and westward to Missouri and Texas is blackish with a pale brown coat of mixed bristles and scales. It feeds on *Solanum nigrum*.

60a (56) Funicle of antennae with but six segments; femora without teeth; elytra glabrous. Fig. 908. .... *Tyloderma fragariae* (Riley)

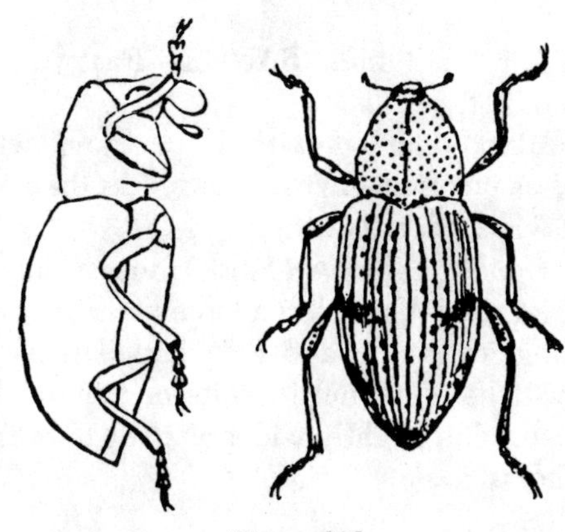

Figure 908

LENGTH: 3.5-4.2 mm. RANGE: Eastern half United States.

Piceous, elytra and legs reddish-brown, the wingcovers marked with blackish and yellowish as pictured; beak longer than head. This "Strawberry Crown Borer" is a destructive pest.

T. aerea Say, smaller (2.2-2.8 mm) and with similar range, but with sides of elytra parallel instead of oval, is a bronzy-black species which seems to live around damp places.

60b Funicle with seven segments, tibiae with strong hook at tip; ventral abdominal segments two to four, practically equal. Fig. 909. .. *Cryptorhynchus pumilus* Boh.

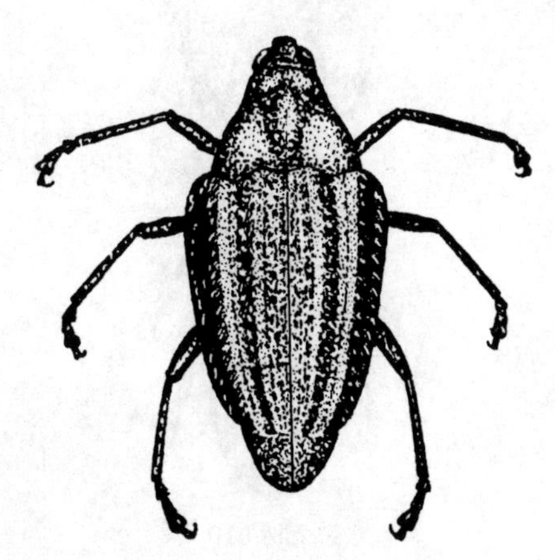

Figure 909

LENGTH: 4-5 mm. RANGE: Eastern half United States.

Black, marked with brown and white scales as pictured. Elytra at base less than one-third wider than thorax.

Cophes obtentus (Hbst.), measuring 5-7 mm and sooty-black, has a large yellowish spot crossing its suture, and a Y-shaped mark on thorax. The scutellum is white or yellow. The elytra are but slightly wider at their base than the thorax.

61a (8) Pygidium covered by elytra; number of segments in funicle often reduced; mentum shortened in the peduncle. 62

61b Pygidium exposed; funicle with six segments, peduncle of mentum elongated. 64

62a (61) Front coxae but narrowly separated; beak no longer than the head; funicle with segments 2-7 transverse, elytral striae deep and with coarse serrated punctures. Fig. 910. .... *Rhyncolus oregonensis* Horn

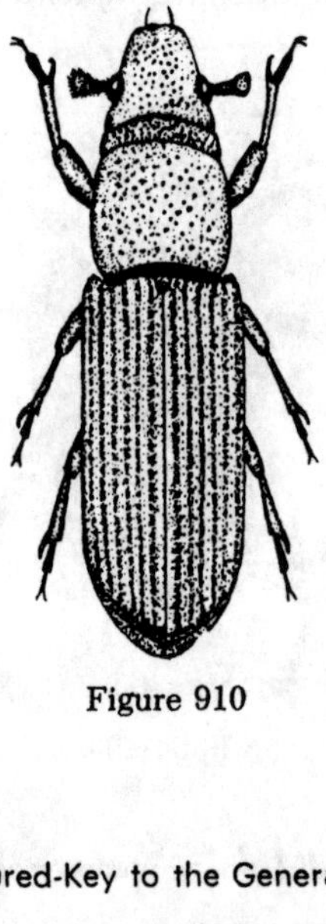

Figure 910

LENGTH: 2.8-3.5 mm. RANGE: much of the United States.

Shining black; legs and antennae dark reddish-brown, club paler, beak as wide as long. Under surface with rather dense coarse punctures.

Phloeophagus apionides Horn, similar in color and size to the above, has a much longer beak which attains its greatest width where the antennae are attached. It ranges from Canada and New England to Ohio and Michigan.

Stenoscelis boevis (Boh.) with a length of 2.8-3.2 mm differs from the above species in having a beak much shorter than the head. It is black, nearly dull, with the elytra and legs brownish and the antennae and tarsi paler; the thorax is nearly one-half wider than long. The cylindrical elytra have their sides parallel and have broad, coarsely punctured serrations. It apparently may be found in much of the eastern part of the United States and Canada and is often common in the dead wood of many species of trees.

62b Front coxae separated by more than half their width; beak usually longer than the head. 63

63a (62) Body depressed; beak dilated at tip; antennae attached near tip of beak. Fig. 911. .... *Cossonus platalea* Say

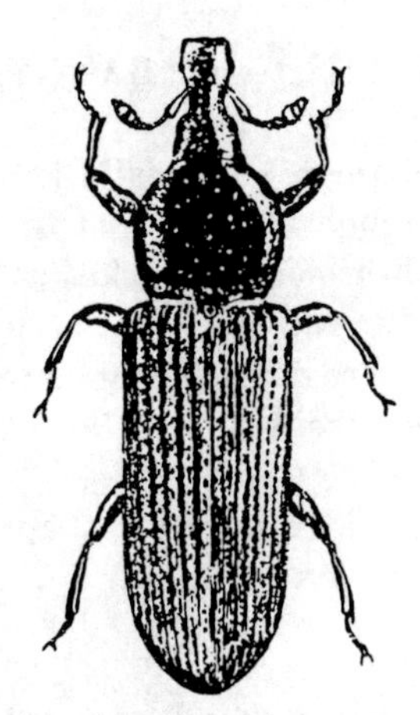
Figure 911

LENGTH: 5.5-6.5 mm. RANGE: Iowa to Virginia and Canada.

Shining black; tarsi and antennae reddish-brown. Disk of thorax flattened with scattered coarse punctures at middle, finer and more numerous at edges.

Himatinum errans Lec., measuring 2 mm, is reddish-brown and is covered with a coarse yellowish pubescence. It ranges from Nova Scotia to Ohio and Florida.

63b Body not flattened; beak not dilated at tip; antennal club large (a). Fig. 912. .... *Caulophilus latinasus* (Say)

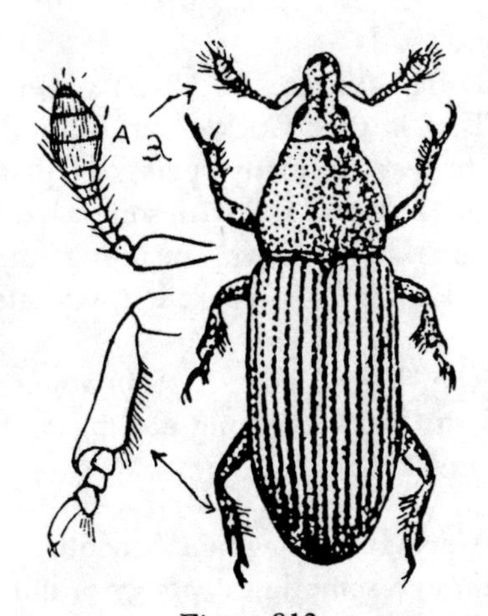

Figure 912

LENGTH: 3 mm. RANGE: Southeastern United States.

Reddish-brown. Thorax as broad as long, with fine even punctures on top. Elytra deep with coarse punctures. It is sometimes a pest of dried cereals and is known as the "Broadnosed Grain Weevil."

A much more slender, tiny little fellow (measuring but 1.5 mm and linear), *Stenomimus pallidus* Boh. is known to range in several of the Central and Middle States, and is found in association with walnut and hickory. It is shining pale reddish-yellow. The thorax is oblong and in width equal to the elytra at their base. The elytral striae are fine with rather coarse, close-set punctures.

64a (61) Large beetles, 20 mm or longer; side pieces of metathorax very wide. Fig. 913. .. *Rhynchophorus cruentatus* (Fab.)

Figure 913

LENGTH: 20-31 mm. RANGE: South Carolina and Florida to Louisiana.

Shining black; thorax often red with black markings; elytra with red spots on black, as pictured; elytral striae deep and punctured; legs with fringe of long yellow hairs on inner side. Some specimens are wholly black and have been given variety significance *zimmermanni.* Fahr.

The larvae feed within the dying trunks of the cabbage palmetto and other palms. The other species are tropical.

64b Smaller, not over 18 mm; side pieces of metathorax not unusually wide. 65

65a (64) Antennal club oval; small slender beetles about 4 mm or less. Fig. 914.

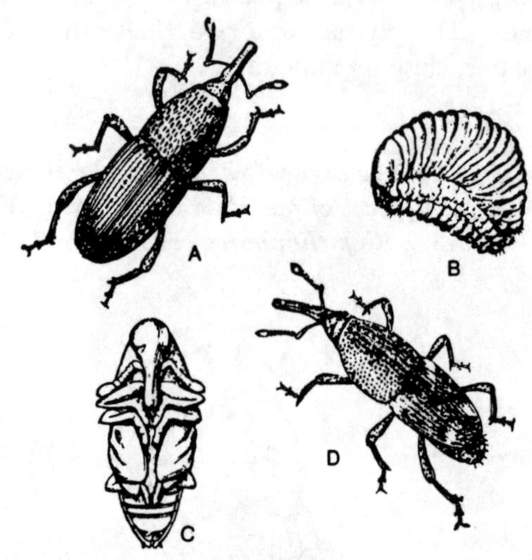

Figure 914

(a) *Sitophilus granarius* (L.)

LENGTH: 3-4 mm. RANGE: cosmopolitan.

This, the "Granary Weevil," is shining chestnut brown to blackish. Thorax with scattered punctures; elytra with punctured deep striae, the intervals smooth. It is flightless but lives in stored grain and travels with its food. The larva (b) and the pupa (c) of this and the following species are highly similar.

(d) *Sitophilus oryza* (L.)

LENGTH: 2.1-2.8 mm. RANGE: cosmopolitan.

Its common name is "Rice Weevil"; its habits are much like those of the above. It is dull reddish-brown to blackish; the elytra often have four reddish spots as pictured. The elytral striae are coarsely punctured, with each puncture bearing a short yellowish hair.

65b Antennal club wedge-shaped, larger, 5 mm or over. 66

**66a (65) Third tarsal segment broad and spongy beneath; red marked with black as pictured. Fig. 915. ....
.. *Rhodobaenus tredecimpunctatus* (Ill.)**

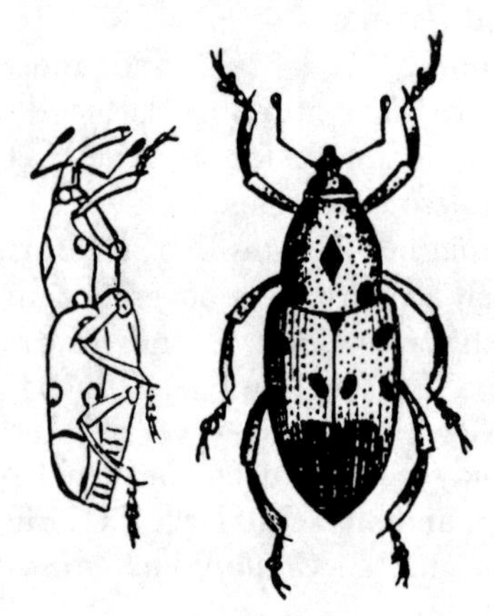

Figure 915

LENGTH: 7-10 mm. RANGE: general.

This is the "Cockle-bur Billbug" but it seems to feed on many species of plants. It is black beneath and red with somewhat variable black markings above, resulting in variety names. The beak is black. It hibernates as an adult.

The "Yucca Weevil" *Scyphophorus yuccae* Horn, 15-17 mm long and black, feeds on yucca in California.

66b Third tarsal segment smooth at middle though sometimes spongy or hairy at the sides. 67

67a (66) **Third segment of hind tarsi but little if at all dilated. 68**

67b **Third segment of all tarsi much dilated, but glabrous at the middle. Fig. 916. .... *Sphenophorus aequalis* Gyll.**

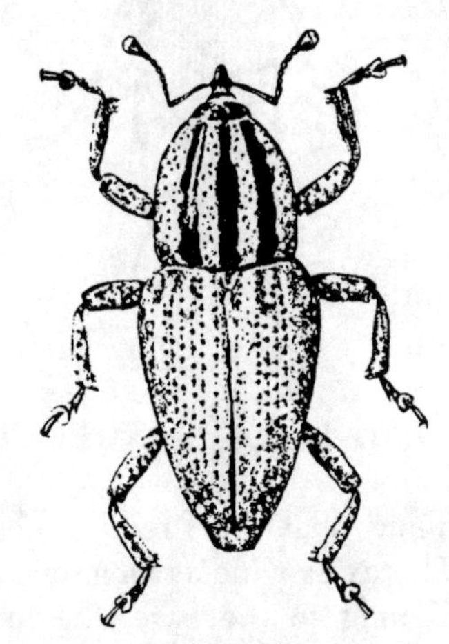

Figure 916

LENGTH: 11-17 mm. RANGE: Most of the southern Canada and United States except New England.

Black with a shiny coating varying in color, light bluish-gray, pale yellowish to reddish-brown, with three darkened elevated streaks on thorax and frequently dark stripes on the elytra. This "Clay-colored Billbug" normally feeds on reeds and bulrushes, but when low places, where these plants grow, are plowed the billbugs frequently attack the young corn plants.

The "Tule Billbug" *S. discolor* Mann. feeding on tules and cat-tails, but attacking grain crops in California is similar in size. It is white beneath and usually black with white streaks above. This is a variety of *aequalis*.

In contrast to *aequalis*, pictured and described above, *S. inaequalis* (Say) is short with wedge-shaped form and has the outer apical angles of the front tibiae prolonged; middle and hind tarsi truncate and the tarsi slender. It ranges from New York to Florida and Texas and measures 6 mm. Further description follows.

Black, often covered with a brownish superficial coat. Beak two-thirds as long as thorax, feebly compressed; base somewhat swollen and finely grooved above. Thorax as long as wide and coarsely punctuate; usually with a large elevated smooth space at middle and two smaller one on either side. Elytra at base as wide as thorax but rapidly narrowing toward apex; striae with coarse, widely spaced punctures.

68a (67) **Third segment of first and second tarsi not at all dilated, wholly glabrous or nearly so. 71**

68b **Third segment of first and second tarsi feebly dilated, with spongy pubescence at the sides. 69**

69a (68) **Thorax with three flattened, elevated, smooth longitudinal lines. 70**

69b **Thorax with only the mid-elevated line distinct and it not extending the entire distance. Fig. 917. .... *Sphenophorus maidis* (Chitt.)**

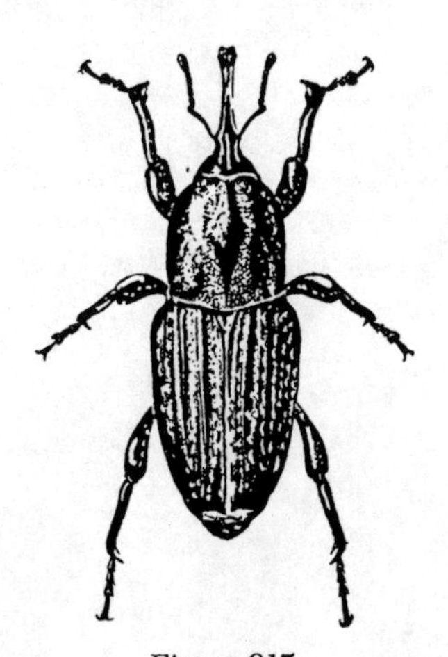

Figure 917

LENGTH: 10-15 mm. RANGE: Southeastern U. S. to Kans. and Tex.

Blackish or dark reddish, without much clay-colored coating; beak three-fifths the length of the thorax, with fine punctures. Vitta (mid-thoracic stripe) but slightly elevated. This "Maize Billbug" often does serious damage to corn.

S. *soltaui* Chitt., 7.5 mm, a rather dull-black species differs from *maidis* in having a fusiform body. It ranges from Ill. to Col. south to Tex.

S. *cariosa* (Oliv.) with a length of 8-11 mm, differs markedly from the above in the much elevated third elytral interval extending thus from the base to the middle of the elytra. It is black with the raised places and the underparts shining. The disk of the thorax has three longitudinal stripes, the outer ones irregular and entire and the middle one widening to a rhomboid figure midway of the thorax then ending abruptly or sometimes continuing as a fine line. It apparently ranges rather generally east of the Mississippi river.

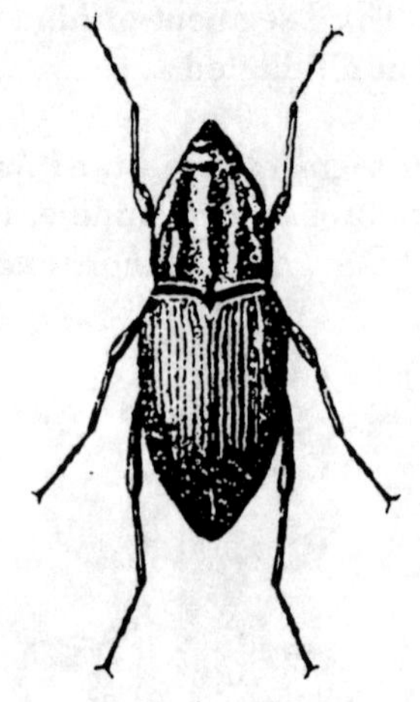

Figure 918

LENGTH: 11-14 mm. RANGE: Central and Western States.

Shining black or reddish-black, quite robust. Thorax as wide as long, outer elevated lines widening to the base, the median one widest at its midpoint.

C. peninsularis Chitt. averaging a bit larger, is black but has a clay covering. It ranges from So. Ga. to Fla.

70a (69) Beak somewhat compressed; all of the elytral intervals about evenly convex, of the same color and about equal in width. Fig. 918. .... *Calendra robustus* Horn

70b Beak nearly cylindrical, a bit narrow at its base; alternating elytral intervals elevated, irregular and sometimes interrupted. Fig. 919. .... *Sphenophorus pertinax* (Oliv.)

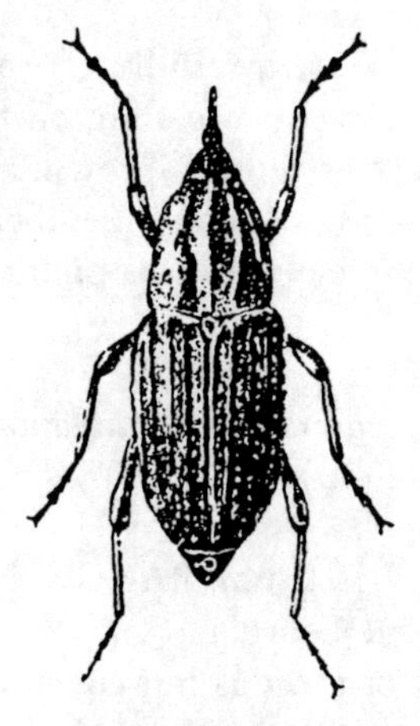

Figure 919

LENGTH: 11-15 mm. RANGE: Eastern U .S. to Iowa and Tex.

Shining black, often with reddish tinge, the interspaces on elytra and thorax covered with a whitish coating.

It breeds in cat-tail and calamus.

71a (68) **Thorax without longitudinal vittae (smooth elevated ridges), surface punctured throughout. Fig. 920.** ....

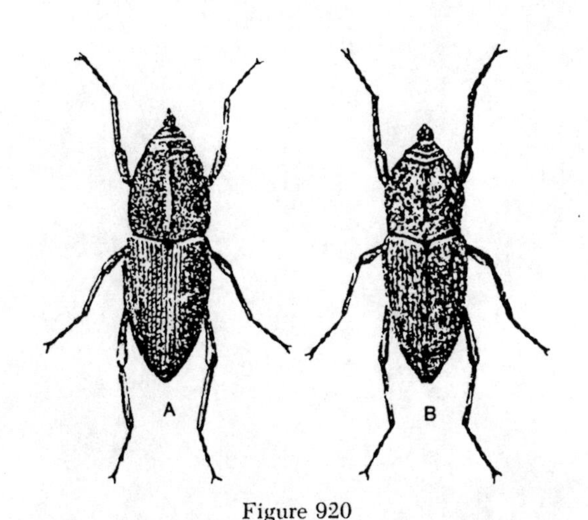

Figure 920

(a) ***Sphenophorus parvulus*** **Gyll.**

LENGTH: 5-6.5 mm. RANGE: Eastern U. S. west to Col. and Tex.

Black with clayey coating of ash-gray; tarsi and antennae reddish-brown. Beak three-fourths as long as thorax. Thorax sometimes with narrow median line. This "Blue-grass Bill-bug" attacks the roots of timothy and blue-grass.

(b) ***Sphenophorus minimus*** **Hart**

LENGTH: 5-6 mm. RANGE: Eastern U. S. to Kan.; also Cal.

Black, usually covered with clay-yellow or ash gray coating.

Beak two-thirds the length of thorax. Thoracic punctures coarse and irregularly placed; elytra widest in front.

71b Thorax with one or more vittae. **72**

72a (71) **Middle thoracic vitta forked in front. Fig. 921.** ....
.... ***Sphenophorus venatus*** **(Say)**

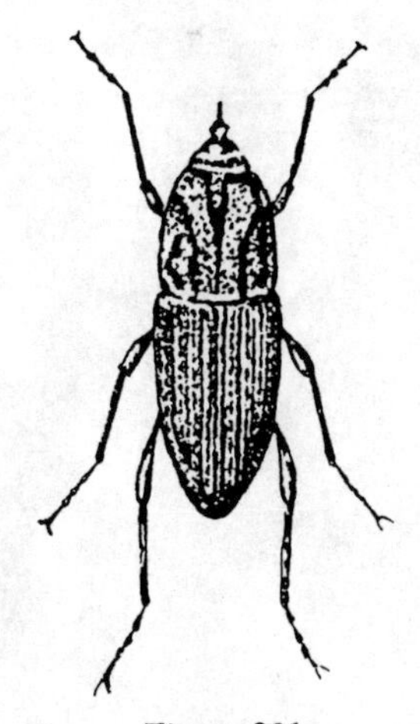

Figure 921

LENGTH: 6-10 mm. RANGE: Probably all of U. S. and southern Canada.

Blackish or reddish; beak three-fourths as long as thorax. Vittae as pictured. Elytra narrowing from base to apex, striae fine with coarse punctures.

72b Median vittae not forked. Fig. 922.

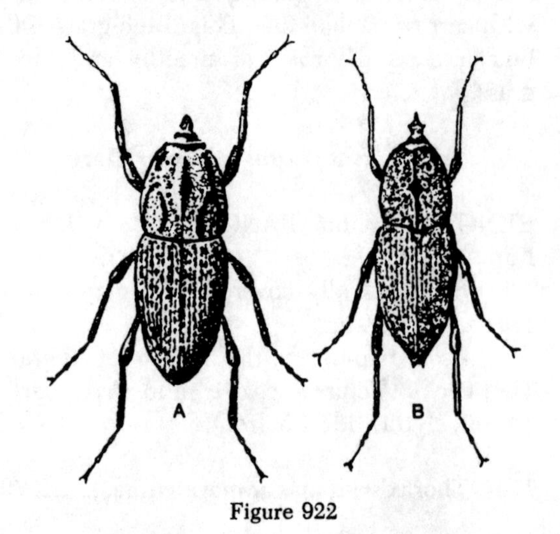

Figure 922

(a) *Sphenophorus zeae* Walsh

LENGTH: 7.5-9 mm. RANGE: Eastern U. S. and Can., west to Col. Also Cal.

This "Timothy Billbug" sometimes attacks corn as well as feeding on timothy. It is shining black or reddish-brown, and lacks the artificial coating of many members of this genus. The thoracic vittae, as pictured, are characteristic.

(b) *Sphenophorus melanocephalus* (Fab.)

LENGTH: 7-10.5 mm. RANGE: New England to Kansas and Georgia.

Black or piceous but covered above with a dense brownish coating. Median vitta ending at half the length of the thorax in a dilated part as pictured.

It might seem from the four pages devoted to this one genus that some partiality were being shown. The fact that not many genera of beetles have as many highly destructive species as this one, seems to justify the amount of attention given it. There are over 70 N. American members of this genus.

General References

Arnett, R. H., Jr. 1962. The beetles of the United States (a Manual for identification). xii+ 1112 pp. (Out-of-print). [Keys to the genera found in the U. S., illustrated, extensive bibliography.]

Arnett, R. H., Jr. 1979- . Checklist of the beetles of Canada, United States, Mexico, Central America, and the West Indies. World Natural History Publications, Marlton, New Jersey. [A continuing series cataloging the species of beetles occurring in North America, including a bibliography.]

Balsbaugh, E. H. and K. L. Hays. 197-. Leaf beetles of Alabama. Bull. 441, Agricultural Experimentation Station, Auburn, AL.

Blatchley, W. S. 1910. An illustrated and descriptive catalogue of the Coleoptera or Beetles (exclusive of the Rhynchophora) known to occur in Indiana. 1385 pp. (Out-of-print). [Useful for most of n.e. U.S.; names outdated.]

Blatchley, W. S. and C. W. Leng. 1916. Rhynchophora or weevils of Northeastern North America. 682 pp. (Out-of-print). [Supplements the preceding work to complete the families for the region.]

Boving, A. G. and F. C. Craighead. 1930. An illustrated synopsis of the principal larval forms of the order Coleoptera. 351 pp. New York Entomological Society, American Museum of Natural History, New York, NY.

Hatch, M. H. 1953-1971. Beetles of the Pacific Northwest. 5 parts. University of Washington Press, Seattle, WA. [Keys to species of the beetles of Oregon, Washington, Idaho, and British Columbia.]

Lindroth, Carl H. 1961-1969. The ground beetles of Canada and Alaska. 6 parts. Entomological Society, Zoological Institute, University, Lund, Sweden.

Linsley, E. G. and J. A. Chemsak. 1961- Cerambycidae of North America. University of California Press, Berkeley, CA. [7 parts have been issued; remainder of the series in press; complete keys and descriptions, with illustrations, of this large family of beetles known from the area.]

Wilcox, John A. 1954. Leaf beetles of Ohio. Bull. 43. Ohio State University, Columbus, OH. Pp. 353-506.

Woodruff, R. E. 1973. The scarab beetles of Florida, part 1. Florida Department of Agriculture and Plant Services, Gainesville, FL. 220 pp. [Keys and descriptions of the beetles of the large family Scarabaeidae for Florida and neighboring regions; approximately half of the species are included in this first part.]

The following organization publishes a journal devoted to beetle study and includes notices of new publications on beetles:

The Coleopterists' Society, c/o Department of Entomology, Smithsonian Institution, Washington, D.C. 20560. [Publishers of "The Coleopterists' Bulletin.]

Supply houses for entomological equipment:

American Biological Supply Co., 1330 Dillon Heights Ave., Baltimore, MD 21228.

BioQuip Products, P. O. Box 61, Santa Monica, CA 90406.

Carolina Biological Supply Co., Burlington, NC 27215.

Macmillan Science Co., Inc., 8200 South Hoyne Ave., Chicago, IL 60620.

Ward's Natural Science Establishment, Inc., P. O. Box 1712, Rochester, NY 14603.

List of Families

Order COLEOPTERA

Suborder ARCHOSTEMATA

1. *Cupedidae* – The reticulated beetles

Suborder ADEPHAGA

2. *Rhysodidae* – The wrinkled bark beetles
3. *Cicindelidae* – The tiger beetles
4. *Carabidae* – The ground beetles
5. *Amphizoidae* – The trout-stream beetles
6. *Haliplidae* – The crawling water beetles
7. *Dytiscidae* – The predaceous diving beetles
8. *Noteridae* – The burrowing water beetles
9. *Gyrinidae* – The whirligig beetles

Suborder MYXOPHAGA

10. *Sphaeriidae* – The minute bog beetles
11. *Hydroscaphidae* – The skiff beetles

Suborder POLYPHAGA

12. *Hydrophylidae* – The water scavenger beetles
13. *Limnebiidae* – The minute moss beetles
14. *Georyssidae* – The minute mud-loving beetles
15. *Ptilidae* – The feather-winged beetles
16. *Limulodidae* – The horse-shoe crab beetles
17. *Dasyceridae* – The minute scavenger beetles
18. *Leptinidae* – The mammal nest beetles
19. *Leiodidae* – The round fungus beetles
20. *Scydmaenidae* – The antlike stone beetles
21. *Silphidae* – The carrion beetles
22. *Leptodiridae* – The small carrion beetles
23. *Scaphidiidae* – The shining fungus beetles
24. *Staphylinidae* – The rove beetles
25. *Pselaphidae* – The antloving beetles
26. *Histeridae* – The clown beetles
27. *Sphaeritidae* – The false clown beetles
28. *Lucanidae* – The stag beetles
29. *Passalidae* – The peg or bess beetles
30. *Scarabaeidae* – The lamellicorn beetles
31. *Eucinetidae* – The plate-thigh beetles
32. *Helodidae* – The marsh beetles
33. *Clambidae* – The minute beetles
34. *Dascillidae* – The soft-bodied plant beetles
35. *Byrrhidae* – The pill beetles
36. *Psephenidae* – The water penny beetles
37. *Brachypsectridae* – The Texas beetle

38. *Ptilodactylidae* – The toed-winged beetles
39. *Rhipiceridae* – The cicada parasite beetles
40. *Callirhipidae* – The cedar beetles
41. *Chelonariidae* – The turtle beetles
42. *Heteroceridae* – The variegated mud-loving beetles
43. *Limnichidae* – The minute marsh-loving beetles
44. *Dryopidae* – The long-toed water beetles
45. *Elmidae* – The drive or riffle beetles
46. *Buprestidae* – The metallic wood-boring beetles
47. *Cebrionidae* – The robust click beetles
48. *Elateridae* – The click beetles
49. *Throscidae* – The false metallic wood-boring beetles
50. *Cerophytidae* – The rare click beetles
51. *Perothopidae* – The beech-tree beetles
52. *Eucnemidae* – The false click beetles
53. *Telegeusidae* – The long-lipped beetles
54. *Phengodidae* – The glowworm beetles
55. *Lampyridae* – The firefly beetles
56. *Cantharidae* – The soldier beetles
57. *Lycidae* – The net-winged beetles
58. *Micromalthidae* – The telephone-pole beetles
59. *Derodontidae* – The tooth-necked fungus beetles
60. *Nosodendronidae* – The wounded-tree beetles
61. *Dermestidae* – The skin beetles and larder beetles
62. *Ptinidae* – The spider beetles
63. *Anobiidae* – The drug-store and death-watch beetles
64. *Bostrichidae* – The horned powder-post beetles
65. *Lyctidae* – The powder-post beetles
66. *Trogossitidae* – The bark-gnawing beetles
67. *Cleridae* – The checkered beetles
68. *Melyridae* – The soft-winged flower beetles
69. *Lymexylonidae* – The ship-timber beetles
70. *Nitidulidae* – The sap-feeding beetles
71. *Rhizophagidae* – The root-eating beetles
72. *Cucujidae* – The flat bark beetles
73. *Sphindidae* – The dry-fungus beetles
74. *Cryptophagidae* – The silken fungus beetles
75. *Languriidae* – The lizard beetles
76. *Erotylidae* – The pleasing fungus beetles
77. *Phalacridae* – The shining flower beetles
78. *Cerylonidae* – The minute bark beetles
79. *Corylophidae* – The minute fungus beetles
80. *Coccinellidae* – The ladybird beetles
81. *Endomychidae* – The handsome fungus beetles
82. *Lathridiidae* – The minute brown scavenger beetles
83. *Biphyllidae* – The false skin beetles
84. *Byturidae* – The fruitworm beetles
85. *Monommatidae* – The opossum beetles
86. *Cephaloidae* – The false longhorned beetles
87. *Tenebrionidae* – The darkling beetles
88. *Alleculidae* – The comb-clawed bark beetles
89. *Lagriidae* – The long-jointed beetles
90. *Salpingidae* – The narrow-waisted bark beetles
91. *Mycetophagidae* – The hairy fungus beetles
92. *Ciidae* – The minute tree-fungus beetles
93. *Colydiidae* – The cylindrical bark beetles
94. *Pyrochroidae* – The fire-colored beetles
95. *Othniidae* – The false tiger beetles
96. *Prostomidae* – The jugular-horned beetles
97. *Oedemeridae* – The pollen feeding beetles
98. *Melandryidae* – The false darkling beetles
99. *Mordellidae* – The tumbling flower beetles

100. *Rhipiphoridae* – The wedge-shaped beetles
101. *Meloidae* – The blister beetles
102. *Anthicidae* – The antlike flower beetles
103. *Pedilidae* – The false antloving flower beetles
104. *Euglenidae* – The antlike leaf beetles
105. *Stylopidae*–The parasitic beetles (Strepsiptera)
106. *Cerambycidae* – The longhorned beetles
107. *Chrysomelidae* – The leaf beetles
108. *Bruchidae* – The pea and bean weevils
109. *Anthribidae* – The fungus weevils
110. *Brentidae* – The straight-snouted weevils.
111. *Platypodidae* – The ambrosia beetles
112. *Scolytidae* – The bark beetles
113. *Curculionidae* – The snout beetles and true weevils

Index and Pictured Glossary

A

Figure 924

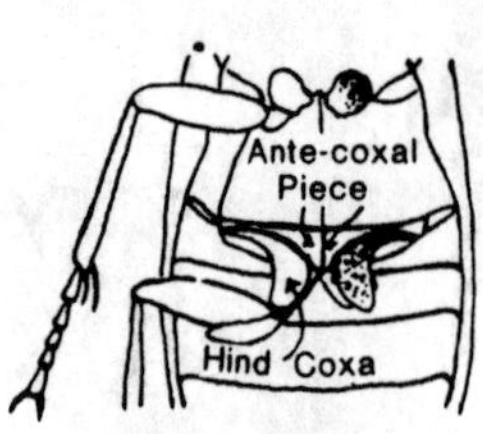

Figure 925

B

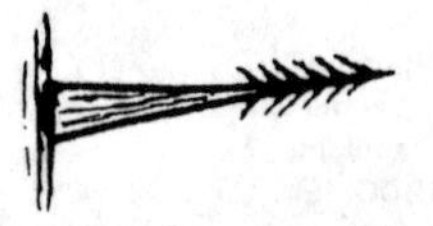

Figure 926

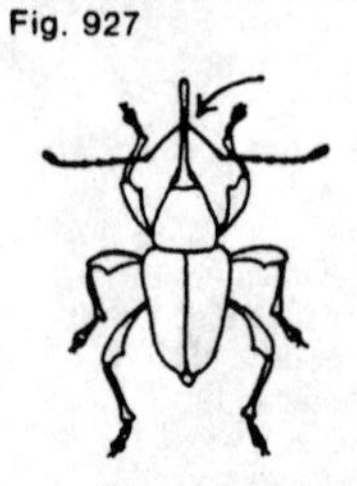

Figure 927

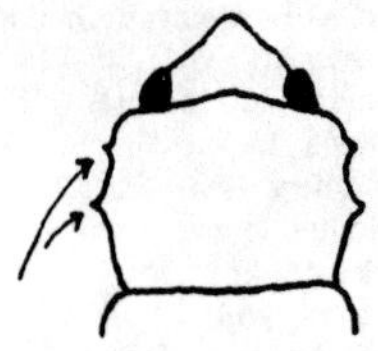

Figure 928

C

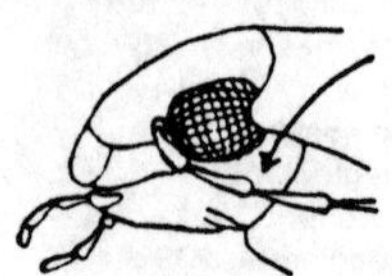

Figure 929

Figure 930

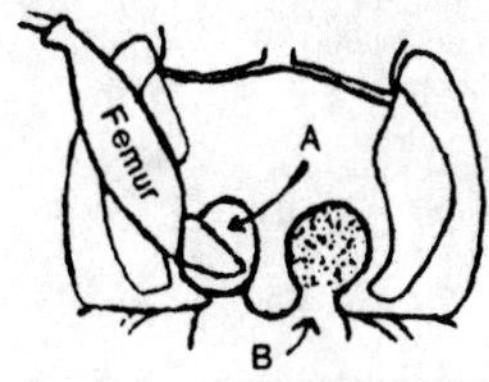

Figure 931

D

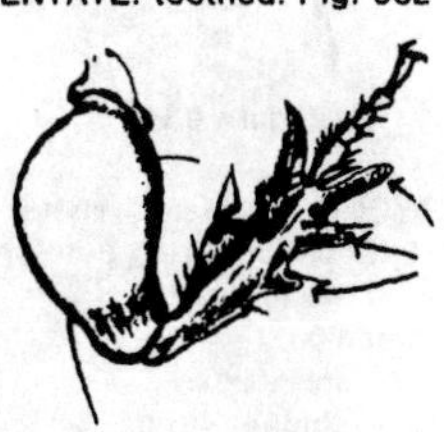

Figure 932

F

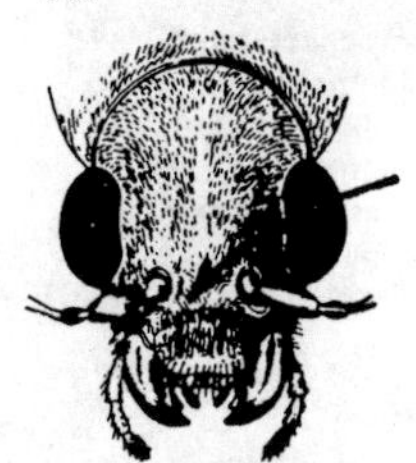

Figure 937

Pĕdicel (2nd joint)
Scāpe (1st joint)
Antĕnnal Suture
Flagĕllum
GENICULATE (Elbowed)

Figure 938

G

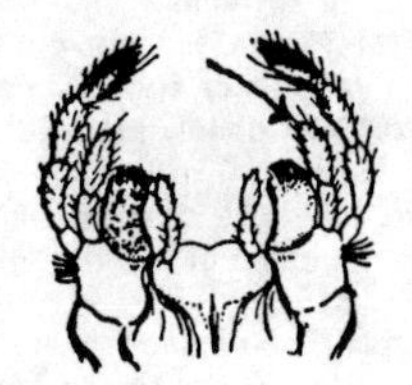

Figure 939

Figure 940

H

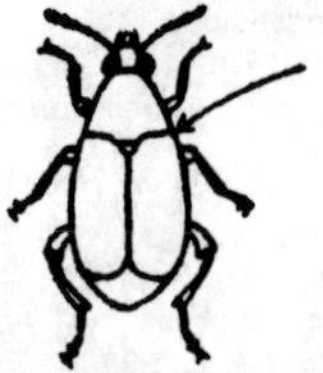

Figure 941

I

J

K

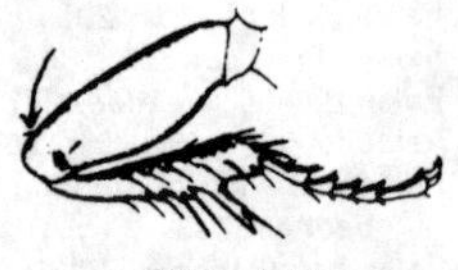

Figure 942

L

Figure 943

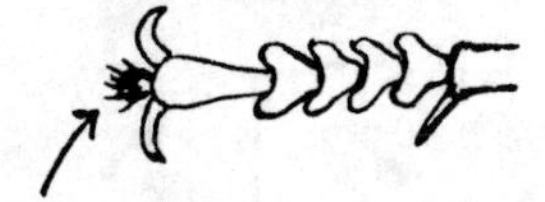

Figure 948

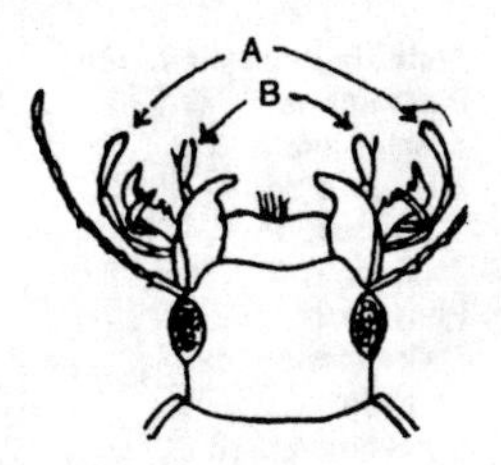

Figure 949

Figure 950

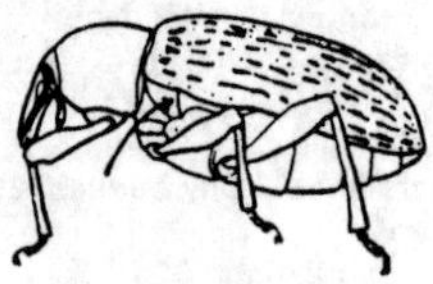

Figure 951

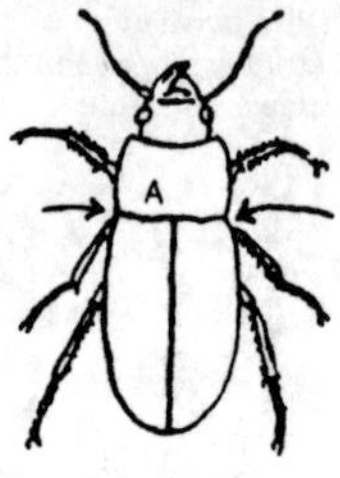

Figure 956

B
A

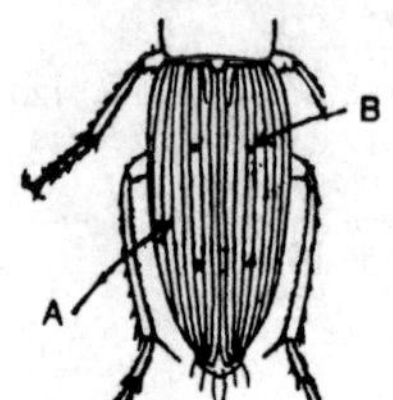

Figure 957

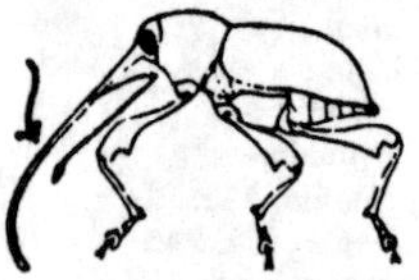

T

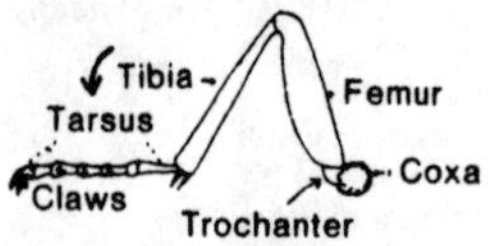

Tibia
Femur
Tarsus
Coxa
Claws
Trochanter

Figure 958

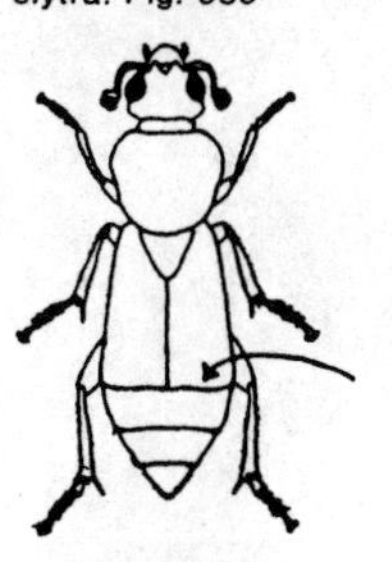

Figure 959

U

V

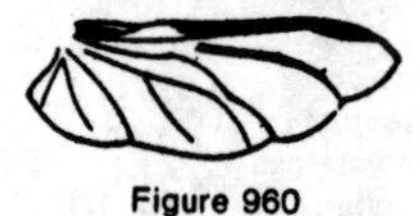

Figure 960